A Unified Algebraic Approach
to
Linear Control Design

A Unified Algebraic Approach to Linear Control Design

ROBERT E. SKELTON
University of California, San Diego

TETSUYA IWASAKI
Tokyo Institute of Technology, Tokyo

KAROLOS M. GRIGORIADIS
University of Houston, Houston

Taylor & Francis
Publishers since 1798

UK Taylor & Francis Ltd, 1 Gunpowder Square, London EC4A 3DE
USA Taylor & Francis Inc., 1900 Frost Road, Suite 101, Bristol, PA 19007

British Library Cataloguing in Publication Data

A catalogue record for this book is available from the British Library

ISBN 0-7484-0592-5

Library of Congress Cataloging in Publication data are available

Cover design by Amanda Barragry

To Judy, Stephanie, Hope, Katie, Grahm and Carson Robert E. Skelton

To Junko and the coming baby Tetsuya Iwasaki

To my parents Karolos M. Grigoriadis

Contents

Series Introduction

Control systems has a long and distinguished tradition stretching back to nineteenth-century dynamics and stability theory. Its establishment as a major engineering discipline in the 1950s arose, essentially, from Second World War driven work on frequency response methods by, amongst others, Nyquist, Bode and Wiener. The intervening 40 years has seen quite unparalleled developments in the underlying theory with applications ranging from the ubiquitous PID controller widely encountered in the process industries through to high-performance/fidelity controllers typical of aerospace applications. This development has been increasingly underpinned by the rapid developments in the, essentially enabling, technology of computing software and hardware.

This view of mathematically model-based systems and control as a mature discipline masks relatively new and rapid developments in the general area of robust control. Here intense research effort is being directed to the development of high-performance controllers which (at least) are robust to specified classes of plant uncertainty. One measure of this effort is the fact that, after a relatively short period of work, 'near world' tests of classes of robust controllers have been undertaken in the aerospace industry. Again, this work is supported by computing hardware and software developments, such as the toolboxes available within numerous commercially marketed controller design/simulation packages.

Recently, there has been increasing interest in the use of so-called intelligent control techniques such as fuzzy logic and neural networks. Basically, these rely on learning (in a prescribed manner) the input–output behaviour of the plant to be controlled. Already, it is clear that there is little to be gained by applying these techniques to cases where mature mathematical model-based approaches yield high-performance control. Instead, their role is (in general terms) almost certainly going to lie in areas where the processes encountered are ill-defined, complex, nonlinear, time-varying and stochastic. A detailed evaluation of their (relative) potential awaits the appearance of a rigorous supporting base (underlying theory and implementation architectures for example) the essential elements of which are beginning to appear in learned journals and conferences.

Elements of control and systems theory/engineering are increasingly finding use outside traditional numerical processing environments. One such general area in which there is increasing interest is intelligent command and control systems which are central, for example, to innovative manufacturing and the management of advanced transportation systems. Another is discrete event systems which mix numeric and logic decision making.

It was in response to these exciting new developments that the present book series of Systems and Control was conceived. It publishes high-quality research texts and reference works in the diverse areas which systems and control now includes. In addition to basic theory, experimental and/or application studies are welcome, as are expository texts where theory, verification and applications come together to provide a unifying coverage of a particular topic or topics.

The book series itself arose out of the seminal text: the 1992 centenary first English translation of Lyapunov's memoir *The General Problem of the Stability of Motion* by A. T. Fuller, and was followed by the 1994 publication of *Advances in Intelligent Control* by C. J. Harris. Since then a number of titles have been published and many more are planned. A full list is given below.

Advances in Intelligent Control, edited by C. J. Harris

Intelligent Control in Biomedicine, edited by D. A. Linkens

Advances in Flight Control, edited by M. B. Tischler

Multiple Model Approaches to Modelling and Control, edited by R. Murray-Smith and T. A. Johansen

Forthcoming

Sliding Mode Control: Theory and Applications, by C. Edwards and S. K. Spurgeon

Neural Network Control of Robot Manipulators and Nonlinear Systems, by F. L. Lewis, S. Jagannathan and A. Yesildirek

Generalized Predictive Control with Applications to Medicine, by M. Mahfouf and D. A. Linkens

Sliding Mode Control in Electro-Mechanical Systems, by V. I. Utkin, J. Guldner and J. Shi

From Process Data to PID Controller Design, by L. Wang and W. R. Cluett

E. ROGERS
J. O'REILLY

Preface

This book provides a unifying point of view of systems and control with a focus on linear systems. With so many available, one may ask why another book on linear systems and control is necessary, especially since each of many directions in control theory has reached a fairly mature state. However, the tools used to develop the existing results are fundamentally different, and there is no unifying point of view. In fact, there is still a wide-ranging debate over which approach is "best". Some of these debates and how they relate to the unifying themes of this book are described in this preface.

Frequency Domain versus State Space Methods

The frequency domain methods do not directly apply to time-varying and nonlinear systems (except for the isolated nonlinearities that allow describing function analysis). On the other hand, the frequency domain, using the classical methods of Evans, Bode, and Nyquist, readily yields simple controllers of low order. The tractable state space optimization theory yields controllers of high order, as opposed to the easily accommodated low-order controllers of classical control. The Youla parametrization [158], when used in optimization problems (see [6]), may yield even higher-order controllers. Many researchers in control theory view the lack of a tractable method for the design of low-order controllers as the most fundamental deficiency of modern control theory. This book presents a parametrization of all stabilizing controllers of low order (equal to or less than the plant order) to aid in the practical problem of designing a simple controller for a high-order complex system.

Deterministic versus Stochastic Methods

The debate over whether to treat the system as deterministic or stochastic has been as heated as the debate over frequency or time domain methods. One argument against stochastic methods is that guarantees of absolute values of signals are not possible; yet, bounds on signals are very practical and important considerations when dealing with real systems with sensor and actuator saturations, and physical limits of stresses in structures, etc. On the other hand, one argument for stochastic analysis is that there are no physical sensors and actuators without some electronic noise on the outputs. In the stochastic literature, *covariance* analysis is a cornerstone in filtering theory and this powerful theory has found many practical uses. In the book by Skelton [125] a step is taken to unify the two points of view (deterministic and stochastic). By offering a deterministic treatment of excitations (initial conditions and impulses), an analysis of the deterministic system is given that is shown to be mathematically equivalent to the covariance analysis of stochastic processes. Thus, a deterministic interpretation of the covariance matrix is given. This book will further extend these ideas to both time-varying and discrete-time systems, so that a background in stochastic processes is not required to read this book.

Control versus Signal Processing

To our knowledge, all controllers that have been implemented in aerospace hardware have been designed in two uncoordinated steps. First, the control analyst designs the controller under the assumption of infinite precision computation (analog or digital). Secondly, the signal processing and computer sciences group within the company implements the given controller in a special set of coordinates chosen for scaling and minimization of computational errors in the flight computer. To integrate these two steps, a characterization of computational errors should be considered in the *initial* design of the controller. In this way it is possible to design controllers to be optimal in the presence of delay, in the presence of finite wordlengths in the control computer, in the digital to analog converter, and in the analog to digital converter. In this regard, this book follows the lead of Williamson [152] and Gevers *et al.* [36] to suggest some degree of integration of the fields of signal processing and control. Furthermore, we shall parametrize all controllers which can stabilize a given system with controllers of a given wordlength and time-delay, assuming a stochastic model of round-off errors.

Modeling versus Control Design

It is well understood by now that the modeling and control design problems are not independent [126]. For this reason, 20 years of model reduction research has left us with few guarantees about closed-loop performance using reduced-order models. Hence, many of these researchers moved to the more specialized subject of "controller reduction", with little additional success. Optimization problems for controllers of fixed order promise a better answer theoretically, but the corresponding nonlinear mathematical program presents computational problems. Alternately, one can separate the low-order controller design into two uncoordinated steps (model reduction then controller design, or controller design then controller reduction), presenting a more tractable but less theoretically satisfying model or controller reduction. This book takes *none* of these three approaches to fixed-order controller design, but introduces a unification of the modeling and control problem in the sense that the controller order is not fixed at the outset, but is guaranteed to be equal to or less than the order of the plant.

Scalar versus Multiobjective Methods

Most of the literature on optimal control deals with a single objective function (or "cost" function). It is not practical to judge the relative merits of a controller by computing a single number. On the other hand multiobjective optimal (or "pareto-optimal") control is computationally demanding. Great improvement in such problems has been made by first *ignoring* optimization, and characterizing all solutions that lie within a set of inequality constraints. The necessary and sufficient conditions for a solution to lie within a set of (multiple) inequality constraints are usually considerably simpler and more tractable than the necessary conditions for optimization. Optimization can often be approached from the point of view of studying "feasibility" of the inequality constraints, by reducing the upper bounds in the inequality constraints until there are no feasible solutions. We take this view and do not focus on optimization, but on the satisfaction of matrix inequality constraints. This gives a multiobjective nature to our problem.

Performance versus Stability

The vast majority of control theory has focused on *stability*, while *performance* guarantees have received much less attention. Being able to guarantee specific bounds on the response is needed in practice. Indeed, "stability" is a mathematical property of a particular "mathematical model" of the system and *never* a guaranteed property of the physical system itself. When writing a book on controller design for real world implementation one must differentiate between the physical system and its mathematical model. Quite often the model of the closed-loop system may have the "asymptotic stability" property, but the physical system *never* does. Even in the absence of external disturbances the control signal produced by a digital computer does not asymptotically go to zero, but to some sort of limit cycle dictated by the computer wordlength. Therefore, the ability to guarantee upper bounds on signals is *more important* than the (admittedly artificial) ability to guarantee asymptotic stability. If stability has been overemphasized in the control literature, then performance has been underemphasized. Lack of stability might be considered a disaster, but in physical systems disaster comes *long before* instability. A stable billion dollar space telescope would be considered a disaster if it fails to meet the pointing accuracy required to make the observations and pictures useful.

The comparison of classical versus modern control methods has been characterized by the following oversimplification: "In classical control one designs for stability, but then must check for performance, whereas in modern control, one designs for performance, but then must check for stability". Both classical and modern approaches have made progress toward integration of stability and performance design objectives, but only for highly specialized definitions of performance and stability margins. Without some assumptions, optimality does not guarantee stability. Hence, it has been popular recently to optimize the H_2 performance scalar (integral squared of the transfer function) subject to an upper bound on the "H_∞ norm" (peak of the transfer function). The H_∞ bound delivers a certain kind of stability margin and the H_2 norm represents a (scalar) performance measure, related to the Root Mean Squared (RMS) behavior of the output signal. The control design procedures in this book can include these kind of design objectives. We also seek a method that can treat time domain "L_∞ norms" (peak of time response), since these represent the physical limits of real time signals, such as sensors and actuators that are subject to saturation.

This book discusses two kinds of robustness: "performance" robustness and "stability" robustness. By these phrases we mean that certain performance and stability guarantees hold in the presence of specified errors in the plant model or in the disturbances.

The 1956 result of Massera [88] states that if a system is asymptotically stable, then there exists a Lyapunov function to prove it so. The practical value of this theorem is that the search for a Lyapunov function is not a waste of time (the set of functions with the properties we seek is never an empty set). The theoretical value of this theorem is that the search for a characterization of all stabilizing controllers would be well served by a characterization of all Lyapunov functions. Moreover for linear systems, only quadratic Lyapunov functions are needed to capture necessary and sufficient conditions for stability. That is the approach of this book, to parametrize (for linear systems) the set of all quadratic Lyapunov functions. This allows us to capture the set of all stabilizing controllers. From this set, all controllers which can meet the (certain matrix inequality) performance requirements are parametrized. The connections between Lyapunov stability and deterministic definitions of RMS performance allows the above unification of the theories for performance and stability guarantees.

Choosing a Design Space

From 1930 to 1955 the two-dimensional space of the *complex plane* was the workhorse of control theory. Design in this space allows placement of poles and zeros. Various other two-dimensional plots have special use in such designs, including the Bode plot, the Nichols plot, the Nyquist plot and the Root Locus. The essential tool here was complex analysis.

In the 1960s *state space* became the popular design space. Because the n-dimensional state space does not lend itself to plotting, the graphical methods made popular for the two-dimensional complex plane played less of a role in this period of control development. Rather, the essential tools used here were optimization, the calculus of variations, ordinary differential equations and Hilbert spaces.

In the 1980s a modern version of complex analysis extended the classical notions of stability margins to Multi-Input Multi-Output systems in a systematic way. In the Single-Input Single-Output (SISO) system the Nyquist plot (for unit feedback systems) could indicate the peak magnitude of the closed-loop frequency response (closed-loop Bode diagram) by adding "M circles" [51] to the Nyquist plot. The M circles are the locus of points in the complex plane of the open-loop transfer function that correspond to the same magnitude in the closed-loop frequency response. The modern H_∞ theory allows the characterization of the peak magnitude of not just one transfer function, but the "magnitude" (norm) of a transfer matrix. The essential tools here were Hardy spaces and complex analysis [29].

In the late 1980s a state space interpretation of H_∞ theory was provided [25], and robust control became a strong focus for research. Guaranteeing an upper bound on the H_∞ norm in the presence of parameter perturbations remains a major focus of control theorists to this day. Objectives of robust H_∞ control are *stability margin* type of goals (in the presence of plant perturbations), and maintaining an H_∞ norm is qualitatively a *stability margin* type of design goal.

In 1977, Nagayasu [92] introduced (in Japanese) the existence condition for solving the state feedback covariance control problem. During the mid 1980s this approach to control, called "covariance control", was developed in the English literature [52]. The objectives here were the assignment of all $\frac{n(n+1)}{2}$ elements of the state covariance matrix. The $\frac{n(n+1)}{2}$ dimensional "covariance space" has some important features. By increasing the design space from n (as in state space) to $\frac{n(n+1)}{2}$ the class of systems that can be studied with the simple tools of *linear algebra* are increased! That is, the class of problems that can be represented as linear problems in the $\frac{n(n+1)}{2}$ dimensional space is larger than the class of problems that are linear in state space. In addition to enlarging the class of control problems that can be treated by linear methods, the covariance control theory needs only the tools of *linear algebra*.

The fundamental contribution of this book is to show that a large class of control problems *reduce to a problem in linear algebra*. In fact, some 18 control problems (nine continuous-time plus nine discrete-time) reduce to a *single* linear algebra problem! Hence linear algebra is the enabling tool that allows students to view the vast majority of all linear system control problems from a common setting. The goal of this book is to show how to use linear algebra to achieve this goal.

It is a pleasure to acknowledge the diligence and patience of Jill Comer and Becky May who typeset much of the manuscript, and to Raymond de Callafon for his review and editing of the manuscript in 1997. Thanks also go to the students of Purdue's course, "Control of Uncertain Systems", who have provided helpful feedback since 1990 when this book was initiated. This book began as a book on covariance control to summarize the covariance control work since 1985. However, the main linear algebra tools which were developed for these purposes have a much broader application than covariance control, so the unifying theme of the book became the driving influence, once we realized that a single problem in linear algebra characterizes more that 20 different and diverse problems in control theory.

CHAPTER ONE

Introduction

To illustrate the concepts of this book, a simple system described by a scalar differential equation is useful

$$\dot{x}(t) = ax(t) + dw(t), \qquad \dot{x} \triangleq \tfrac{d}{dt}[x(t)]$$
$$y(t) = cx(t) \tag{1.1}$$

where $x(t)$ is the state, $w(t)$ is the external input to the system, and $y(t)$ is the output of interest. Scalars a, d and c are given constants. The solution of this differential equation is

$$y(t) = ce^{at}x(0) + \int_0^t ce^{a(t-\tau)}dw(\tau)d\tau. \tag{1.2}$$

Consider two possible excitations due to the impulsive disturbance in $w(t)$ and the nonzero initial state $x(0)$ indexed by $i = 1, 2$ where

$$\{w(t) \;=\; w\delta(t), \quad x(0) = 0\} \text{ for } i = 1 \tag{1.3a}$$
$$\{w(t) \;=\; 0, \quad x(0) = x_0\} \text{ for } i = 2. \tag{1.3b}$$

Let $y(i, t)$ denote the response $y(t)$ in the event of excitation i. Then

$$y(1, t) \;=\; ce^{at}dw \tag{1.4a}$$
$$y(2, t) \;=\; ce^{at}x_0. \tag{1.4b}$$

If the differential equation (1.1) describes the time history of the error due to the excitations $w(t)$ and $x(0)$ from a certain desirable system state, or equivalently, if the variable y dictates the error, performance of the system can be measured by

$$Y \triangleq \sum_{i=1}^{2} \int_0^\infty y^2(i, t)dt \tag{1.5}$$

$$X \triangleq \sum_{i=1}^{2} \int_0^\infty x^2(i, t)dt \tag{1.6}$$

where it can readily be shown that X and Y satisfy the following:

$$Y = c^2 X, \; 0 = 2Xa + d^2 W + X_0 \tag{1.7}$$

where $W \triangleq w^2$ and $X_0 \triangleq x_0^2$. Note that we do not presume to know the actual initial condition x_0, or the impulsive disturbance w, only their magnitudes. Actually, the presumption about the excitations is even less specific than uncertainties about x_0 and w. Note in (1.7) that our performance measure X is equivalent over a wide class of excitations characterized by $d^2 W + X_0 = constant$.

Let us refer to $y(t)$ as *first-order* information about system (1.1), and to Y as *second-order* information. The study of *first-order* information occupies much of the control literature. However, it is the premise of this book that *second-order* information can be more informative, and indeed, that many essential properties of (1.1) can be characterized by X, the second-order information about the state, x. We will later motivate the use of the word "covariance" as a label for X.

These ideas will be illustrated in this chapter, but for motivational illustrations we will use the scalar system (1.1), or its control problem counterpart:

$$\begin{aligned} \dot{x}(t) &= ax(t) + bu(t) + dw(t) \\ y(t) &= cx(t), \qquad u(t) = gy(t) \end{aligned} \tag{1.8}$$

where $u(t)$ is the control input and g is the feedback gain to be determined. The subsequent chapters will develop these ideas and the mathematical design tools for the general Multi-Input Multi-Output (MIMO) case.

1.1 OUTPUT PERFORMANCE AND SECOND-ORDER INFORMATION

From (1.1)–(1.5), define a certain norm (size) of $y(i, t)$ by

$$||y||_{\mathcal{L}_2} \triangleq \left[\sum_{i=1}^{2} \int_0^\infty y^2(i, t)dt \right]^{1/2}. \tag{1.9}$$

The quantity (1.9) is closely related to a Root Mean Square (RMS) value, $\left[\frac{1}{T}\int_0^T y^2(t)dt\right]^{1/2}$ for some finite T, which is frequently used as a measure of system performance. However, the infinite horizon ($T \to \infty$) is often more important in engineering problems. More motivation for our use of performance measure (1.9) will be added later. Quite often in practice the $||y||_{\mathcal{L}_2}$ value of the output is of interest, rather than the actual output trajectory $y(t)$. For example, in telescope or antenna pointing problems, pointing control accuracies need not exceed the resolution of the film or image processing equipment. As long as a light ray is controlled such that it remains within the grain size of the film (the faster the film, the larger the grain size), the quality of the picture is limited only by the properties of the film and optics and *not* by the control system. This is the ultimate objective of the control designer, to design the control system so that the controller is not the limiting factor in the total system performance.

Physical systems are never asymptotically stable, nor completely controllable, nor completely observable. That is, there is always a limit on our ability to control the system state (we cannot take it exactly to zero), or to observe (measure) what the complete system is actually doing. Nor can we measure with infinite precision. There are periodic behaviors (limit cycles) in the output due to round-off in finite precision computers, or drifts and nonlinearities in the actuators and sensors. But our inability to control the output to zero or to know exactly what the output is doing might not prevent us from *placing bounds* on their RMS or $||y||_{\mathcal{L}_2}$ values. We will show that *bounds* on the absolute value of first-order information (signals) can be related explicitly to second-order information. This is the fundamental advantage of working with second-order information; and we will show that "performance on the average", or "bounds on the absolute values" constitutes a more realistic objective for control design than the "drive the output zero" requirement associated with *first-order* system properties. We will develop a complete theory for assigning a specific value of X, which fixes the values $||y||_{\mathcal{L}_2} = Y^{1/2} = (c^2X)^{1/2}$, or provide upper bounds on the same. We will also be able to fix upper bounds on the absolute value $|y(t)|$, for certain disturbances.

In this book on the control of second-order information, we begin with linear systems with a detailed study of their stability properties.

1.2 STABILITY, POLE LOCATIONS AND SECOND-ORDER INFORMATION

Consider the system (1.1). The system is said to be stable if $x(t)$ approaches zero as $t \to \infty$ for all initial x_0 with $w(t) = 0$. In view of (1.2), the system is stable if and only if $a < 0$. We argue here that second-order information is closely connected to stability. Indeed, a given set of first-order data $x(t)$, $0 \le t < \infty$, might not reflect any stability properties of a. See from (1.2) and (1.4) that $x(t)$ can be zero for some initial condition ($x_0 = -dw$), *independently* of the properties of a.

On the other hand, a given positive value of X is *equivalent* to stability of a. To see this, compute X in (1.7) to obtain for system (1.1),

$$2aX = -(d^2W + X_0). \tag{1.10}$$

Hence $a < 0$ is equivalent to $X > 0$ provided $d^2W + X_0 > 0$. Furthermore, Y in (1.5) can be computed from the *experimental* first-order data $y(i, t)$, without knowledge of model data a, d, c. In our attempt to develop control techniques that can be tested in practice, and improved by redesign in the field, this fact cannot be overemphasized. If data is available, Y can be computed *directly* from (1.5), or, equivalently, from the model-based equation (1.7).

Now consider the control problem to choose g in (1.8). To apply (1.7) to model (1.8), instead of model (1.1), replace a in (1.7) by $a + bgc$ and solve (1.7) for g,

$$g = -\frac{2aX + d^2W + X_0}{2bcX}. \tag{1.11}$$

Having noted already the equivalence of $X > 0$ and $a + bgc < 0$ (and such equivalence *does not* depend on the specific numerical values of the positive number $d^2W + X_0$, hence stability does not depend on initial conditions), we can regard (1.11) as a parametrization of all stabilizing controllers, $u = gy$. The parametrization is explicit in terms of an arbitrary positive number X, where for design purposes, X should be chosen according to a *desired* constraint $Y = c^2X = ||y||_{\mathcal{L}_2} < \overline{Y}$.

Example 1.2.1 Consider a vertically fired rocket [125], where m = *mass*, g = *gravity constant*, v = *speed*, f_c = *control force*, f_n = *impulsive disturbance* such as arising from startup transients in the rocket firing. In the form of (1.8), the model is given by the following system parameters:

$$a = 0,\ b = m^{-1} = d,\ c = 1,\ u \triangleq f_c - mg,\ w = f_n,\ y = x = v - \bar{v}$$
$$f_n(t) = w\delta(t),\ W = w^2,\ X_0 = (v(0) - \bar{v})^2$$

where $w(t)$ is the assumed impulsive error in the thrust, $\bar{v}$ is the desired (constant) velocity and $v(0)$ is the assumed initial velocity. (We specify here only the *magnitude* of the initial error $X_0 = (v(0) - \bar{v})^2$, we do not presume to know the actual initial condition.) Then the control $u = g(v - \bar{v})$ that regulates the speed to the desired value $\bar{v}$, with a guaranteed error $||y||_{\mathcal{L}_2} = ||(v - \bar{v})||_{\mathcal{L}_2} = 0.01$ is, from, (1.11),

$$g = -\frac{m}{2}10^4\left[\frac{W}{m^2} + X_0\right].$$

Note that the *evaluation* of the $||y||_{\mathcal{L}_2}$ performance of any controller requires knowledge of a set of excitation events (which we loosely call "disturbances", w and $v(0)$ in the above example). Since we are controlling only second-order information, an entire set of various disturbances can lead to the same controller. In this example the controller is invariant under the infinite number of disturbances $(w,\ v(0))$ satisfying

$$\frac{w^2}{m^2} + (v(0) - \bar{v})^2 = constant.$$

In *first-order* analysis the disturbance environment must be specified *exactly* to solve the differential equations for $y(t)$. However, in second-order analysis the disturbance environment need not be exactly specified. It is reasonable and easier in practice to specify a possible set of disturbances, rather than pinning our analytical predictions on a single specific disturbance.

Note that the eigenvalues of the closed-loop system (of course there's only one $a + bgc$ in this example) can also be related to the second-order information X or Y

$$a + bgc = -\frac{d^2W + X_0}{2X} \tag{1.12}$$

and the performance constraint $X \leq \overline{X}$ guarantees a stability margin

$$a + bgc \leq -\frac{d^2W + X_0}{2\overline{X}}. \tag{1.13}$$

Hence, both *transient* properties (eigenvalues), and *steady-state* properties $||y||_{\mathcal{L}_2}$ can be related to second-order information X.

For the general MIMO system, we will develop a parametrization of all stabilizing controllers (state feedback, output feedback, or dynamic controllers) in much the same spirit as (1.11). These results will also be easy to use for assigning poles to a region (by choosing X appropriately, as in (1.13)).

1.3 STABILITY ROBUSTNESS AND SECOND-ORDER INFORMATION

The "stability robustness" issue addresses this concern: Given that the nominal closed-loop system (1.8) is stable ($a+bgc < 0$), what can be guaranteed about the stability of $a+\Delta a+bgc$, where Δa represents the uncertainty in a (of course, we are oversimplifying everything in this motivational chapter; the b, c and d may also change in general). Y changes to $Y + \Delta Y$ when a changes to $a + \Delta a$. Suppose we desire a specific $||y||_{\mathcal{L}_2}$ performance bound on the output, say, $||y||_{\mathcal{L}_2}{}^2 = Y + \Delta Y \leq 0.02$, for all uncertainty Δa in the range

$$-0.1 \leq \frac{\Delta a}{|a|} \leq 0.1 \tag{1.14}$$

and we also desire stability of $a+bgc+\Delta a$ over this range. From (1.10), with $a+\Delta a+bgc$ replacing a, and $X + \Delta X$ replacing X,

$$Y + \Delta Y = c^2(X + \Delta X) = -\frac{c^2(d^2W + X_0)}{2(a + \Delta a + bgc)} \leq 0.02. \tag{1.15}$$

From (1.14) and (1.15), it can be shown that the controller (1.11) will satisfy both the stability and the performance constraint for this range of choices of X

$$X \leq \frac{d^2W + X_0}{0.2|a| + 50(d^2W + X_0)c^2} \tag{1.16}$$

in the design equation (1.11). The text will relate stability robustness properties of general linear systems to the second-order information X and Y, and parametrize robustly stabilizing controllers in terms of X as in (1.11).

1.4 DISTURBANCE ATTENUATION AND SECOND-ORDER INFORMATION

Suppose now we desire to bound the *first-order* property $|y(t)| \leq \sqrt{0.02}$ subject to zero initial state $x(0) = 0$ and arbitrary disturbances $w(t)$ with a known upper bound β on the "energy" of the disturbance

$$\int_0^\infty w^2(t)dt \leq \beta. \tag{1.17}$$

By this definition, Figure 1.1 shows three disturbances with the same energy level, β,

$$\int_0^\infty w_j^2(t)dt = 2, \qquad j = 1, 2, 3$$

$$w_1(t) = \sqrt{2}e^{-(1/2)t}, \qquad w_2(t) = \begin{cases} \sqrt{2}, & 1 \leq t \leq 2 \\ 0 \end{cases}$$

$$w_3(t) = \begin{cases} \frac{\sqrt{3}}{2}(t-3), & 3 \leq t \leq 5 \\ 0 \end{cases}.$$

The text will show for the general case what we now state for the scalar case,

$$y^2(t) \leq Y\left[\int_0^\infty w^2(\tau)d\tau\right], \quad 0 \leq t \leq \infty. \tag{1.18}$$

Hence, in the presence of an arbitrary disturbance $w(t)$, the peak value of $y(t)$ over all time is bounded by the right-hand side of (1.18), where Y is computed by (1.7) with $X_0 = 0$ *as if* $w(t)$ were impulsive with intensity $W = 1$. Hence, Y describes a disturbance robustness feature of a linear system. In fact, the inequality in (1.18) becomes arbitrarily close to

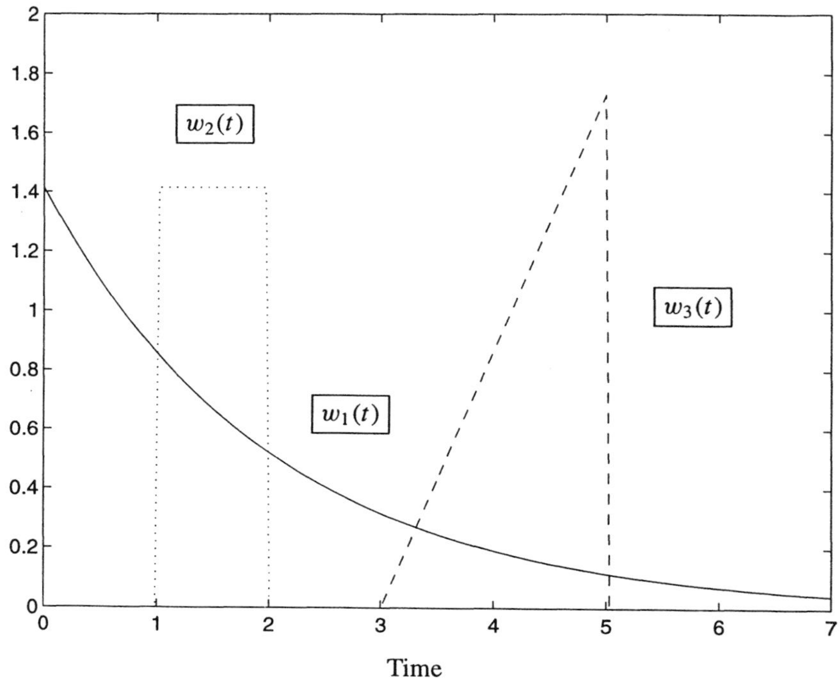

Figure 1.1 Energy equivalent signals.

equality for a special disturbance $\bar{w}(t)$ [19]. This special "worst-case" disturbance is useful in practice, since the engineer can guarantee that if $y(t)$ has acceptable peak values with disturbance $\bar{w}(t)$ satisfying (1.17), then $y(t)$ will have acceptable peak values *for all* other disturbances whose energy level is bounded by the same value β. Such disturbances will be shown.

By choosing the second-order information c^2X appropriately, desired disturbance robustness properties are obtained, guaranteeing upper bounds on the peak value of the signal $y(t)$ over all time. In systems with sensor and actuator saturations this signal bounding capability is extremely important.

1.5 STABILITY MARGINS MEASURED BY H_∞ NORMS

For the closed-loop system (1.8), the peak value of the frequency response is a reliable measure of relative stability (for a unity feedback system the M circles on the Nyquist plot indicate this magnitude),

$$H_\infty \triangleq \max_\omega \left|c(j\omega - (a+bgc))^{-1}d\right| = \max_\omega \frac{|cd|}{\{\omega^2 + (a+bgc)^2\}^{1/2}} = \frac{|cd|}{|a+bgc|}.$$

It will be shown below that the number we define here as H_∞ is less than or equal to a specified number $\gamma > 0$ if $X_\infty > 0$ solves

$$0 = 2(a + bgc)X_\infty + \gamma^{-2}c^2X_\infty^2 + d^2. \tag{1.19}$$

Suppose $c \neq 0$, $d \neq 0$ and the equation (1.19) has a positive real solution X_∞. This is equivalent to stability of $a + bgc$ and

$$\frac{|cd|}{|a + bgc|} \leq \gamma$$

since the positive solution of (1.19) is

$$X_\infty = \frac{a + bgc}{\gamma^{-2}c^2}\left[-1 + \sqrt{1 - \left(\frac{|cd|}{\gamma|a + bgc|}\right)^2}\right].$$

Solving (1.19) for g, all stabilizing gains yielding the H_∞ norm less than or equal to γ are

$$g = -\frac{2aX_\infty + d^2 + \gamma^{-2}c^2X_\infty^2}{2bcX_\infty}, \quad X_\infty > 0. \tag{1.20}$$

Notice that the linear algebra problem that solves the "covariance" control problem for g, yielding (1.11), also solves (1.19) for g, simply by replacing the forcing term $(d^2W + X_0)$ in (1.7) (where $a \rightarrow a + bgc$) by $(d^2 + \gamma^{-2}c^2X_\infty^2)$. Compare (1.20) with (1.11) to see this point. Hence, almost all essential results of this book come from the solution of the covariance control problem (1.11). If $W = 1$ and $X_0 = 0$, then $d^2W + X_0 < d^2 + \gamma^{-2}c^2X_\infty^2$ and hence $X < X_\infty$ if X_∞ is positive real. Hence X_∞ is an upper bound on the covariance of the system. By choosing the same controller to satisfy both a performance constraint ($X < \overline{X}$) from (1.7), and a stability margin constraint (1.19), for some $\gamma > 0$, the competing objectives of stability versus performance can be satisfied or traded. Solving *either* (1.7) or (1.19) for the controller g is a problem in linear algebra. To prepare for these problems the next chapter is devoted to a review of linear algebra and matrix methods.

1.6 COMPUTATIONAL ERRORS

One of the fundamental deficiencies of existing control theory is the lack of treatment of the finite precision computational issues. Suppose (1.1) represents a dynamic system under measurement feedback control, controlled by analog computation, which introduces computational errors of the form

$$\begin{aligned} \dot{x} &= ax + b(u + e_u) + w \\ u &= g(y + e_y) \\ y &= cx \end{aligned}$$

where e_y and e_u are the errors introduced by the sensor and actuator errors, respectively (in digital control e_y and e_u would be the errors introduced by the finite wordlength of the Analog to Digital (A/D) converters and the Digital to Analog (D/A) converters). The magnitude of the errors e_u and e_y depend upon the quality and accuracy of the sensor and actuator hardware. In this illustration these "magnitudes" E_u and E_y are the square of the intensities of impulsive errors e_u and e_y. W is the square of the intensity of the impulsive $w(t)$. Applying these three impulses one at a time, the closed-loop system

$$\dot{x} = (a + bgc)x + be_u + bge_y + w$$

yields the norm $||x||_{\mathcal{L}_2}$ of the state

$$X \triangleq ||x||^2_{\mathcal{L}_2} = \sum_{i=1}^{3} \int_0^{\infty} x^2(i, t)dt \tag{1.21}$$

satisfying

$$0 = 2X(a + bgc) + b^2 E_u + b^2 g^2 E_y + W. \tag{1.22}$$

Completing the square and factoring yields the equation, linear in g,

$$bg + \frac{cX}{E_y} = \pm\frac{1}{\sqrt{E_y}}\left[\frac{c^2X^2}{E_y} - 2aX - b^2E_u - W\right]^{1/2}.$$

The X corresponding to a real solution g follows from

$$\frac{c^2X^2}{E_y} - 2aX - b^2E_u - W \geq 0.$$

Hence the smallest admissible $X = X^o > 0$ is

$$X^o \triangleq \frac{1}{c^2}\left[aE_y + \sqrt{a^2E_y^2 + E_yc^2(W + b^2E_u)}\,\right] \tag{1.23}$$

corresponding to the value $g = g^o$,

$$g^o = -\frac{cX^o}{bE_y} \tag{1.24}$$

Note that the optimal performance is achieved at a *finite* g^o as opposed to the *infinite* gain predicted by the standard optimal control theory which ignores computational errors (hence, an infinite precision sensor, $E_y = 0$), where, from (1.22), X is arbitrarily small if $|g|$ is arbitrarily large. For an infinite precision actuator $E_u = 0$,

$$X^o(E_u = 0) = \frac{1}{c^2}\left[aE_y + \sqrt{a^2E_y^2 + c^2E_yW}\,\right] > 0.$$

Hence, an accurate sensor is more important than an accurate actuator, in the sense that X^o can be taken to zero with a perfect sensor, but not with a perfect actuator. Two observations are important here. In the presence of computational errors in the controller:

1 There exist lower bounds on performance which cannot be predicted by the usual infinite precision assumptions for controller implementation.
2 The performance bounds are a function of the precision with which the controller (sensors, actuators, control computer) is implemented.

These concepts will be developed in the discrete-time case to include the effects of A/D, and D/A conversion, computational delay, and computer round-off.

CHAPTER 1 SUMMARY

The tools of this text will allow (but not be limited to) the following types of control design specifications, related to our example (1.8): We seek a g to satisfy some combinations of the following:

(i) Disturbance attenuation requirement for finite energy disturbances: We require

$$|y(t)| \leq \epsilon_1, \quad \forall t \geq 0$$

for all $\int_0^\infty w^2(t)dt \leq \epsilon_2$, for given ϵ_1 and ϵ_2.

(ii) Disturbance attenuation requirement for impulsive disturbances: We require

$$\begin{aligned} ||y||_{\mathcal{L}_2} &\leq \epsilon_4 \\ ||u||_{\mathcal{L}_2} &\leq \epsilon_5 \end{aligned}$$

in response to $w(t) = w\delta(t)$ with $|w| \leq \epsilon_6$, for given ϵ_4, ϵ_5 and ϵ_6.

(iii) Robust stability requirement: For a given ϵ_3 we require that the closed-loop system remain stable when a changes to $a + \Delta a$, where the perturbation Δa satisfies

$$|\Delta a| \leq \epsilon_3.$$

(iv) The poles of the closed-loop system should lie in a specified region in the complex plane (e.g. a circle with center given at $-\epsilon_6$ and radius given by $0 < \epsilon_7 < \epsilon_6$, or a specified rectangular region).

(v) The H_∞ norm of a specified closed-loop transfer function should be (the frequency response should have a peak) less than a specified number ϵ_8.

(vi) Controllers of low (less than plant) order to accomplish (i)–(v) are required (only constant gain state feedback is considered in the motivational examples of this introductory chapter).

(vii) Performance should be guaranteed in the presence of computational errors in the controller and A/D, D/A converters.

Such design tools are developed for both continuous-time and discrete-time systems. Beginning with these elementary motivations, we seek to develop a unified approach to control built upon second-order information as the fundamentally important data. The advantages of this approach to education and the theory of control are several:

1 Using second-order properties in the design goals allows one to parametrize all stabilizing controllers in terms of physically meaningful data X.

2 The RMS performance of multiple outputs (Y is a matrix in this case) can be simultaneously controlled to preassigned values, or upper bounds

$$Y \leq \bar{Y}.$$

This lends a multiobjective capability to the theory of Multi-Input, Multi-Output systems.

3 In the presence of a class of uncertainties in the model data, stability and performance bounds can be incorporated into the control design specifications. We refer to these two design objectives as *stability robustness* and *performance robustness*.

4 All stabilizing controllers of *fixed order* can be parametrized in terms of second-order information. This text will develop tools to design low-order controllers with guaranteed stability and RMS performances on multiple inputs and outputs.

5 Pole assignment can be accomplished by proper choices of the second-order information, and poles can be assigned to a *region*.
6 Disturbance attenuation properties can be accomplished by proper choices of the second-order information.
7 The extensions to time-varying systems are straightforward, but are not contained in this book.

CHAPTER TWO

Linear Algebra Review

This chapter documents, for later use, certain results from linear algebra. Some common notations are listed in Table 2.1, and a more fundamental review of linear algebra appears in Appendix A. Many results of this chapter are taken from [153].

Table 2.1 Common notations

j	$\triangleq$	imaginary unit ($\sqrt{-1}$)
$\mathbf{I}_n$	$\triangleq$	$n \times n$ identity matrix
$\mathbf{A}^T$	$\triangleq$	transpose of matrix $\mathbf{A}$
$\mathbf{A}^*$	$\triangleq$	complex conjugate transpose of $\mathbf{A}$
$\mathbf{A}^+$	$\triangleq$	the Moore–Penrose inverse of $\mathbf{A}$
$tr\mathbf{A}$	$\triangleq$	$\sum_{i=1}^{n} a_{ii}$ (trace of $\mathbf{A}$)
$\lambda_i[\mathbf{A}]$	$\triangleq$	eigenvalue λ_i of matrix $\mathbf{A}$
$\mathbf{A} > \mathbf{0}$	$\triangleq$	$\lambda_i[\mathbf{A}] > 0, \ \forall i$ ($\mathbf{A}$ "symmetric positive definite")
$\mathbf{A} \geq \mathbf{0}$	$\triangleq$	$\lambda_i[\mathbf{A}] \geq 0, \ \forall i$ ($\mathbf{A}$ "symmetric positive semidefinite")
$\mathbf{A} = \mathbf{E}\mathbf{\Lambda}\mathbf{E}^{-1}$	$\triangleq$	spectral decomposition of $\mathbf{A}$ $\mathbf{\Lambda} \triangleq$ Jordan form $\mathbf{E} \triangleq$ modal matrix (of eigenvectors or generalized eigenvectors)
$\mathbf{A} = \mathbf{U}\mathbf{\Sigma}\mathbf{V}^*$	$\triangleq$	singular value decomposition (SVD) of $\mathbf{A}$
$\Vert\mathbf{A}\Vert_F = tr[\mathbf{A}\mathbf{A}^T]^{1/2}$	$\triangleq$	Frobenius norm of $\mathbf{A}$
$\Vert\mathbf{A}\Vert = \bar{\sigma}[\mathbf{A}]$	$\triangleq$	maximum singular value of $\mathbf{A}$
$\underline{\sigma}[\mathbf{A}]$	$\triangleq$	minimum (nonzero) singular value of $\mathbf{A}$
$\mathbf{A} = \mathbf{F}\mathbf{F}^*$	$\triangleq$	factoral decomposition of $\mathbf{A} \geq \mathbf{0}$

2.1 SINGULAR VALUE DECOMPOSITION

The modal data of a square matrix are its eigenvalues and eigenvectors. The matrix decomposition $\mathbf{A} = \mathbf{E}\mathbf{\Lambda}\mathbf{E}^{-1}$ (where the eigenvalues of $\mathbf{A}$ appear on the diagonal of $\mathbf{\Lambda}$ and the eigenvectors are the columns of $\mathbf{E}$) is called the spectral decomposition of $\mathbf{A}$. If $\mathbf{A}$ is a Hermitian matrix, then the eigenvalues and the eigenvectors of the spectral decomposition of $\mathbf{A}$ can be chosen to be real, and the eigenvectors can be chosen to be orthonormal. The matrix decompositions in this section require modal data computations for Hermitian matrices. In this regard, this section is a special case of spectral decomposition. In another way it is more general, since we shall *not* restrict $\mathbf{A}$ to be square.

A square unitary matrix $\mathbf{U}$ is defined by the property $\mathbf{U}^*\mathbf{U} = \mathbf{I} = \mathbf{U}\mathbf{U}^*$, where the * superscript denotes complex conjugate transpose. The main result is as follows.

Theorem 2.1.1 *Let* $\mathbf{A} \in \mathcal{C}^{k\times n}$ *be a matrix of rank* r. *Then there exist unitary matrices* $\mathbf{U}$ *and* $\mathbf{V}$ *such that*

$$\mathbf{A} = \mathbf{U}\mathbf{\Sigma}\mathbf{V}^* \tag{2.1}$$

where $\mathbf{U}$ *satisfies*

$$\mathbf{A}\mathbf{A}^*\mathbf{U} = \mathbf{U}\mathbf{\Sigma}\mathbf{\Sigma}^* \tag{2.2}$$

and $\mathbf{V}$ *satisfies*

$$\mathbf{A}^*\mathbf{A}\mathbf{V} = \mathbf{V}\mathbf{\Sigma}^*\mathbf{\Sigma} \tag{2.3}$$

where $\mathbf{\Sigma}$ *has the canonical structure*

$$\mathbf{\Sigma} = \begin{bmatrix} \mathbf{\Sigma}_0 & \mathbf{0} \\ \mathbf{0} & \mathbf{0} \end{bmatrix}, \qquad \mathbf{\Sigma}_0 = \mathrm{diag}\,(\sigma_1, \ \ldots, \ \sigma_r) > \mathbf{0}. \tag{2.4}$$

The numbers σ_i, $i = 1, \ \ldots, \ r$ *are called the nonzero singular values of* $\mathbf{A}$.

Proof. Since $\mathbf{A}^*\mathbf{A} \geq \mathbf{0}$, its eigenvalues are all real and non-negative. Let σ_i^2, $i = 1, \ldots, n$ be the eigenvalues of $\mathbf{A}^*\mathbf{A}$. For convenience, we arrange σ_i such that

$$\sigma_1^2 \geq \sigma_2^2 \geq \ldots \geq \sigma_n^2 \geq 0.$$

Suppose that the rank of $\mathbf{A}$ is r. Then, the rank of $\mathbf{A}^*\mathbf{A}$ is r, and we have

$$\sigma_1^2 \geq \sigma_2^2 \geq \ldots \geq \sigma_r^2 > 0, \qquad \sigma_{r+1} = \ldots = \sigma_n = 0.$$

Let $\mathbf{v}_i$, $i = 1, \ldots, n$ be the orthonormal eigenvectors of $\mathbf{A}^*\mathbf{A}$ associated with σ_i^2. Define

$$\mathbf{V} = \left[\mathbf{v}_1 \ \mathbf{v}_2 \ldots \mathbf{v}_r \,\middle|\, \mathbf{v}_{r+1} \ldots \mathbf{v}_n\right] = \left[\ \mathbf{V}_1 \ \middle| \ \mathbf{V}_2\ \right].$$

Then from the spectral decomposition theorem

$$\mathbf{V}^*\mathbf{A}^*\mathbf{A}\mathbf{V} = \begin{bmatrix} \mathbf{\Sigma}_0^2 & \mathbf{0} \\ \mathbf{0} & \mathbf{0} \end{bmatrix}$$

where $\mathbf{\Sigma}_0^2 = \mathrm{diag}(\sigma_1^2, \ldots, \sigma_r^2)$. This provides

$$\mathbf{V}_2^*\mathbf{A}^*\mathbf{A}\mathbf{V}_2 = \mathbf{0} \tag{2.5}$$

and

$$\mathbf{V}_1^*\mathbf{A}^*\mathbf{A}\mathbf{V}_1 = \mathbf{\Sigma}_0^2 \tag{2.6}$$

or equivalently

$$\mathbf{\Sigma}_0^{-1}\mathbf{V}_1^*\mathbf{A}^*\mathbf{A}\mathbf{V}_1\mathbf{\Sigma}_0^{-1} = \mathbf{I}.$$

Define the $k \times r$ matrix $\mathbf{U}_1$ by

$$\mathbf{U}_1 \triangleq \mathbf{A}\mathbf{V}_1\mathbf{\Sigma}_0^{-1} \tag{2.7}$$

which is column unitary since $\mathbf{U}_1^*\mathbf{U}_1 = \mathbf{I}$. Hence, there exists matrix $\mathbf{U}_2$ such that $\mathbf{U} = [\mathbf{U}_1 \ \ \mathbf{U}_2]$ is unitary. Then

$$\mathbf{U}^*\mathbf{A}\mathbf{V} = \begin{bmatrix} \mathbf{U}_1^*\mathbf{A}\mathbf{V}_1 & \mathbf{U}_1^*\mathbf{A}\mathbf{V}_2 \\ \mathbf{U}_2^*\mathbf{A}\mathbf{V}_1 & \mathbf{U}_2^*\mathbf{A}\mathbf{V}_2 \end{bmatrix}.$$

Now (2.7) implies that $\mathbf{A}\mathbf{V}_1 = \mathbf{U}_1\mathbf{\Sigma}_0$ and (2.5) implies that $\mathbf{A}\mathbf{V}_2 = \mathbf{0}$. Hence

$$\mathbf{U}^*\mathbf{A}\mathbf{V} = \begin{bmatrix} \mathbf{\Sigma}_0 & \mathbf{0} \\ \mathbf{0} & \mathbf{0} \end{bmatrix} \quad \text{or} \quad \mathbf{A} = \mathbf{U}\begin{bmatrix} \mathbf{\Sigma}_0 & \mathbf{0} \\ \mathbf{0} & \mathbf{0} \end{bmatrix}\mathbf{V}^*. \qquad \square$$

For more on singular value decomposition, see [50], [153] or many other books on linear algebra.

2.2 MOORE–PENROSE INVERSE

It has been shown that any $n \times m$ matrix $\mathbf{A}$ can be expressed as a *singular value decomposition*

$$\mathbf{A} = \mathbf{U}\mathbf{\Sigma}\mathbf{V}^* \tag{2.8}$$

where $\mathbf{U}$ and $\mathbf{V}$ are $n \times n$ and $m \times m$ unitary matrices respectively, and $\mathbf{\Sigma}$ is an $n \times m$ matrix given by

$$\mathbf{\Sigma} = \begin{bmatrix} \mathbf{\Sigma}_0 & \mathbf{0} \\ \mathbf{0} & \mathbf{0} \end{bmatrix}; \quad \mathbf{\Sigma}_0 = \operatorname{diag}(\sigma_1, \sigma_2, \ldots, \sigma_r) \tag{2.9}$$

where $\sigma_k > 0$ for $k = 1, 2, \ldots, r$ are the *singular values* of $\mathbf{A}$, and r is the rank of $\mathbf{A}$.

Given the $n \times m$ matrix $\mathbf{A}$ in (2.8) and (2.9), define the $m \times n$ matrix

$$\mathbf{A}^+ \triangleq \mathbf{V}\begin{bmatrix} \mathbf{\Sigma}_0^{-1} & \mathbf{0} \\ \mathbf{0} & \mathbf{0} \end{bmatrix}\mathbf{U}^*. \tag{2.10}$$

From (2.8), (2.9) and (2.10) the $n \times n$ matrix $\mathbf{A}\mathbf{A}^+$ is given by

$$\mathbf{A}\mathbf{A}^+ = \mathbf{U}\begin{bmatrix} \mathbf{I}_r & \mathbf{0} \\ \mathbf{0} & \mathbf{0} \end{bmatrix}\mathbf{U}^*$$

and the $m \times m$ matrix $\mathbf{A}^+\mathbf{A}$ is given by

$$\mathbf{A}^+\mathbf{A} = \mathbf{V}\begin{bmatrix} \mathbf{I}_r & \mathbf{0} \\ \mathbf{0} & \mathbf{0} \end{bmatrix}\mathbf{V}^*.$$

Exercise 2.2.1 Show that $\mathbf{A}^+$ in (2.10) satisfies

$$\begin{aligned} \mathbf{A}\mathbf{A}^+\mathbf{A} = \mathbf{A}; &\qquad \mathbf{A}^+\mathbf{A}\mathbf{A}^+ = \mathbf{A}^+ \\ (\mathbf{A}\mathbf{A}^+)^* = \mathbf{A}\mathbf{A}^+; &\qquad (\mathbf{A}^+\mathbf{A})^* = \mathbf{A}^+\mathbf{A}. \end{aligned} \tag{2.11}$$

Definition 2.2.1 A matrix $\mathbf{A}^{+}$ which satisfies properties (2.11) is called the *Moore–Penrose inverse* of $\mathbf{A}$.

Theorem 2.2.1 *For every real $n \times m$ matrix $\mathbf{A}$, there exists a unique $m \times n$ matrix $\mathbf{A}^{+}$, the Moore–Penrose inverse of $\mathbf{A}$, which satisfies (2.11). Moreover, if $\mathbf{A}$ has the singular value decomposition (2.8), (2.9) then $\mathbf{A}^{+}$ is given by (2.10).*

Proof. By construction (and Exercise 2.2.1) we have shown that (2.11) has at least *one* solution as given by (2.10). We must now show that this solution is *unique*. First consider a solution $\mathbf{A}^{(1)}$ of the equation

$$\mathbf{A}\mathbf{A}^{(1)}\mathbf{A} = \mathbf{A}.$$

Substitution of $\mathbf{A}$ using the SVD formula in (2.8) and (2.9) implies $\mathbf{A}^{(1)}$ is of the form

$$\mathbf{A}^{(1)} = \mathbf{V}\begin{bmatrix} \boldsymbol{\Sigma}_0^{-1} & \mathbf{Z}_{12} \\ \mathbf{Z}_{21} & \mathbf{Z}_{22} \end{bmatrix}\mathbf{U}^{*}$$

where the matrices $\mathbf{Z}_{12}$ of dimension $r \times (n-r)$, $\mathbf{Z}_{21}$ of dimension $(m-r) \times r$ and $\mathbf{Z}_{22}$ of dimension $(m-r) \times (n-r)$ are arbitrary. Now

$$\mathbf{A}^{(1)}\mathbf{A}\mathbf{A}^{(1)} = \mathbf{V}\begin{bmatrix} \boldsymbol{\Sigma}_0^{-1} & \mathbf{Z}_{12} \\ \mathbf{Z}_{21} & \mathbf{Z}_{21}\boldsymbol{\Sigma}_0\mathbf{Z}_{12} \end{bmatrix}\mathbf{U}^{*}.$$

Hence $\mathbf{A}^{(1)}\mathbf{A}\mathbf{A}^{(1)} = \mathbf{A}^{(1)}$ implies

$$\mathbf{Z}_{22} = \mathbf{Z}_{21}\boldsymbol{\Sigma}_0\mathbf{Z}_{12}.$$

Also, $\mathbf{A}\mathbf{A}^{(1)} = (\mathbf{A}\mathbf{A}^{(1)})^{*}$ implies $\mathbf{Z}_{12} = \mathbf{0}$, and $(\mathbf{A}^{(1)}\mathbf{A})^{*} = \mathbf{A}^{(1)}\mathbf{A}$ implies $\mathbf{Z}_{21} = \mathbf{0}$. Therefore, if *all* four conditions hold in (2.11), $\mathbf{A}^{(1)} = \mathbf{A}^{+}$ is unique. □

Exercise 2.2.2 Consider an $n \times m$ matrix $\mathbf{A}$. Show that

(i) If $\mathbf{A}$ has full row rank n, that is, $\text{rank}(\mathbf{A}\mathbf{A}^{*}) = n$, then

$$\mathbf{A}^{+} = \mathbf{A}^{*}(\mathbf{A}\mathbf{A}^{*})^{-1}.$$

(ii) If $\mathbf{A}$ has full column rank m, that is, $\text{rank}(\mathbf{A}^{*}\mathbf{A}) = m$, then

$$\mathbf{A}^{+} = (\mathbf{A}^{*}\mathbf{A})^{-1}\mathbf{A}^{*}.$$

Exercise 2.2.3 Show that if $\mathbf{A}^{*}\mathbf{A} = \mathbf{I}$, then $\mathbf{A}^{+} = \mathbf{A}^{*}$.

Exercise 2.2.4 Let $\mathbf{X} = \mathbf{X}^{*}$, $\det[\mathbf{X}] \neq 0$, $\mathbf{A}^{*}\mathbf{A} = \mathbf{I}$. Then show that $[\mathbf{A}\mathbf{X}]^{+} = \mathbf{X}^{-1}\mathbf{A}^{*}$.

2.3 SOLUTIONS OF SELECTED LINEAR ALGEBRA PROBLEMS

This section records the solution of certain linear matrix equations and linear matrix inequalities that will provide the solution of *every* control problem to be discussed later in the book. Indeed, the major point of this book is to show that the following linear algebra results solve a large variety of linear control design problems.

2.3.1 AXB = Y

For the given set of matrices $\mathbf{A}, \mathbf{B}, \mathbf{Y}$, we consider in this section all solutions $\mathbf{X}$ to the linear algebraic equations of the form $\mathbf{AXB} = \mathbf{Y}$.

Theorem 2.3.1 *Let $\mathbf{A}$ be an $n_1 \times n_2$ matrix, $\mathbf{X}$ be an $n_2 \times n_3$ matrix, $\mathbf{B}$ be an $n_3 \times n_4$ matrix and $\mathbf{Y}$ be an $n_1 \times n_4$ matrix. Then the following statements are equivalent:*

(i) The equation

$$\mathbf{AXB} = \mathbf{Y} \tag{2.12}$$

has a solution $\mathbf{X}$.

(ii) $\mathbf{A}$, $\mathbf{B}$ and $\mathbf{Y}$ satisfy

$$\mathbf{AA}^+\mathbf{YB}^+\mathbf{B} = \mathbf{Y}. \tag{2.13}$$

(iii) $\mathbf{A}$, $\mathbf{B}$ and $\mathbf{Y}$ satisfy

$$(\mathbf{I} - \mathbf{AA}^+)\mathbf{Y} = \mathbf{0}, \;\; \mathbf{Y}(\mathbf{I} - \mathbf{B}^+\mathbf{B}) = \mathbf{0}. \tag{2.14}$$

In this case all solutions are

$$\mathbf{X} = \mathbf{A}^+\mathbf{YB}^+ + \mathbf{Z} - \mathbf{A}^+\mathbf{AZBB}^+ \tag{2.15}$$

where $\mathbf{Z}$ is an arbitrary $n_2 \times n_3$ matrix and $\mathbf{A}^+$ denotes the Moore–Penrose inverse of $\mathbf{A}$.

Proof. The implication (i) $\Rightarrow$ (ii) can be verified by multiplying both sides of (2.12) by $\mathbf{AA}^+$ from the left and by $\mathbf{B}^+\mathbf{B}$ from the right. To prove the converse, suppose (2.13) holds. Then using (2.15)

$$\begin{aligned} \mathbf{AXB} &= \mathbf{A}[\mathbf{A}^+\mathbf{YB}^+ + \mathbf{Z} - \mathbf{A}^+\mathbf{AZBB}^+]\mathbf{B} \\ &= \mathbf{AA}^+\mathbf{YB}^+\mathbf{B} + \mathbf{AZB} - \mathbf{AA}^+\mathbf{AZBB}^+\mathbf{B} \\ &= \mathbf{Y} \end{aligned}$$

which holds by virtue of the pseudo-inverse property $\mathbf{AA}^+\mathbf{A} = \mathbf{A}$ and (2.13). Thus we have (i) $\Rightarrow$ (ii) and it has been shown that any $\mathbf{X}$ given by (2.15) is a solution of (2.12).

To prove that any solution $\mathbf{X}$ to (2.12) can be generated by (2.15), we must show that for any solution of (2.12), there exists a $\mathbf{Z}$ satisfying (2.15). That is, solve

$$\mathbf{X} = \mathbf{A}^+(\mathbf{AXB})\,\mathbf{B}^+ + \mathbf{Z} - \mathbf{A}^+\mathbf{AZBB}^+,$$

for $\mathbf{Z}$. Obviously, a choice $\mathbf{Z} = \mathbf{X}$ works.

To prove the equivalence of (ii) and (iii), suppose (2.13) holds. Replace $\mathbf{Y}$ in (2.14) by the left-hand side of (2.13) to get

$$(\mathbf{I} - \mathbf{AA}^+)(\mathbf{AA}^+\mathbf{YB}^+\mathbf{B}) = \mathbf{0}, \;\; \mathbf{AA}^+\mathbf{YB}^+\mathbf{B}(\mathbf{I} - \mathbf{B}^+\mathbf{B}) = \mathbf{0}.$$

Hence (2.13) implies (2.14). Now suppose (2.14) holds, then using $\mathbf{AA}^+\mathbf{Y} = \mathbf{Y}$ and $\mathbf{YB}^+\mathbf{B} = \mathbf{Y}$, we have

$$(\mathbf{AA}^+\mathbf{Y})\mathbf{B}^+\mathbf{B} = \mathbf{YBB}^+ = \mathbf{Y}.$$

This completes the proof. □

Corollary 2.3.1 *Consider an $m \times n$ matrix* $\mathbf{A}$ *and a vector* $\mathbf{y} \in \mathcal{C}^m$. *Then the following statements are equivalent:*

(i) There exists a vector $\mathbf{x} \in \mathcal{C}^n$ *such that*

$$\mathbf{A}\mathbf{x} = \mathbf{y}. \tag{2.16}$$

(ii) $\mathbf{A}$ *and* $\mathbf{y}$ *satisfy*

$$(\mathbf{I} - \mathbf{A}\mathbf{A}^+)\mathbf{y} = \mathbf{0}. \tag{2.17}$$

In this case all solution vectors $\mathbf{x}$ *are given by*

$$\mathbf{x} = \mathbf{A}^+\mathbf{y} + (\mathbf{I} - \mathbf{A}^+\mathbf{A})\mathbf{z}$$

where $\mathbf{z}$ *is an arbitrary vector in* $\mathcal{C}^n$.

Proof. The proof follows immediately from Theorem 2.3.1 by setting $\mathbf{B} = 1$. □

Exercise 2.3.1 Show that in terms of the SVD of $\mathbf{A}$,

$$\mathbf{A} = [\mathbf{U}_1 \;\; \mathbf{U}_2]\begin{bmatrix} \boldsymbol{\Sigma} & \mathbf{0} \\ \mathbf{0} & \mathbf{0} \end{bmatrix}\begin{bmatrix} \mathbf{V}_1^* \\ \mathbf{V}_2^* \end{bmatrix}$$

all solutions of (2.16) are given by

$$\mathbf{x} = [\mathbf{V}_1 \;\; \mathbf{V}_2]\begin{bmatrix} \boldsymbol{\Sigma}^{-1}\mathbf{U}_1^*\mathbf{y} \\ \mathbf{z}_1 \end{bmatrix} \tag{2.18}$$

where $\mathbf{z}_1$ is an arbitrary vector of dimension $p = n - r$, where $\boldsymbol{\Sigma} \in \mathcal{R}^{r \times r}$ and r is the rank of $\mathbf{A}$.

Example 2.3.1 Use SVD to solve $\mathbf{A}\mathbf{x} = y$ where y is a scalar, and

$$\mathbf{A} = [\, 1 \quad 1 \,].$$

The SVD of $\mathbf{A}$ is

$$\mathbf{A}_1 = \mathbf{U}_1\boldsymbol{\Sigma}_1\mathbf{V}_1^T$$

where

$$\mathbf{U}_1 = 1; \quad \boldsymbol{\Sigma}_1 = [\, \boldsymbol{\Sigma}_0 \quad \mathbf{0} \,], \quad \boldsymbol{\Sigma}_0 = \sqrt{2}; \quad \mathbf{V}_1 = \frac{1}{\sqrt{2}}\begin{bmatrix} 1 & 1 \\ 1 & -1 \end{bmatrix}.$$

Condition (2.17) is satisfied. Hence, all solutions $\mathbf{x}$ are parametrized as

$$\mathbf{x} = \frac{1}{\sqrt{2}}\begin{bmatrix} 1 & 1 \\ 1 & -1 \end{bmatrix}\begin{bmatrix} \frac{1}{\sqrt{2}}y \\ z_1 \end{bmatrix}$$

where z_1 is an arbitrary real number.

The following result provides a solution to a Frobenius norm minimization problem.

Theorem 2.3.2 *Consider the matrices* $\mathbf{A}$, $\mathbf{X}$, $\mathbf{B}$ *and* $\mathbf{Y}$ *with dimensions as in Theorem 2.3.1 and the Frobenius norm minimization problem*

$$\min \|\mathbf{AXB} - \mathbf{Y}\|_F.$$

with respect to the matrix $\mathbf{X}$. *Then, the minimum is achieved by any* $\mathbf{X}$ *belonging to the following class of matrices generated by an arbitrary matrix* $\mathbf{Z}$*:*

$$\mathbf{X}_{\text{opt}} = \mathbf{A}^+\mathbf{Y}\mathbf{B}^+ + \mathbf{Z} - \mathbf{A}^+\mathbf{AZBB}^+. \tag{2.19}$$

Proof. Let $\mathbf{X}$ be an arbitrary matrix and define $\mathbf{\Theta} \triangleq \mathbf{X} - \mathbf{X}_{\text{opt}}$. Then define

$$\begin{aligned} f(\mathbf{X}) &\triangleq \|\mathbf{AXB} - \mathbf{Y}\|_F^2 - \|\mathbf{AX}_{\text{opt}}\mathbf{B} - \mathbf{Y}\|_F^2 \\ &= tr(2\mathbf{A\Theta B A}_0^T) + \|\mathbf{A\Theta B}\|_F^2 \end{aligned}$$

where

$$\mathbf{A}_0 \triangleq \mathbf{AX}_{\text{opt}}\mathbf{B} - \mathbf{Y}.$$

Note that using the expression (2.19), we have

$$\begin{aligned} tr(\mathbf{A\Theta BA}_0^T) &= tr(\mathbf{\Theta B}(\mathbf{AX}_{\text{opt}}\mathbf{B} - \mathbf{Y})^T\mathbf{A}) \\ &= tr(\mathbf{\Theta B}(\mathbf{B}^+\mathbf{BYAA}^+ - \mathbf{Y})\mathbf{A}) \\ &= tr(\mathbf{\Theta}(\mathbf{BB}^+\mathbf{BYAA}^+\mathbf{A} - \mathbf{BYA})) \\ &= 0 \end{aligned}$$

where the properties (2.11) have been used. Therefore, we have $f(\mathbf{X}) \geq 0$ for any $\mathbf{X}$, with equality holding when $\mathbf{X} = \mathbf{X}_{\text{opt}}$. □

Notice that if the equation $\mathbf{AXB} = \mathbf{Y}$ is consistent, that is, if condition (2.13) is satisfied, then the minimum value of $\|\mathbf{AXB} - \mathbf{Y}\|_F$ is zero and the parametrization (2.19) provides all solutions of $\mathbf{AXB} = \mathbf{Y}$.

2.3.2 AX = C, XB = D

Theorem 2.3.3 *Let matrices* $\mathbf{A}$, $\mathbf{B}$, $\mathbf{C}$ *and* $\mathbf{D}$ *be given. Then the following statements are equivalent:*

(i) There exists a common solution $\mathbf{X}$ *to the two linear matrix equations*

$$\mathbf{AX} = \mathbf{C}, \qquad \mathbf{XB} = \mathbf{D}. \tag{2.20}$$

(ii) The following three conditions hold

$$\mathbf{AA}^+\mathbf{C} = \mathbf{C} \tag{2.21a}$$

$$\mathbf{DB}^+\mathbf{B} = \mathbf{D} \tag{2.21b}$$

$$\mathbf{AD} = \mathbf{CB}. \tag{2.21c}$$

In this case, all solutions are given by

$$\mathbf{X} = \mathbf{A}^+\mathbf{C} + \mathbf{DB}^+ - \mathbf{A}^+\mathbf{ADB}^+ + (\mathbf{I} - \mathbf{A}^+\mathbf{A})\mathbf{Z}(\mathbf{I} - \mathbf{BB}^+)$$

where $\mathbf{Z}$ *is arbitrary.*

Proof. Necessity of conditions (2.21) is easy to establish since (2.21a) and (2.21b) correspond to the solvability conditions of the first and second equation (2.20), respectively and (2.21c) is required since

$$\mathbf{AXB} = \mathbf{CB} = \mathbf{AD}.$$

To prove sufficiency of the conditions (2.21) as well as the expression for all solutions, consider the general solution of the first equation $\mathbf{AX} = \mathbf{C}$

$$\mathbf{X} = \mathbf{A}^+\mathbf{C} + (\mathbf{I} - \mathbf{A}^+\mathbf{A})\mathbf{Y}$$

where $\mathbf{Y}$ is arbitrary. For a solution to exist, (2.21a) is a necessary and sufficient condition. Substituting the expression for $\mathbf{X}$ into the second equation $\mathbf{XB} = \mathbf{D}$ we obtain

$$(\mathbf{I} - \mathbf{A}^+\mathbf{A})\mathbf{YB} = \mathbf{D} - \mathbf{A}^+\mathbf{CB}. \tag{2.22}$$

This equation has a solution for $\mathbf{Y}$ if and only if the two conditions corresponding to equations (2.14) are satisfied. These provide

$$(\mathbf{D} - \mathbf{A}^+\mathbf{CB})(\mathbf{I} - \mathbf{B}^+\mathbf{B}) = \mathbf{0}.$$

That is,

$$\mathbf{DB}^+\mathbf{B} = \mathbf{D}$$

and

$$\left[\mathbf{I} - (\mathbf{I} - \mathbf{A}^+\mathbf{A})(\mathbf{I} - \mathbf{A}^+\mathbf{A})^+\right](\mathbf{D} - \mathbf{A}^+\mathbf{CB}) = \mathbf{0}$$

or

$$\mathbf{A}^+\mathbf{A}(\mathbf{D} - \mathbf{A}^+\mathbf{CB}) = \mathbf{0}.$$

That is,

$$\mathbf{A}^+\mathbf{AD} = \mathbf{A}^+\mathbf{CB}.$$

Pre-multiplying by $\mathbf{A}$ and using (2.21a) we obtain (2.21c).

The general solution of (2.22) with respect to $\mathbf{Y}$ is

$$\mathbf{Y} = (\mathbf{I} - \mathbf{A}^+\mathbf{A})(\mathbf{D} - \mathbf{A}^+\mathbf{CB})\mathbf{B}^+ + \mathbf{Z} - (\mathbf{I} - \mathbf{A}^+\mathbf{A})\mathbf{ZBB}^+.$$

Hence, the general solution $\mathbf{X}$ of the equations (2.20) is given by

$$\begin{aligned}
\mathbf{X} &= \mathbf{A}^+\mathbf{C} + (\mathbf{I} - \mathbf{A}^+\mathbf{A})\mathbf{Y} \\
&= \mathbf{A}^+\mathbf{C} + (\mathbf{I} - \mathbf{A}^+\mathbf{A})(\mathbf{D} - \mathbf{A}^+\mathbf{CB})\mathbf{B}^+ \\
&\quad +(\mathbf{I} - \mathbf{A}^+\mathbf{A})\mathbf{Z} - (\mathbf{I} - \mathbf{A}^+\mathbf{A})\mathbf{ZBB}^+ \\
&= \mathbf{A}^+\mathbf{C} + \mathbf{DB}^+ - \mathbf{A}^+\mathbf{CBB}^+ - \mathbf{A}^+\mathbf{ADB}^+ + \mathbf{A}^+\mathbf{AA}^+\mathbf{CBB}^+ \\
&\quad +(\mathbf{I} - \mathbf{A}^+\mathbf{A})\mathbf{Z}(\mathbf{I} - \mathbf{BB}^+)
\end{aligned}$$

which proves the general solution of the theorem. □

2.3.3 $AX = C$, $X = X^*$

Theorem 2.3.4 *Let matrices* $\mathbf{A}$ *and* $\mathbf{C}$ *be given. Then the following statements are equivalent:*

(i) There exists a Hermitian solution $\mathbf{X} = \mathbf{X}^*$ *to* $\mathbf{AX} = \mathbf{C}$.

(ii) The following two conditions hold

$$\mathbf{CA}^* = (\mathbf{CA}^*)^*, \quad (\mathbf{I} - \mathbf{AA}^+)\mathbf{C} = \mathbf{0}.$$

In this case, all such solutions are given by

$$\begin{aligned} \mathbf{X} &= \mathbf{A}^+\mathbf{C} + \mathbf{C}^*\mathbf{A}^{+*} - \mathbf{A}^+\mathbf{CA}^*\mathbf{A}^{+*} \\ &\quad +(\mathbf{I} - \mathbf{A}^+\mathbf{A})\boldsymbol{\Theta}(\mathbf{I} - \mathbf{A}^+\mathbf{A}) \\ \boldsymbol{\Theta} &= \boldsymbol{\Theta}^* \ \textit{arbitrary}. \end{aligned} \tag{2.23}$$

Proof. We must prove that $\mathbf{AX} = \mathbf{C}$ and $\mathbf{XA}^* = \mathbf{C}^*$ have a common solution. From Theorem 2.3.3 there exists such a solution if and only if

$$(\mathbf{I} - \mathbf{AA}^+)\mathbf{C} = \mathbf{0}$$

$$\mathbf{C}^*(\mathbf{I} - \mathbf{A}^{*+}\mathbf{A}^*) = \mathbf{0}$$

$$\mathbf{AC}^* = \mathbf{CA}^*$$

where the first and second conditions are redundant. This proves the necessary and sufficient conditions for existence. All solutions are from Theorem 2.3.3

$$\mathbf{X} = \mathbf{A}^+\mathbf{C} + \mathbf{C}^*\mathbf{A}^{*+} - \mathbf{A}^+\mathbf{AC}^*\mathbf{A}^{*+} + (\mathbf{I} - \mathbf{A}^+\mathbf{A})\boldsymbol{\Theta}(\mathbf{I} - \mathbf{A}^*\mathbf{A}^{*+}).$$

This proves (2.23) since $\mathbf{AC}^* = \mathbf{CA}^*$ and $(\mathbf{A}^+\mathbf{A}) = (\mathbf{A}^+\mathbf{A})^*$. □

2.3.4 AX = C, XB = D, X = X*

Theorem 2.3.5 *Let matrices* **A**, **B**, **C** *and* **D** *be given, and suppose (2.21) holds. Then the following statements are equivalent.*

(i) There exists a common Hermitian solution ($\mathbf{X} = \mathbf{X}^*$) *to (2.20).*

(ii) The conditions

$$\begin{aligned} \mathbf{RP}^* &= \mathbf{PR}^* \\ \mathbf{AA}^+\mathbf{C} &= \mathbf{C} \end{aligned} \tag{2.24}$$

$$(\mathbf{I} - \mathbf{TT}^+)(\mathbf{D}^* - \mathbf{B}^*\mathbf{A}^+\mathbf{C}) = \mathbf{0}$$

hold, where

$$\mathbf{T} = \mathbf{B}^*(\mathbf{I} - \mathbf{A}^+\mathbf{A}), \quad \mathbf{P} \triangleq \begin{bmatrix} \mathbf{A} \\ \mathbf{B}^* \end{bmatrix}, \quad \mathbf{R} \triangleq \begin{bmatrix} \mathbf{C} \\ \mathbf{D}^* \end{bmatrix}.$$

In this case, all Hermitian solutions are

$$\begin{aligned} \mathbf{X} &= \mathbf{P}^+\mathbf{R} + \mathbf{R}^*\mathbf{P}^{+*} - \mathbf{P}^+\mathbf{RP}^*\mathbf{P}^{+*} \\ &\quad +(\mathbf{I} - \mathbf{P}^+\mathbf{P})\boldsymbol{\Theta}(\mathbf{I} - \mathbf{P}^+\mathbf{P}), \\ \boldsymbol{\Theta} &= \boldsymbol{\Theta}^* \ \textit{arbitrary}. \end{aligned} \tag{2.25}$$

Proof. Equations (2.20) have a common Hermitian solution if and only if

$$\mathbf{PX} = \mathbf{R} \tag{2.26}$$

has a Hermitian solution. According to Theorem 2.3.4, this is equivalent to $\mathbf{RP}^* = (\mathbf{RP}^*)^*$ (which provides the first of the conditions (2.24)), and the equations

$$\mathbf{AX} = \mathbf{C}, \qquad \mathbf{B}^*\mathbf{X} = \mathbf{D}^* \tag{2.27}$$

being consistent. The first one of these equations is solvable for $\mathbf{X}$ if and only if

$$\mathbf{AA}^+\mathbf{C} = \mathbf{C} \tag{2.28}$$

which is the second condition in (2.24) and the general solution is

$$\mathbf{X} = \mathbf{A}^+\mathbf{C} + (\mathbf{I} - \mathbf{A}^+\mathbf{A})\mathbf{Z}$$

where $\mathbf{Z}$ is an arbitrary matrix of appropriate dimension. Substituting this expression in the second equation $\mathbf{B}^*\mathbf{X} = \mathbf{D}^*$ results in

$$\mathbf{TZ} = \mathbf{D}^* - \mathbf{B}^*\mathbf{A}^+\mathbf{C}.$$

This equation has a solution for $\mathbf{Z}$ if and only if the third condition in (2.24) is satisfied. The expression for the general solution (2.25) is obtained by applying the general solution of Theorem 2.3.4. □

By replacing $\mathbf{X}$ by $j\mathbf{X}$ and $\mathbf{C}$ by $j\mathbf{C}$, and $\mathbf{\Theta}$ by $j\mathbf{\Theta}$, $j \triangleq \sqrt{-1}$, Theorem 2.3.4 can be used to find skew-Hermitian solutions to linear algebra problems (note that $j\mathbf{X}$ is Hermitian if and only if $\mathbf{X}$ is skew-Hermitian).

2.3.5 AX = C, X = –X*

Corollary 2.3.2 *Let matrices* $\mathbf{A}$ *and* $\mathbf{C}$ *be given. Then the following statements are equivalent:*

(i) There exists a skew-Hermitian solution $\mathbf{X}$ *to the equation*

$$\mathbf{AX} = \mathbf{C}, \qquad \mathbf{X} = -\mathbf{X}^*. \tag{2.29}$$

(ii) The following two conditions hold

$$\mathbf{CA}^* = -\mathbf{AC}^*, \qquad (\mathbf{I} - \mathbf{AA}^+)\mathbf{C} = \mathbf{0}. \tag{2.30}$$

In this case, all solutions are given by

$$\mathbf{X} = \mathbf{A}^+\mathbf{C} - \mathbf{C}^*\mathbf{A}^{+*} - \mathbf{A}^+\mathbf{CA}^*\mathbf{A}^{+*} + (\mathbf{I} - \mathbf{A}^+\mathbf{A})\mathbf{\Theta}(\mathbf{I} - \mathbf{A}^+\mathbf{A}) \tag{2.31}$$

$$\mathbf{\Theta} = -\mathbf{\Theta}^* \quad \textit{arbitrary} \tag{2.32}$$

where $\mathbf{\Theta}$ *is an arbitrary skew-Hermitian matrix.*

Replacing $\mathbf{X}$, $\mathbf{C}$, $\mathbf{D}$ in Theorem 2.3.3 by $j\mathbf{X}$, $j\mathbf{C}$, $j\mathbf{D}$ leads to the following result.

2.3.6 XB = D, X = –X*

Corollary 2.3.3 *Let matrices* **B** *and* **D** *be given. Then the following statements are equivalent:*

(i) There exists a skew-Hermitian solution to the equation

$$\mathbf{XB} = \mathbf{D}, \qquad \mathbf{X} = -\mathbf{X}^*.$$

(ii) The following conditions hold

$$\mathbf{D}(\mathbf{I} - \mathbf{B}^+\mathbf{B}) = \mathbf{0}$$

$$\mathbf{B}^*\mathbf{D} = -\mathbf{D}^*\mathbf{B}.$$

In this case, all solutions are given by

$$\mathbf{X} = -\mathbf{B}^{*+}\mathbf{D}^* + \mathbf{DB}^+ + \mathbf{B}^{*+}\mathbf{D}^*\mathbf{BB}^+ + (\mathbf{I} - \mathbf{BB}^+)\mathbf{S}(\mathbf{I} - \mathbf{BB}^+)$$

for arbitrary $\mathbf{S} = -\mathbf{S}^*$.

2.3.7 AX = C, XB = D, X = –X*

Corollary 2.3.4 *Let matrices* **A**, **B**, **C** *and* **D** *be given. Then the following statements are equivalent:*

(i) There exists a common skew-Hermitian solution **X** *to the equations (2.20).*
(ii) The condition (2.21) holds and in addition

$$\begin{aligned} (\mathbf{I} - \mathbf{PP}^+)\mathbf{R} &= \mathbf{0} \\ \mathbf{RP}^* &= -\mathbf{PR}^* \end{aligned} \tag{2.33}$$

where

$$\mathbf{P} \triangleq \begin{bmatrix} \mathbf{A} \\ \mathbf{B}^* \end{bmatrix}, \quad \mathbf{R} \triangleq \begin{bmatrix} \mathbf{C} \\ -\mathbf{D}^* \end{bmatrix}.$$

In this case all skew-Hermitian solutions are

$$\mathbf{X} = \mathbf{P}^+\mathbf{R} - \mathbf{R}^*\mathbf{P}^{+*} - \mathbf{P}^+\mathbf{RP}^*\mathbf{P}^{+*} + (\mathbf{I} - \mathbf{P}^+\mathbf{P})\boldsymbol{\Theta}(\mathbf{I} - \mathbf{P}^+\mathbf{P}) \tag{2.34}$$

where $\boldsymbol{\Theta}$ *is an arbitrary skew-Hermitian matrix.*

2.3.8 AX + (AX)* + Q = 0

Theorem 2.3.6 *Let* **A** *and* **Q** *be given where* $\mathbf{Q} = \mathbf{Q}^*$. *Then the following statements are equivalent:*

(i) There exists a matrix **X** *satisfying*

$$\mathbf{AX} + (\mathbf{AX})^* + \mathbf{Q} = \mathbf{0}. \tag{2.35}$$

(ii) **A** *and* **Q** *satisfy*

$$(\mathbf{I} - \mathbf{AA}^+)\mathbf{Q}(\mathbf{I} - \mathbf{AA}^+) = \mathbf{0}. \tag{2.36}$$

In this case, all such $\mathbf{X}$ *are given by*

$$\mathbf{X} = -\frac{1}{2}\mathbf{A}^+\mathbf{Q}(2\mathbf{I} - \mathbf{A}\mathbf{A}^+) + \mathbf{A}^+\mathbf{S}\mathbf{A}\mathbf{A}^+ + (\mathbf{I} - \mathbf{A}^+\mathbf{A})\mathbf{Z} \tag{2.37}$$

where $\mathbf{Z}$ *is arbitrary and* $\mathbf{S}$ *is an arbitrary skew-Hermitian matrix.*

Proof. First note that (2.35) holds if and only if

$$\mathbf{A}\mathbf{X} = -\frac{1}{2}(\mathbf{Q} + \hat{\mathbf{S}})$$

holds for some skew-Hermitian $\hat{\mathbf{S}} = -\hat{\mathbf{S}}^*$. Using Theorem 2.3.1, the above equation is solvable for $\mathbf{X}$ if and only if

$$(\mathbf{I} - \mathbf{A}\mathbf{A}^+)(\mathbf{Q} + \hat{\mathbf{S}}) = \mathbf{0}, \tag{2.38}$$

in which case, all solutions are

$$\mathbf{X} = -\frac{1}{2}\mathbf{A}^+(\mathbf{Q} + \hat{\mathbf{S}}) + (\mathbf{I} - \mathbf{A}^+\mathbf{A})\mathbf{Z} \tag{2.39}$$

where $\mathbf{Z}$ is arbitrary. From Corollary 2.3.2, there exists a skew-Hermitian matrix $\hat{\mathbf{S}}$ satisfying (2.38) if and only if

$$-(\mathbf{I} - \mathbf{A}\mathbf{A}^+)\mathbf{Q}(\mathbf{I} - \mathbf{A}\mathbf{A}^+)^* = (\mathbf{I} - \mathbf{A}\mathbf{A}^+)\mathbf{Q}(\mathbf{I} - \mathbf{A}\mathbf{A}^+)^*,$$
$$-[\mathbf{I} - (\mathbf{I} - \mathbf{A}\mathbf{A}^+)(\mathbf{I} - \mathbf{A}\mathbf{A}^+)^+](\mathbf{I} - \mathbf{A}\mathbf{A}^+)\mathbf{Q} = \mathbf{0}$$

or equivalently,

$$(\mathbf{I} - \mathbf{A}\mathbf{A}^+)\mathbf{Q}(\mathbf{I} - \mathbf{A}\mathbf{A}^+) = \mathbf{0}$$

holds. In this case, all such $\hat{\mathbf{S}}$ are given by

$$\hat{\mathbf{S}} = -(\mathbf{I} - \mathbf{A}\mathbf{A}^+)\mathbf{Q}\mathbf{A}\mathbf{A}^+ + \mathbf{Q}(\mathbf{I} - \mathbf{A}\mathbf{A}^+) + \mathbf{A}\mathbf{A}^+\bar{\mathbf{S}}\mathbf{A}\mathbf{A}^+ \tag{2.40}$$

where $\bar{\mathbf{S}}$ is an arbitrary skew-Hermitian matrix. Substituting (2.40) into (2.39) and defining $\mathbf{S} \triangleq -\frac{1}{2}\bar{\mathbf{S}}$ yields (2.37). □

2.3.9 AXBC + (AXBC)* + Q = 0

Theorem 2.3.7 *Let matrices* $\mathbf{A}$, $\mathbf{B}$, $\mathbf{C}$ *and* $\mathbf{Q}$ *be given, where* $\mathbf{Q}$ *is Hermitian and* $\mathbf{C}$ *is a square invertible matrix. Then the following statements are equivalent:*

(i) There exists a matrix $\mathbf{X}$ *satisfying*

$$\mathbf{A}\mathbf{X}\mathbf{B}\mathbf{C} + (\mathbf{A}\mathbf{X}\mathbf{B}\mathbf{C})^* + \mathbf{Q} = \mathbf{0}. \tag{2.41}$$

(ii) The following three conditions hold

$$(\mathbf{I} - \mathbf{A}\mathbf{A}^+)\mathbf{Q}(\mathbf{I} - \mathbf{A}\mathbf{A}^+) = \mathbf{0} \tag{2.42}$$

$$(\mathbf{I} - \mathbf{B}^+\mathbf{B})\mathbf{C}^{-*}\mathbf{Q}\mathbf{C}^{-1}(\mathbf{I} - \mathbf{B}^+\mathbf{B}) = \mathbf{0} \tag{2.43}$$

$$(\mathbf{I} - \mathbf{D}\mathbf{D}^+)(\mathbf{I} - \mathbf{B}^+\mathbf{B})\mathbf{C}^{-*}\mathbf{Q} = \mathbf{0} \tag{2.44}$$

where

$$\mathbf{D} \triangleq (\mathbf{I} - \mathbf{B}^+\mathbf{B})\mathbf{C}^{-*}\mathbf{A}\mathbf{A}^+ \tag{2.45}$$

and $\mathbf{C}^{-*} \triangleq (\mathbf{C}^{-1})^*$.

In this case, all such matrices $\mathbf{X}$ *are given by*

$$\mathbf{X} = -\frac{1}{2}\mathbf{A}^{+}(\mathbf{Q}+\mathbf{S})\mathbf{C}^{-1}\mathbf{B}^{+} + \mathbf{Z} - \mathbf{A}^{+}\mathbf{A}\mathbf{Z}\mathbf{B}\mathbf{B}^{+} \tag{2.46}$$

where $\mathbf{Z}$ *is arbitrary and*

$$\begin{gathered}\mathbf{S} \triangleq [\mathbf{P}^{+}\mathbf{R} + (\mathbf{I} - \mathbf{P}^{+}\mathbf{P})\boldsymbol{\Theta}](\mathbf{I} - \mathbf{P}^{+}\mathbf{P}) - (\mathbf{P}^{+}\mathbf{R})^{*} \\ \mathbf{P} \triangleq \begin{bmatrix} \mathbf{I} - \mathbf{A}\mathbf{A}^{+} \\ (\mathbf{I} - \mathbf{B}^{+}\mathbf{B})\mathbf{C}^{-*} \end{bmatrix}, \quad \mathbf{R} \triangleq \begin{bmatrix} -(\mathbf{I} - \mathbf{A}\mathbf{A}^{+}) \\ (\mathbf{I} - \mathbf{B}^{+}\mathbf{B})\mathbf{C}^{-*} \end{bmatrix}\mathbf{Q}\end{gathered} \tag{2.47}$$

where $\boldsymbol{\Theta} = -\boldsymbol{\Theta}^{*}$ *is an arbitrary skew-Hermitian matrix.*

Proof. The equality (2.41) holds if and only if

$$\mathbf{AXBC} = -\frac{1}{2}(\mathbf{Q}+\mathbf{S}), \quad \mathbf{S} = -\mathbf{S}^{*}$$

holds for some skew-Hermitian matrix $\mathbf{S}$. From Theorem 2.3.1, the above equation is solvable for $\mathbf{X}$ if and only if

$$\begin{aligned}(\mathbf{I} - \mathbf{A}\mathbf{A}^{+})(\mathbf{Q}+\mathbf{S}) &= \mathbf{0} \\ (\mathbf{Q}+\mathbf{S})\mathbf{C}^{-1}(\mathbf{I} - \mathbf{B}^{+}\mathbf{B}) &= \mathbf{0}\end{aligned}$$

hold, in which case, all solutions $\mathbf{X}$ are given by (2.46). Rearranging, we have

$$\begin{aligned}(\mathbf{I} - \mathbf{A}\mathbf{A}^{+})\mathbf{S} &= -(\mathbf{I} - \mathbf{A}\mathbf{A}^{+})\mathbf{Q}, \\ \mathbf{S}\mathbf{C}^{-1}(\mathbf{I} - \mathbf{B}^{+}\mathbf{B}) &= -\mathbf{Q}\mathbf{C}^{-1}(\mathbf{I} - \mathbf{B}^{+}\mathbf{B}).\end{aligned}$$

Using Corollary 2.3.4, there exists a skew-Hermitian matrix $\mathbf{S}$ satisfying the above equations if and only if

$$\begin{aligned}&\left[\mathbf{I} - (\mathbf{I} - \mathbf{A}\mathbf{A}^{+})(\mathbf{I} - \mathbf{A}\mathbf{A}^{+})^{+}\right](\mathbf{I} - \mathbf{A}\mathbf{A}^{+})\mathbf{Q} = \mathbf{0}, \\ &(\mathbf{I} - \mathbf{D}\mathbf{D}^{+})\left[(\mathbf{I} - \mathbf{B}^{+}\mathbf{B})\mathbf{C}^{-*}\mathbf{Q} + (\mathbf{I} - \mathbf{B}^{+}\mathbf{B})\mathbf{C}^{-*}(\mathbf{I} - \mathbf{A}\mathbf{A}^{+})\mathbf{Q}\right] = \mathbf{0},\end{aligned}$$

hold and

$$\begin{bmatrix} -(\mathbf{I} - \mathbf{A}\mathbf{A}^{+})\mathbf{Q}(\mathbf{I} - \mathbf{A}\mathbf{A}^{+}) & -(\mathbf{I} - \mathbf{A}\mathbf{A}^{+})\mathbf{Q}\mathbf{C}^{-1}(\mathbf{I} - \mathbf{B}^{+}\mathbf{B}) \\ (\mathbf{I} - \mathbf{B}^{+}\mathbf{B})\mathbf{C}^{-*}\mathbf{Q}(\mathbf{I} - \mathbf{A}\mathbf{A}^{+}) & (\mathbf{I} - \mathbf{B}^{+}\mathbf{B})\mathbf{C}^{-*}\mathbf{Q}\mathbf{C}^{-1}(\mathbf{I} - \mathbf{B}^{+}\mathbf{B}) \end{bmatrix}$$

is a skew-Hermitian matrix, or equivalently, (2.42)–(2.44) hold, where we used the identity

$$(\mathbf{I} - \mathbf{D}\mathbf{D}^{+})(\mathbf{I} - \mathbf{B}^{+}\mathbf{B})\mathbf{C}^{-*}\mathbf{A}\mathbf{A}^{+} = (\mathbf{I} - \mathbf{D}\mathbf{D}^{+})\mathbf{D} = \mathbf{0}.$$

In this case, all such $\mathbf{S}$ are given by (2.47). □

2.3.10 AX = B, XX* = I

Theorem 2.3.8 *Let* $\mathbf{A} \in \mathcal{C}^{a\times b}$ *and* $\mathbf{B} \in \mathcal{C}^{a\times c}$ *be given matrices, where* $c \geq b$. *Then the following statements are equivalent:*

(i) There exists $\mathbf{X}$ *satisfying*

$$\mathbf{AX} = \mathbf{B}, \qquad \mathbf{XX}^* = \mathbf{I}. \tag{2.48}$$

(ii) $\mathbf{A}$ *and* $\mathbf{B}$ *satisfy*

$$\mathbf{AA}^* = \mathbf{BB}^*. \tag{2.49}$$

In this case, all such $\mathbf{X}$ *are given by*

$$\mathbf{X} = \begin{bmatrix} \mathbf{V}_{A1} & \mathbf{V}_{A2} \end{bmatrix} \begin{bmatrix} \mathbf{I} & \mathbf{0} \\ \mathbf{0} & \mathbf{U} \end{bmatrix} \begin{bmatrix} \mathbf{V}_{B1}^* \\ \mathbf{V}_{B2}^* \end{bmatrix} \tag{2.50}$$

where $\mathbf{U}$ *is an arbitrary matrix such that* $\mathbf{UU}^* = \mathbf{I}$ *and* $\mathbf{V}_{A1}$, $\mathbf{V}_{A2}$, $\mathbf{V}_{B1}$ *and* $\mathbf{V}_{B2}$ *are defined from the SVDs of* $\mathbf{A}$ *and* $\mathbf{B}$ *as follows*

$$\mathbf{A} = \begin{bmatrix} \mathbf{U}_{A1} & \mathbf{U}_{A2} \end{bmatrix} \begin{bmatrix} \mathbf{\Sigma}_A & \mathbf{0} \\ \mathbf{0} & \mathbf{0} \end{bmatrix} \begin{bmatrix} \mathbf{V}_{A1}^* \\ \mathbf{V}_{A2}^* \end{bmatrix} = \mathbf{U}_A \mathbf{\Sigma}_A \mathbf{V}_A^* \tag{2.51}$$

$$\mathbf{B} = \begin{bmatrix} \mathbf{U}_{A1} & \mathbf{U}_{A2} \end{bmatrix} \begin{bmatrix} \mathbf{\Sigma}_A & \mathbf{0} \\ \mathbf{0} & \mathbf{0} \end{bmatrix} \begin{bmatrix} \mathbf{V}_{B1}^* \\ \mathbf{V}_{B2}^* \end{bmatrix} = \mathbf{U}_A \mathbf{\Sigma}_A \mathbf{V}_B^*. \tag{2.52}$$

Proof. Square both sides of $\mathbf{AX} = \mathbf{B}$ to see that

$$\mathbf{AX}(\mathbf{AX})^* = \mathbf{AXX}^*\mathbf{A}^* = \mathbf{AA}^* = \mathbf{BB}^*.$$

This proves necessity of (2.49). For sufficiency, recall from the SVD of $\mathbf{A}$ that $\mathbf{U}_A$ satisfies

$$\mathbf{AA}^*\mathbf{U}_A = \mathbf{U}_A \begin{bmatrix} \mathbf{\Sigma}_A{}^2 & \mathbf{0} \\ \mathbf{0} & \mathbf{0} \end{bmatrix}. \tag{2.53}$$

But from (2.49) and (2.53) it is clear that we can choose $\mathbf{U}_A = \mathbf{U}_B$, $\mathbf{\Sigma}_A = \mathbf{\Sigma}_B$. Hence (2.52). Now define

$$\mathbf{Z}^* \triangleq \begin{bmatrix} \mathbf{Z}_1^* \\ \mathbf{Z}_2^* \end{bmatrix} \triangleq \begin{bmatrix} \mathbf{V}_{A1}^* \\ \mathbf{V}_{A2}^* \end{bmatrix} \mathbf{X}. \tag{2.54}$$

Then (2.48) is equivalent to

$$\begin{aligned} &\begin{bmatrix} \mathbf{U}_{A1} & \mathbf{U}_{A2} \end{bmatrix} \begin{bmatrix} \mathbf{\Sigma}_A & \mathbf{0} \\ \mathbf{0} & \mathbf{0} \end{bmatrix} \begin{bmatrix} \mathbf{Z}_1^* \\ \mathbf{Z}_2^* \end{bmatrix} = \begin{bmatrix} \mathbf{U}_{A1} & \mathbf{U}_{A2} \end{bmatrix} \begin{bmatrix} \mathbf{\Sigma}_A & \mathbf{0} \\ \mathbf{0} & \mathbf{0} \end{bmatrix} \mathbf{V}_B^* \\ &\mathbf{Z}^*\mathbf{Z} = \mathbf{I} \end{aligned} \tag{2.55}$$

which is equivalent to

$$\mathbf{\Sigma}_A \mathbf{Z}_1^* = \mathbf{\Sigma}_A \mathbf{V}_{B1}^*, \quad \mathbf{Z}_2^*\mathbf{Z}_2 = \mathbf{I}, \ \mathbf{Z}_2^*\mathbf{Z}_1 = \mathbf{0}, \tag{2.56}$$

which is equivalent to (using the fact $\mathbf{V}_{B2}^*\mathbf{V}_{B1} = \mathbf{0}$)

$$\mathbf{Z}_1 = \mathbf{V}_{B1}, \quad \mathbf{Z}_2 = \mathbf{V}_{B2}\mathbf{U}^*, \quad \mathbf{UU}^* = \mathbf{I}, \tag{2.57}$$

which is equivalent to (since $\mathbf{X} = \mathbf{V}_A\mathbf{Z}^*$)

$$\mathbf{X} = \mathbf{V}_A \begin{bmatrix} \mathbf{I} & \mathbf{0} \\ \mathbf{0} & \mathbf{U} \end{bmatrix} \mathbf{V}_B^* \qquad \mathbf{UU}^* = \mathbf{I}. \tag{2.58}$$

This completes the proof of Theorem 2.3.8. □

2.3.11 (AX + B)R(AX + B)* = Q

Theorem 2.3.9 *Let matrices* $\mathbf{A}$, $\mathbf{B}$, $\mathbf{R}$ *and* $\mathbf{Q}$ *be given. Suppose* $\mathbf{Q} = \mathbf{Q}^* \in \mathcal{C}^{n\times n}$, $\mathbf{R} = \mathbf{R}^* \in \mathcal{C}^{r\times r}$ *and* $\mathbf{R} > \mathbf{0}$. *Then the following statements are equivalent:*

(i) There exists a matrix $\mathbf{X}$ *such that*

$$(\mathbf{AX}+\mathbf{B})\mathbf{R}(\mathbf{AX}+\mathbf{B})^* = \mathbf{Q}. \tag{2.59}$$

(ii) The following conditions hold:

$$\mathbf{Q} \geq \mathbf{0}, \qquad \text{rank}(\mathbf{Q}) \leq r, \tag{2.60}$$

$$(\mathbf{I}-\mathbf{AA}^+)(\mathbf{Q}-\mathbf{BRB}^*)(\mathbf{I}-\mathbf{AA}^+) = \mathbf{0}. \tag{2.61}$$

In this case, all such $\mathbf{X}$ *are given by*

$$\mathbf{X} = \mathbf{A}^+(\mathbf{LUR}^{-1/2}-\mathbf{B}) + (\mathbf{I}-\mathbf{A}^+\mathbf{A})\mathbf{Z} \tag{2.62}$$

where $\mathbf{Z}$ *is arbitrary and*

$$\begin{aligned}
&\mathbf{LL}^* = \mathbf{Q}, \quad \mathbf{L} \in \mathcal{C}^{n\times r}\\
&\mathbf{U} \triangleq \mathbf{V}_L \begin{bmatrix} \mathbf{I} & \mathbf{0} \\ \mathbf{0} & \mathbf{U}_F \end{bmatrix} \mathbf{V}_R^*\\
&(\mathbf{I}-\mathbf{AA}^+)\mathbf{L} = \mathbf{U}_L \begin{bmatrix} \mathbf{\Sigma}_L & \mathbf{0} \\ \mathbf{0} & \mathbf{0} \end{bmatrix} \mathbf{V}_L^* \qquad (SVD)\\
&(\mathbf{I}-\mathbf{AA}^+)\mathbf{BR}^{1/2} = \mathbf{U}_L \begin{bmatrix} \mathbf{\Sigma}_L & \mathbf{0} \\ \mathbf{0} & \mathbf{0} \end{bmatrix} \mathbf{V}_R^* \qquad (SVD)
\end{aligned} \tag{2.63}$$

where $\mathbf{U}_F$ *is an arbitrary matrix such that* $\mathbf{U}_F\mathbf{U}_F^* = \mathbf{I}$.

Proof. Since the left-hand side of (2.59) is positive semidefinite with rank less than or equal to r, (2.60) is necessary. Pre- and post-multiplying both sides of (2.59) by $\mathbf{I}-\mathbf{AA}^+$, we have (2.61). This proves the necessity of (2.60) and (2.61). To prove sufficiency, suppose (2.60) and (2.61) hold. Then there exists $\mathbf{L} \in \mathcal{C}^{n\times r}$ such that $\mathbf{Q} = \mathbf{LL}^*$. Now,

$$(\mathbf{AX}+\mathbf{B})\mathbf{R}(\mathbf{AX}+\mathbf{B})^* = \mathbf{LL}^*$$

holds if and only if

$$(\mathbf{AX}+\mathbf{B})\mathbf{R}^{1/2} = \mathbf{LU}, \qquad \mathbf{UU}^* = \mathbf{I}$$

or equivalently,

$$\mathbf{AX} = \mathbf{LUR}^{-1/2} - \mathbf{B}, \qquad \mathbf{UU}^* = \mathbf{I}$$

holds for some orthogonal matrix $\mathbf{U}$. The above equation is solvable for $\mathbf{X}$ if and only if

$$(\mathbf{I}-\mathbf{AA}^+)(\mathbf{LUR}^{-1/2}-\mathbf{B}) = \mathbf{0}$$

and all solutions $\mathbf{X}$ are given by (2.62). Rearranging, we have

$$(\mathbf{I}-\mathbf{AA}^+)\mathbf{LU} = (\mathbf{I}-\mathbf{AA}^+)\mathbf{BR}^{1/2}, \qquad \mathbf{UU}^* = \mathbf{I}.$$

From Theorem 2.3.8, the above equation is solvable for an orthogonal matrix $\mathbf{U}$ if and only if

$$(\mathbf{I}-\mathbf{AA}^+)\mathbf{LL}^*(\mathbf{I}-\mathbf{AA}^+) = (\mathbf{I}-\mathbf{AA}^+)\mathbf{BRB}^*(\mathbf{I}-\mathbf{AA}^+)$$

or equivalently, (2.61) holds. In this case, all such $\mathbf{U}$ are given by (2.63). □

2.3.12 $\mu \mathbf{BB}^* - \mathbf{Q} > 0$

In the sequel, we shall need the following definition. For a matrix $\mathbf{B} \in \mathcal{C}^{n \times m}$ with rank r, let $\mathbf{B}^{\perp} \in \mathcal{C}^{(n-r)\times n}$ be any matrix such that $\mathbf{B}^{\perp}\mathbf{B} = \mathbf{0}$ and $\mathbf{B}^{\perp}\mathbf{B}^{\perp *} > \mathbf{0}$. Note that such a matrix $\mathbf{B}^{\perp}$ exists if and only if $\mathbf{B}$ has linearly dependent rows ($n > r$), and the set of all such matrices can be captured by $\mathbf{B}^{\perp} = \mathbf{T}\mathbf{U}_2^*$, where $\mathbf{T}$ is an arbitrary nonsingular matrix and $\mathbf{U}_2$ is from the SVD

$$\mathbf{B} = \begin{bmatrix} \mathbf{U}_1 & \mathbf{U}_2 \end{bmatrix} \begin{bmatrix} \boldsymbol{\Sigma}_1 & \mathbf{0} \\ \mathbf{0} & \mathbf{0} \end{bmatrix} \begin{bmatrix} \mathbf{V}_1^* \\ \mathbf{V}_2^* \end{bmatrix}. \tag{2.64}$$

Theorem 2.3.10 (Finsler's Theorem) *Let matrices $\mathbf{B} \in \mathcal{C}^{n\times m}$ and $\mathbf{Q} \in \mathcal{C}^{n\times n}$ be given. Suppose* rank $(\mathbf{B}) < n$ *and* $\mathbf{Q} = \mathbf{Q}^*$. *Let $(\mathbf{B}_r, \mathbf{B}_\ell)$ be any full rank factor of $\mathbf{B}$, i.e. $\mathbf{B} = \mathbf{B}_\ell \mathbf{B}_r$, and define $\mathbf{D} \triangleq (\mathbf{B}_r\mathbf{B}_r^*)^{-1/2}\mathbf{B}_\ell^+$. Then the following statements are equivalent:*

(i) There exists a scalar μ such that

$$\mu \mathbf{BB}^* - \mathbf{Q} > \mathbf{0}. \tag{2.65}$$

(ii) The following condition holds

$$\mathbf{P} \triangleq \mathbf{B}^{\perp}\mathbf{Q}\mathbf{B}^{\perp *} < \mathbf{0}. \tag{2.66}$$

If the above statements hold, then all scalars μ satisfying (2.65) are given by

$$\mu > \mu_{\min} \triangleq \lambda_{\max}\left[\mathbf{D}(\mathbf{Q} - \mathbf{Q}\mathbf{B}^{\perp *}\mathbf{P}^{-1}\mathbf{B}^{\perp}\mathbf{Q})\mathbf{D}^*\right]. \tag{2.67}$$

Proof. Let $\mathbf{T}$ be a square nonsingular matrix defined by

$$\mathbf{T} \triangleq \begin{bmatrix} \mathbf{D} \\ \mathbf{B}^{\perp} \end{bmatrix}. \tag{2.68}$$

By a congruence transformation with $\mathbf{T}$, (2.65) is equivalent to

$$\begin{bmatrix} \mu\mathbf{I} - \mathbf{DQD}^* & -\mathbf{DQB}^{\perp *} \\ -\mathbf{B}^{\perp}\mathbf{QD}^* & -\mathbf{B}^{\perp}\mathbf{QB}^{\perp *} \end{bmatrix} > \mathbf{0} \tag{2.69}$$

or equivalently,

$$\mathbf{P} \triangleq \mathbf{B}^{\perp}\mathbf{QB}^{\perp *} < \mathbf{0} \tag{2.70}$$

$$\mu\mathbf{I} - \mathbf{DQD}^* + \mathbf{DQB}^{\perp *}\mathbf{P}^{-1}\mathbf{B}^{\perp}\mathbf{QD}^* > \mathbf{0} \tag{2.71}$$

which proves the necessity of (2.66). Now, to prove sufficiency, suppose $\mathbf{P} < \mathbf{0}$. Clearly, there exists a μ satisfying (2.71) and all such μ are given by (2.67). □

The existence condition (2.66) is known as Finsler's Theorem (see references in [65], [105]). Theorem 2.3.10 provides not only the existence condition but also all acceptable values of μ. Note that $\mu_{\min} \leq 0$ if and only if $\mathbf{Q} \leq \mathbf{0}$ since $\mu_{\min} \leq 0$ is equivalent to

$$\mathbf{D}(\mathbf{Q} - \mathbf{QB}^{\perp *}\mathbf{P}^{-1}\mathbf{B}^{\perp}\mathbf{Q})\mathbf{D}^* \leq \mathbf{0}$$

which is equivalent to $\mathbf{TQT}^* \leq \mathbf{0}$ since $\mathbf{P} < \mathbf{0}$.

The following result can be verified by a similar procedure to the proof of Theorem 2.3.10.

Corollary 2.3.5 *Let matrices* $\mathbf{B}$ *and* $\mathbf{Q} = \mathbf{Q}^*$ *be given. Then the following statements are equivalent:*

(i) There exists a symmetric matrix $\mathbf{X}$ *such that*

$$\mathbf{Q} + \mathbf{B}\mathbf{X}\mathbf{B}^* > \mathbf{0}.$$

(ii) One of the following conditions hold

$$\mathbf{B}^{\perp}\mathbf{Q}\mathbf{B}^{\perp *} > \mathbf{0} \quad or \quad \mathbf{B}\mathbf{B}^* > \mathbf{0}.$$

Suppose (ii) holds and $\mathbf{B}^*\mathbf{B} > \mathbf{0}$*, but* $\mathbf{B}\mathbf{B}^*$ *is singular. Then all matrices* $\mathbf{X}$ *satisfying the condition in (i) are given by*

$$\mathbf{X} > \mathbf{B}^{+}[\mathbf{Q}\mathbf{B}^{\perp *}(\mathbf{B}^{\perp}\mathbf{Q}\mathbf{B}^{\perp *})^{-1}\mathbf{B}^{\perp}\mathbf{Q} - \mathbf{Q}]\mathbf{B}^{+*}.$$

2.3.13 (A + BXC)R(A + BXC)* < Q

Theorem 2.3.11 *Let matrices* $\mathbf{A} \in \mathcal{C}^{n\times \ell}$, $\mathbf{B} \in \mathcal{C}^{n\times m}$, $\mathbf{C} \in \mathcal{C}^{k\times \ell}$, $\mathbf{R} \in \mathcal{C}^{\ell\times \ell}$ *and* $\mathbf{Q} \in \mathcal{C}^{n\times n}$ *be given. Suppose* $\mathbf{B}^*\mathbf{B} > \mathbf{0}$, $\mathbf{C}\mathbf{C}^* > \mathbf{0}$, $\mathbf{R} > \mathbf{0}$ *and* $\mathbf{Q} > \mathbf{0}$. *Then the following statements are equivalent:*

(i) There exists a matrix $\mathbf{X}$ *such that*

$$(\mathbf{A} + \mathbf{B}\mathbf{X}\mathbf{C})\mathbf{R}(\mathbf{A} + \mathbf{B}\mathbf{X}\mathbf{C})^* < \mathbf{Q}. \tag{2.72}$$

(ii) The following two conditions hold

$$\mathbf{B}^{\perp}(\mathbf{Q} - \mathbf{A}\mathbf{R}\mathbf{A}^*)\mathbf{B}^{\perp *} > \mathbf{0} \quad or \quad \mathbf{B}\mathbf{B}^* > \mathbf{0}$$

$$\mathbf{C}^{*\perp}(\mathbf{R}^{-1} - \mathbf{A}^*\mathbf{Q}^{-1}\mathbf{A})\mathbf{C}^{*\perp *} > \mathbf{0} \quad or \quad \mathbf{C}^*\mathbf{C} > \mathbf{0}.$$

If the above statements hold, then all matrices $\mathbf{X}$ *satisfying (2.72) are given by*

$$\mathbf{X} = -(\mathbf{B}^*\mathbf{\Phi}\mathbf{B})^{-1}\mathbf{B}^*\mathbf{\Phi}\mathbf{A}\mathbf{R}\mathbf{C}^*(\mathbf{C}\mathbf{R}\mathbf{C}^*)^{-1} + (\mathbf{B}^*\mathbf{\Phi}\mathbf{B})^{-1/2}\mathbf{L}\mathbf{\Psi}^{1/2}$$

where $\mathbf{L}$ *is an arbitrary matrix such that* $\|\mathbf{L}\| < 1$ *and*

$$\mathbf{\Phi} \triangleq (\mathbf{Q} - \mathbf{A}\mathbf{R}\mathbf{A}^* + \mathbf{A}\mathbf{R}\mathbf{C}^*(\mathbf{C}\mathbf{R}\mathbf{C}^*)^{-1}\mathbf{C}\mathbf{R}\mathbf{A}^*)^{-1}$$

$$\mathbf{\Psi} \triangleq \mathbf{R}_c - \mathbf{R}_c\mathbf{C}\mathbf{R}\mathbf{A}^*(\mathbf{\Phi} - \mathbf{\Phi}\mathbf{B}(\mathbf{B}^*\mathbf{\Phi}\mathbf{B})^{-1}\mathbf{B}^*\mathbf{\Phi})\mathbf{A}\mathbf{R}\mathbf{C}^*\mathbf{R}_c$$

$$\mathbf{R}_c \triangleq (\mathbf{C}\mathbf{R}\mathbf{C}^*)^{-1}.$$

Proof. After expanding and completing the square, the inequality (2.72) can equivalently be written as

$$(\mathbf{B}\mathbf{X} + \mathbf{A}\mathbf{R}\mathbf{C}^*\mathbf{R}_c)\mathbf{R}_c^{-1}(\mathbf{B}\mathbf{X} + \mathbf{A}\mathbf{R}\mathbf{C}^*\mathbf{R}_c)^* < \mathbf{\Phi}^{-1}$$

where $\mathbf{\Phi}$ and $\mathbf{R}_c$ are defined above. Note that the assumptions $\mathbf{R} > \mathbf{0}$ and $\mathbf{C}\mathbf{C}^* > \mathbf{0}$ imply that $\mathbf{R}_c > \mathbf{0}$. Then, by the Schur complement formula, the above inequality and $\mathbf{R}_c > \mathbf{0}$ are equivalent to

$$(\mathbf{X}^*\mathbf{B}^* + \mathbf{R}_c\mathbf{C}\mathbf{R}\mathbf{A}^*)\mathbf{\Phi}(\mathbf{B}\mathbf{X} + \mathbf{A}\mathbf{R}\mathbf{C}^*\mathbf{R}_c) < \mathbf{R}_c$$

and $\mathbf{\Phi} > \mathbf{0}$. After expanding, completing the square with respect to $\mathbf{X}$ yields

$$(\mathbf{X} + \mathbf{\Phi}_B\mathbf{B}^*\mathbf{\Phi}\mathbf{A}\mathbf{R}\mathbf{C}^*\mathbf{R}_c)^*\mathbf{\Phi}_B^{-1}(\mathbf{X} + \mathbf{\Phi}_B\mathbf{B}^*\mathbf{\Phi}\mathbf{A}\mathbf{R}\mathbf{C}^*\mathbf{R}_c) < \mathbf{\Psi} \tag{2.73}$$

where $\mathbf{\Phi}_B \triangleq (\mathbf{B}^*\mathbf{\Phi}\mathbf{B})^{-1}$. Since the left-hand side is non-negative, we have $\mathbf{\Psi} > \mathbf{0}$. Thus, $\mathbf{\Phi} > \mathbf{0}$ and $\mathbf{\Psi} > \mathbf{0}$ are necessary for the existence of $\mathbf{X}$ satisfying (2.72). To prove the converse, suppose $\mathbf{\Phi} > \mathbf{0}$ and $\mathbf{\Psi} > \mathbf{0}$. In this case, $\mathbf{\Phi}_B^{-1/2}$ and $\mathbf{\Psi}^{-1/2}$ exist, and (2.73) can be equivalently written

$$\|\mathbf{L}\| < 1, \quad \mathbf{L} \triangleq \mathbf{\Phi}_B{}^{-1/2}(\mathbf{X} + \mathbf{\Phi}_B\mathbf{B}^*\mathbf{\Phi}\mathbf{A}\mathbf{R}\mathbf{C}^*\mathbf{R}_c)\mathbf{\Psi}^{-1/2}.$$

Solving for $\mathbf{X}$, we have

$$\mathbf{X} = -\mathbf{\Phi}_B\mathbf{B}^*\mathbf{\Phi}\mathbf{A}\mathbf{R}\mathbf{C}^*\mathbf{R}_c + \mathbf{\Phi}_B^{1/2}\mathbf{L}\mathbf{\Psi}^{1/2}, \quad \|\mathbf{L}\| < 1.$$

By construction, the above $\mathbf{X}$ solves (2.72) and in fact, any solution $\mathbf{X}$ can be generated by the above formula.

Finally, we shall show that the existence conditions $\mathbf{\Phi} > \mathbf{0}$ and $\mathbf{\Psi} > \mathbf{0}$ are equivalent to statement (ii). To this end, note that $\mathbf{\Phi} > \mathbf{0}$ holds if and only if there exists $\mathbf{V} > \mathbf{0}$ such that

$$\mathbf{Q} - \mathbf{A}\mathbf{R}\mathbf{A}^* + \mathbf{A}\mathbf{R}\mathbf{C}^*(\mathbf{C}\mathbf{R}\mathbf{C}^* + \mathbf{V})^{-1}\mathbf{C}\mathbf{R}\mathbf{A}^* > \mathbf{0},$$

or equivalently, using the matrix inversion lemma,

$$\mathbf{Q} - \mathbf{A}(\mathbf{R}^{-1} + \mathbf{C}^*\mathbf{V}^{-1}\mathbf{C})^{-1}\mathbf{A}^* > \mathbf{0},$$

or equivalently, using the Schur complement formula,

$$\mathbf{R}^{-1} + \mathbf{C}^*\mathbf{V}^{-1}\mathbf{C} - \mathbf{A}^*\mathbf{Q}^{-1}\mathbf{A} > \mathbf{0}.$$

From Finsler's Theorem, there exists $\mathbf{V} > \mathbf{0}$ satisfying the above inequality if and only if

$$\mathbf{C}^{*\perp}(\mathbf{R}^{-1} - \mathbf{A}^*\mathbf{Q}^{-1}\mathbf{A})\mathbf{C}^{*\perp *} > \mathbf{0} \quad \text{or} \quad \mathbf{C}^*\mathbf{C} > \mathbf{0}$$

holds. The equivalence between $\mathbf{\Psi} > \mathbf{0}$ and the first condition in statement (ii) can be shown by a similar procedure. This completes the proof. □

For the special case where $\mathbf{C} = \mathbf{I}$, Theorem 2.3.11 reduces to the following.

Corollary 2.3.6 *Let matrices* $\mathbf{A}$, $\mathbf{B}$, $\mathbf{Q}$ *and* $\mathbf{R}$ *be given. Suppose* $\mathbf{Q} = \mathbf{Q}^*$, $\mathbf{R} = \mathbf{R}^* > \mathbf{0}$ *and* $\mathbf{B}^*\mathbf{B} > \mathbf{0}$. *Then the following statements are equivalent:*

(i) There exists a matrix $\mathbf{X}$ *such that*

$$(\mathbf{A} + \mathbf{B}\mathbf{X})\mathbf{R}(\mathbf{A} + \mathbf{B}\mathbf{X})^* < \mathbf{Q}. \tag{2.74}$$

(ii) $\mathbf{Q} > \mathbf{0}$ *and*

$$\mathbf{B}^{\perp}(\mathbf{Q} - \mathbf{A}\mathbf{R}\mathbf{A}^*)\mathbf{B}^{\perp *} > \mathbf{0} \quad \textit{or} \quad \mathbf{B}\mathbf{B}^* > \mathbf{0}.$$

If the above statements hold, then all matrices $\mathbf{X}$ *satisfying (2.74) are given by*

$$\mathbf{X} = -(\mathbf{B}^*\mathbf{Q}^{-1}\mathbf{B})^{-1}\mathbf{B}^*\mathbf{Q}^{-1}\mathbf{A} + (\mathbf{B}^*\mathbf{Q}^{-1}\mathbf{B})^{-1/2}\mathbf{L}\mathbf{\Psi}^{1/2}$$

where $\mathbf{L}$ *is an arbitrary matrix such that* $\|\mathbf{L}\| < 1$ *and*

$$\mathbf{\Psi} \triangleq \mathbf{R}^{-1} - \mathbf{A}^*\mathbf{Q}^{-1}\mathbf{A} + \mathbf{A}^*\mathbf{Q}^{-1}\mathbf{B}(\mathbf{B}^*\mathbf{Q}^{-1}\mathbf{B})^{-1}\mathbf{B}^*\mathbf{Q}^{-1}\mathbf{A}.$$

The following result is also a special case of Theorem 2.3.11 where $\mathbf{B} = \mathbf{I}$ and $\mathbf{C} = \mathbf{I}$.

Corollary 2.3.7 *Let matrices* $\mathbf{A}$, $\mathbf{B}$, $\mathbf{C}$, $\mathbf{Q}$ *and* $\mathbf{R}$ *be given, where all the matrices except* $\mathbf{B}$ *are symmetric. Then the following statements are equivalent.*

(i) There exists a matrix $\mathbf{X}$ *such that*

$$\begin{bmatrix} \mathbf{Q} & \mathbf{X} \\ \mathbf{X}^* & \mathbf{R} \end{bmatrix} > \begin{bmatrix} \mathbf{A} & \mathbf{B} \\ \mathbf{B}^* & \mathbf{C} \end{bmatrix}. \tag{2.75}$$

(ii) $\mathbf{V} \triangleq \mathbf{Q} - \mathbf{A} > \mathbf{0}$ *and* $\mathbf{W} \triangleq \mathbf{R} - \mathbf{C} > \mathbf{0}$.

Suppose the above statements hold. Then all matrices $\mathbf{X}$ *satisfying (2.75) are given by*

$$\mathbf{X} = \mathbf{B} + \mathbf{V}^{1/2}\mathbf{L}\mathbf{W}^{1/2},$$

where $\mathbf{L}$ *is an arbitrary matrix such that* $\|\mathbf{L}\| < 1$.

Proof. Suppose (i) holds. Then the necessity of (ii) is obvious. To prove the converse, suppose (ii) holds. Then using the Schur complement formula, (2.75) is equivalent to

$$(\mathbf{X} - \mathbf{B})\mathbf{W}^{-1}(\mathbf{X} - \mathbf{B})^* < \mathbf{V}.$$

Now the result follows as a special case of Theorem 2.3.11. □

2.3.14 BXC + (BXC)* + Q < 0

Theorem 2.3.12 *Let matrices* $\mathbf{B} \in \mathcal{C}^{n\times m}$, $\mathbf{C} \in \mathcal{C}^{k\times n}$ *and* $\mathbf{Q} = \mathbf{Q}^* \in \mathcal{C}^{n\times n}$ *be given. Then the following statements are equivalent:*

(i) There exists a matrix $\mathbf{X}$ *satisfying*

$$\mathbf{BXC} + (\mathbf{BXC})^* + \mathbf{Q} < \mathbf{0}. \tag{2.76}$$

(ii) The following two conditions hold

$$\mathbf{B}^{\perp}\mathbf{Q}\mathbf{B}^{\perp *} < \mathbf{0} \quad or \quad \mathbf{B}\mathbf{B}^* > \mathbf{0}$$

$$\mathbf{C}^{*\perp}\mathbf{Q}\mathbf{C}^{*\perp *} < \mathbf{0} \quad or \quad \mathbf{C}^*\mathbf{C} > \mathbf{0}.$$

Suppose the above statements hold. Let r_b *and* r_c *be the ranks of* $\mathbf{B}$ *and* $\mathbf{C}$, *respectively, and* $(\mathbf{B}_\ell, \mathbf{B}_r)$ *and* $(\mathbf{C}_\ell, \mathbf{C}_r)$ *be any full rank factors of* $\mathbf{B}$ *and* $\mathbf{C}$, *i.e.* $\mathbf{B} = \mathbf{B}_\ell\mathbf{B}_r$, $\mathbf{C} = \mathbf{C}_\ell\mathbf{C}_r$. *Then all matrices* $\mathbf{X}$ *in statement (i) are given by*

$$\mathbf{X} = \mathbf{B}_r^+\mathbf{K}\mathbf{C}_\ell^+ + \mathbf{Z} - \mathbf{B}_r^+\mathbf{B}_r\mathbf{Z}\mathbf{C}_\ell\mathbf{C}_\ell^+$$

where $\mathbf{Z}$ *is an arbitrary matrix and*

$$\mathbf{K} \triangleq -\mathbf{R}^{-1}\mathbf{B}_\ell^*\boldsymbol{\Phi}\mathbf{C}_r^*(\mathbf{C}_r\boldsymbol{\Phi}\mathbf{C}_r^*)^{-1} + \mathbf{S}^{1/2}\mathbf{L}(\mathbf{C}_r\boldsymbol{\Phi}\mathbf{C}_r^*)^{-1/2}$$

$$\mathbf{S} \triangleq \mathbf{R}^{-1} - \mathbf{R}^{-1}\mathbf{B}_\ell^*[\boldsymbol{\Phi} - \boldsymbol{\Phi}\mathbf{C}_r^*(\mathbf{C}_r\boldsymbol{\Phi}\mathbf{C}_r^*)^{-1}\mathbf{C}_r\boldsymbol{\Phi}]\mathbf{B}_\ell\mathbf{R}^{-1}$$

where $\mathbf{L}$ *is an arbitrary matrix such that* $\|\mathbf{L}\| < 1$ *and* $\mathbf{R}$ *is an arbitrary positive definite matrix such that*

$$\boldsymbol{\Phi} \triangleq (\mathbf{B}_\ell\mathbf{R}^{-1}\mathbf{B}_\ell^* - \mathbf{Q})^{-1} > \mathbf{0}.$$

Proof. Suppose statement (i) holds. Then $\mathbf{K} \triangleq \mathbf{B}_r\mathbf{X}\mathbf{C}_\ell$ satisfies

$$\mathbf{B}_\ell\mathbf{K}\mathbf{C}_r + (\mathbf{B}_\ell\mathbf{K}\mathbf{C}_r)^* + \mathbf{Q} < \mathbf{0}.$$

Let a matrix $\mathbf{R} > \mathbf{0}$ be such that

$$\mathbf{B}_\ell\mathbf{K}\mathbf{C}_r + (\mathbf{B}_\ell\mathbf{K}\mathbf{C}_r)^* + \mathbf{Q} + \mathbf{C}_r^*\mathbf{K}^*\mathbf{R}\mathbf{K}\mathbf{C}_r < \mathbf{0}$$

or equivalently,

$$(\mathbf{B}_\ell\mathbf{R}^{-1} + \mathbf{C}_r^*\mathbf{K}^*)\mathbf{R}(\mathbf{R}^{-1}\mathbf{B}_\ell^* + \mathbf{K}\mathbf{C}_r) < \mathbf{B}_\ell\mathbf{R}^{-1}\mathbf{B}_\ell^* - \mathbf{Q} =: \boldsymbol{\Phi}^{-1}.$$

Note that such a matrix $\mathbf{R} > \mathbf{0}$ always exists and a choice of such $\mathbf{R}$ is $\mathbf{R} = \varepsilon\mathbf{I}$ for sufficiently small $\varepsilon > 0$. Now, using Corollary 2.3.6, there exists a matrix $\mathbf{K}$ satisfying the above inequality if and only if $\boldsymbol{\Phi} > \mathbf{0}$ and either $\mathbf{C}_r^{*\perp}\mathbf{Q}\mathbf{C}_r^{*\perp*} < \mathbf{0}$ or $\mathbf{C}_r^*\mathbf{C}_r > \mathbf{0}$, in which case all such matrices $\mathbf{K}$ are given by

$$\mathbf{K} = -\mathbf{R}^{-1}\mathbf{B}_\ell^*\boldsymbol{\Phi}\mathbf{C}_r^*(\mathbf{C}_r\boldsymbol{\Phi}\mathbf{C}_r^*)^{-1} + \mathbf{S}^{1/2}\mathbf{L}(\mathbf{C}_r\boldsymbol{\Phi}\mathbf{C}_r^*)^{-1/2}$$

where $\mathbf{L}$ is an arbitrary matrix such that $\|\mathbf{L}\| < 1$ and

$$\mathbf{S} \triangleq \mathbf{R}^{-1} - \mathbf{R}^{-1}\mathbf{B}_\ell^*[\boldsymbol{\Phi} - \boldsymbol{\Phi}\mathbf{C}_r^*(\mathbf{C}_r\boldsymbol{\Phi}\mathbf{C}_r^*)^{-1}\mathbf{C}_r\boldsymbol{\Phi}]\,\mathbf{B}_\ell\mathbf{R}^{-1} > \mathbf{0}.$$

Using Finsler's Theorem (Corollary 2.3.5), $\boldsymbol{\Phi} > \mathbf{0}$ holds for some $\mathbf{R} > \mathbf{0}$ if and only if $\mathbf{B}_\ell^\perp\mathbf{Q}\mathbf{B}_\ell^{\perp*} < \mathbf{0}$ or $\mathbf{B}_\ell\mathbf{B}_\ell^* > \mathbf{0}$. It is easy to verify that $\mathbf{C}_r^{*\perp}\mathbf{Q}\mathbf{C}_r^{*\perp*} < \mathbf{0}$ and $\mathbf{B}_\ell^\perp\mathbf{Q}\mathbf{B}_\ell^{\perp*} < \mathbf{0}$ are equivalent to $\mathbf{C}^{*\perp}\mathbf{Q}\mathbf{C}^{*\perp*} < \mathbf{0}$ and $\mathbf{B}^\perp\mathbf{Q}\mathbf{B}^{\perp*} < \mathbf{0}$, respectively. Finally, note that $\mathbf{K} = \mathbf{B}_r\mathbf{X}\mathbf{C}_\ell$ holds if and only if

$$\mathbf{X} = \mathbf{B}_r^+\mathbf{K}\mathbf{C}_\ell^+ + \mathbf{Z} - \mathbf{B}_r^+\mathbf{B}_r\mathbf{Z}\mathbf{C}_\ell\mathbf{C}_\ell^+$$

holds for some $\mathbf{Z}$. Thus we have established the necessity of statement (ii). Sufficiency and the explicit formula for $\mathbf{X}$ follow by construction. This completes the proof. □

Note that all solutions $\mathbf{X}$ to (2.76) can be captured by the freedoms in the choice of parameters $\mathbf{Z}$, $\mathbf{L}$ and $\mathbf{R}$. For the special case where $\mathbf{B}$ and $\mathbf{C}$ have full column and rows, respectively, the freedom due to $\mathbf{Z}$ disappears since $\mathbf{B}_r^+\mathbf{B}_r = \mathbf{I}$ and $\mathbf{C}_\ell\mathbf{C}_\ell^+ = \mathbf{I}$. In view of the above proof, the freedom $\mathbf{R} > \mathbf{0}$ can be restricted to have the structure $\mathbf{R} = r\mathbf{I}$ for some scalar $r > 0$ without loss of generality. The following result shows another special case where the freedom $\mathbf{R} > \mathbf{0}$ (as well as $\mathbf{Z}$) disappears.

Corollary 2.3.8 *Let matrices* $\mathbf{B}$, $\mathbf{C}$ *and* $\mathbf{Q} = \mathbf{Q}^*$ *be given. Suppose the conditions in statement (ii) of Theorem 2.3.12 hold. If we further assume that*

$$\mathbf{B}^*\mathbf{B} > \mathbf{0}, \quad \mathbf{C}\mathbf{B}^{\perp*}\mathbf{B}^\perp\mathbf{C}^* > \mathbf{0},$$

then all matrices $\mathbf{X}$ *satisfying*

$$\mathbf{B}\mathbf{X}\mathbf{C} + (\mathbf{B}\mathbf{X}\mathbf{C})^* + \mathbf{Q} < \mathbf{0}$$

are given by

$$\mathbf{X} = \mathbf{X}_1 + \mathbf{X}_2\mathbf{L}\mathbf{X}_3$$

where $\mathbf{L}$ *is an arbitrary matrix such that* $\|\mathbf{L}\| < 1$ *and*

$$\mathbf{X}_1 \triangleq (\mathbf{C}_1^* - \mathbf{Q}_{12}\mathbf{Q}_{22}^{-1}\mathbf{C}_2^*)(\mathbf{C}_2\mathbf{Q}_{22}^{-1}\mathbf{C}_2^*)^{-1}$$

$$\mathbf{X}_2 \triangleq (\mathbf{Q}_{12}\mathbf{Q}_{22}^{-1}\mathbf{Q}_{12}^* - \mathbf{Q}_{11} - \mathbf{X}_1\mathbf{X}_3^{-2}\mathbf{X}_1^*)^{1/2}$$

$$\mathbf{X}_3 \triangleq (-\mathbf{C}_2\mathbf{Q}_{22}^{-1}\mathbf{C}_2^*)^{-1/2}$$

$$\begin{bmatrix} \mathbf{Q}_{11} & \mathbf{Q}_{12} \\ \mathbf{Q}_{12}^* & \mathbf{Q}_{22} \end{bmatrix} \triangleq \begin{bmatrix} \mathbf{B}^+ \\ \mathbf{B}^\perp \end{bmatrix} \mathbf{Q} \begin{bmatrix} \mathbf{B}^{+*} & \mathbf{B}^{\perp *} \end{bmatrix}$$

$$\begin{bmatrix} \mathbf{C}_1 & \mathbf{C}_2 \end{bmatrix} \triangleq \mathbf{C} \begin{bmatrix} \mathbf{B}^{+*} & \mathbf{B}^{\perp *} \end{bmatrix}.$$

Proof. By a congruent transformation, we have

$$\begin{bmatrix} \mathbf{B}^+ \\ \mathbf{B}^\perp \end{bmatrix} (\mathbf{BXC} + (\mathbf{BXC})^* + \mathbf{Q}) \begin{bmatrix} \mathbf{B}^{+*} & \mathbf{B}^{\perp *} \end{bmatrix} < \mathbf{0},$$

or equivalently, using the definitions given above,

$$\begin{bmatrix} \mathbf{Q}_{11} + \mathbf{XC}_1 + \mathbf{C}_1^*\mathbf{X}^* & \mathbf{Q}_{12} + \mathbf{XC}_2 \\ \mathbf{Q}_{12}^* + \mathbf{C}_2^*\mathbf{X}^* & \mathbf{Q}_{22} \end{bmatrix} < \mathbf{0}.$$

Since statement (ii) in Theorem 2.3.12 holds, we have $\mathbf{Q}_{22} < \mathbf{0}$ and hence the above inequality is equivalent to

$$\mathbf{Q}_{11} + \mathbf{XC}_1 + \mathbf{C}_1^*\mathbf{X}^* - (\mathbf{Q}_{12} + \mathbf{XC}_2)\mathbf{Q}_{22}^{-1}(\mathbf{Q}_{12} + \mathbf{XC}_2)^* < \mathbf{0}.$$

Now, by supposition,

$$\mathbf{C}_2\mathbf{C}_2^* = \mathbf{CB}^{\perp *}\mathbf{B}^\perp\mathbf{C}^* > \mathbf{0},$$

and hence $\mathbf{C}_2\mathbf{Q}_{22}^{-1}\mathbf{C}_2^* < \mathbf{0}$. Thus, after expanding, we can complete the square as follows:

$$(\mathbf{X} - \mathbf{X}_1)\mathbf{X}_3^{-2}(\mathbf{X} - \mathbf{X}_1)^* < \mathbf{X}_2^2$$

where matrices $\mathbf{X}_1$, $\mathbf{X}_2$ and $\mathbf{X}_3$ are defined above. Then the result follows as a special case of Corollary 2.3.6. This completes the proof. □

Note that the result of Theorem 2.3.11 follows as a special case of Theorem 2.3.12. For $\mathbf{R} > \mathbf{0}$,

$$(\mathbf{A} + \mathbf{BXC})\mathbf{R}(\mathbf{A} + \mathbf{BXC})^* < \mathbf{Q}$$

is equivalent to

$$\begin{bmatrix} \mathbf{B} \\ \mathbf{0} \end{bmatrix} (-\mathbf{X}) \begin{bmatrix} \mathbf{0} & \mathbf{C} \end{bmatrix} + \begin{bmatrix} \mathbf{0} \\ \mathbf{C}^* \end{bmatrix} (-\mathbf{X})^* \begin{bmatrix} \mathbf{B}^* & \mathbf{0} \end{bmatrix} - \begin{bmatrix} \mathbf{Q} & \mathbf{A} \\ \mathbf{A}^* & \mathbf{R}^{-1} \end{bmatrix} < \mathbf{0}.$$

Thus, applying Theorem 2.3.12, we have the existence conditions in statement (ii) of Theorem 2.3.11. If $\mathbf{B}^*\mathbf{B} > \mathbf{0}$ and $\mathbf{CC}^* > \mathbf{0}$, then the assumptions in Corollary 2.3.8 are satisfied, i.e.

$$\begin{bmatrix} \mathbf{B} \\ \mathbf{0} \end{bmatrix}^* \begin{bmatrix} \mathbf{B} \\ \mathbf{0} \end{bmatrix} = \mathbf{B}^*\mathbf{B} > \mathbf{0}$$

$$\begin{bmatrix} \mathbf{0} & \mathbf{C} \end{bmatrix} \begin{bmatrix} \mathbf{B} \\ \mathbf{0} \end{bmatrix}^{\perp *} \begin{bmatrix} \mathbf{B} \\ \mathbf{0} \end{bmatrix}^{\perp} \begin{bmatrix} \mathbf{0} \\ \mathbf{C}^* \end{bmatrix} = \mathbf{CC}^* > \mathbf{0}.$$

Hence, the result of Corollary 2.3.8 can be applied to obtain an explicit formula for $\mathbf{X}$, where all the freedoms are captured by $\|\mathbf{L}\| < 1$, as is the case for the formula given in Theorem 2.3.11.

Theorem 2.3.12 reduces to the following when specialized to the case $\mathbf{C} = \mathbf{I}$.

Corollary 2.3.9 *Let matrices* $\mathbf{B}$ *and* $\mathbf{Q} = \mathbf{Q}^*$ *be given. Then the following statements are equivalent:*

(i) There exists a matrix $\mathbf{X}$ *satisfying*

$$\mathbf{BX} + (\mathbf{BX})^* + \mathbf{Q} < \mathbf{0}.$$

(ii) The following condition holds

$$\mathbf{B}^{\perp}\mathbf{Q}\mathbf{B}^{\perp *} < \mathbf{0} \quad or \quad \mathbf{BB}^* > \mathbf{0}.$$

Suppose the above statements hold and further assume that $\mathbf{B}^*\mathbf{B} > \mathbf{0}$. *Then all matrices* $\mathbf{X}$ *in statement (i) are given by*

$$\mathbf{X} = -\rho\mathbf{B}^* + \sqrt{\rho}\mathbf{L}\boldsymbol{\Omega}^{1/2}$$

where $\mathbf{L}$ *is any matrix such that* $\|\mathbf{L}\| < 1$ *and* $\rho > \mathbf{0}$ *is any scalar such that*

$$\boldsymbol{\Omega} \triangleq \rho\mathbf{BB}^* - \mathbf{Q} > \mathbf{0}.$$

CHAPTER 2 SUMMARY

Chapter 2 defines the notation, techniques, and virtually all the mathematical tools that are needed to derive all results of this book. Linear algebra concepts have become increasingly important in the state space analysis and design of control systems [127]. A detailed presentation of the singular value decomposition and its connections with control theory can be found in [75]. The Moore–Penrose generalized inverse was defined in [103]. Solvability conditions for linear matrix equations can be found in [108], [3] and [72] although many results in this chapter are new in this area. Most of the solvability conditions of linear and quadratic matrix inequalities are new although the basic Theorem 2.3.10 dates back to Finsler [28]. More theoretical details in matrix methods are available in the books [50], [41], [153].

Theorem 2.3.12 is the most important result in this book. Almost all control problems in this book can be analytically solved by this theorem, i.e. approximately 20 different control problems all reduce to this problem of linear algebra. The point of the book is to show how to rearrange these control problems so that they take the form of Theorem 2.3.12.

CHAPTER THREE

Analysis of First-Order Information

This chapter reviews the classical analysis of state space models including the system "abilities"; observability, controllability and stability.

3.1 SOLUTIONS OF LINEAR DIFFERENTIAL EQUATIONS

We consider a linear, time-varying dynamic system of the form

$$\begin{aligned} \dot{\mathbf{x}}(t) &= \mathbf{A}(t)\mathbf{x}(t) + \mathbf{B}(t)\mathbf{u}(t) \\ \mathbf{y}(t) &= \mathbf{C}(t)\mathbf{x}(t) + \mathbf{D}(t)\mathbf{u}(t) \end{aligned} \tag{3.1}$$

where $\mathbf{A}(t)$, $\mathbf{B}(t)$, $\mathbf{C}(t)$ and $\mathbf{D}(t)$ are the system matrices which may be functions of time t, and $\mathbf{x}(t)$ is the state vector, $\mathbf{u}(t)$ is the control input, and $\mathbf{y}(t)$ is the output. The solution of (3.1) is given by

$$\mathbf{x}(t) = \mathbf{\Phi}(t, t_0)\mathbf{x}(t_0) + \int_{t_0}^{t} \mathbf{\Phi}(t, \sigma)\mathbf{B}(\sigma)\mathbf{u}(\sigma)d\sigma$$

where $\mathbf{\Phi}(t, t_0)$ is called the state transition matrix and is generated by solving the differential equation

$$\frac{d}{dt}\mathbf{\Phi}(t, t_0) = \mathbf{A}(t)\mathbf{\Phi}(t, t_0), \qquad \mathbf{\Phi}(t_0, t_0) = \mathbf{I}.$$

If $\mathbf{A}$ is a constant matrix, then

$$\mathbf{\Phi}(\tau + t_0, t_0) = \mathbf{\Phi}(\tau, 0) = e^{\mathbf{A}\tau} \triangleq \sum_{i=0}^{\infty} \frac{\mathbf{A}^i \tau^i}{i!} \tag{3.2}$$

and

$$\mathbf{x}(t) = e^{\mathbf{A}(t-t_0)}\mathbf{x}(t_0) + \int_{t_0}^{t} e^{\mathbf{A}(t-\sigma)}\mathbf{B}\mathbf{u}(\sigma)d\sigma. \tag{3.3}$$

Various properties of $e^{\mathbf{A}t}$ may be found in a first course in linear systems [11, 125, 67], such as the following.

Theorem 3.1.1 *Let* $\mathbf{A}$ *be a constant* $n \times n$ *matrix. Then*

$$e^{\mathbf{A}t} = \mathcal{L}^{-1}[s\mathbf{I} - \mathbf{A}]^{-1} \tag{3.4a}$$

$$e^{\mathbf{A}t} = \mathbf{E}e^{\Lambda t}\mathbf{E}^{-1} \tag{3.4b}$$

$$e^{\mathbf{A}t} = \sum_{i=0}^{n-1} \mathbf{A}^i \alpha_i(t) \tag{3.4c}$$

where $\mathcal{L}^{-1}$ *is the inverse Laplace transformation operator,* $\mathbf{A} = \mathbf{E}\Lambda\mathbf{E}^{-1}$ *is the spectral decomposition of* $\mathbf{A}$*, and the functions* $\alpha_i(t)$ *are computed from the inverse Laplace transforms of*

$$\begin{aligned}
\alpha_{n-1}(s) &= |s\mathbf{I} - \mathbf{A}|^{-1} \\
\alpha_{n-2}(s) &= |s\mathbf{I} - \mathbf{A}|^{-1}(s + a_{n-1}) \\
\alpha_{n-3}(s) &= |s\mathbf{I} - \mathbf{A}|^{-1}(s^2 + a_{n-1}s + a_{n-2}) \\
&\vdots \\
\alpha_0(s) &= |s\mathbf{I} - \mathbf{A}|^{-1}(s^{n-1} + a_{n-1}s^{n-2} + \cdots a_2 s + a_1)
\end{aligned} \tag{3.5a}$$

where

$$|s\mathbf{I} - \mathbf{A}| = s^n + a_{n-1}s^{n-1} + a_{n-2}s^{n-2} + \cdots a_1 s + a_0.$$

The above theorem allows one to compute the *first-order* information $\mathbf{y}(t)$ for a linear system. To illustrate how difficult and unreliable such calculations can be, consider four examples for the computation of $e^{\mathbf{A}t}$:

$$\mathbf{A}_1 = \begin{bmatrix} -1 & 1 \\ 0 & -1 \end{bmatrix}, \quad \mathbf{A}_2 = \begin{bmatrix} -1+\epsilon & 1 \\ 0 & -1 \end{bmatrix},$$

$$\mathbf{A}_3 = \begin{bmatrix} -1 & 1 \\ 0 & -1+\epsilon \end{bmatrix}, \quad \mathbf{A}_4 = \begin{bmatrix} -1 & 1+\epsilon \\ 0 & -1 \end{bmatrix}.$$

Matrix $\mathbf{A}_1$ is *defective* (meaning that it does not have linearly independent eigenvectors). Matrices $\mathbf{A}_2, \mathbf{A}_3, \mathbf{A}_4$ might represent various effects of computational errors in an attempt to study the first-order behavior of $\dot{\mathbf{x}} = \mathbf{A}_1\mathbf{x}$. While both $\mathbf{A}_1$ and $\mathbf{A}_4$ are defective, matrices $\mathbf{A}_2$ and $\mathbf{A}_3$ are nondefective, for any (even arbitrarily small) $\epsilon \neq 0$. Hence $\mathbf{y}(t) = \mathbf{C}e^{\mathbf{A}t}\mathbf{x}(0)$ for each $\mathbf{A}_1, \mathbf{A}_2$ above yields, for $\mathbf{C} = [1 \quad 0]$, $\mathbf{x}(0) = [0 \ 1]^T$,

$$\begin{aligned}
y^1(t) &= \mathbf{C}e^{\mathbf{A}_1 t}\mathbf{x}(0) \\
&= \mathbf{C}\mathbf{E}_1 e^{\Lambda_1 t}\mathbf{E}_1^{-1}\mathbf{x}(0), \quad \mathbf{E}_1 = \begin{bmatrix} 1 & 0 \\ 0 & 1 \end{bmatrix}, \quad \Lambda_1 = \begin{bmatrix} -1 & 1 \\ 0 & -1 \end{bmatrix} \\
&= te^{-t} \\
y^2(t) &= \mathbf{C}e^{\mathbf{A}_2 t}\mathbf{x}(0) \\
&= \mathbf{C}\mathbf{E}_2 e^{\Lambda_2 t}\mathbf{E}_2^{-1}\mathbf{x}(0), \quad \mathbf{E}_2 = \begin{bmatrix} 1 & 1 \\ -\epsilon & 0 \end{bmatrix}, \quad \Lambda_2 = \begin{bmatrix} -1 & 0 \\ 0 & -1+\epsilon \end{bmatrix} \\
&= \left(\frac{e^{\epsilon t} - 1}{\epsilon}\right)e^{-t}.
\end{aligned}$$

The error $y^2(t) - y^1(t)$ is

$$y^2(t) - y^1(t) = \left[\frac{e^{\epsilon t} - 1}{\epsilon} - t\right]e^{-t}.$$

Hence, small modeling errors can drastically change the character of the first-order response of a linear system, and great care is required to obtain a good model for control design.

3.2 SOLUTIONS OF LINEAR DIFFERENCE EQUATIONS

Consider a linear discrete-time dynamic system

$$\begin{aligned} \mathbf{x}_{k+1} &= \mathbf{A}_k\mathbf{x}_k + \mathbf{B}_k\mathbf{u}_k \\ \mathbf{y}_k &= \mathbf{C}_k\mathbf{x}_k + \mathbf{D}_k\mathbf{u}_k \end{aligned} \tag{3.6}$$

where $\mathbf{A}_k$, $\mathbf{B}_k$, $\mathbf{C}_k$, $\mathbf{D}_k$, $\mathbf{x}_k$, $\mathbf{y}_k$, $\mathbf{u}_k$ denote matrices and vectors that are functions of the time index k, i.e. at time t_k, $\mathbf{A}_k \triangleq \mathbf{A}(t_k)$, etc. The solution of (3.6) for $\mathbf{x}_k$ is

$$\begin{aligned} \mathbf{x}_k &= \mathbf{\Phi}_{k0}\mathbf{x}_0 + \sum_{i=1}^{k} \mathbf{\Phi}_{ki}\mathbf{B}_{i-1}\mathbf{u}_{i-1} \\ \mathbf{\Phi}_{kk} &\hat{=} \mathbf{I}\ \mathbf{\Phi}_{ki} \triangleq \Pi_{\alpha=1}^{k-1}\mathbf{A}_\alpha. \end{aligned}$$

If $\mathbf{A}$ and $\mathbf{B}$ are constant then $\mathbf{A}_\alpha = \mathbf{A}$ for all α and this solution reduces to

$$\mathbf{x}_k = \mathbf{A}^k\mathbf{x}_0 + \sum_{i=1}^{k} \mathbf{A}^{k-i}\mathbf{B}\mathbf{u}_{i-1}. \tag{3.7}$$

3.3 CONTROLLABILITY AND OBSERVABILITY OF CONTINUOUS-TIME SYSTEMS

3.3.1 Controllability

Consider the system

$$\begin{aligned} \dot{\mathbf{x}}(t) &= \mathbf{A}(t)\mathbf{x}(t) + \mathbf{B}(t)\mathbf{u}(t) \\ \mathbf{y}(t) &= \mathbf{C}(t)\mathbf{x}(t). \end{aligned} \tag{3.8}$$

Definition 3.3.1 System (3.8) is said to be completely state controllable at time $t = t_0$ if there exists a time $t_f > t_0$ and a control $\mathbf{u}(t)$, $t \in [t_0,\ t_f]$ such that the state is transferred from an arbitrary initial state $\mathbf{x}(t_0) = \mathbf{x}_0$ to an arbitrarily specified $\mathbf{x}(t_f) = \mathbf{x}_f$ in a finite time $t_f < \infty$.

Suppose we wish to know whether (3.8) is completely controllable at t_0. This is equivalent to asking whether there exists a $\mathbf{u}(\sigma)$, $\sigma \in [t_0,\ t_f]$ such that

$$\int_{t_0}^{t_f} \mathbf{\Phi}(t_f, \sigma)\,\mathbf{B}(\sigma)\mathbf{u}(\sigma)d\sigma = \tilde{\mathbf{x}} \tag{3.9}$$

for some $t_f < \infty$ and for any specified $\tilde{\mathbf{x}} \triangleq \mathbf{x}_f - \mathbf{\Phi}(t_f, t_0)\mathbf{x}_0$. Since every element of the vector $\tilde{\mathbf{x}}$ is arbitrary, the rows of the matrix

$$\mathbf{R}(\sigma) \triangleq \mathbf{\Phi}(t_f, \sigma)\mathbf{B}(\sigma) \tag{3.10}$$

must be linearly independent on the interval $\sigma \in [t_0, \; t_f]$. This is equivalent to

$$\mathbf{X}(t_f) \triangleq \int_{t_0}^{t_f} \mathbf{R}(\sigma)\mathbf{R}^T(\sigma)d\sigma > \mathbf{0}, \tag{3.11}$$

which is equivalent to

$$\begin{aligned} \dot{\mathbf{X}}(t) &= \mathbf{X}(t)\mathbf{A}^T(t) + \mathbf{A}(t)\mathbf{X}(t) + \mathbf{B}(t)\mathbf{B}^T(t) \\ \mathbf{X}(t_0) &= \mathbf{0} \\ \mathbf{X}(t_f) &> \mathbf{0}. \end{aligned} \tag{3.12}$$

Equation (3.12) may be derived from (3.11) by replacing t_f by t and differentiating $\mathbf{X}(t)$ with respect to t.

Theorem 3.3.1 *The system $\dot{\mathbf{x}}(t) = \mathbf{A}(t)\mathbf{x}(t) + \mathbf{B}(t)\mathbf{u}(t)$ is completely state controllable at time t_0 if and only if there exists $t_f < \infty$ such that (3.12) holds.*

Exercise 3.3.1 Suppose system (3.8) is given. Using similar steps as (3.8)–(3.10) prove that the output $\mathbf{y}(t_f)$ can be taken to an arbitrary value for some $t_f < \infty$ if and only if

$$\mathbf{C}(t_f) \int_{t_0}^{t_f} \Phi(t_f, \sigma)\mathbf{B}(\sigma)\mathbf{B}^T(\sigma)\Phi^T(t_f, \sigma)d\sigma \mathbf{C}^T(t_f) > \mathbf{0},$$

or equivalently

$$\begin{aligned} &\dot{\mathbf{X}}(t) = \mathbf{X}(t)\mathbf{A}^T(t) + \mathbf{A}(t)\mathbf{X}(t) + \mathbf{B}(t)\mathbf{B}^T(t) \\ &\mathbf{X}(t_0) = \mathbf{0} \\ &\mathbf{C}(t_f)\mathbf{X}(t_f)\mathbf{C}^T(t_f) > \mathbf{0}. \end{aligned} \tag{3.13}$$

Now suppose hereafter that **A**, **B**, **C** are constant matrices. Equation (3.11) can also be written as

$$\begin{aligned} \mathbf{X}(t_f) &= \int_{t_0}^{t_f} e^{\mathbf{A}(t_f-\sigma)}\mathbf{B}\mathbf{B}^T e^{\mathbf{A}^T(t_f-\sigma)}d\sigma \\ &= -\int_{t_f-t_0}^{0} e^{\mathbf{A}\tau}\mathbf{B}\mathbf{B}^T e^{\mathbf{A}^T\tau}d\tau \\ &= \int_{0}^{t_f-t_0} e^{\mathbf{A}\tau}\mathbf{B}\mathbf{B}^T e^{\mathbf{A}^T\tau}d\tau. \end{aligned}$$

Since the integrand $e^{\mathbf{A}\tau}\mathbf{B}\mathbf{B}^T e^{\mathbf{A}^T\tau} \geq 0$, it follows that $\mathbf{X}(t_2) \geq \mathbf{X}(t_1)$ if $t_2 \geq t_1$ (for any given t_0). Hence, the existence of a t_f such that $\mathbf{X}(t_f) > 0$ does not depend upon the choice of t_0. Now suppose $\mathbf{X}(t_f) > 0$ for some $t_f < \infty$. Then $\mathbf{X}(\tilde{t}) > 0$ for every $\tilde{t} \geq t_f$, including the limiting case $t_f = \infty$. Likewise, if $\mathbf{X}(\infty)$ is not positive definite then $\mathbf{X}(\tilde{t})$ is not positive definite for any $t_f < \infty$. This proves the following.

Corollary 3.3.1 *The linear time-invariant system $\dot{\mathbf{x}} = \mathbf{A}\mathbf{x} + \mathbf{B}\mathbf{u}$ is completely state controllable if and only if there exists some $t_f < \infty$ such that*

$$\mathbf{X}(t_f) = \int_{0}^{t_f} e^{\mathbf{A}\tau}\mathbf{B}\mathbf{B}^T e^{\mathbf{A}^T\tau}d\tau > \mathbf{0}$$

*or equivalently (3.12) holds for $t_0 = 0$ and **A**, **B** constant.*

Note that $\mathbf{X}(t_f)$ always exists for $t_f < \infty$ but might not exist for $t_f = \infty$. This is illustrated by example.

Example 3.3.1 Computing $\mathbf{X}(t_f)$ for $\mathbf{A} = \begin{bmatrix} 1 & 0 \\ 0 & -1 \end{bmatrix}$, $\mathbf{B} = \begin{bmatrix} 1 \\ 1 \end{bmatrix}$ yields

$$\begin{aligned}
\mathbf{X}(t_f) &= \int_0^{t_f} e^{\mathbf{A}\tau}\mathbf{B}\mathbf{B}^T e^{\mathbf{A}^T\tau} d\tau \\
&= \int_0^{t_f} \begin{bmatrix} e^{\tau} & 0 \\ 0 & e^{-\tau} \end{bmatrix} \begin{bmatrix} 1 & 1 \\ 1 & 1 \end{bmatrix} \begin{bmatrix} e^{\tau} & 0 \\ 0 & e^{-\tau} \end{bmatrix} d\tau \\
&= \begin{bmatrix} \frac{1}{2}(e^{2t_f} - 1) & t_f \\ t_f & \frac{1}{2}(1 - e^{-2t_f}) \end{bmatrix}
\end{aligned}$$

which is positive definite for every $0 < t_f < \infty$, but $\mathbf{X}(t_f)$ does not exist for $t_f = \infty$.

This example shows that t_f cannot be taken as infinity in Corollary 3.3.1, without some assumptions. Now define $\mathbf{X}(\infty)$ by (if it exists),

$$\mathbf{X}(\infty) = \int_0^{\infty} e^{\mathbf{A}\tau}\mathbf{B}\mathbf{B}^T e^{\mathbf{A}^T\tau} d\tau. \tag{3.14}$$

This matrix is called the controllability Gramian. Using (3.4c) write (3.14) as

$$\mathbf{X}(\infty) = \begin{bmatrix}\mathbf{B} & \mathbf{AB} & \cdots & \mathbf{A}^{n-1}\mathbf{B}\end{bmatrix} \Omega \begin{bmatrix}\mathbf{B} & \mathbf{AB} & \cdots & \mathbf{A}^{n-1}\mathbf{B}\end{bmatrix}^T \tag{3.15}$$

where

$$\boldsymbol{\Omega} \triangleq \int_0^{\infty} \begin{bmatrix} \alpha_0(t)\mathbf{I} \\ \alpha_1(t)\mathbf{I} \\ \vdots \\ \alpha_{n-1}(t)\mathbf{I} \end{bmatrix} \begin{bmatrix}\alpha_0(t)\mathbf{I} & \alpha_1(t)\mathbf{I} & \cdots & \alpha_{n-1}(t)\mathbf{I}\end{bmatrix} dt.$$

Suppose the integral in $\boldsymbol{\Omega}$ is finite (i.e. $\boldsymbol{\Omega}$ exists). The matrix $\boldsymbol{\Omega}$ is positive definite if $\alpha_i(t)$ $(i = 0, \ldots, n-1)$ are linearly independent functions on the interval $t \in [0, \infty]$. This is true for any $\mathbf{A}$ and may be proved from (3.5a), but we omit this proof. Hence, from (3.15), see that rank$[\mathbf{B}\ \ \mathbf{AB}\ \ \cdots\ \mathbf{A}^{n-1}\mathbf{B}] = n$ is necessary and sufficient for $\mathbf{X}(\infty) > \mathbf{0}$. We define $\mathbf{A}$ to be an asymptotically stable matrix if all eigenvalues of $\mathbf{A}$ are in the open left half plane. Notice from (3.5) that for an asymptotically stable matrix $\mathbf{A}$,

$$\lim_{t\to\infty} \alpha_i(t) = 0,$$

and matrix $\boldsymbol{\Omega}$ is finite if $\mathbf{A}$ is asymptotically stable. Hence, if $\mathbf{A} \in \mathcal{R}^{n\times n}$ is asymptotically stable then $\boldsymbol{\Omega}$ exists and is positive definite, and thus $\mathbf{X}(\infty) > \mathbf{0}$ is equivalent to rank$[\mathbf{B}\ \ \mathbf{AB}\ \ \ldots\ \ \mathbf{A}^{n-1}\mathbf{B}] = n$. This discussion gives a necessary and sufficient condition for $\mathbf{X}(\infty) > \mathbf{0}$, under the assumption that $\mathbf{A}$ is asymptotically stable. Finally, we must remove the stability assumption.

Corollary 3.3.2 $\mathbf{X}(\infty)$ *exists if and only if the controllable modes of* $\dot{\mathbf{x}} = \mathbf{Ax} + \mathbf{Bu}$ *are asymptotically stable. If* $\mathbf{X}(\infty)$ *exists, then* $\mathbf{X}(\infty) > 0$ *if and only if* $(\mathbf{A}, \mathbf{B})$ *is a controllable pair.*

Proof. We will only prove the case for nondefective $\mathbf{A}$. Describe the time-invariant system $\dot{\mathbf{x}} = \mathbf{A}\mathbf{x} + \mathbf{B}\mathbf{u}$ in its modal coordinates, i.e.

$$\dot{\mathbf{x}} = \begin{bmatrix} \lambda_1 & & & \\ & \lambda_2 & & \\ & & \ddots & \\ & & & \lambda_n \end{bmatrix} \mathbf{x} + \begin{bmatrix} \mathbf{b}_1^* \\ \mathbf{b}_2^* \\ \vdots \\ \mathbf{b}_n^* \end{bmatrix} \mathbf{u}.$$

The "modes" are characterized by the eigenvalue, eigenvector pairs and a "controllable mode" i that is "asymptotically stable" corresponds to $\{Real[\lambda_i] < 0,\ \mathbf{b}_i \neq \mathbf{0}\}$. Hence $\mathbf{X}_{ij}(\infty) = \int_0^\infty e^{\lambda_i \tau}\mathbf{b}_i^*\mathbf{b}_j e^{\lambda_j \tau} d\tau$ exists if and only if $\mathbf{b}_k = \mathbf{0}$ whenever $Real[\lambda_k] > 0$. □

Exercise 3.3.2 (i) Show that $\mathbf{X}(\infty)$ exists but is not positive definite for the pair

$$\mathbf{A} = \begin{bmatrix} 1 & 0 \\ 0 & -1 \end{bmatrix}, \quad \mathbf{B} = \begin{bmatrix} 0 \\ 1 \end{bmatrix}.$$

(ii) Show that $X(\infty)$ exists and is positive definite for the pair

$$\mathbf{A} = \begin{bmatrix} -2 & 0 \\ 0 & -1 \end{bmatrix}, \quad \mathbf{B} = \begin{bmatrix} 1 \\ 1 \end{bmatrix}.$$

(iii) Show that $\mathbf{X}(\infty)$ does not exist but the pair

$$\mathbf{A} = \begin{bmatrix} 1 & 0 \\ 0 & -1 \end{bmatrix}, \quad \mathbf{B} = \begin{bmatrix} 1 \\ 1 \end{bmatrix}$$

is controllable.

Note that if $\mathbf{X}(\infty)$ exists it satisfies $\mathbf{X}(\infty) = \mathbf{X}$, where

$$\mathbf{0} = \mathbf{X}\mathbf{A}^T + \mathbf{A}\mathbf{X} + \mathbf{B}\mathbf{B}^T. \tag{3.16}$$

However, all $\mathbf{X}$ that satisfy (3.16) might not be matrices of the form (3.14) (note that (3.14) is always a positive semidefinite matrix).

Exercise 3.3.3 For $\mathbf{A} = \begin{bmatrix} 1 & 0 \\ 0 & -1 \end{bmatrix}$, $\mathbf{B} = \begin{bmatrix} 0 \\ 1 \end{bmatrix}$, (3.16) yields

$$\mathbf{X} = \begin{bmatrix} 0 & X_{12} \\ X_{12} & 1/2 \end{bmatrix}$$

for arbitrary X_{12} (say $X_{12} = 1$), and $\mathbf{X}$ is not a positive semidefinite matrix for $X_{12} \neq 0$. Hence this $\mathbf{X} \neq \mathbf{X}(\infty)$ and $\mathbf{X}(\infty)$ does not exist.

Corollary 3.3.3 *The matrix X solving (3.16) is unique if and only if there are no two eigenvalues of* $\mathbf{A}$ *that are symmetrically located about the $j\omega$ axis.*

Proof. The left eigenvectors $\mathbf{l}_i$ of $\mathbf{A}$ satisfy $\mathbf{l}_i^*\mathbf{A} = \lambda_i \mathbf{l}_i^*$. Multiply (3.16) from the left by $\mathbf{l}_i^*$ and from the right by $\mathbf{l}_k$ to get

$$0 = \mathbf{l}_i^*\mathbf{X}\mathbf{A}^T\mathbf{l}_k + \mathbf{l}_i^*\mathbf{A}\mathbf{X}\mathbf{l}_k + \mathbf{l}_i^*\mathbf{B}\mathbf{B}^T\mathbf{l}_k.$$

But using $\mathbf{l}_i^*\mathbf{A} = \lambda_i \mathbf{l}_i^*$ (for the scalar $\lambda^* = \bar{\lambda}$)

$$0 = \mathbf{l}_i^*\mathbf{X}\mathbf{l}_k(\bar{\lambda}_k + \lambda_i) + \mathbf{l}_i^*\mathbf{B}\mathbf{B}^T\mathbf{l}_k$$

yields unique values for the elements of the transformed $\hat{\mathbf{X}}$: $\hat{\mathbf{X}}_{ik} = [\mathbf{E}^{-1}\mathbf{X}\mathbf{E}^{-*}]_{ik} = \mathbf{l}_i^*\mathbf{X}\mathbf{l}_k$, $\mathbf{E}^{-*} = [\mathbf{l}_1 \cdots \mathbf{l}_n]$,

$$\hat{\mathbf{X}}_{ik} \triangleq \mathbf{l}_i^*\mathbf{X}\mathbf{l}_k = -\frac{\mathbf{l}_i^*\mathbf{B}\mathbf{B}^T\mathbf{l}_k}{\lambda_i + \bar{\lambda}_k} \quad \forall i, k$$

if and only if $\lambda_i + \bar{\lambda}_k \neq 0$ for all i and k. Or, equivalently, since the eigenvalues occur in complex conjugate pairs, $\lambda_i + \lambda_k \neq 0$ for all i and k. □

Corollary 3.3.3 suggests one computational procedure for solving Lyapunov equations of the form (3.16), but the procedure is not efficient because the transformation to Jordan form is computationally difficult. Instead of transforming to a diagonal form, it is much easier to transform $\mathbf{A}$ to a triangular form

$$\mathbf{U}^*\mathbf{A}\mathbf{U} = \begin{bmatrix} \lambda_1 & \lambda_{12} & \cdots & \lambda_{1n} \\ 0 & \lambda_2 & \cdots & \lambda_{2n} \\ \vdots & \vdots & \ddots & \vdots \\ 0 & 0 & & \lambda_n \end{bmatrix} \tag{3.17}$$

since there exists a unitary $\mathbf{U}$ to do this. Assume that $\mathbf{A}$ in (3.16) is already in upper triangular form. Then the solution of (3.16) can be expressed in terms of its columns defined by

$$\mathbf{X}_i = \begin{bmatrix} X_{1i} \\ X_{2i} \\ \vdots \\ X_{ii} \end{bmatrix} \quad i = 1, 2, \ldots, n \tag{3.18}$$

by computing sequentially for $i = n, n-1, \ldots, 2, 1$,

$$\begin{bmatrix} X_{1i} \\ X_{2i} \\ \vdots \\ X_{ii} \end{bmatrix} = -(\lambda_i \mathbf{I}_i + \mathbf{A}_i)^{-1} \left\{ \begin{bmatrix} Q_{1i} \\ Q_{2i} \\ \vdots \\ Q_{ii} \end{bmatrix} + \begin{bmatrix} \lambda_{1,i+1} & \cdots & \lambda_{1n} \\ \lambda_{2,i+1} & \cdots & \lambda_{2n} \\ \vdots & & \\ \lambda_{i,i+1} & \cdots & \lambda_{in} \end{bmatrix} \begin{bmatrix} X_{i,i+1} \\ X_{i,i+2} \\ \vdots \\ X_{in} \end{bmatrix} + \sum_{k=i+1}^{n} \lambda_{ik} \begin{bmatrix} X_{1k} \\ X_{2k} \\ \vdots \\ X_{ik} \end{bmatrix} \right\}, \tag{3.19}$$

where $\mathbf{A}_i$ is the $i \times i$ upper left-hand corner of $\mathbf{A}$. The proof is easy to show by construction. See [153] for a proof. Note that $\lambda_i \mathbf{I} + \mathbf{A}_i$ is triangular and therefore easy to invert. Hence (3.19) yields the unique solution to (3.16) if $\lambda_i + \lambda_j \neq 0$ for any i, j.

Corollary 3.3.4 *If the controllable modes of* $\dot{\mathbf{x}} = \mathbf{A}\mathbf{x} + \mathbf{B}\mathbf{u}$ *are asymptotically stable, the following statements are equivalent:*

(i) The system is completely state controllable.
(ii) $\int_0^\infty e^{\mathbf{A}\tau}\mathbf{B}\mathbf{B}^T e^{\mathbf{A}^T\tau}d\tau > 0.$
(iii) $\mathbf{X} > \mathbf{0},\ \mathbf{0} = \mathbf{A}\mathbf{X} + \mathbf{X}\mathbf{A}^T + \mathbf{B}\mathbf{B}^T$
(iv) $\text{rank}[\mathbf{B}\ \ \mathbf{A}\mathbf{B}\ \cdots\ \mathbf{A}^{n-1}\mathbf{B}] = n$ *(dimension of* $\mathbf{x}$*)*

3.3.2 Observability

Now consider the system

$$\dot{\mathbf{x}}(t) = \mathbf{A}(t)\mathbf{x}(t), \qquad \mathbf{y}(t) = \mathbf{C}(t)\mathbf{x}(t). \tag{3.20}$$

Suppose we wish to determine $\mathbf{x}(t_0)$ given the data $\mathbf{y}(t)$, $t_0 \le t \le t_f$. Note that knowledge of $\mathbf{x}(t_0)$ is equivalent to knowledge of $\mathbf{x}(t)$ for any t, since $\mathbf{\Phi}(t, t_0)$ is invertible and $\mathbf{x}(t) = \mathbf{\Phi}(t, t_0)\mathbf{x}(t_0)$. From (3.20),

$$\mathbf{y}(t) = \mathbf{C}(t)\mathbf{\Phi}(t, t_0)\mathbf{x}(t_0). \tag{3.21}$$

Definition 3.3.2 *The system (3.20) is said to be completely observable at time* $t_f > t_0$ *if the data* $\mathbf{y}(t),\ t \in [t_0,\ t_f]$ *yields a unique solution* $\mathbf{x}(t_0)$ *to (3.21).*

Now consider $\mathbf{y}(t) = \mathbf{C}(t)\mathbf{x}(t)$ and some given data $\mathbf{y}(t)$ over an interval $t_0 \le t \le t_f$. In order for $\mathbf{x}(t_0)$ to have a unique solution in (3.21) the columns of $\mathbf{C}(t)\mathbf{\Phi}(t, t_0)$ must be linearly independent on the interval $[t_0,\ t_f]$. This means that $\mathbf{K}(t_0) \triangleq \int_{t_0}^{t_f} \mathbf{\Phi}^T(\sigma, t_0)\mathbf{C}^T(\sigma)\mathbf{C}(\sigma)\mathbf{\Phi}(\sigma, t_0)d\sigma > \mathbf{0}$ (where $\dot{\mathbf{\Phi}}(t, t_0) = \mathbf{A}(t)\mathbf{\Phi}(t, t_0)$, $\mathbf{\Phi}(t_0, t_0) = \mathbf{I}$), or equivalently

$$\left.\begin{array}{rcl} -\dot{\mathbf{K}}(t) & = & \mathbf{K}(t)\mathbf{A}(t) + \mathbf{A}^T(t)\mathbf{K}(t) + \mathbf{C}^T(t)\mathbf{C}(t) \\ \mathbf{K}(t_f) & = & \mathbf{0} \\ \mathbf{K}(t_0) & > & 0 \ \text{for some } t_0 < t_f. \end{array}\right\} \tag{3.22}$$

These results are summarized as follows.

Theorem 3.3.2 *The system* $\dot{\mathbf{x}}(t) = \mathbf{A}(t)\mathbf{x}(t)$, $\mathbf{y}(t) = \mathbf{C}(t)\mathbf{x}(t)$ *is completely observable at time* t_f *if and only if there exists* $0 < t_0 < t_f$ *such that* $\mathbf{K}(t_0) > 0$, *where*

$$\begin{array}{rcl} -\dot{\mathbf{K}}(t) & = & \mathbf{K}(t)\mathbf{A} + \mathbf{A}^T\mathbf{K}(t) + \mathbf{C}^T\mathbf{C} \\ \mathbf{K}(t_f) & = & \mathbf{0}. \end{array} \tag{3.23}$$

The time-invariant cases follow in a natural way from the above theorems by setting $\dot{\mathbf{K}}(t)$ to zero. The matrix $\mathbf{K}$ below is called the observability Gramian

$$\mathbf{K} = \int_0^\infty e^{A^T\tau}C^T C e^{A\tau}dt.$$

Corollary 3.3.5 *If the observable modes of* $\dot{\mathbf{x}} = \mathbf{A}\mathbf{x}$, $\mathbf{y} = \mathbf{C}\mathbf{x}$ *are asymptotically stable, the following statements are equivalent:*

(i) The system is completely observable.
(ii) $\int_0^\infty e^{\mathbf{A}^T\tau}\mathbf{C}^T\mathbf{C}e^{\mathbf{A}\tau}d\tau > \mathbf{0}.$
(iii) $\mathbf{K} > \mathbf{0},\ \mathbf{0} = \mathbf{K}\mathbf{A} + \mathbf{A}^T\mathbf{K} + \mathbf{C}^T\mathbf{C}.$
(iv) $\text{rank}[\mathbf{C}^T\ \ \mathbf{A}^T\mathbf{C}^T\ \cdots\ \mathbf{A}^{n-1T}\mathbf{C}^T] = n$ *(dimension of* $\mathbf{x}$*).*

Exercise 3.3.4

1 Show that $(\mathbf{A}, \mathbf{BB}^T)$ is a controllable pair if and only if $(\mathbf{A}, \mathbf{B})$ is a controllable pair.
2 Show that, for any $\mathbf{W} > \mathbf{0}$, $(\mathbf{A}, \mathbf{BWB}^T)$ is a controllable pair if and only if $(\mathbf{A}, \mathbf{B})$ is controllable.

3.4 CONTROLLABILITY AND OBSERVABILITY OF DISCRETE-TIME SYSTEMS

3.4.1 Controllability

Consider now the discrete-time system (3.6).

Definition 3.4.1 The system (3.6) is called "output controllable at time k_0" if there exists an integer k_f and a sequence $\{\mathbf{u}_{k_0}, \mathbf{u}_{k_0+1}, \mathbf{u}_{k_0+2}, \ldots, \mathbf{u}_{k_f}\}$ such that $\mathbf{y}_{k_f} = \mathbf{y}_f$ for an arbitrarily specified $\mathbf{y}_f$, for any given initial state $\mathbf{x}_{k_0}$.

When $(\mathbf{A}, \mathbf{B}, \mathbf{C}, \mathbf{D})$ are constant matrices, the "at time $k_0 = 0$" can be deleted in the definition and $k_0 = 0$ can be substituted without loss. When $\mathbf{C} = \mathbf{I}$, of course $\mathbf{y}$ is replaced by $\mathbf{x}$, and $\mathbf{C}$, $\mathbf{D}$ need not be stated, where output controllability reduces to state controllability in the definition.

Theorem 3.4.1 *These statements are equivalent:*

(i) The matrix time-varying triple $(\mathbf{A}_k, \mathbf{B}_k, \mathbf{C}_k, \mathbf{D}_k)$ is output controllable at time k_0.
(ii) There exists $k_f > k_0$ such that

$$\mathbf{X}_{k+1} = \mathbf{A}_k\mathbf{X}_k\mathbf{A}_k^T + \mathbf{B}_k\mathbf{B}_k^T \tag{3.24}$$

$$\mathbf{X}_{k_0} = \mathbf{0}$$

$$\mathbf{C}_{k_f}\mathbf{X}_{k_f}\mathbf{C}_{k_f}{}^T + \mathbf{D}_f\mathbf{D}_f^T > \mathbf{0}. \tag{3.25}$$

Theorem 3.4.2 *Let $\mathbf{A}, \mathbf{B}, \mathbf{C}$ be constant and suppose $\mathbf{X}$ exists satisfying*

$$\mathbf{X} = \mathbf{AXA}^T + \mathbf{BB}^T. \tag{3.26}$$

Then the following two statements are equivalent:

(i) The system (3.6) is output controllable.
(ii) $\mathbf{CXC}^T + \mathbf{DD}^T > \mathbf{0}$.

The following statements are also equivalent:

(i) The system (3.6) is state controllable.
(ii) $\mathbf{X} > \mathbf{0}$.

The solution to (3.26), if it exists, is

$$\mathbf{X} = \sum_{i=0}^{\infty} \mathbf{A}^i\mathbf{BB}^T(\mathbf{A}^T)^i \tag{3.27}$$

as proved by direct substitution into (3.26). From (3.27)

$$\mathbf{X} = \boldsymbol{\Omega}\boldsymbol{\Omega}^T, \quad \boldsymbol{\Omega} \triangleq [\mathbf{B} \;\; \mathbf{AB} \;\; \mathbf{A}^2\mathbf{B} \cdots]$$

making it clear that rank $\mathbf{X}$ = rank $\boldsymbol{\Omega}$.

The Cayley–Hamilton theorem [11, 125, 67] states that for any real square matrix $\mathbf{A}$ with characteristic equation

$$\lambda^m + a_{m-1}\lambda^{m-1} + \cdots a_1\lambda + a_0 = 0,$$

the following holds:

$$\mathbf{A}^m + a_{m-1}\mathbf{A}^{m-1} + \cdots a_1\mathbf{A} + a_0\mathbf{I} = 0.$$

Due to the Cayley–Hamilton theorem

$$\text{rank}[\mathbf{B}\ \ \mathbf{AB}\ \ \cdots\ \ \mathbf{A}^{n-1}\mathbf{B}] = \text{rank}[\mathbf{B}\ \ \mathbf{AB}\ \ \cdots\ \ \mathbf{A}^{i}\mathbf{B}]$$

for any $i > n - 1$, since $\mathbf{A}^n$ equals some linear combination of lower powers of $\mathbf{A}$. Hence no new linearly independent columns in $\boldsymbol{\Omega}$ are added beyond the $\mathbf{A}^{n-1}\mathbf{B}$ column block.

Theorem 3.4.3 *If* $\mathbf{X}$ *in (3.27) exists, then the following statements are equivalent:*

(i) The matrix pair $(\mathbf{A}, \mathbf{B})$ *is (state) controllable.*
(ii) rank $\mathbf{X}$ = rank$[\mathbf{B}\ \mathbf{AB}\ \cdots\ \mathbf{A}^{n-1}\ \mathbf{B}] = n$ *(= dimension of* $\mathbf{A}$*).*
(iii) $\mathbf{X} > \mathbf{0}$.

Exercise 3.4.1 Define $\mathbf{X}_k$ as the matrix

$$\mathbf{X}_k \triangleq \sum_{i=j+1}^{k} \mathbf{A}^{k-i}\mathbf{B}\mathbf{B}^T(\mathbf{A}^T)^{k-i}$$

and show that

$$\mathbf{X}_{k+1} = \mathbf{A}\mathbf{X}_k\mathbf{A}^T + \mathbf{B}\mathbf{B}.$$

Corollary 3.4.1 *There exists a unique solution* $\mathbf{X}$ *to (3.26) if and only if* $\lambda_i[\mathbf{A}] \neq (\lambda_j[\mathbf{A}])^{-1}$ *for all* i, j.

Proof. Pre- and post-multiply (3.26) by the matrix of left eigenvectors of $\mathbf{A}$ (where $\lambda\mathbf{l}^* = \mathbf{l}^*\mathbf{A}$, $\mathbf{E}^{-*} = [\mathbf{l}_1\mathbf{l}_2\cdots\mathbf{l}_n]$), as follows,

$$\begin{aligned}
\mathbf{E}^{-1}\mathbf{X}\mathbf{E}^{-*} &= \mathbf{E}^{-1}(\mathbf{A}\mathbf{X}\mathbf{A}^* + \mathbf{B}\mathbf{B}^*)\mathbf{E}^{-*} \\
&= \mathbf{E}^{-1}\mathbf{A}\mathbf{E}\mathbf{E}^{-1}\mathbf{X}\mathbf{E}^{-*}\mathbf{E}^*\mathbf{A}^*\mathbf{E}^{-*} + \mathbf{E}^{-1}\mathbf{B}\mathbf{B}^*\mathbf{E}^{-*} \\
\mathcal{X}_{ij} &= (\Lambda\mathcal{X}\bar{\Lambda})_{ij} + l_i^*\mathbf{B}\mathbf{B}^*l_i \\
&= \lambda_i\mathcal{X}_{ij}\bar{\lambda}_j \\
\mathcal{X}_{ij} &= (1 - \lambda_i\bar{\lambda}_j)^{-1}l_i^*\mathbf{B}\mathbf{B}^*l_j.
\end{aligned}$$

Hence the ij element of the transformed matrix $\mathbf{X}$ is unique if $\lambda_i\lambda_j \neq 1$. □

Now solve (3.26) by exploiting the symmetric structure of $\mathbf{X}$ and the ith column of dimension i, as in (3.18), assuming that $\mathbf{A}$ has been transformed to an upper triangular form (3.17). Then the solution of (3.26) is given recursively for $i = n, n-1, \ldots, 2, 1$ by

$$\mathbf{X}_i = (\mathbf{I}_i - \lambda_i\mathbf{A}_i)^{-1}\left\{\mathbf{Q}_i + \sum_{k=i+1}^{n}\mathbf{A}_{ik}\mathbf{X}_k\lambda_{ik} + \sum_{k=i}^{n-1}\tilde{\mathbf{A}}_{i,n-k}\tilde{\mathbf{X}}_k\lambda_{ik}\right\} \tag{3.28}$$

where $\mathbf{A}_{ik}$ is the $i \times k$ upper left corner of $\mathbf{A}$, $\tilde{\mathbf{A}}_{i,n-k}$ is the $i \times (n-k)$ upper right corner of $\mathbf{A}$ and the ith column of $\mathbf{X}$ is denoted by

$$\mathbf{X}_{icol} = \begin{bmatrix} \mathbf{X}_i \\ \tilde{\mathbf{X}}_i \end{bmatrix}, \quad \mathbf{X}_i = \begin{bmatrix} X_{ij} \\ \vdots \\ X_{ii} \end{bmatrix}, \quad \tilde{\mathbf{X}}_i = \begin{bmatrix} X_{i,i+1} \\ \vdots \\ X_{i,n} \end{bmatrix}.$$

See [153] for a proof of (3.28), or easily verify by construction.

3.4.2 Observability

Define the system

$$\begin{aligned} \mathbf{x}_k &= \mathbf{A}_k \mathbf{x}_k + \mathbf{B}_k \mathbf{u}_k \\ \mathbf{y}_k &= \mathbf{C}_k \mathbf{x}_k + \mathbf{D}_k \mathbf{u}_k. \end{aligned} \tag{3.29}$$

Definition 3.4.2 We say that (3.29) is observable at time t_p if there exists a time $q \leq p$ such that knowledge of $\{\mathbf{u}(k), \mathbf{y}(k)\ q \leq k \leq p\}$ allows a unique solution for $\mathbf{x}(q)$.

From (3.29) write

$$\begin{bmatrix} \mathbf{y}_q \\ \mathbf{y}_{q+1} \\ \vdots \\ \mathbf{y}_p \end{bmatrix} = \begin{bmatrix} \mathbf{C}_q \\ \mathbf{C}_{q+1}\mathbf{A}_q \\ \vdots \\ \mathbf{C}_p \mathbf{A}_{p-1} \cdots \mathbf{A}_q \end{bmatrix} \mathbf{x}_q + \begin{bmatrix} \mathbf{D}_q & & & \\ \mathbf{C}_{q+1}\mathbf{B}_q & \mathbf{D}_{q+1} & & \\ \vdots & & \ddots & \\ \mathbf{C}_p \mathbf{A}_{p-2} \cdots \mathbf{A}_{q-1}\mathbf{B}_q & \cdots & \cdots & \mathbf{D}_p \end{bmatrix} \begin{bmatrix} \mathbf{u}(q) \\ \vdots \\ \vdots \\ \mathbf{u}(p) \end{bmatrix}$$

or, simply,

$$\tilde{\mathbf{y}}(p, q) = \tilde{\mathbf{C}}(p, q)\mathbf{x}_q + \tilde{\mathbf{B}}(p, q)\tilde{\mathbf{u}}(p, q). \tag{3.30}$$

Hence, observability at time p is equivalent to the existence of a unique $\mathbf{x}_q$ satisfying (3.30), given the matrix $\tilde{\mathbf{C}}(p, q)$, and the vector $\hat{\mathbf{y}}(p, q) \triangleq \tilde{\mathbf{y}}(p, q) - \tilde{\mathbf{B}}(p, q)\tilde{\mathbf{u}}(p, q)$. This linear algebra problem has solution

$$\mathbf{x}_q = \tilde{\mathbf{C}}^+(p, q)\hat{\mathbf{Y}}(p, q) + (\mathbf{I} - \tilde{\mathbf{C}}^+(p, q)\tilde{\mathbf{C}}(p, q))\mathbf{z}$$

if the following existence condition holds

$$(\mathbf{I} - \tilde{\mathbf{C}}(p, q)\tilde{\mathbf{C}}^+(p, q))\hat{\mathbf{y}}(p, q) = \mathbf{0}. \tag{3.31}$$

The solution $\mathbf{x}_q$ is *unique* if the columns of $\tilde{\mathbf{C}}(p, q)$ are linearly independent, in which case

$$\mathbf{I} - \tilde{\mathbf{C}}^+(p, q)\tilde{\mathbf{C}}(p, q) = \mathbf{0}. \tag{3.32}$$

There exists at least one solution for $\mathbf{x}(q)$ if (3.31) holds, and there exists a solution for *arbitrary* input/output data $\hat{\mathbf{Y}}(p, q)$ if and only if the rows of $\tilde{\mathbf{C}}(p, q)$ are linearly independent so that

$$\mathbf{I} - \tilde{\mathbf{C}}(p, q)\tilde{\mathbf{C}}^+(p, q) = \mathbf{0}. \tag{3.33}$$

Since $\tilde{\mathbf{C}} \in \mathcal{R}^{n_y(p-q+1)\times n_x}$, uniqueness requires $n_y(p-q+1) \geq n_x$. Specifically, we require a left inverse of $\tilde{\mathbf{C}}(p,q)$, or equivalently

$$\mathbf{P}(p,q) \triangleq \tilde{\mathbf{C}}^T(p,q)\tilde{\mathbf{C}}(p,q) > 0$$

or

$$\mathbf{P}(p,q) = \sum_{i=0}^{p-q} \mathbf{A}^{Ti}\mathbf{C}^T\mathbf{C}\mathbf{A}^i. \tag{3.34}$$

Since $\mathbf{P}(p_2,q_2) \geq \mathbf{P}(p_1,q_1)$ whenever $p_2 - q_2 \geq p_1 - q_1$, we test observability in the time-invariant case by the condition

$$\mathbf{P} \triangleq \sum_{i=0}^{\infty} \mathbf{A}^{Ti}\mathbf{C}^T\mathbf{C}\mathbf{A}^i > 0. \tag{3.35}$$

Since observability is a function only of the matrix pair $(\mathbf{A},\mathbf{C})$ we may say, relative to (3.29), that the "matrix pair $(\mathbf{A},\mathbf{C})$ is observable" (or not).

Exercise 3.4.2 Show that $\mathbf{P}(p,q)$ in (3.34) satisfies

$$\mathbf{P} = \mathbf{A}^T\mathbf{P}\mathbf{A} + \mathbf{C}^T\mathbf{C}. \tag{3.36}$$

Exercise 3.4.3 For the system given in (3.29), suppose $\mathbf{C} = [0 \quad 1]$. Find out how much data $q\ u(k),\ y(k),\ q \leq k \leq p$ is required to uniquely compute the initial state $\mathbf{x}(0)$.

Theorem 3.4.4 *The following statements are equivalent:*

(i) The time-varying system (3.29) is observable at time p.
(ii) There exists q such that

$$\begin{aligned} \mathbf{P}_k &= \mathbf{A}_k^T\mathbf{P}_{k+1}\mathbf{A}_k + \mathbf{C}_k^T\mathbf{C}_k, \quad \mathbf{P}_p = \mathbf{0} \\ \mathbf{P}_q &> \mathbf{0}. \end{aligned} \tag{3.37}$$

If $(\mathbf{A},\ \mathbf{C})$ is a pair of constant matrices, and if $\mathbf{P}_q$ exists from (3.37) for $p = \infty$ the following statements are equivalent:
(iii) The time-invariant system (3.29) is observable.
(iv) $\mathbf{P} = \mathbf{A}^T\mathbf{P}\mathbf{A} + \mathbf{C}^T\mathbf{C}, \quad \mathbf{P} > \mathbf{0}$.

Note for the time-invariant case, and from the Cayley–Hamilton theorem that

$$\begin{aligned} \operatorname{rank}\mathbf{P} &= \operatorname{rank}\sum_{i=0}^{\infty}\mathbf{A}^{iT}\mathbf{C}^T\mathbf{C}\mathbf{A}^i = \operatorname{rank}[\mathbf{C}^T \quad \mathbf{A}^T\mathbf{C}\cdots][\mathbf{C}^T \quad \mathbf{A}^T\mathbf{C}^T\cdots]^T \\ &= \operatorname{rank}[\mathbf{C}^T \quad \mathbf{A}^T\mathbf{C}^T \cdots \mathbf{A}^{n-1T}\mathbf{C}^T]. \end{aligned}$$

Hence observability is equivalent to

$$\operatorname{rank}[\mathbf{C}^T \quad \mathbf{A}^T\mathbf{C}^T \cdots \mathbf{A}^{n-1T}\mathbf{C}^T] = n.$$

The solution of (3.36) is unique if $\lambda_i[\mathbf{A}] \neq (\lambda_j[\mathbf{A}])^{-1}$ for any i, j. The solution of (3.36) follows from the same algorithm (3.28) by substitutions $\mathbf{A} \to \mathbf{A}^T,\ \mathbf{C} \to \mathbf{B}^T$.

3.5 LYAPUNOV STABILITY OF LINEAR SYSTEMS

The early work of Lyapunov [84] remains to this day one of the most powerful methodologies for stability analysis. No other stability method can treat such a large class of problems: nonlinear systems, time-varying systems, linear systems. In fact, Massera [88] has pointed out that under mild assumptions, a Lyapunov function always exists for proving asymptotic stability of a solution, if the solution is indeed asymptotically stable. This result extended the Lyapunov method beyond the sufficient conditions of Lyapunov's work, to include discussions of necessity as well. In fact, for linear systems, this opens the door for our characterization of the class of all quadratic Lyapunov functions that can be used to prove stability of a given stable system. This result can then be easily extended beyond analysis to parametrize the set of all plant parameters and controller parameters that can stabilize the system. Hence, relatively new tests are given for a plant to be stabilizable by a controller of fixed order (e.g. state feedback, output feedback). All these results follow in subsequent chapters. This chapter states the relevant Lyapunov stability theory for linear systems.

3.5.1 Continuous-Time Systems

For the linear time-invariant system

$$\dot{\mathbf{x}} = \mathbf{A}\mathbf{x} \qquad \mathbf{y} = \mathbf{C}\mathbf{x} \tag{3.38}$$

the solution for any τ is

$$\mathbf{y}(t) = \mathbf{C}e^{\mathbf{A}(t-\tau)}\mathbf{x}(\tau). \tag{3.39}$$

Theorem 3.5.1 *The following result characterizes the stability of nonlinear systems The null solution of the system* $\dot{\mathbf{x}} = (\mathbf{x}, \mathbf{t})$ *is asymptotically stable in the sense of Lyapunov if there exists a scalar function* $\mathcal{V}(0, t) = 0$, $\mathcal{V}(\infty, t) = \infty$, $\mathcal{V}(\mathbf{x}, t) > 0$ *for all* $\mathbf{x}$ *such that either (a)* $\dot{\mathcal{V}}(\mathbf{x}, t) \leq 0$ *and* $\frac{d^i}{dt^i}\mathcal{V}(\mathbf{x}, t) = 0$ *for all* $i > 0$ *implies* $\mathbf{x} = \mathbf{0}$, *or (b)* $\dot{\mathcal{V}}(\mathbf{x}, t) < 0$.

Generally, stability is a property of a *solution*. For linear systems stability of the null solution $\mathbf{x}(t) = \mathbf{0}$ is equivalent to the stability of any other solution since from (3.39) stability will depend only on $\mathbf{A}$ and not $\mathbf{x}(\tau)$. By a slight abuse of language in the linear systems of this book we refer simply to the "stability of the system" (3.38).

We will now show how to construct a Lyapunov function for any linear system, where $\mathbf{y}$ in (3.38) has no physical significance in this discussion. We choose any $\mathbf{C}$ such that (3.38) is observable. This means that

$$\int_0^\infty e^{\mathbf{A}^T\sigma}\mathbf{C}^T\mathbf{C}e^{\mathbf{A}\sigma}d\sigma > \mathbf{0}. \tag{3.40}$$

Define

$$\mathcal{V}(\mathbf{x}(t)) \triangleq \int_t^\infty \mathbf{y}^T(\sigma)\mathbf{y}(\sigma)d\sigma.$$

Then from (3.38) and (3.40)

$$\dot{\mathcal{V}}(\mathbf{x}(t)) = -\mathbf{y}^T(t)\mathbf{y}(t) = -\mathbf{x}^T(t)\mathbf{C}^T\mathbf{C}\mathbf{x}(t). \tag{3.41}$$

From (3.40) and (3.41), using $\mathbf{x}(\sigma) = e^{\mathbf{A}(\sigma - t)}\mathbf{x}(t)$, the Lyapunov function is

$$\begin{aligned}\mathcal{V}(\mathbf{x}(t)) &= \int_0^\infty \mathbf{x}^T(\sigma)\mathbf{C}^T\mathbf{C}\mathbf{x}(\sigma)d\sigma \\ &= \mathbf{x}^T(t)\int_t^\infty e^{\mathbf{A}^T(\sigma-t)}\mathbf{C}^T\mathbf{C}e^{\mathbf{A}(\sigma-t)}d\,\sigma\,\mathbf{x}(t) \\ &= \mathbf{x}^T(t)\mathbf{P}\mathbf{x}(t), \quad \mathbf{P} \triangleq \int_0^\infty e^{\mathbf{A}^T\tau}\mathbf{C}^T\mathbf{C}e^{\mathbf{A}\tau}d\tau\end{aligned}$$

where $\mathbf{P} > \mathbf{0}$ by assumption of observability of $(\mathbf{A}, \mathbf{C})$.

Exercise 3.5.1 Show that if $\mathbf{P}$ exists, it satisfies the equation

$$\mathbf{0} = \mathbf{PA} + \mathbf{A}^T\mathbf{P} + \mathbf{C}^T\mathbf{C}. \tag{3.42}$$

Exercise 3.5.2 Show that for some $\mathbf{\Omega}$

$$\mathbf{P} \triangleq \int_0^\infty e^{\mathbf{A}^T\tau}\mathbf{C}^T\mathbf{C}e^{\mathbf{A}\tau}d\tau = [\mathbf{C}^T \;\; \mathbf{A}^T\mathbf{C}^T \;\; \cdots]\mathbf{\Omega}[\mathbf{C}^T \;\; \mathbf{A}^T\mathbf{C}^T \;\; \cdots]^T$$

and that rank $\mathbf{P}$ = rank $[\mathbf{C}^T \quad \mathbf{A}^T\mathbf{C}^T \cdots \mathbf{A}^{n-1T}\mathbf{C}^T]$. Note that since the integrand $e^{\mathbf{A}^T\sigma}\mathbf{C}^T\mathbf{C}e^{\mathbf{A}\sigma}$ is a non-negative definite matrix, then $\int_t^{t_2} e^{\mathbf{A}^T\sigma}\mathbf{C}^T\mathbf{C}e^{\mathbf{A}\sigma}d\sigma \geq \int_t^{t_1} e^{\mathbf{A}^T\sigma}\mathbf{C}^T\mathbf{C}e^{\mathbf{A}\sigma}d\sigma$ if $t_2 \geq t_1$, and observability at any $t_f < \infty$ implies observability at $t_f = \infty$. Note also that observability (at any time) is guaranteed if one chooses a nonsingular $\mathbf{C}$. Hence $\mathcal{V}(\mathbf{x}(t)) > 0 \; \forall \; \mathbf{x} \neq \mathbf{0}$, where $\mathbf{P} > \mathbf{0}$ is parametrized by $\mathbf{C}$ such that $(\mathbf{A}, \; \mathbf{C})$ is observable. Now, $\dot{\mathcal{V}} \leq 0$ obviously from (3.41), and $\dot{\mathcal{V}} \equiv 0$ is equivalent to $\mathbf{y}(t) \equiv \mathbf{0}$. Hence, the question "does $\dot{\mathcal{V}} \equiv 0$ imply $\mathbf{x} = \mathbf{0}$?" reduces to "does $\mathbf{y}(t) \equiv \mathbf{0}$ imply $\mathbf{x} = \mathbf{0}$?" But this is true if and only if all state variables are observable in $\mathbf{y}(t)$. This is guaranteed by the $(\mathbf{A}, \; \mathbf{C})$ observable assumption. Hence, we have the following conclusion.

Theorem 3.5.2 *The following are equivalent statements:*

(i) The system $\dot{\mathbf{x}} = \mathbf{A}\mathbf{x}$ is asymptotically stable in the sense of Lyapunov.
(ii) The eigenvalues of $\mathbf{A}$ lie in the open left half plane.
(iii) If $(\mathbf{A}, \; \mathbf{C})$ is an observable pair, there exists $\mathbf{P} > \mathbf{0}$ satisfying $\mathbf{0} = \mathbf{PA} + \mathbf{A}^T\mathbf{P} + \mathbf{C}^T\mathbf{C}$.

Proof that (ii) and (iii) are equivalent follows by multiplying the Lyapunov equation by $\mathbf{e}_i^*$ from the left and by $\mathbf{e}_i$ from the right, where $\mathbf{A}\mathbf{e}_i = \lambda_i\mathbf{e}_i$, to get

$$\begin{aligned}0 &= \mathbf{e}_i^*\mathbf{PA}\mathbf{e}_i + \mathbf{e}_i^*\mathbf{A}^*\mathbf{P}\mathbf{e}_i + \mathbf{e}_i^*\mathbf{C}^*\mathbf{C}\mathbf{e}_i \\ &= \mathbf{e}_i^*\mathbf{P}\mathbf{e}_i(\lambda_i + \overline{\lambda}_i) + \mathbf{e}_i^*\mathbf{C}^*\mathbf{C}\mathbf{e}_i.\end{aligned}$$

Since $\mathbf{e}_i^*\mathbf{C}^*\mathbf{C}\mathbf{e}_i > \mathbf{0}$ (note that $\mathbf{C}\mathbf{e}_i \neq \mathbf{0}$ because of observability of every mode), $\mathbf{e}_i^*\mathbf{P}\mathbf{e}_i > \mathbf{0}$, then $\lambda_i + \overline{\lambda}_i = 2Real[\lambda_i] < 0$.

Now consider another Lyapunov function, for $t > 0$,

$$\mathcal{V}(\mathbf{x}(t)) = \mathbf{x}^*(t)\mathbf{X}^{-1}\mathbf{x}(t) \tag{3.43}$$

where $\mathbf{X}$ is defined by

$$\mathbf{X} \triangleq \int_0^\infty e^{\mathbf{A}\sigma}\mathbf{B}\mathbf{B}^T e^{\mathbf{A}^T\sigma}d\sigma$$

which satisfies (if $\mathbf{X}$ exists)

$$\mathbf{0} = \mathbf{X}\mathbf{A}^T + \mathbf{A}\mathbf{X} + \mathbf{B}\mathbf{B}^T. \tag{3.44}$$

Then

$$\begin{aligned} \dot{\mathcal{V}}(\mathbf{x}(t)) &= \dot{\mathbf{x}}^T(t)\mathbf{X}^{-1}\mathbf{x}(t) + \mathbf{x}^T(t)\mathbf{X}^{-1}(t)\dot{\mathbf{x}}(t) \\ &= \mathbf{x}^T(t)\mathbf{A}^T\mathbf{X}^{-1}\mathbf{x}(t) + \mathbf{x}^T(t)\mathbf{X}^{-1}\mathbf{A}\mathbf{x}(t) \\ &= \mathbf{x}^T(t)\mathbf{X}^{-1}[\mathbf{X}\mathbf{A}^T + \mathbf{A}\mathbf{X}]\mathbf{X}^{-1}\mathbf{x}(t). \end{aligned} \tag{3.45}$$

Hence, $\mathcal{V}(\mathbf{x}(t)) > 0$, if $\mathbf{X} > \mathbf{0}$ satisfies (3.44), and $\dot{\mathcal{V}}(\mathbf{x}(t)) \leq 0$, $\dot{\mathcal{V}}(\mathbf{x}(t)) \equiv 0$ implies $\mathbf{x} = \mathbf{0}$ if $(\mathbf{A}, \mathbf{B})$ is a controllable pair.

Theorem 3.5.3 *The following statements are equivalent:*

(i) The system $\dot{\mathbf{x}} = \mathbf{A}\mathbf{x}$ *is asymptotically stable in the sense of Lyapunov.*
(ii) The eigenvalues of $\mathbf{A}$ *lie in the open left half plane.*
(iii) If $(\mathbf{A},\ \mathbf{B})$ *is a controllable pair, then there exists* $\mathbf{X} > \mathbf{0}$*, satisfying*

$$\mathbf{0} = \mathbf{X}\mathbf{A}^T + \mathbf{A}\mathbf{X} + \mathbf{B}\mathbf{B}^T. \tag{3.46}$$

(iv) If $(\mathbf{A},\ \mathbf{B})$ *is a stabilizable pair, then there exists* $\mathbf{X} \geq \mathbf{0}$*, satisfying (3.46).*

Obviously $(\mathbf{A}, \mathbf{B})$ or equivalently $(\mathbf{A}, \mathbf{B}\mathbf{B}^T)$ is controllable for any nonsingular $\mathbf{B}$, and $(\mathbf{A}, \mathbf{C})$ is observable for any nonsingular $\mathbf{C}$. Hence Theorems 3.5.2 and 3.5.3 readily lead to the following.

Corollary 3.5.1 *The following statements are equivalent:*

(i) $\dot{\mathbf{x}} = \mathbf{A}\mathbf{x}$ *is asymptotically stable.*
(ii) There exists $\mathbf{P} > \mathbf{0}$ *satisfying*

$$\mathbf{P}\mathbf{A} + \mathbf{A}^T\mathbf{P} < \mathbf{0}.$$

(iii) There exists $\mathbf{X} > \mathbf{0}$ *satisfying*

$$\mathbf{X}\mathbf{A}^T + \mathbf{A}\mathbf{X} < \mathbf{0}. \tag{3.47}$$

One final statement is important for the case when only stability, rather than asymptotic stability, is of interest.

Corollary 3.5.2 *The following statements are equivalent:*

(i) $\dot{\mathbf{x}} = \mathbf{A}\mathbf{x}$ *is at least stable.*
(ii) There exists $\mathbf{P} > \mathbf{0}$ *satisfying*

$$\mathbf{P}\mathbf{A} + \mathbf{A}^T\mathbf{P} \leq \mathbf{0}.$$

(iii) There exists $\mathbf{X} > \mathbf{0}$ *satisfying*

$$\mathbf{X}\mathbf{A}^T + \mathbf{A}\mathbf{X} \leq \mathbf{0}.$$

Example 3.5.1 Show that the eigenvalues of $\mathbf{A}$ lie in the open left half plane if and only if $\mathbf{X} > 0$ satisfies

$$\mathbf{X}\mathbf{A}^T + \mathbf{A}\mathbf{X} < \mathbf{0}.$$

Solution. Define some left eigenvector of $\mathbf{A}$ by $\mathbf{l}$. Then $\mathbf{l}^*\mathbf{A} = \lambda\mathbf{l}^*$ where λ is an eigenvalue of $\mathbf{A}$. For negative definiteness:

$$\mathbf{l}^*(\mathbf{X}\mathbf{A}^T + \mathbf{A}\mathbf{X})\mathbf{l} < \mathbf{0} \qquad \forall \mathbf{l} \neq \mathbf{0}$$

and using $\mathbf{l}^*\mathbf{A} = \lambda\mathbf{l}^*$

$$\mathbf{l}^*\mathbf{X}\mathbf{l}\bar{\lambda} + \lambda\mathbf{l}^*\mathbf{X}\mathbf{l} < 0$$

$$(\lambda + \bar{\lambda})\,\mathbf{l}^*\mathbf{X}\mathbf{l} < 0$$

$$(Re\,\lambda)\mathbf{l}^*\mathbf{X}\mathbf{l} < 0.$$

But since $\mathbf{X} > \mathbf{0}$, this is equivalent to

$$Re\,\lambda < 0.$$

Note that $2Re\,\lambda \leq 0$ results when the inequality $\mathbf{X}\mathbf{A}^T + \mathbf{A}\mathbf{X} \leq \mathbf{0}$ is used.

3.5.2 Discrete-Time Systems

Stability theorems for the discrete case follow in a similar manner as for the continuous-time case. Hence, results are merely summarized without proof below. Consider a linear system described by

$$\mathbf{x}_{k+1} = \mathbf{A}\mathbf{x}_k, \qquad \mathbf{y}_k = \mathbf{C}\mathbf{x}_k$$

where $\mathbf{C}$ is any such that $(\mathbf{A},\ \mathbf{C})$ is observable. This means

$$\mathbf{P} \triangleq \sum_{i=k}^{\infty}(\mathbf{A}^T)^{i-k}\mathbf{C}^T\mathbf{C}\mathbf{A}^{i-k} > \mathbf{0}.$$

Define

$$\begin{aligned}
\mathcal{V}(\mathbf{x}_k) &\triangleq \sum_{i=k}^{\infty}\mathbf{y}_i^T\mathbf{y}_i \\
&= \sum_{i=k}^{\infty}\mathbf{x}_k^T(\mathbf{A}^T)^{i-k}\mathbf{C}^T\mathbf{C}\mathbf{A}^{i-k}\mathbf{x}_k \\
&= \mathbf{x}_k^T\left[\sum_{i=k}^{\infty}(\mathbf{A}^T)^{i-k}\mathbf{C}^T\mathbf{C}\mathbf{A}^{i-k}\right]\mathbf{x}_k \\
&= \mathbf{x}_k^T\mathbf{P}\mathbf{x}_k.
\end{aligned}$$

Then

$$\begin{aligned}
\mathcal{V}(\mathbf{x}_{k+1}) - \mathcal{V}(\mathbf{x}_k) &= -\mathbf{y}_k^T\mathbf{y}_k \\
&= -\mathbf{x}_k^T\mathbf{C}^T\mathbf{C}\mathbf{x}_k.
\end{aligned}$$

Note also that

$$\mathcal{V}(\mathbf{x}_{k+1}) - \mathcal{V}(\mathbf{x}_k) = \mathbf{x}_{k+1}^T\mathbf{P}\mathbf{x}_{k+1} - \mathbf{x}_k^T\mathbf{P}\mathbf{x}_k.$$

It is straightforward to show by substitution of $\mathbf{x}_{k+1} = \mathbf{A}\mathbf{x}_k$ that $\mathbf{P}$ satisfies the linear matrix equation

$$\mathbf{P} = \mathbf{A}^T\mathbf{P}\mathbf{A} + \mathbf{C}^T\mathbf{C}.$$

Consider now another Lyapunov function

$$\mathcal{V}(\mathbf{x}_k) = \mathbf{x}_k^T\mathbf{X}^{-1}\mathbf{x}_k \qquad k > 0$$

where $\mathbf{X}$ is defined by

$$\mathbf{X} \triangleq \sum_{i=0}^{\infty} \mathbf{A}^i\mathbf{B}\mathbf{B}^T(\mathbf{A}^T)^i$$

which satisfies

$$\mathbf{X} = \mathbf{A}\mathbf{X}\mathbf{A}^T + \mathbf{B}\mathbf{B}^T.$$

Theorem 3.5.4 *The following statements are equivalent:*

(i) All eigenvalues of $\mathbf{A}$ *lie in the open unit circle.*
(ii) $\mathbf{P} - \mathbf{A}^T\mathbf{P}\mathbf{A} > \mathbf{0} \quad \mathbf{P} > \mathbf{0}.$
(iii) $\mathbf{X} - \mathbf{A}\mathbf{X}\mathbf{A}^T > \mathbf{0}, \quad \mathbf{X} > \mathbf{0}.$

Exercise 3.5.3 Show that the eigenvalues of $\mathbf{A}$ lie inside the open unit disk if and only if

$$-\mathbf{X} + \mathbf{A}\mathbf{X}\mathbf{A}^T < \mathbf{0}, \qquad \mathbf{X} > \mathbf{0}.$$

CHAPTER 3 SUMMARY

This chapter contains a solution to the state space equations, and a discussion of the classical tests for observability and controllability of continuous- and discrete-time systems. But since controllability and observability have nothing to do with stability, we need more "abilities" of linear systems to capture the essential performance. In later chapters, we will be interested in characterizing the set of all observability and controllability Gramians (3.14)–(3.23) which can be assigned by feedback control. More detailed expositions on the state space solution and the observability and controllability of continuous- and discrete-time systems can be found in traditional linear systems textbooks such as [67], [11], [125]. Lyapunov stability theory is much more general than the linear system analysis presented here. For asymptotically stable solutions of linear or nonlinear systems, a Lyapunov function always exists which will prove asymptotic stability. For nonlinear systems, such Lyapunov functions are hard to find and there is no general procedure to construct a Lyapunov function for nonlinear systems. However, such searches for Lyapunov functions are not necessary for linear systems, since quadratic functions always work. This chapter parametrizes all quadratic Lyapunov functions that a linear system can have. There are two subdivisions of this set: Those Lyapunov matrices satisfying equality constraints (3.46) for a controllable pair $(\mathbf{A}, \mathbf{B})$; and those Lyapunov matrices satisfying inequality constraints (3.47). When the plant matrix $\mathbf{A}$ contains adjustable parameters such as control parameters, then further chapters can take the next step of parametrizing the set of all control parameters and Lyapunov matrices that a stable system can have. A detailed exposition of Lyapunov stability theory for linear systems can be found in [151]. For discrete-time systems see [78]. Lyapunov techniques are used extensively for control design of nonlinear systems in the very important work of Corless, Leitmann, and others [2, 17, 18, 110, 16, 15].

CHAPTER FOUR

Second-Order Information in Linear Systems

4.1 THE DETERMINISTIC COVARIANCE MATRIX FOR CONTINUOUS-TIME SYSTEMS

Consider the time-varying system

$$\begin{aligned} \dot{\mathbf{x}}(t) &= \mathbf{A}(t)\mathbf{x}(t) + \mathbf{D}(t)\mathbf{w}(t) \qquad \mathbf{w}(t) \in \mathcal{R}^{n_w}, \quad \mathbf{x}(t) \in \mathcal{R}^{n_x} \\ \mathbf{y}(t) &= \mathbf{C}(t)\mathbf{x}(t). \end{aligned} \tag{4.1}$$

Define $\mathbf{x}(i, t, \tau)$, $\mathbf{y}(i, t, \tau)$ as the state (output) response at time t due to the ith excitation applied at time $\tau \leq t$. The admissible excitation events are $r = n_x + n_w$ in number (n_w impulses and n_x initial conditions):

$$\left.\begin{aligned} w_\alpha(t) &= w_\alpha \delta(t - \tau), \quad & \alpha &= 1, \ldots, n_w \\ x_\beta(\tau) &= x_{\beta 0}, & \beta &= 1, \ldots, n_x \end{aligned}\right\} \tag{4.2}$$

where $i = \alpha$ for $i \leq n_w$ and $i = n_w + \beta$ for $i > n_w$, and where w_α is the strength of the impulse applied in the αth input channel.

With zero initial conditions $\mathbf{x}(0) = \mathbf{0}$, and unit intensities $w_\alpha = 1, \alpha = 1, \ldots, n_w$, the vectors $\mathbf{y}(i, t, \tau)$ associated with system (4.1) are simply the columns of the impulse response matrix

$$\mathbf{C}(t)\mathbf{\Phi}(t, \tau)\mathbf{D}(\tau) = [\mathbf{y}(1, t, \tau), \mathbf{y}(2, t, \tau), \ldots, \mathbf{y}(n_w, t, \tau)] \tag{4.3}$$

where $\mathbf{\Phi}(t, \tau)$ is the state transition matrix for $\mathbf{A}(t)$. For the time-invariant case we can take $\tau = 0$ and write

$$\mathbf{C}e^{\mathbf{A}t}\mathbf{D} = [\mathbf{y}(1, t), \mathbf{y}(2, t), \ldots, \mathbf{y}(n_w, t)]. \tag{4.4}$$

Define

$$\mathbf{X}(t) \triangleq \sum_{i=1}^{r} \int_0^t \mathbf{x}(i, t, \tau)\mathbf{x}^T(i, t, \tau)d\tau, \qquad r = n_x + n_w. \tag{4.5}$$

Consider that, from (4.1) and (4.2),

$$\mathbf{x}(i,\ t,\ \tau) = \begin{cases} e^{\mathbf{A}(t-\tau)}\mathbf{x}_{\beta 0}, & i = n_w + \beta \\ e^{\mathbf{A}(t-\tau)}\mathbf{D}\mathbf{w}_{i0}, & i \le n_w \end{cases}$$

where

$$\mathbf{x}_{\beta 0} = \begin{bmatrix} 0 \\ \vdots \\ x_{\beta 0} \\ \vdots \\ 0 \end{bmatrix}, \quad or\ \mathbf{w}_{i0} = \begin{bmatrix} 0 \\ \vdots \\ w_{i0} \\ \vdots \\ 0 \end{bmatrix}$$

and

$$\dot{\mathbf{x}}(i,\ t,\ \tau) = \mathbf{A}e^{\mathbf{A}(t-\tau)}\mathbf{d},\ \mathbf{d} \triangleq \mathbf{x}_{\beta 0}\ or\ \mathbf{D}\mathbf{w}_{i0}.$$

Then from (4.5) it is easy to show that

$$\dot{\mathbf{X}}(t) = \sum_{i=1}^{r} \mathbf{x}(i,\ t,\ t)\mathbf{x}^T(i,\ t,\ t) + \mathbf{A}\int_0^t \sum_{i=1}^{r} \mathbf{x}(i,\ t,\ \tau)\mathbf{x}^T(i,\ t,\ \tau)d\tau$$

$$+ \int_0^t \sum_{i=1}^{r} \mathbf{x}(i,\ t,\ \tau)\mathbf{x}^T(i,\ t,\ \tau)d\tau\mathbf{A}^T.$$

and

$$\sum_{i=1}^{r} \mathbf{x}(i,\ t,\ t)\mathbf{x}^T(i,\ t,\ t) = \mathbf{D}\mathbf{W}_0\mathbf{D}^T + \mathbf{X}_0$$

where

$$\mathbf{W}_0 \triangleq \text{diag}\,[\ldots\ w_i^2\ \ldots] > \mathbf{0}, \quad \mathbf{X}_0 \triangleq \text{diag}\,[\ldots\ \mathbf{x}_i^2(0)\ \ldots] > \mathbf{0}. \tag{4.6}$$

Hence $\mathbf{X}(t)$ satisfies the differential equation

$$\dot{\mathbf{X}}(t) = \mathbf{X}(t)\mathbf{A}^T + \mathbf{A}\mathbf{X}(t) + \mathbf{D}\mathbf{W}_0\mathbf{D}^T + \mathbf{X}_0, \quad \mathbf{X}(0) = \mathbf{0}. \tag{4.7}$$

Definition 4.1.1 For system $\dot{\mathbf{x}} = \mathbf{A}\mathbf{x} + \mathbf{D}\mathbf{w}$, the deterministic covariance, or simply "D-covariance" of the state is defined by (4.5) and satisfies (4.7).

In the stability theory of Theorem 3.5.3 we are free to choose $\mathbf{B}$. Now comparing (3.46) with (4.7) leads to the following conclusion by choosing $\mathbf{B}\mathbf{B}^T$ in (3.46) as $\mathbf{B}\mathbf{B}^T \triangleq \mathbf{D}\mathbf{W}_0\mathbf{D}^T + \mathbf{X}_0$. Note that for *any* $\mathbf{A}$ the pair $(\mathbf{A},\ \mathbf{B})$ is controllable with this choice, since $\mathbf{X}_0 > \mathbf{0}$ guarantees that $\mathbf{B}\mathbf{B}^T > \mathbf{0}$.

In the time-invariant case, it is not necessary to let the time of excitations be variable as in (4.5), so τ can be taken as 0, and integrating over t instead yields the steady-state answer

$$\mathbf{X} \triangleq \sum_{i=1}^{r} \int_0^\infty \mathbf{x}(i,\ t)\mathbf{x}^T(i,\ t)dt = \int_0^\infty e^{\mathbf{A}t}(\mathbf{D}\mathbf{W}_0\mathbf{D}^T + \mathbf{X}_0)e^{\mathbf{A}^T t}dt \tag{4.8}$$

where

$$\mathbf{x}(i,\ t) = e^{\mathbf{A}t}\mathbf{x}_{i0} \quad or \quad e^{\mathbf{A}t}\mathbf{D}\mathbf{w}_{i0}$$

$$\mathbf{0} = \mathbf{X}\mathbf{A}^T + \mathbf{A}\mathbf{X} + \mathbf{D}\mathbf{W}_0\mathbf{D}^T + \mathbf{X}_0, \quad \mathbf{X}_0 > \mathbf{0}. \tag{4.9}$$

Theorem 4.1.1 *Let* $(\mathbf{A}, \mathbf{D}, \mathbf{W}_0, \mathbf{X}_0)$ *be constant, with* $\mathbf{X}_0 > \mathbf{0}$. *These statements are equivalent:*

(i) The eigenvalues of $\mathbf{A}$ *all lie in the open left half plane.*
(ii) The D-covariance satisfying (4.9) is positive definite $\mathbf{X} > \mathbf{0}$.

The proof follows immediately from Theorem 3.5.3.

Define

$$\mathbf{Y}(t) \triangleq \sum_{i=1}^{r} \int_0^t \mathbf{y}(i,\ t,\ \tau)\mathbf{y}^T(i,\ t,\ \tau)d\tau = \mathbf{C}(t).\mathbf{X}(t)\mathbf{C}^T(t). \tag{4.10}$$

The outputs $\mathbf{y}(i,\ t,\ \tau)$ denote the response at time t, applying only the ith excitation at time τ and where $\mathbf{X}(t)$ satisfies (4.7), or, in the time-invariant case

$$\mathbf{Y} = \sum_{i=1}^{r} \int_0^\infty \mathbf{y}(i,\ t)\mathbf{y}^T(i,\ t)dt = \mathbf{CXC}^T \tag{4.11}$$

where $\mathbf{X}$ satisfies (4.9). The matrices (4.10) and (4.11) have yet another physical significance.

Theorem 4.1.2 *Let an unknown strictly proper linear system (time-invariant) have zero initial conditions and let impulses (of any finite intensity* w_i, $w_i(t) = w_i\delta(t)$*) be applied at the input, one at a time. Compute the following integral* $\mathbf{Y}$ *from the response* $\mathbf{y}(i, t)$ *of these* n_w *experiments*

$$\mathbf{Y} = \sum_{i=1}^{n_w} \int_0^\infty \mathbf{y}(i,\ t)\mathbf{y}^T(i,\ t)dt. \tag{4.12}$$

There exists some input $\mathbf{w}(t)$ *(not impulses) to take the output* $\mathbf{y}(t)$ *to an arbitrarily specified value* $\mathbf{y}(t_f) = \bar{\mathbf{y}}$ *in a finite time* $t_f < \infty$ *if and only if the calculation* (4.12) *yields* $\mathbf{Y} > \mathbf{0}$.

The proof of this theorem follows from the output controllability results of (3.13) requiring

$$\mathbf{CXC}^T > \mathbf{0}$$
$$\mathbf{0} = \mathbf{XA}^T + \mathbf{AX} + \mathbf{DW}_0\mathbf{D}^T.$$

The important conclusion from Theorem 4.1.2 is that the necessary and sufficient condition for output controllability can be stated in terms of a *physically meaningful* matrix (4.12) (which is obtained from the physical impulse responses rather than knowledge of the internal model). Consider, for example, the time-invariant case (4.8), where the square root of the diagonal elements of $\mathbf{X}$ represent the (loosely called RMS) value of the state variables, as defined by

$$x_k(RMS) \triangleq [X_{kk}]^{1/2} = \left[\sum_{i=1}^{r} \int_0^\infty x_k^2(i,\ t)dt\right]^{1/2}, \quad k = 1,\ \dots,\ n_x.$$

Likewise

$$y_k(RMS) \triangleq [Y_{kk}]^{1/2} = \left[\mathbf{CXC}^T\right]_{kk}^{1/2}, \quad k = 1,\ \dots,\ n_y. \tag{4.13}$$

Hence, a parametrization of all stable linear systems given in terms of all attainable RMS values of the state variables x_k, $k = 1,\ \dots,\ n_x$ provides an explicit connection between stability and RMS performance. Later in the text we exploit these relationships to show, for a single-input system, an explicit one-to-one correspondence between the n_x coefficients of the characteristic polynomial and the n_x RMS values of the state variables.

4.2 MODELS FOR CONTROL DESIGN (CONTINUOUS-TIME)

Consider now that (4.1) represents a closed-loop system to be designed. Then all results of the previous section apply by replacing the matrices **A**, **D**, **C** by their closed-loop equivalents. To develop the matrices let subscripts p and c denote, respectively, plant and controller matrices. Then the system of interest is described by

$$\begin{array}{rrcllll} \textit{PLANT}: & \dot{\mathbf{x}}_p(t) & = & \mathbf{A}_p\mathbf{x}_p(t) & +\mathbf{B}_p\mathbf{u}(t) & +\mathbf{D}_p\mathbf{w}(t) \\ \textit{OUTPUT}: & \mathbf{y}(t) & = & \mathbf{C}_p\mathbf{x}_p(t) & +\mathbf{B}_y\mathbf{u}(t) & +\mathbf{D}_y\mathbf{w}(t) \\ \textit{MEASUREMENT}: & \mathbf{z}(t) & = & \mathbf{M}_p\mathbf{x}_p(t) & & +\mathbf{D}_z\mathbf{w}(t) \\ \textit{CONTROLLER}: & \dot{\mathbf{x}}_c(t) & = & \mathbf{A}_c\mathbf{x}_c(t) & +\mathbf{B}_c\mathbf{z}(t) & \\ & \mathbf{u}(t) & = & \mathbf{C}_c\mathbf{x}_c(t) & +\mathbf{D}_c\mathbf{z}(t) & \end{array} \tag{4.14}$$

or, by assembling these equations in compact form,

$$\begin{aligned} \begin{bmatrix} \dot{\mathbf{x}}_p(t) \\ \dot{\mathbf{x}}_c(t) \end{bmatrix} &= \begin{bmatrix} \mathbf{A}_p + \mathbf{B}_p\mathbf{D}_c\mathbf{M}_p & \mathbf{B}_p\mathbf{C}_c \\ \mathbf{B}_c\mathbf{M}_p & \mathbf{A}_c \end{bmatrix}\begin{bmatrix} \mathbf{x}_p(t) \\ \mathbf{x}_c(t) \end{bmatrix} + \begin{bmatrix} \mathbf{D}_p + \mathbf{B}_p\mathbf{D}_c\mathbf{D}_z \\ \mathbf{B}_c\mathbf{D}_z \end{bmatrix}\mathbf{w}(t) \\ \mathbf{y}(t) &= \begin{bmatrix} \mathbf{C}_p + \mathbf{B}_y\mathbf{D}_c\mathbf{M}_p & \mathbf{B}_y\mathbf{C}_c \end{bmatrix}\begin{bmatrix} \mathbf{x}_p(t) \\ \mathbf{x}_c(t) \end{bmatrix} + \begin{bmatrix} \mathbf{D}_y + \mathbf{B}_y\mathbf{D}_c\mathbf{D}_z \end{bmatrix}\mathbf{w}(t) \end{aligned}$$

or simply

$$\begin{aligned} \dot{\mathbf{x}}(t) &= \mathbf{A}_{c\ell}\mathbf{x}(t) + \mathbf{B}_{c\ell}\mathbf{w}(t) \\ \mathbf{y}(t) &= \mathbf{C}_{c\ell}\mathbf{x}(t) + \mathbf{D}_{c\ell}\mathbf{w}(t) \end{aligned} \tag{4.15}$$

where

$$\begin{aligned} \mathbf{A}_{c\ell} &\triangleq \mathbf{A} + \mathbf{BGM}, \quad & \mathbf{B}_{c\ell} = \mathbf{D} + \mathbf{BGE} \\ \mathbf{C}_{c\ell} &\triangleq \mathbf{C} + \mathbf{HGM}, \quad & \mathbf{D}_{c\ell} = \mathbf{F} + \mathbf{HGE} \end{aligned} \tag{4.16}$$

$$\begin{aligned} &\mathbf{A} \triangleq \begin{bmatrix} \mathbf{A}_p & \mathbf{0} \\ \mathbf{0} & \mathbf{0} \end{bmatrix}, \quad \mathbf{B} \triangleq \begin{bmatrix} \mathbf{B}_p & \mathbf{0} \\ \mathbf{0} & \mathbf{I}_{n_c} \end{bmatrix}, \quad \mathbf{M} \triangleq \begin{bmatrix} \mathbf{M}_p & \mathbf{0} \\ \mathbf{0} & \mathbf{I}_{n_c} \end{bmatrix} \\ &\mathbf{E} \triangleq \begin{bmatrix} \mathbf{D}_z \\ \mathbf{0} \end{bmatrix}, \quad \mathbf{H} \triangleq \begin{bmatrix} \mathbf{B}_y & \mathbf{0} \end{bmatrix}, \quad \mathbf{D} \triangleq \begin{bmatrix} \mathbf{D}_p \\ \mathbf{0} \end{bmatrix}, \quad \mathbf{G} \triangleq \begin{bmatrix} \mathbf{D}_c & \mathbf{C}_c \\ \mathbf{B}_c & \mathbf{A}_c \end{bmatrix} \\ &\mathbf{F} \triangleq \mathbf{D}_y, \quad \mathbf{x} \triangleq \begin{bmatrix} \mathbf{x}_p \\ \mathbf{x}_c \end{bmatrix}, \quad \mathbf{C} \triangleq \begin{bmatrix} \mathbf{C}_p & \mathbf{0} \end{bmatrix} \end{aligned} \tag{4.17}$$

and the vector dimensions are: $\mathbf{x}_p \in \mathcal{R}^{n_p}, \mathbf{x}_c \in \mathcal{R}^{n_c}, \mathbf{y}_p \in \mathcal{R}^{n_y}$,

$$\begin{aligned} &\mathbf{z} \in \mathcal{R}^{n_z}, \quad \mathbf{w} \in \mathcal{R}^{n_w}, \quad \mathbf{u} \in \mathcal{R}^{n_u} \\ &\mathbf{x} \in \mathcal{R}^{n_x}, \quad n_x = n_p + n_c. \end{aligned}$$

Applying the definition (4.5) to system (4.15) we can see from (4.7) that $\mathbf{X}(t)$ satisfies

$$\begin{aligned}\dot{\mathbf{X}}(t) &= \mathbf{X}(t)(\mathbf{A}+\mathbf{BGM})^T + (\mathbf{A}+\mathbf{BGM})\mathbf{X}(t) \\ &\quad +(\mathbf{D}+\mathbf{BGE})\mathbf{W}_0(\mathbf{D}+\mathbf{BGE})^T + \mathbf{X}_0 \\ \mathbf{X}(0) &= \mathbf{0}\end{aligned} \tag{4.18}$$

where the initial conditions are arranged as follows

$$\mathbf{X}_{po} = \begin{bmatrix} x_{p_1}^2(0) & & \\ & \ddots & \\ & & x_{p_{np}}^2(0)\end{bmatrix} > \mathbf{0}, \quad \mathbf{X}_{co} \triangleq \begin{bmatrix} x_{c_1}^2(0) & & \\ & \ddots & \\ & & x_{c_{nc}}^2(0)\end{bmatrix} > \mathbf{0}$$

$$\mathbf{X}_0 = \begin{bmatrix} \mathbf{X}_{po} & \mathbf{0} \\ \mathbf{0} & \mathbf{X}_{co}\end{bmatrix} > \mathbf{0},$$

and the impulsive elements $w_i(t)$ have strengths w_{p_i} and

$$\mathbf{W}_0 = \begin{bmatrix} w_1^2 & & \\ & \ddots & \\ & & w_{n_w}^2\end{bmatrix} > \mathbf{0}.$$

If the steady-state exists for the time-invariant case then, $\lim_{t\to\infty} \mathbf{X}(t) \triangleq \mathbf{X}$ satisfies [125]

$$\mathbf{0} = \mathbf{X}(\mathbf{A}+\mathbf{BGM})^T + (\mathbf{A}+\mathbf{BGM})\mathbf{X} + (\mathbf{D}+\mathbf{BGE})\mathbf{W}_0(\mathbf{D}+\mathbf{BGE})^T + \mathbf{X}_0. \tag{4.19}$$

4.3 STOCHASTIC INTERPRETATIONS

Let $\mathbf{w}_p(t)$, $\mathbf{v}(t)$, and $\mathbf{x}_p(0)$, $\mathbf{x}_c(0)$ in (4.14) be random with statistics described as follows [77]

$$\mathcal{E}\begin{bmatrix}\mathbf{w}(t)\\ \mathbf{x}(0)\end{bmatrix} = 0, \quad \mathcal{E}\begin{bmatrix}\mathbf{w}(t)\\ \mathbf{x}(0)\end{bmatrix}\begin{bmatrix}\mathbf{w}(\tau)\\ \mathbf{x}(0)\end{bmatrix}^T = \begin{bmatrix}\mathbf{W}_s(t)\delta(t-\tau) & \mathbf{0}\\ \mathbf{0} & \mathbf{X}^0\end{bmatrix} \tag{4.20}$$

where

$$\mathbf{W}_s = \begin{bmatrix}\mathbf{W}_p & \mathbf{W}_{pv}\\ \mathbf{W}_{pv}^T & \mathbf{V}\end{bmatrix} > 0, \quad \mathbf{X}^0 > 0 \tag{4.21}$$

and $\mathcal{E}[\cdot]$ is the expectation operator (an integral over the sample space weighted by the probability density function; see any book on stochastic processes, such as (4.1)–(4.2). The matrix $\mathbf{X}^0$ is the covariance of the random initial state $\mathbf{x}(0)$. The matrix $\mathbf{W}_0$ is not necessarily diagonal and is the intensity matrix of the white noise process $\mathbf{w}(t)$. $\mathbf{W}_{pv}$ is the correlation between the random processes $\mathbf{w}_p(t)$, $\mathbf{v}(t)$. Define the covariance matrix $\mathbf{X}_s(t)$ for the state of (4.15) by

$$\mathbf{X}_s(t) \triangleq \mathcal{E}\left[\mathbf{x}(t)\mathbf{x}^T(t)\right]. \tag{4.22}$$

Then $\mathbf{X}_s(t)$ satisfies

$$\begin{aligned}\dot{\mathbf{X}}_s(t) &= \mathbf{X}_s(t)(\mathbf{A}+\mathbf{BGM})^T + (\mathbf{A}+\mathbf{BGM})\mathbf{X}_s(t) \\ &\quad + (\mathbf{D}+\mathbf{BGE})\mathbf{W}_s(\mathbf{D}+\mathbf{BGE})^T \\ \mathbf{X}_s(0) &= \mathbf{X}^0.\end{aligned} \tag{4.23}$$

It is interesting to compare the deterministic theory (4.18) with the stochastic theory (4.23), especially the conditions under which they represent the same mathematical problem.

Theorem 4.3.1 *Define $\mathbf{X}_s(t)$ by (4.22) when (4.15) is excited by random variables with statistics (4.20) and (4.21). Define $\mathbf{X}(t)$ by (4.5) where $\mathbf{x}(i, t, \tau)$ refers to the response of (4.15) under the deterministic conditions similar to (4.2). Then $\mathbf{X}_s(t) = \mathbf{X}(t)$ for all $t \geq 0$ if:*

(i) the stochastic process $\mathbf{w}(t)$ in (4.20) and (4.21) is an independent process ($\mathbf{W}_s(t)$ is diagonal) with constant intensity ($\mathbf{W}_s$ constant);
(ii) $\mathcal{E}[w_i(t)w_i(\tau)] = w_i^2\delta(t-\tau)$ (the intensity of the white noise process $w_i(t)$ is equal in magnitude to the square of the intensity of the impulse applied in the deterministic case);
(iii) $\mathbf{X}^0 = \mathbf{X}_0 = \mathbf{0}$ (the initial conditions are zero in both the deterministic and stochastic problems).

The difference between the stochastic and deterministic problems deserve comment. In the stochastic problem (4.23) the matrix $\mathbf{W}_s$ can be time-varying and an arbitrary positive definite matrix, whereas in the deterministic problem (4.18) $\mathbf{W}_0$ is constant diagonal. This difference seems to be insignificant, since a time-varying scaling and transformation of the disturbance input $\mathbf{w}(t)$ could be applied to eliminate this difference. That is, one can choose $\mathbf{\Theta}(t)$ such that

$$\mathbf{w}_s = \mathbf{w}(stochastic) = \mathbf{\Theta}(t)\mathbf{w}(new) = \mathbf{\Theta}(t)w_{sn}$$

where $\mathbf{w}_{sn}$ is intended to have constant diagonal intensities $\mathbf{W}_{sn}$:

$$\begin{aligned}\mathcal{E}\left[\mathbf{w}_s(t)\mathbf{w}_s^T(\tau)\right] &= \mathbf{\Theta}(t)\mathcal{E}[\mathbf{w}_{sn}(t)\mathbf{w}_{sn}^T(\tau)]\mathbf{\Theta}^T(t) = \mathbf{W}_s(t)\delta(t-\tau)\\ &= \mathbf{\Theta}(t)\mathbf{W}_{sn}\mathbf{\Theta}^T(t) = \mathbf{W}_s(t).\end{aligned}$$

The initial state enters the two problems in substantially different ways. In the stochastic case (4.23) has a positive definite initial condition, whereas (4.18) has zero initial condition. But (4.18) has an extra forcing term $\mathbf{X}_0$ to "compensate" for the zero initial condition. A scalar case illustrates the essential differences in the treatment of initial conditions. Let $\mathbf{W}_0 = \mathbf{W}_s$. For the *scalar* equation (4.18) (stable time-invariant case)

$$X(t) = \frac{(D+BGE)^2W_0 + X_0}{-2(A+BGM)}[1 - e^{2(A+BGM)t}]$$

and for the scalar equation (4.23)

$$X_s(t) = \left(X^0 + \frac{(D+BGE)^2W_s}{2(A+BGM)}\right)e^{2(A+BGM)t} - \frac{(D+BGE)^2W_s}{2(A+BGM)}$$

yielding an initial difference ($t = 0$),

$$X_s(0) - X(0) = X_s(0) = X^0$$

and a steady-state difference

$$X_s(\infty) - X(\infty) = \frac{X_0}{2(A+BGM)} < 0.$$

Hence the deterministic problem is conservative in comparison with variances from the stochastic analysis. That is, $X(\infty) > X_s(\infty)$, meaning that the RMS value of the state $\sqrt{X(\infty)}$ is larger than the standard deviation $\sqrt{X_s(\infty)}$ associated with the stochastic problem, and this difference is zero when X_0 in the deterministic problem is zero.

The final comparison is perhaps the most important one. Consider the steady-state covariance $\mathbf{X}_s$ and the D-covariance $\mathbf{X}$, associated with (4.15). $\mathbf{X}_s$ and $\mathbf{X}$ satisfy

$$\begin{aligned}\mathbf{0} &= \mathbf{X}_s\mathbf{A}_{c\ell}^T + \mathbf{A}_{c\ell}\mathbf{X}_s + \mathbf{B}_{c\ell}\mathbf{W}_s\mathbf{B}_{c\ell}^T \\ \mathbf{0} &= \mathbf{X}\mathbf{A}_{c\ell}^T + \mathbf{A}_{c\ell}\mathbf{X} + \mathbf{B}_{c\ell}\mathbf{W}_0\mathbf{B}_{c\ell}^T + \mathbf{X}_0,\end{aligned}$$

respectively. Due to $\mathbf{X}_0 > 0$, asymptotic stability of $\mathbf{A}_{c\ell}$ is equivalent to $\mathbf{X} > 0$, regardless of properties of the matrix $\mathbf{D}$. On the other hand asymptotic stability of $\mathbf{A}_{c\ell}$ is not equivalent to $\mathbf{X}_s > 0$ because $(\mathbf{A}_{c\ell}, \mathbf{B}_{c\ell})$ might not be controllable, whereas $(\mathbf{A}_{c\ell}, \mathbf{B}_{c\ell}\mathbf{W}_0\mathbf{B}_{c\ell}^T + \mathbf{X}_0)$ is always controllable. *If* $(\mathbf{A} + \mathbf{BGM}, \mathbf{D} + \mathbf{BGE})$ is a controllable pair, *then* $\mathbf{X} > 0$ is equivalent to asymptotic stability of $\mathbf{A} + \mathbf{BGM}$. It is an unfortunate deficiency of the stochastic problem (4.23) that stability of $\mathbf{A} + \mathbf{BGM}$ is not equivalent to the condition $\mathbf{X}_s > \mathbf{0}$. Furthermore, to make matters worse, the controllability condition *never* holds in the physical world we seek to model. That is, *physical* systems are *never* completely controllable nor observable, even though the (simplified) mathematical model might be. See [125] for a more complete discussion of the uncontrollability and unobservability of physical systems.

Hence, to capture the set of all stabilizing controllers of fixed order, one can parametrize the set of all $\mathbf{G}(t)$ and $\mathbf{X}(t) > \mathbf{0}$ satisfying (4.18), but, because $(\mathbf{A}, \mathbf{D})$ might not be controllable one cannot capture the set of all stabilizing controllers by parametrizing all $\mathbf{G}(t)$ and $\mathbf{X}(t) > 0$ satisfying (4.23). This gives the deterministic approach a big advantage.

However, the language of (4.23) ("covariance analysis") is more familiar, and indeed covariance analysis is the *cornerstone* of several fields of systems theory (filtering theory, state estimation, identification, linear quadratic Gaussian optimal control, etc.). So, we should not dismiss the stochastic problem, but rather modify it slightly to suit our needs. Hereafter, whenever we refer to the stochastic problem (4.23) we shall assume that $\mathbf{D}$ is a nonsingular matrix. This is equivalent to adding noise sources that can excite all state variables, which is the effect of the $\mathbf{X}_0$ term in (4.18).

We must choose a name for matrix $\mathbf{X}$ (appearing in (4.5), (4.8) and (4.18)). In order to simplify our language we will call it the "D-covariance" (deterministic covariance) matrix, since it is the deterministic equivalent of the familiar stochastic definition of covariance, (4.20)–(4.22). When the context makes it clear whether we are discussing deterministic or stochastic problems, we may drop the "D" and just use the word "covariance".

4.4 THE DISCRETE SYSTEM D-COVARIANCE

Consider the time-varying systems

$$\mathbf{x}_{k+1} = \mathbf{A}\mathbf{x}_k + \mathbf{D}\mathbf{w}_k. \tag{4.24}$$

Define $\mathbf{x}(i, k, j)$ as the state at time k due to the ith excitation applied at time $j \leq k$. The admissible set of excitation events are $r = n_x + n_v$ in number (n_w pulses and n_x initial conditions)

$$\left.\begin{aligned} w_\alpha(k) &= w_\alpha\delta_{kj} \qquad & \alpha &= 1, \ldots, n_w \\ x_\beta(j) &= x_{\beta 0} \qquad & \beta &= 1, \ldots, n_x \end{aligned}\right\} \tag{4.25}$$

where for convention we (arbitrarily) assign $i = \alpha$ for $i \leq n_w$ and $i = n_w + \beta$ for $i > n_w$, and when w_α is the magnitude of the pulse applied in the αth input channel. Define

$$\mathbf{X}_k \triangleq \sum_{i=1}^{r} \sum_{j=0}^{k-1} \mathbf{x}(i, k, j)\mathbf{x}^T(i, k, j). \tag{4.26}$$

Consider that from (4.24), (4.25)

$$\mathbf{x}(i, k, j) = \begin{cases} \mathbf{A}^{k-j}\mathbf{x}_{\beta 0} & i = n_w + \beta \\ \mathbf{A}^{k-j-1}\mathbf{D}\mathbf{w}_{i0} & i \leq n_w \end{cases}$$

where

$$\mathbf{w}_{i0} = \begin{bmatrix} 0 \\ \vdots \\ w_i \\ \vdots \\ 0 \end{bmatrix}, \quad \text{when } \mathbf{w}(k) = \mathbf{w}_{i0}\delta_{kj}, \quad i \leq n_w$$

$$\mathbf{x}_{\beta 0} = \begin{bmatrix} 0 \\ \vdots \\ x_{\beta 0} \\ \vdots \\ 0 \end{bmatrix} \quad \text{when } i = n_w + \beta, \ \beta = 1, \ldots, n_x$$

and

$$\mathbf{x}(i, k+1, j) = \mathbf{A}^{k-j}\mathbf{d}, \ \mathbf{d} \triangleq \mathbf{A}\mathbf{x}_{\beta 0} \text{ or } \mathbf{D}\mathbf{w}_{i0}.$$

Then from (4.26) the reader should show that $\mathbf{X}_k$ satisfies

$$\mathbf{X}_{k+1} = \mathbf{A}\mathbf{X}_k\mathbf{A}^T + \mathbf{D}\mathbf{W}\mathbf{D}^T + \mathbf{X}_0 \tag{4.27}$$

where

$$\mathbf{W} = \text{diag}\left[\cdots w_\alpha^2 \cdots\right] > \mathbf{0}, \ \mathbf{X}_0 = \text{diag}\left[\cdots x_{\beta 0}^2 \cdots\right] > \mathbf{0}.$$

Definition 4.4.1 The matrix defined by (4.26), satisfying (4.27), is called the "D-covariance" for the system (4.24).

Theorem 4.4.1 *Let* $(\mathbf{A},\mathbf{D})$ *be constant. The eigenvalues of* $\mathbf{A}$ *all lie in the open unit disk if and only if the steady-state D-covariance exists satisfying*

$$\mathbf{X} = \mathbf{A}\mathbf{X}\mathbf{A}^T + \mathbf{D}\mathbf{W}\mathbf{D}^T + \mathbf{X}_0, \quad \mathbf{X} > \mathbf{0}. \tag{4.28}$$

Further motivation for selecting the second-order information (the covariance matrix) as the design space for linear systems comes from the fact that a larger class of systems can be treated with linear methods in the space of covariance matrices than in the state space. Consider an example. Let a certain system be described by

$$x(k+1) = \frac{1}{2}x(k) + x(k)u(k). \tag{4.29}$$

If $u(k)$ is a zero mean white noise (with covariance U) uncorrelated with $x(k)$, then the state covariance satisfies the linear equation,

$$X(k+1) = \frac{1}{4}X(k) + X(k)U. \tag{4.30}$$

Hence, in the space of covariances (4.29) is a linear system, whereas in the space of the state (4.29) is a nonlinear system. Hence, developing a control design theory based on second-order information (covariances), where the design space is the space of all symmetric matrices, will allow a larger class of systems to be treated with linear methods.

4.5 MODELS FOR CONTROL DESIGN (DISCRETE-TIME)

The discrete-time equivalent of the previous section is presented now. The systems of interest are described by

$$\begin{array}{rrcllll} \mathit{PLANT}: & \mathbf{x}_p(k+1) & = & \mathbf{A}_p\mathbf{x}_p(k) & +\mathbf{B}_p\mathbf{u}(k) & +\mathbf{D}_p\mathbf{w}(k) \\ \mathit{OUTPUT}: & \mathbf{y}(k) & = & \mathbf{C}_p\mathbf{x}_p(k) & +\mathbf{B}_y\mathbf{u}(k) & +\mathbf{D}_y\mathbf{w}(k) \\ \mathit{MEASUREMENT}: & \mathbf{z}(k) & = & \mathbf{M}_p\mathbf{x}_p(k) & & +\mathbf{D}_z\mathbf{w}(k) \\ \mathit{CONTROLLER}: & \mathbf{x}_c(k+1) & = & \mathbf{A}_c\mathbf{x}_c(k) & +\mathbf{B}_c\mathbf{z}(k) & \\ & \mathbf{u}(k) & = & \mathbf{C}_c\mathbf{x}_c(k) & +\mathbf{D}_c\mathbf{z}(k) & \end{array} \tag{4.31}$$

or, by assembling these equations in matrix form

$$\begin{aligned} \begin{bmatrix} \mathbf{x}_p(k+1) \\ \mathbf{x}_c(k+1) \end{bmatrix} &= \begin{bmatrix} \mathbf{A}_p + \mathbf{B}_p\mathbf{D}_c\mathbf{M}_p & \mathbf{B}_p\mathbf{C}_c \\ \mathbf{B}_c\mathbf{M}_p & \mathbf{A}_c \end{bmatrix} \begin{bmatrix} \mathbf{x}_p(k) \\ \mathbf{x}_c(k) \end{bmatrix} + \begin{bmatrix} \mathbf{D}_p + \mathbf{B}_p\mathbf{D}_c\mathbf{D}_z \\ \mathbf{B}_c\mathbf{D}_z \end{bmatrix} \mathbf{w}(k) \\ \mathbf{y}(k) &= \begin{bmatrix} \mathbf{C}_p + \mathbf{B}_y\mathbf{D}_c\mathbf{M}_p & \mathbf{B}_y\mathbf{C}_c \end{bmatrix} \begin{bmatrix} \mathbf{x}_p(k) \\ \mathbf{x}_c(k) \end{bmatrix} + \begin{bmatrix} \mathbf{D}_y + \mathbf{B}_y\mathbf{D}_c\mathbf{D}_z \end{bmatrix} \mathbf{w}(k) \end{aligned}$$

or simply

$$\begin{aligned} \mathbf{x}(k+1) &= \mathbf{A}_{c\ell}\mathbf{x}(k) + \mathbf{B}_{c\ell}\mathbf{w}(k) \\ \mathbf{y}(k) &= \mathbf{C}_{c\ell}\mathbf{x}(k) + \mathbf{D}_{c\ell}\mathbf{w}(k) \end{aligned} \tag{4.32}$$

where

$$\mathbf{A}_{c\ell} \triangleq \mathbf{A} + \mathbf{BGM}, \qquad \mathbf{B}_{c\ell} \triangleq \mathbf{D} + \mathbf{BGE}$$
$$\mathbf{C}_{c\ell} \triangleq \mathbf{C} + \mathbf{HGM}, \qquad \mathbf{D}_{c\ell} \triangleq \mathbf{F} + \mathbf{HGE}$$

$$\mathbf{A} \triangleq \begin{bmatrix} \mathbf{A}_p & \mathbf{0} \\ \mathbf{0} & \mathbf{0} \end{bmatrix}, \quad \mathbf{B} \triangleq \begin{bmatrix} \mathbf{B}_p & \mathbf{0} \\ \mathbf{0} & \mathbf{I}_{n_c} \end{bmatrix}, \quad \mathbf{M} \triangleq \begin{bmatrix} \mathbf{M}_p & \mathbf{0} \\ \mathbf{0} & \mathbf{I}_{n_c} \end{bmatrix}, \quad \mathbf{G} \triangleq \begin{bmatrix} \mathbf{D}_c & \mathbf{C}_c \\ \mathbf{B}_c & \mathbf{A}_c \end{bmatrix}$$

$$\mathbf{E} \triangleq \begin{bmatrix} \mathbf{D}_z \\ \mathbf{0} \end{bmatrix}, \quad \mathbf{H} \triangleq \begin{bmatrix} \mathbf{B}_y & \mathbf{0} \end{bmatrix}, \quad \mathbf{D} \triangleq \begin{bmatrix} \mathbf{D}_p \\ \mathbf{0} \end{bmatrix},$$

$$\mathbf{F} \triangleq \mathbf{D}_y, \quad \mathbf{x} \triangleq \begin{bmatrix} \mathbf{x}_p \\ \mathbf{x}_c \end{bmatrix}, \quad \mathbf{C} \triangleq \begin{bmatrix} \mathbf{C}_p & \mathbf{0} \end{bmatrix}$$

and the vector dimensions are the same as in the continuous-time case (4.15). Consider the equation

$$\begin{aligned} \mathbf{X}(k+1) &= (\mathbf{A} + \mathbf{BGM})\mathbf{X}(k)(\mathbf{A} + \mathbf{BGM})^T \\ &+ (\mathbf{D} + \mathbf{BGE})\mathbf{W}_0(\mathbf{D} + \mathbf{BGE})^T + \mathbf{X}_0 \\ \mathbf{X}(0) &= \mathbf{0} \end{aligned} \tag{4.33}$$

where

$$\mathbf{X}_0 \triangleq \begin{bmatrix} \mathbf{X}_{p0} & \mathbf{0} \\ \mathbf{0} & \mathbf{X}_{c0} \end{bmatrix}, \quad \mathbf{X}_{p0} \triangleq \begin{bmatrix} x_{p_1 0}^2 & & \\ & \ddots & \\ & & x_{p_{n_p} 0}^2 \end{bmatrix}$$

$$\mathbf{X}_{c0} \triangleq \begin{bmatrix} x_{c_1 0}^2 & & \\ & \ddots & \\ & & x_{c_{n_c} 0}^2 \end{bmatrix}, \quad \mathbf{W}_0 = \begin{bmatrix} w_1^2 & & \\ & \ddots & \\ & & w_{n_w}^2 \end{bmatrix},$$

where w_i is the magnitude of a pulse in the ith channel $w_i(k) = w_i \delta_{k\tau}$ applied at time $k = \tau$, and $x_{p_i 0}$ is the initial condition at $k = m$, $x_{p_i}(m) \triangleq x_{p_i 0}$. Then $\mathbf{X}(k)$ is defined by

$$\mathbf{X}(k) = \sum_{i=1}^{r} \sum_{m=0}^{k} \mathbf{x}(i,\ k,\ m)\mathbf{x}^T(i,\ k,\ m), \quad r = n_x + n_w \tag{4.34}$$

where $\mathbf{x}(i,\ k,\ m)$ denotes the response of (4.32) when only the ith excitation is applied at time m, from the admissible set of excitations

$$\left\{ \begin{array}{ll} w_\alpha(k) = w_\alpha \delta_{km} & \alpha = 1,\ 2,\ \ldots,\ n_w \\ x_\beta(m) = x_{\beta o} & \beta = 1,\ \ldots,\ n_x \end{array} \right\}$$

for a total of $r = n_w + n_x$ excitations.

In the time-invariant case $m = 0$ and the steady-state value of $\mathbf{X}(k)$ becomes

$$\mathbf{X} = \sum_{i=1}^{r} \sum_{k=0}^{\infty} \mathbf{x}(i,\ k)\mathbf{x}^T(i,\ k) \triangleq \lim_{k \to \infty} \mathbf{X}(k) \tag{4.35}$$

and satisfies, if it exists,

$$\mathbf{X} = (\mathbf{A} + \mathbf{BGM})\mathbf{X}(\mathbf{A} + \mathbf{BGM})^T + (\mathbf{D} + \mathbf{BGE})\mathbf{W}_0(\mathbf{D} + \mathbf{BGE})^T + \mathbf{X}_0\,. \tag{4.36}$$

The stochastic equivalent of (4.33) is as follows:

$$\begin{aligned} \mathbf{X}_s(k+1) &= (\mathbf{A} + \mathbf{BGM})\mathbf{X}_s(k)(\mathbf{A} + \mathbf{BGM})^T \\ &+ (\mathbf{D} + \mathbf{BGE})\mathbf{W}_s(\mathbf{D} + \mathbf{BGE})^T \\ \mathbf{X}(0) &= \mathbf{X}^0 \end{aligned} \tag{4.37}$$

where

$$\mathbf{X}_s(k) = \mathcal{E}\mathbf{x}(k)\mathbf{x}(k)^T$$

$$\mathcal{E}\begin{bmatrix} \mathbf{w}(k) \\ \mathbf{x}(0) \end{bmatrix} = 0$$

$$\mathcal{E}\begin{bmatrix} \mathbf{w}(k) \\ \mathbf{x}(0) \end{bmatrix}\begin{bmatrix} \mathbf{w}(j) \\ \mathbf{x}(0) \end{bmatrix}^T = \begin{bmatrix} \mathbf{W}_s(k)\delta_{kj} & \mathbf{0} \\ \mathbf{0} & \mathbf{X}^0 \end{bmatrix}.$$

Similar arguments hold, comparing stochastic and deterministic interpretations of (4.33), (4.37), as in the continuous-time case.

4.6 SYSTEM PERFORMANCE ANALYSIS

4.6.1 Continuous-Time Systems

Consider the linear time-invariant continuous-time system

$$\begin{bmatrix} \dot{\mathbf{x}}(t) \\ \mathbf{y}(t) \end{bmatrix} = \begin{bmatrix} \mathbf{A} & \mathbf{B} \\ \mathbf{C} & \mathbf{D} \end{bmatrix} \begin{bmatrix} \mathbf{x}(t) \\ \mathbf{w}(t) \end{bmatrix} \tag{4.38}$$

where $\mathbf{x}$ is the state, $\mathbf{w}$ is the disturbance, and $\mathbf{y}$ is the output of interest. Suppose that the system is considered to have "good" performance if $\mathbf{y}$ is "small" regardless of the disturbance $\mathbf{w}$. The purpose of this section is to define quantitative measures of system performance and to provide (computable) characterizations of the performance measures.

A standard way to quantify system performance is to consider the *system gain* Γ:

$$\Gamma \triangleq \sup_{\mathbf{w}} \frac{\text{size}(\mathbf{y})}{\text{size}(\mathbf{w})},$$

or equivalently,

$$\Gamma \triangleq \sup_{\mathbf{w}} \{\text{size}(\mathbf{y}) : \text{size}(\mathbf{w}) \leq 1\}.$$

The quantity Γ measures the size of the output signal $\mathbf{y}$ in response to the worst-case disturbance $\mathbf{w}$ with zero initial state. Thus, the smaller the system gain, the better the system performance. The definition of system gain Γ given above is still not concrete enough; we need to specify *how to measure the size of signals* $\mathbf{w}$ *and* $\mathbf{y}$. Clearly, different ways of measuring the size lead to different performance measures. Several measures for the size of a signal are summarized below.

The size of a square integrable (vector) signal $\mathbf{v}$ may be measured by

$$\|\mathbf{v}\|_{\mathcal{L}_2} \triangleq \left(\int_0^\infty \|\mathbf{v}(t)\|^2 dt \right)^{1/2},$$

where $\|\cdot\|$ is the Euclidean norm of a (constant) vector; for a vector $\mathbf{x}$, $\|\mathbf{x}\| \triangleq \sqrt{\mathbf{x}^T\mathbf{x}}$. The quantity $\|\mathbf{v}\|_{\mathcal{L}_2}$ is called the $\mathcal{L}_2$ norm of signal $\mathbf{v}$. It is also referred to as the energy of signal $\mathbf{v}$, in the control literature. The size of magnitude-bounded signal $\mathbf{v}$ can be measured by

$$\|\mathbf{v}\|_{\mathcal{L}_\infty} \triangleq \sup_{t \geq 0} \|\mathbf{v}(t)\|.$$

The quantity $\|\mathbf{v}\|_{\mathcal{L}_\infty}$ is called the $\mathcal{L}_\infty$ norm of signal $\mathbf{v}$. Note that, if $\mathbf{v}$ is a scalar signal, then $\|\mathbf{v}\|_{\mathcal{L}_\infty}$ is simply the peak value. The size of an impulsive signal $\mathbf{v}(t) = \mathbf{v}_0\delta(t)$ may be quantified as $\|\mathbf{v}_0\|$, where $\delta(\cdot)$ is the Dirac's delta function.

Now, using these notions of the signal size, we define the following performance measures (system gains) for system (4.38):

Impulse-to-Energy Gain:

$$\Gamma_{ie} \triangleq \sup_{\substack{\mathbf{w}(t) = \mathbf{w}_0\delta(t) \\ \|\mathbf{w}_0\| \leq 1}} \|\mathbf{y}\|_{\mathcal{L}_2}.$$

Energy-to-Peak Gain:

$$\Gamma_{ep} \triangleq \sup_{\|\mathbf{w}\|_{\mathcal{L}_2}\leq 1} \|\mathbf{y}\|_{\mathcal{L}_\infty}.$$

Energy-to-Energy Gain:

$$\Gamma_{ee} \triangleq \sup_{\|\mathbf{w}\|_{\mathcal{L}_2}\leq 1} \|\mathbf{y}\|_{\mathcal{L}_2}.$$

To analyze system performance, we would like to compute the system gains defined above. The following results are useful for this purpose. We use $\|\mathbf{A}\|$ to denote the spectral norm (the maximum singular value) of a matrix $\mathbf{A}$.

Theorem 4.6.1 *Consider system (4.38). The impulse-to-energy gain Γ_{ie} is finite if the system is strictly proper ($\mathbf{D} = \mathbf{0}$) and asymptotically stable. In this case, Γ_{ie} is given by*

$$\Gamma_{ie} = \|\mathbf{B}^T\mathbf{Y}\mathbf{B}\|^{1/2}, \quad \mathbf{Y}\mathbf{A} + \mathbf{A}^T\mathbf{Y} + \mathbf{C}^T\mathbf{C} = \mathbf{0}, \tag{4.39}$$

or alternatively,

$$\Gamma_{ie} = \inf_{\mathbf{P}}\{\|\mathbf{B}^T\mathbf{P}\mathbf{B}\|^{1/2} \ : \ \mathbf{P}\mathbf{A} + \mathbf{A}^T\mathbf{P} + \mathbf{C}^T\mathbf{C} < \mathbf{0}\ \}. \tag{4.40}$$

Proof. Recall that the state trajectory of the system (4.38) in response to $\mathbf{w}$ with zero initial state is given by

$$\mathbf{x}(t) = \int_0^t e^{\mathbf{A}(t-\tau)}\mathbf{B}\mathbf{w}(\tau)d\tau.$$

Using this, for impulsive disturbance $\mathbf{w}(t) = \mathbf{w}_0\delta(t)$, we have

$$\mathbf{y}(t) \ = \ \mathbf{C}e^{\mathbf{A}t}\mathbf{B}\mathbf{w}_0.$$

Hence, the $\mathcal{L}_2$ norm of the output signal $\mathbf{y}$ is given by

$$\begin{aligned} \|\mathbf{y}\|^2_{\mathcal{L}_2} &= \int_0^\infty \mathbf{w}_0^T\mathbf{B}^T e^{\mathbf{A}^T t}\mathbf{C}^T\mathbf{C}e^{\mathbf{A}t}\mathbf{B}\mathbf{w}_0 dt \\ &= \mathbf{w}_0^T\mathbf{B}^T\mathbf{Y}\mathbf{B}\mathbf{w}_0 \end{aligned} \tag{4.41}$$

where

$$\mathbf{Y} \triangleq \int_0^\infty e^{\mathbf{A}^T t}\mathbf{C}^T\mathbf{C}e^{\mathbf{A}t}dt.$$

Since the system is asymptotically stable, matrix $\mathbf{Y}$ defined above exists and can be computed as the solution to the Lyapunov equation in (4.39). Now, the impulse-to-energy gain Γ_{ie} is given by maximizing the right-hand side of (4.41) over $\mathbf{w}_0$ subject to $\|\mathbf{w}_0\| \leq 1$. Clearly, the worst-case direction $\mathbf{w}_0$ of the impulsive disturbance is given by the (unit) eigenvector of $\mathbf{B}^T\mathbf{Y}\mathbf{B}$ corresponding to the largest eigenvalue. Hence

$$\max_{\|\mathbf{w}_0\|\leq 1} \mathbf{w}_0^T\mathbf{B}^T\mathbf{Y}\mathbf{B}\mathbf{w}_0 = \|\mathbf{B}^T\mathbf{Y}\mathbf{B}\|,$$

and we have (4.39).

Now we prove (4.40). Define $\mathbf{R} \triangleq \mathbf{P} - \mathbf{Y}$ and consider the Lyapunov inequality in (4.40):

$$(\mathbf{R} + \mathbf{Y})\mathbf{A} + \mathbf{A}^T(\mathbf{R} + \mathbf{Y}) + \mathbf{C}^T\mathbf{C} < \mathbf{0}.$$

Using the Lyapunov equation in (4.39), we have

$$\mathbf{R}\mathbf{A} + \mathbf{A}^T\mathbf{R} < \mathbf{0}.$$

Since $\mathbf{A}$ is stable, it follows that $\mathbf{R} > \mathbf{0}$. Thus we have $\mathbf{P} > \mathbf{Y} \geq \mathbf{0}$. Note that inequality $\mathbf{P} > \mathbf{Y}$ is tight, i.e. for any given $\varepsilon > 0$, there exists $\mathbf{P}$ satisfying the Lyapunov inequality in (4.40) such that $\|\mathbf{P} - \mathbf{Y}\| \leq \varepsilon$. Hence, the smallest (infimum) value of $\|\mathbf{B}^T\mathbf{P}\mathbf{B}\|^{1/2}$ is given by $\|\mathbf{B}^T\mathbf{Y}\mathbf{B}\|^{1/2}$, which is equal to Γ_{ie} due to (4.39). This completes the proof. □

Theorem 4.6.2 *Consider system (4.38). The energy-to-peak gain Γ_{ep} is finite if the system is strictly proper ($\mathbf{D} = \mathbf{0}$) and asymptotically stable. In this case, Γ_{ep} is given by*

$$\Gamma_{ep} = \|\mathbf{C}\mathbf{X}\mathbf{C}^T\|^{1/2}, \quad \mathbf{A}\mathbf{X} + \mathbf{X}\mathbf{A}^T + \mathbf{B}\mathbf{B}^T = \mathbf{0}, \tag{4.42}$$

or alternatively,

$$\Gamma_{ep} = \inf_{\mathbf{Q}}\{\|\mathbf{C}\mathbf{Q}\mathbf{C}^T\|^{1/2} : \mathbf{A}\mathbf{Q} + \mathbf{Q}\mathbf{A}^T + \mathbf{B}\mathbf{B}^T < \mathbf{0}\}. \tag{4.43}$$

Proof. [154, 19] Let $\mathbf{Q}$ be any symmetric matrix satisfying the Lyapunov inequality in (4.43). Note that $\mathbf{Q} > \mathbf{0}$ since $\mathbf{A}$ is stable. After multiplying the Lyapunov inequality in (4.43) by $\mathbf{Q}^{-1}$ from the left and the right, the use of the Schur complement formula yields

$$\Phi \triangleq \begin{bmatrix} \mathbf{Q}^{-1}\mathbf{A} + \mathbf{A}^T\mathbf{Q}^{-1} & \mathbf{Q}^{-1}\mathbf{B} \\ \mathbf{B}^T\mathbf{Q}^{-1} & -\mathbf{I} \end{bmatrix} < \mathbf{0}.$$

Let $\mathbf{x}$ and $\mathbf{w}$ be any signals that satisfy the state equation (4.38). Define $\mathbf{v}^T \triangleq [\ \mathbf{x}^T\ \ \mathbf{w}^T\ ]$ and consider

$$\begin{aligned}
\mathbf{v}^T(t)\Phi\mathbf{v}(t) &= \mathbf{x}^T(t)\mathbf{Q}^{-1}(\mathbf{A}\mathbf{x}(t) + \mathbf{B}\mathbf{w}(t)) + (\mathbf{A}\mathbf{x}(t) + \mathbf{B}\mathbf{w}(t))^T\mathbf{Q}^{-1}\mathbf{x}(t) - \mathbf{w}^T(t)\mathbf{w}(t) \\
&= \frac{d}{dt}(\mathbf{x}^T(t)\mathbf{Q}^{-1}\mathbf{x}(t)) - \|\mathbf{w}(t)\|^2 < 0.
\end{aligned}$$

Integrating from $t = 0$ to τ with zero initial state,

$$\mathbf{x}^T(\tau)\mathbf{Q}^{-1}\mathbf{x}(\tau) < \int_0^\tau \|\mathbf{w}(t)\|^2 dt \leq \|\mathbf{w}\|^2_{\mathcal{L}_2}.$$

Using the Schur complement formula, we have

$$\begin{bmatrix} \|\mathbf{w}\|^2_{\mathcal{L}_2} & \mathbf{x}^T(\tau) \\ \mathbf{x}(\tau) & \mathbf{Q} \end{bmatrix} > \mathbf{0},$$

which implies that

$$\begin{bmatrix} 1 & \mathbf{0} \\ \mathbf{0} & \mathbf{C} \end{bmatrix}\begin{bmatrix} \|\mathbf{w}\|^2_{\mathcal{L}_2} & \mathbf{x}^T(\tau) \\ \mathbf{x}(\tau) & \mathbf{Q} \end{bmatrix}\begin{bmatrix} 1 & \mathbf{0} \\ \mathbf{0} & \mathbf{C}^T \end{bmatrix} = \begin{bmatrix} \|\mathbf{w}\|^2_{\mathcal{L}_2} & \mathbf{y}^T(\tau) \\ \mathbf{y}(\tau) & \mathbf{C}\mathbf{Q}\mathbf{C}^T \end{bmatrix} \geq \mathbf{0}.$$

Using the Schur complement formula again,

$$\frac{1}{\|\mathbf{w}\|_{\mathcal{L}_2}^2}\mathbf{y}(\tau)\mathbf{y}^T(\tau) \le \mathbf{CQC}^T \le \|\mathbf{CQC}^T\|\mathbf{I}.$$

Hence

$$\mathbf{y}^T(\tau)\mathbf{y}(\tau) \le \|\mathbf{w}\|_{\mathcal{L}_2}^2\|\mathbf{CQC}^T\| \le \|\mathbf{CQC}^T\|$$

for all disturbances such that $\|\mathbf{w}\|_{\mathcal{L}_2}^2 \le 1$. Since the above inequality holds for all $\tau \ge 0$, and for all $\mathbf{Q}$ satisfying the Lyapunov inequality in (4.43), we have established that

$$\Gamma_{ep} \le \inf_{\mathbf{Q}}\{\|\mathbf{CQC}^T\|^{1/2} : \mathbf{AQ} + \mathbf{QA}^T + \mathbf{BB}^T < \mathbf{0}\} = \|\mathbf{CXC}^T\|^{1/2} \tag{4.44}$$

where $\mathbf{X}$ is the solution to the Lyapunov equation in (4.42), and the last equality can be verified in a similar manner to the proof of Theorem 4.6.1.

Now we show that inequality $\le$ in (4.44) is tight. Consider the disturbance signal given by

$$\mathbf{w}_T(t) \triangleq \begin{cases} \lambda_T^{-1/2}\mathbf{B}^T e^{\mathbf{A}^T(T-t)}\mathbf{C}^T\mathbf{v}_T & (0 \le t \le T) \\ \mathbf{0} & (T < t) \end{cases} \tag{4.45}$$

where $\mathbf{v}_T$ is the unit eigenvector of $\mathbf{CX}_T\mathbf{C}^T$ corresponding to the largest eigenvalue λ_T, where

$$\mathbf{X}_T \triangleq \int_0^T e^{\mathbf{A}t}\mathbf{BB}^T e^{\mathbf{A}^T t}dt.$$

Note that

$$\begin{aligned} \|\mathbf{w}_T\|_{\mathcal{L}_2}^2 &= \int_0^T \frac{1}{\lambda_T}\mathbf{v}_T^T\mathbf{C}e^{\mathbf{A}(T-t)}\mathbf{BB}^T e^{\mathbf{A}^T(T-t)}\mathbf{C}^T\mathbf{v}_T dt \\ &= \frac{1}{\lambda_T}\mathbf{v}_T^T(\mathbf{CX}_T\mathbf{C}^T)\mathbf{v}_T = 1, \end{aligned}$$

for any fixed $T > 0$. We show that the $\mathcal{L}_\infty$ norm of the output signal $\mathbf{y}_T$, in response to this disturbance $\mathbf{w}_T$, approaches $\|\mathbf{CXC}^T\|^{1/2}$ when T approaches infinity. To this end, first note that

$$\begin{aligned} \mathbf{y}_T(T) &= \int_0^T \mathbf{C}e^{\mathbf{A}(T-\tau)}\mathbf{B}(\lambda_T^{-1/2}\mathbf{B}^T e^{\mathbf{A}^T(T-\tau)}\mathbf{C}^T\mathbf{v}_T)d\tau \\ &= \lambda_T^{-1/2}\mathbf{CX}_T\mathbf{C}^T\mathbf{v}_T. \end{aligned}$$

Hence,

$$\|\mathbf{y}_T(T)\|^2 = \frac{1}{\lambda_T}\mathbf{v}_T^T(\mathbf{CX}_T\mathbf{C}^T)^2\mathbf{v}_T = \lambda_T.$$

Taking the limit,

$$\lim_{T\to\infty}\|\mathbf{y}_T(T)\|^2 = \lim_{T\to\infty}\lambda_T = \|\mathbf{CXC}^T\|.$$

Thus we have shown that $\Gamma_{ep} \ge \|\mathbf{CXC}^T\|^{1/2}$ using a particular disturbance $\mathbf{w}$ in (4.45). From this inequality and (4.44), we conclude the result. □

System gains Γ_{ie} and Γ_{ep} are related to the $\mathcal{H}_2$ norm of the transfer matrix $\mathbf{T}(s) \triangleq \mathbf{C}(s\mathbf{I}-\mathbf{A})^{-1}\mathbf{B}$, defined in the frequency domain by

$$\|\mathbf{T}\|_{\mathcal{H}_2} \triangleq tr\left(\frac{1}{2\pi}\int_{-\infty}^{\infty}\mathbf{T}(j\omega)\mathbf{T}(j\omega)^* d\omega\right)^{1/2}$$

or equivalently,

$$\|\mathbf{T}\|_{\mathcal{H}_2} \triangleq tr\left(\frac{1}{2\pi}\int_{-\infty}^{\infty}\mathbf{T}(j\omega)^*\mathbf{T}(j\omega) d\omega\right)^{1/2}.$$

Using the fact that

$$\mathbf{X} \triangleq \frac{1}{2\pi}\int_{-\infty}^{\infty}(j\omega\mathbf{I}-\mathbf{A})^{-1}\mathbf{B}\mathbf{B}^T(-j\omega\mathbf{I}-\mathbf{A}^T)^{-1}d\omega$$

$$\mathbf{Y} \triangleq \frac{1}{2\pi}\int_{-\infty}^{\infty}(-j\omega\mathbf{I}-\mathbf{A}^T)^{-1}\mathbf{C}^T\mathbf{C}(j\omega\mathbf{I}-\mathbf{A})^{-1}d\omega$$

satisfy

$$\mathbf{A}\mathbf{X}+\mathbf{X}\mathbf{A}^T+\mathbf{B}\mathbf{B}^T=\mathbf{0}$$

$$\mathbf{Y}\mathbf{A}+\mathbf{A}^T\mathbf{Y}+\mathbf{C}^T\mathbf{C}=\mathbf{0}$$

it is easy to see that the $\mathcal{H}_2$ norm of $\mathbf{T}(s)$ is given by

$$\|\mathbf{T}\|_{\mathcal{H}_2}^2 = tr(\mathbf{C}\mathbf{X}\mathbf{C}^T) = tr(\mathbf{B}^T\mathbf{Y}\mathbf{B}).$$

Note that, if we replace $tr(\cdot)$ by $\|\cdot\|$, then we obtain the characterizations of Γ_{ie} and Γ_{ep} in Theorems 4.6.1 and 4.6.2. Since $tr(\mathbf{A}) = \|\mathbf{A}\|$ if $\mathbf{A}$ is a (non-negative) scalar, we see that $\|\mathbf{T}\|_{\mathcal{H}_2} = \Gamma_{ie} = \Gamma_{ep}$ for single-input, single-output systems. A time-domain interpretation of the $\mathcal{H}_2$ norm is given by

$$\|\mathbf{T}\|_{\mathcal{H}_2}^2 = \sum_{i=1}^{n_w}\|\mathbf{y}(i,\cdot)\|_{\mathcal{L}_2}^2$$

where $\mathbf{y}(i,\cdot)$ is the output signal in response to an impulsive disturbance with unit intensity applied to the ith disturbance channel, and n_w is the number of such channels (i.e. the dimension of $\mathbf{w}$).

Finally, we shall give a computable characterization of the energy-to-energy gain Γ_{ee}.

Theorem 4.6.3 *Consider system (4.38) and let a scalar $\gamma > 0$ be given. Suppose the system is asymptotically stable. Then the following statements are equivalent:*

(i) $\Gamma_{ee} < \gamma$.

(ii) $\mathbf{R} \triangleq \gamma^2\mathbf{I}-\mathbf{D}^T\mathbf{D} > \mathbf{0}$ *and there exists a symmetric matrix* $\mathbf{Y} > 0$ *such that*

$$\mathbf{Y}\mathbf{A}+\mathbf{A}^T\mathbf{Y}+(\mathbf{Y}\mathbf{B}+\mathbf{C}^T\mathbf{D})\mathbf{R}^{-1}(\mathbf{Y}\mathbf{B}+\mathbf{C}^T\mathbf{D})^T+\mathbf{C}^T\mathbf{C}=\mathbf{0}$$

and $\mathbf{A}+\mathbf{B}\mathbf{R}^{-1}(\mathbf{Y}\mathbf{B}+\mathbf{C}^T\mathbf{D})^T$ *is asymptotically stable.*

(iii) $\mathbf{R} \triangleq \gamma^2\mathbf{I}-\mathbf{D}^T\mathbf{D} > \mathbf{0}$ *and there exists a symmetric matrix* $\mathbf{P} > 0$ *such that*

$$\mathbf{P}\mathbf{A}+\mathbf{A}^T\mathbf{P}+(\mathbf{P}\mathbf{B}+\mathbf{C}^T\mathbf{D})\mathbf{R}^{-1}(\mathbf{P}\mathbf{B}+\mathbf{C}^T\mathbf{D})^T+\mathbf{C}^T\mathbf{C}<\mathbf{0}.$$

(iv) There exists a symmetric matrix $\mathbf{P} > 0$ *such that*

$$\begin{bmatrix} \mathbf{PA}+\mathbf{A}^T\mathbf{P} & \mathbf{PB} & \mathbf{C}^T \\ \mathbf{B}^T\mathbf{P} & -\gamma\mathbf{I} & \mathbf{D}^T \\ \mathbf{C} & \mathbf{D} & -\gamma\mathbf{I} \end{bmatrix} < \mathbf{0}. \tag{4.46}$$

Proof. We shall prove (iii) $\Leftrightarrow$ (iv) $\Rightarrow$ (i) only. A complete proof may be found in [25, 150].
First note that statement (iii) is equivalent to

$$\mathbf{\Phi} \triangleq \begin{bmatrix} \mathbf{PA}+\mathbf{A}^T\mathbf{P} & \mathbf{PB} \\ \mathbf{B}^T\mathbf{P} & -\gamma^2 I \end{bmatrix} + \begin{bmatrix} \mathbf{C}^T \\ \mathbf{D}^T \end{bmatrix} \begin{bmatrix} \mathbf{C} & \mathbf{D} \end{bmatrix} < \mathbf{0}$$

where we used the Schur complement formula. Another use of the Schur complement formula for $\mathbf{\Phi}/\gamma < \mathbf{0}$ yields statement (iv). Let $\mathbf{x}$ and $\mathbf{w}$ be any signals that satisfy the state equation (4.38). Then we have $\mathbf{v}^T(t)\mathbf{\Phi}\mathbf{v}(t) < 0$ for $\mathbf{v} \triangleq [\ \mathbf{x}^T\ \mathbf{w}^T\ ]^T$, or equivalently,

$$\frac{d}{dt}(\mathbf{x}^T(t)\mathbf{P}\mathbf{x}(t)) - \gamma^2\mathbf{w}^T(t)\mathbf{w}(t) + \mathbf{z}^T(t)\mathbf{z}(t) < 0.$$

Integrating from $t = 0$ to ∞ with zero initial state, and using the stability property $\lim_{t\to\infty}\mathbf{x}(t) = \mathbf{0}$, we have

$$\|\mathbf{z}\|^2_{\mathcal{L}_2} < \gamma^2\|\mathbf{w}\|^2_{\mathcal{L}_2}.$$

Thus we conclude $\Gamma_{ee} < \gamma$. □

The energy-to-energy gain Γ_{ee} has a significant frequency domain interpretation; it is equal to the $\mathcal{H}_\infty$ norm of the transfer matrix $\mathbf{T}(s) \triangleq \mathbf{C}(s\mathbf{I}-\mathbf{A})^{-1}\mathbf{B}+\mathbf{D}$:

$$\|\mathbf{T}\|_{\mathcal{H}_\infty} \triangleq \sup_\omega \|\mathbf{T}(j\omega)\|.$$

Hence, $\|\mathbf{T}\|_{\mathcal{H}_\infty} < \gamma$ is equivalent to each of statements (i), (ii) and (iii) in Theorem 4.6.3. The fact that $\Gamma_{ee} = \|\mathbf{T}\|_{\mathcal{H}_\infty}$ may be proved by using the Parseval's equality [83]. The $\mathcal{H}_\infty$ norm is related to robustness to norm-bounded perturbations as will be shown later.

Example 4.6.1 Consider the following second-order system:

$$P(s) = \frac{k}{s^2 + 2\zeta\omega s + \omega^2}$$

where ω is the natural frequency and ζ is the damping. A state space realization for this system is given by

$$\begin{bmatrix} \mathbf{A} & \mathbf{B} \\ \mathbf{C} & \mathbf{D} \end{bmatrix} = \left[\begin{array}{cc|c} -2\zeta\omega & -\omega^2 & 1 \\ 1 & 0 & 0 \\ \hline 0 & k & 0 \end{array}\right].$$

We consider the following two cases:

$$P_1(s):\ \zeta = 0.1,\ \omega = 1,\ k = 1$$

$$P_2(s):\ \zeta = 1,\ \omega = 3,\ k = 20.$$

The system gains Γ_{ie}, Γ_{ep} and Γ_{ee} are computed by solving the Lyapunov equations in (4.39), (4.42) and the linear matrix inequality in (4.46), for $\mathbf{Y}$, $\mathbf{X}$ and $\mathbf{P}$, respectively. In particular, to compute the energy-to-energy gain Γ_{ee}, the scalar γ is minimized subject to (4.46); in this way, Γ_{ee} can be found as the minimum value of γ. The results are summarized in Tables 4.1 and 4.2.

Table 4.1 System gains for $P_1(s)$ and $P_2(s)$

	Γ_{ie}	Γ_{ep}	Γ_{ee}
$P_1(s)$	2.500	2.500	5.026
$P_2(s)$	3.703	3.703	2.222

Table 4.2 Solutions to (4.39), (4.42) and (4.46)

	$\mathbf{Y}$	$\mathbf{X}$	$\mathbf{P}$
$P_1(s)$	$\begin{bmatrix} 2.500 & 0.500 \\ 0.500 & 2.600 \end{bmatrix}$	$\begin{bmatrix} 2.500 & 0 \\ 0 & 2.500 \end{bmatrix}$	$\begin{bmatrix} 1.005 & 0.101 \\ 0.101 & 1.005 \end{bmatrix}$
$P_2(s)$	$\begin{bmatrix} 3.704 & 22.222 \\ 22.222 & 166.667 \end{bmatrix}$	$\begin{bmatrix} 0.0833 & 0 \\ 0 & 0.0093 \end{bmatrix}$	$\begin{bmatrix} 13.334 & 19.985 \\ 19.985 & 120.007 \end{bmatrix}$

Note that, for single-input, single-output systems, the impulse-to-energy gain and the energy-to-peak gain have exactly the same value, which is equal to the $\mathcal{H}_2$ norm. Recall also that the energy-to-energy gain Γ_{ee} is equal to the $\mathcal{H}_\infty$ norm. From Table 4.1, we see that the system $P_1(s)$ has the smaller $\mathcal{H}_2$ norm and the larger $\mathcal{H}_\infty$ norm than $P_2(s)$. This fact is also evident from the Bode plots of $P_1(s)$ and $P_2(s)$ shown in Figure 4.1.

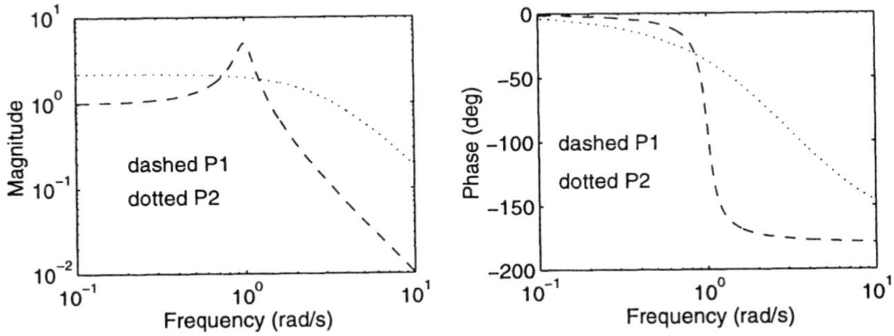

Figure 4.1 Bode plots of $P_1(s)$ and $P_2(s)$.

The $\mathcal{H}_\infty$ norm and the $\mathcal{H}_2$ norm correspond to the peak value of the magnitude plot and the area under the magnitude plot, respectively. The system $P_1(s)$ has a low damping, and hence its magnitude plot has a sharp peak, resulting in a larger system gain Γ_{ee}.

4.6.2 Discrete-Time Systems

Consider the linear time-invariant discrete-time system

$$\begin{bmatrix} \mathbf{x}(k+1) \\ \mathbf{y}(k) \end{bmatrix} = \begin{bmatrix} \mathbf{A} & \mathbf{B} \\ \mathbf{C} & \mathbf{D} \end{bmatrix} \begin{bmatrix} \mathbf{x}(k) \\ \mathbf{w}(k) \end{bmatrix} \tag{4.47}$$

where $\mathbf{x}$ is the state, $\mathbf{y}$ and $\mathbf{w}$ are the output of interest and the disturbance input, respectively. The purpose of this section is to develop system performance analysis results which are analogous to the continuous-time counterpart presented in the previous subsection. To this end, we shall first give several ways of measuring the size of discrete-time signals (sequences).

The size of square summable signal $\mathbf{v}$ can be measured by the ℓ_2 norm:

$$\|\mathbf{v}\|_{\ell_2} \triangleq \left(\sum_{k=0}^{\infty} \|\mathbf{v}(k)\|^2 \right)^{1/2}.$$

The size of magnitude-bounded signal $\mathbf{v}$ may be measured by the ℓ_∞ norm:

$$\|\mathbf{v}\|_{\ell_\infty} \triangleq \sup_{k \geq 0} \|\mathbf{v}(k)\|.$$

The size of pulse signal $\mathbf{v}(k) = \mathbf{v}_0 \delta(k)$ may be quantified as $\|\mathbf{v}_0\|$, where $\delta(\cdot)$ is the Kronecker's delta; $\delta(0) = 1$ and $\delta(k) = 0$ for all $k \neq 0$.

System gains for the discrete-time system (4.47) can be defined using the above signal-size measures as follows.

Pulse-to-Energy Gain:

$$\Upsilon_{pe} \triangleq \sup_{\mathbf{w}(k)=\mathbf{w}_0\delta(k),\ \|\mathbf{w}_0\| \leq 1} \|\mathbf{y}\|_{\ell_2}.$$

Energy-to-Peak Gain:

$$\Upsilon_{ep} \triangleq \sup_{\|\mathbf{w}\|_{\ell_2} \leq 1} \|\mathbf{y}\|_{\ell_\infty}.$$

Energy-to-Energy Gain:

$$\Upsilon_{ee} \triangleq \sup_{\|\mathbf{w}\|_{\ell_2} \leq 1} \|\mathbf{y}\|_{\ell_2}.$$

Next we characterize these system gains in terms of algebraic conditions.

Theorem 4.6.4 *Consider system (4.47). Suppose the system is asymptotically stable. Then the pulse-to-energy gain Υ_{pe} is given by*

$$\Upsilon_{pe} = \|\mathbf{B}^T\mathbf{Y}\mathbf{B} + \mathbf{D}^T\mathbf{D}\|^{1/2}, \quad \mathbf{Y} = \mathbf{A}^T\mathbf{Y}\mathbf{A} + \mathbf{C}^T\mathbf{C} \tag{4.48}$$

or alternatively,

$$\Upsilon_{pe} = \inf_{\mathbf{P}} \{ \|\mathbf{B}^T\mathbf{P}\mathbf{B} + \mathbf{D}^T\mathbf{D}\|^{1/2} : \mathbf{P} > \mathbf{A}^T\mathbf{P}\mathbf{A} + \mathbf{C}^T\mathbf{C} \}. \tag{4.49}$$

Proof. Recall that the solution of the difference equation (4.47) is given by

$$\mathbf{x}(k) = \mathbf{A}^k\mathbf{x}(0) + \sum_{i=0}^{k-1} \mathbf{A}^{k-i-1}\mathbf{B}\mathbf{w}(i).$$

For pulse disturbance $\mathbf{w}(k) = \mathbf{w}_0\delta(k)$ with zero initial state, we have

$$\mathbf{y}(k) = \begin{cases} \mathbf{D}\mathbf{w}_0 & (k = 0) \\ \mathbf{C}\mathbf{A}^{k-1}\mathbf{B}\mathbf{w}_0 & (k \geq 1) \end{cases}.$$

The ℓ_2 norm of this output signal is given by

$$\begin{aligned} \|\mathbf{y}\|_{\ell_2}^2 &= \mathbf{w}_0^T\mathbf{D}^T\mathbf{D}\mathbf{w}_0 + \sum_{k=1}^{\infty} \mathbf{w}_0^T\mathbf{B}^T(\mathbf{A}^T)^{k-1}\mathbf{C}^T\mathbf{C}\mathbf{A}^{k-1}\mathbf{B}\mathbf{w}_0 \\ &= \mathbf{w}_0^T(\mathbf{B}^T\mathbf{Y}\mathbf{B} + \mathbf{D}^T\mathbf{D})\mathbf{w}_0 \end{aligned}$$

where

$$\mathbf{Y} \triangleq \sum_{\ell=0}^{\infty}(\mathbf{A}^T)^{\ell}\mathbf{C}^T\mathbf{C}\mathbf{A}^{\ell}.$$

This $\mathbf{Y}$ satisfies the Lyapunov equation in (4.48). The worst-case direction of the pulse disturbance, which maximize $\|\mathbf{y}\|_{\ell_2}$, is given by the (unit) eigenvector of $\mathbf{B}^T\mathbf{Y}\mathbf{B} + \mathbf{D}^T\mathbf{D}$ corresponding to the largest eigenvalue. In this case, the worst-case ℓ_2 norm of the output is $\|\mathbf{y}\|_{\ell_2}^2 = \|\mathbf{B}^T\mathbf{Y}\mathbf{B} + \mathbf{D}^T\mathbf{D}\|$. This proves (4.48). Finally, the other characterization (4.49) can be proved in a similar manner to the proof of Theorem 4.6.1, and hence is omitted. □

Theorem 4.6.5 *Consider system (4.47) and suppose the system is asymptotically stable. Then the energy-to-peak gain Υ_{ep} is given by*

$$\Upsilon_{ep} = \|\mathbf{C}\mathbf{X}\mathbf{C}^T + \mathbf{D}\mathbf{D}^T\|^{1/2}, \quad \mathbf{X} = \mathbf{A}\mathbf{X}\mathbf{A}^T + \mathbf{B}\mathbf{B}^T, \tag{4.50}$$

or alternatively,

$$\Upsilon_{ep} = \inf_{\mathbf{Q}}\{\|\mathbf{C}\mathbf{Q}\mathbf{C}^T + \mathbf{D}\mathbf{D}^T\|^{1/2} : \mathbf{Q} > \mathbf{A}\mathbf{Q}\mathbf{A}^T + \mathbf{B}\mathbf{B}^T\}. \tag{4.51}$$

Proof. [19, 163] Consider the Lyapunov inequality in (4.51). Noting that $\mathbf{Q} > \mathbf{0}$ due to stability assumption, and using the Schur complement formula, we have

$$\Phi \triangleq \begin{bmatrix} \mathbf{Q}^{-1} & \mathbf{0} \\ \mathbf{0} & \mathbf{I} \end{bmatrix} - \begin{bmatrix} \mathbf{A}^T \\ \mathbf{B}^T \end{bmatrix} \mathbf{Q}^{-1} \begin{bmatrix} \mathbf{A} & \mathbf{B} \end{bmatrix} > \mathbf{0}.$$

Let $\mathbf{w}$ be any square summable signal, and $\mathbf{x}$ be the solution to the state equation (4.51) for this disturbance. Then we have $\mathbf{v}^T(k)\Phi\mathbf{v}(k) > \mathbf{0}$ for $\mathbf{v} \triangleq [\ \mathbf{x}^T\ \ \mathbf{w}^T\ ]^T$, or equivalently,

$$\mathbf{x}^T(k)\mathbf{Q}^{-1}\mathbf{x}(k) - \mathbf{x}^T(k+1)\mathbf{Q}^{-1}\mathbf{x}(k+1) + \|\mathbf{w}(k)\|^2 > 0.$$

Taking the summation over $k = 0, 1, \ldots, n-1$, we have

$$\mathbf{x}^T(n)\mathbf{Q}^{-1}\mathbf{x}(n) < \sum_{k=0}^{n-1} \|\mathbf{w}(k)\|^2$$

where we used $\mathbf{x}(0) = \mathbf{0}$. Using the Schur complement formula,

$$\begin{bmatrix} \sum_{k=0}^{n-1} \|\mathbf{w}(k)\|^2 & \mathbf{x}^T(n) \\ \mathbf{x}(n) & \mathbf{Q} \end{bmatrix} > \mathbf{0},$$

which implies that

$$\begin{bmatrix} 1 & \mathbf{0} \\ \mathbf{0} & \mathbf{C} \end{bmatrix} \begin{bmatrix} \sum_{k=0}^{n-1} \|\mathbf{w}(k)\|^2 & \mathbf{x}^T(n) \\ \mathbf{x}(n) & \mathbf{Q} \end{bmatrix} \begin{bmatrix} 1 & \mathbf{0} \\ \mathbf{0} & \mathbf{C}^T \end{bmatrix} + \begin{bmatrix} \mathbf{w}^T(n) \\ \mathbf{D} \end{bmatrix} \begin{bmatrix} \mathbf{w}(n) & \mathbf{D}^T \end{bmatrix} \geq \mathbf{0}$$

or equivalently,

$$\begin{bmatrix} \sum_{k=0}^{n-1} \|\mathbf{w}(k)\|^2 & \mathbf{y}^T(n) \\ \mathbf{y}(n) & \mathbf{CQC}^T + \mathbf{DD}^T \end{bmatrix} \geq \mathbf{0}.$$

Using the Schur complement formula again,

$$\mathbf{y}^T(n)\mathbf{y}(n) \leq \left(\sum_{k=0}^{n-1} \|\mathbf{w}(k)\|^2 \right) (\mathbf{CQC}^T + \mathbf{DD}^T) \leq \|\mathbf{w}\|_{\ell_2}^2 \|\mathbf{CQC}^T + \mathbf{DD}^T\| \mathbf{I}.$$

Hence, after a few more manipulations, we have

$$\|\mathbf{y}(n)\|^2 \leq \|\mathbf{w}\|_{\ell_2}^2 \|\mathbf{CQC}^T + \mathbf{DD}^T\|$$

for all $n \geq 1$ and $\mathbf{Q}$ satisfying the Lyapunov inequality in (4.51). Note that, for $n = 0$,

$$\|\mathbf{y}(n)\|^2 = \mathbf{w}^T(0)\mathbf{D}^T\mathbf{D}\mathbf{w}(0) \leq \|\mathbf{DD}^T\| \|\mathbf{w}(0)\|^2 \leq \|\mathbf{CQC}^T + \mathbf{DD}^T\|.$$

Thus we conclude that

$$\Upsilon_{ep} \leq \inf_{\mathbf{Q}} \{ \|\mathbf{CQC}^T + \mathbf{DD}^T\|^{1/2} : \mathbf{Q} > \mathbf{AQA}^T + \mathbf{BB}^T \} = \|\mathbf{CXC}^T + \mathbf{DD}^T\|^{1/2} \quad (4.52)$$

where $\mathbf{X}$ is the solution to the Lyapunov equation in (4.50). In the above, the last equality can be verified by a similar argument to that used in the proof of Theorem 4.6.1.

We now prove that $\Upsilon_{ep} \geq \|\mathbf{CXC}^T + \mathbf{DD}^T\|^{1/2}$ by exhibiting a worst-case disturbance. Consider

$$w_n(k) \triangleq \begin{cases} \lambda_n^{-1/2}\mathbf{B}^T(\mathbf{A}^T)^{n-k}\mathbf{C}^T\mathbf{v}_n & (k = 0, 1, \ldots, n) \\ \lambda_n^{-1/2}\mathbf{D}^T\mathbf{v}_n & (k = n+1) \\ 0 & (k = n+1, n+2, \ldots) \end{cases} \quad (4.53)$$

where $\mathbf{v}_n$ is the eigenvector of $\mathbf{CX}_n\mathbf{C}^T + \mathbf{DD}^T$ corresponding to the largest eigenvalue λ_n, where

$$\mathbf{X}_n \triangleq \sum_{k=0}^{n} \mathbf{A}^k\mathbf{BB}^T(\mathbf{A}^T)^k.$$

For any fixed integer $n \geq 0$, we have

$$\begin{aligned} \|\mathbf{w}_n\|_{\ell_2}^2 &= \sum_{k=0}^{n} \frac{1}{\lambda_n}\mathbf{v}_n^T\mathbf{CA}^{n-k}\mathbf{BB}^T(\mathbf{A}^T)^{n-k}\mathbf{C}^T\mathbf{v}_n + \frac{1}{\lambda_n}\mathbf{v}_n^T\mathbf{DD}^T\mathbf{v}_n \\ &= \frac{1}{\lambda_n}\mathbf{v}_n^T(\mathbf{CX}_n\mathbf{C}^T + \mathbf{DD}^T)\mathbf{v}_n = 1. \end{aligned}$$

With this unit energy disturbance, the output signal $\mathbf{y}_n(k)$ at $k = n + 1$ is given by

$$\begin{aligned} \mathbf{y}_n(n+1) &= \mathbf{C}\left(\sum_{k=0}^{n} \mathbf{A}^{n-k}\mathbf{B}(\lambda_n^{-1/2}\mathbf{B}^T(\mathbf{A}^T)^{n-k}\mathbf{C}^T\mathbf{v}_n)\right) + \mathbf{D}\left(\lambda_n^{-1/2}\mathbf{D}^T\mathbf{v}_n\right) \\ &= \lambda_n^{-1/2}(\mathbf{C}\mathbf{X}_n\mathbf{C}^T + \mathbf{D}\mathbf{D}^T)\mathbf{v}_n. \end{aligned}$$

Therefore,

$$\|\mathbf{y}_n(n+1)\|^2 = \frac{1}{\lambda_n}\mathbf{v}_n^T(\mathbf{C}\mathbf{X}_n\mathbf{C}^T + \mathbf{D}\mathbf{D}^T)^2\mathbf{v}_n = \lambda_n.$$

Taking the limit,

$$\lim_{n\to\infty} \|\mathbf{y}_n(n+1)\|^2 = \|\mathbf{C}\mathbf{X}\mathbf{C}^T + \mathbf{D}\mathbf{D}^T\|.$$

Thus, we established that $\Upsilon_{ep} \geq \|\mathbf{C}\mathbf{X}\mathbf{C}^T + \mathbf{D}\mathbf{D}^T\|^{1/2}$. From this inequality and (4.52), we conclude the result. □

The energy-to-energy gain Υ_{ee} can be characterized by the following algebraic conditions.

Theorem 4.6.6 *Consider system (4.47) and let a scalar $\gamma > 0$ be given. Suppose the system is asymptotically stable. Then the following statements are equivalent.*

(i) $\Upsilon_{ee} < \gamma$.

(ii) There exists a solution $\mathbf{Y} = \mathbf{Y}^T \geq \mathbf{0}$ to the Riccati equation

$$\mathbf{Y} = \mathbf{A}^T\mathbf{Y}\mathbf{A} + (\mathbf{A}^T\mathbf{Y}\mathbf{B} + \mathbf{C}^T\mathbf{D})(\gamma^2\mathbf{I} - \mathbf{B}^T\mathbf{Y}\mathbf{B} - \mathbf{D}^T\mathbf{D})^{-1}(\mathbf{A}^T\mathbf{Y}\mathbf{B} + \mathbf{C}^T\mathbf{D})^T + \mathbf{C}^T\mathbf{C}$$

such that

$$\mathbf{R} \triangleq \gamma^2\mathbf{I} - \mathbf{B}^T\mathbf{Y}\mathbf{B} - \mathbf{D}^T\mathbf{D} > \mathbf{0}$$

and all the eigenvalues of $\mathbf{A} + \mathbf{B}\mathbf{R}^{-1}(\mathbf{B}^T\mathbf{Y}\mathbf{A} + \mathbf{D}^T\mathbf{C})$ have (strictly) negative real parts.

(iii) There exists a symmetric matrix $\mathbf{P} > \mathbf{0}$ such that

$$\mathbf{P} > \mathbf{A}^T\mathbf{P}\mathbf{A} + (\mathbf{A}^T\mathbf{P}\mathbf{B} + \mathbf{C}^T\mathbf{D})(\gamma^2\mathbf{I} - \mathbf{B}^T\mathbf{P}\mathbf{B} - \mathbf{D}^T\mathbf{D})^{-1}(\mathbf{A}^T\mathbf{P}\mathbf{B} + \mathbf{C}^T\mathbf{D})^T + \mathbf{C}^T\mathbf{C},$$

$$\gamma^2\mathbf{I} - \mathbf{B}^T\mathbf{P}\mathbf{B} - \mathbf{C}^T\mathbf{C} > \mathbf{0}.$$

(iv) There exists a symmetric matrix $\mathbf{P} > \mathbf{0}$ such that

$$\begin{bmatrix} \mathbf{P} & \mathbf{0} \\ \mathbf{0} & \gamma^2\mathbf{I} \end{bmatrix} > \begin{bmatrix} \mathbf{A} & \mathbf{B} \\ \mathbf{C} & \mathbf{D} \end{bmatrix}^T \begin{bmatrix} \mathbf{P} & \mathbf{0} \\ \mathbf{0} & \mathbf{I} \end{bmatrix} \begin{bmatrix} \mathbf{A} & \mathbf{B} \\ \mathbf{C} & \mathbf{D} \end{bmatrix}. \tag{4.54}$$

Proof. Here, we shall give a simple proof for (iii) ⇔ (iv) ⇒ (i). Detailed proofs for other implications may be found in [32, 96].

First note that the Riccati inequality in statement (iii) and the linear matrix inequality in statement (iv) are related by the Schur complement formula, and hence equivalent. To prove (iv) ⇒ (i), consider any square summable disturbance signal $\mathbf{w}$ (not identically zero) and the corresponding state $\mathbf{x}$ satisfying the difference equation (4.47) with zero initial state. Define a vector signal $\mathbf{v} \triangleq [\ \mathbf{x}^T\ \ \mathbf{w}^T\ ]^T$. Then multiplying inequality (4.54) by $\mathbf{v}^T(k)$ from the left and by $\mathbf{v}(k)$ from the right, we have

$$\mathbf{x}^T(k)\mathbf{P}\mathbf{x}(k) + \gamma^2\|\mathbf{w}(k)\|^2 > \mathbf{x}^T(k+1)\mathbf{P}\mathbf{x}(k+1) + \|\mathbf{y}(k)\|^2.$$

Taking the summation from $k = 0$ to ∞ with zero initial state,

$$\gamma^2 \|\mathbf{w}\|_{\ell_2}^2 > \|\mathbf{y}\|_{\ell_2}^2,$$

where we used the stability property $\lim_{k\to\infty} \mathbf{x}(k) = 0$. Thus, we conclude that $\Upsilon_{ee} < \gamma$. □

As in the continuous-time case, the energy-to-energy gain Υ_{ee} has a frequency domain interpretation:

$$\Upsilon_{ee} = \|\mathbf{T}\|_{\mathcal{H}_\infty} \triangleq \max_{0\le\theta\le 2\pi} \|\mathbf{T}(e^{j\theta})\|$$

where

$$\mathbf{T}(z) \triangleq \mathbf{C}(z\mathbf{I} - \mathbf{A})^{-1}\mathbf{B} + \mathbf{D}.$$

The quantity $\|\mathbf{T}\|_{\mathcal{H}_\infty}$ is called the $\mathcal{H}_\infty$ norm of the (discrete-time) transfer matrix $\mathbf{T}(z)$. We will show in the next section that $\|\mathbf{T}\|_{\mathcal{H}_\infty}$ is related to robustness of the system (4.47) subject to perturbation $\mathbf{\Delta}$ that affects the system dynamics by $\mathbf{w} = \mathbf{\Delta y}$.

Example 4.6.2 Consider the stable continuous-time system described by

$$\hat{\mathbf{A}} = \begin{bmatrix} -1 & 0 \\ 1 & -2 \end{bmatrix}, \quad \hat{\mathbf{B}} = \begin{bmatrix} 1 & 0 \\ 0 & 2 \end{bmatrix}.$$

This system is discretized by using the zero-order hold with the sampling time $T = 0.5$ to get

$$\mathbf{A} = \begin{bmatrix} 0.9048 & 0 \\ 0.0861 & 0.8187 \end{bmatrix}, \quad \mathbf{B} = \begin{bmatrix} 0.0952 & 0 \\ 0.0045 & 0.1813 \end{bmatrix}.$$

Suppose that $\mathbf{C}$ and $\mathbf{D}$ matrices are given by $\mathbf{C} = \mathbf{I}_2$ and $\mathbf{D} = \mathbf{0}$, respectively. In this case, the energy-to-peak gain Υ_{ep} can be computed by solving the Lyapunov equation in (4.50). The solution $\mathbf{X}$ is found to be

$$\mathbf{X} = \begin{bmatrix} 0.2449 & 0.0848 \\ 0.0848 & 0.5024 \end{bmatrix}$$

and the energy-to-peak gain is computed as $\Upsilon_{ep} = 0.7265$. Note that the $\mathcal{H}_2$ norm of this system is given by 0.8645 and is different from Υ_{ep} since there is more than one output.

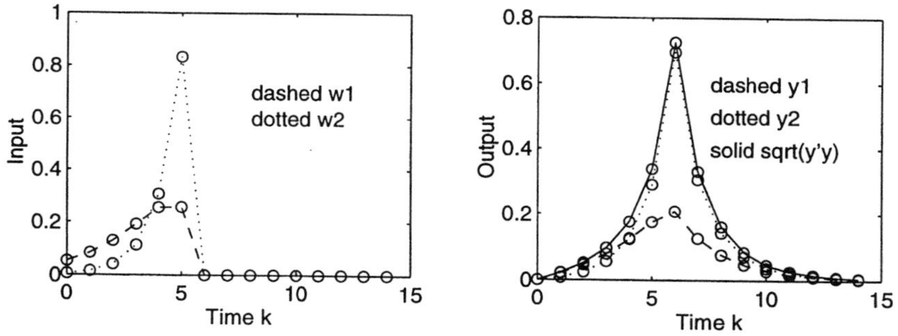

Figure 4.2 Υ_{ep} worst-case disturbance and system responses.

Table 4.3 Pulse-to-energy gains and the worst-case disturbance directions

	Υ_{pe}	$\mathbf{w}_0$
Case: $\mathbf{C}$	0.7065	$\begin{bmatrix} 0.3844 \\ 0.9232 \end{bmatrix}$
Case: $\mathbf{C}_1$	0.4949	$\begin{bmatrix} 1 \\ 0 \end{bmatrix}$
Case: $\mathbf{C}_2$	0.6929	$\begin{bmatrix} -0.1985 \\ -0.9801 \end{bmatrix}$

The worst-case disturbance $\mathbf{w}(k)$ is computed using the formula in (4.53) with $n = 5$, and is plotted together with the system response to this disturbance in Figure 4.2. The solid line in the figure exhibits the time history of $\|\mathbf{y}(k)\| = (\mathbf{y}(k)^T\mathbf{y}(k))^{1/2}$. Note that the peak value of $\|\mathbf{y}(k)\|$ occurs at $k = 6$ with $\|\mathbf{y}(6)\| = 0.7259$ which is very close to Υ_{ep}. By using a larger n, the peak value of $\|\mathbf{y}(k)\|$ can be made arbitrarily close to Υ_{ep}. From Figure 4.2, we see that the worst-case disturbance $\mathbf{w}$ is a monotonically increasing function up to the time $k = 5$, just before the output peak occurs.

Now we compute the pulse-to-energy gain of the system. Three cases are considered: $\mathbf{y} = \mathbf{x}$, $\mathbf{y} = x_1$, and $\mathbf{y} = x_2$, or equivalently, we consider $\mathbf{C} = \mathbf{I}_2$, $\mathbf{C}_1 = [\ 1\ \ 0\ ]$, and $\mathbf{C}_2 = [\ 0\ \ 1\ ]$, with $\mathbf{D} = \mathbf{0}$ for all three cases. For each case, the pulse-to-energy gain Υ_{pe} is computed by solving the Lyapunov equation in (4.48). The result is summarized in Table 4.3 together with the worst-case direction of the pulse disturbance $\mathbf{w}_0$. Recall from the proof of Theorem 4.6.4 that the worst-case direction is given by the eigenvector of $\mathbf{B}^T\mathbf{Y}\mathbf{B}+\mathbf{D}^T\mathbf{D}$ where $\mathbf{Y}$ is the solution to the Lyapunov equation (4.48).

The trajectory of the pulse response for each case is plotted in the phase plane (x_1–x_2 plane) in Figure 4.3. The dashed lines indicate the directions of the eigenvectors of $\mathbf{A}$:

$$\lambda_1 = 0.8187 \text{ with } e_1 = \begin{bmatrix} 0 \\ 1 \end{bmatrix} \text{ and } \lambda_2 = 0.9048 \text{ with } e_2 = \begin{bmatrix} 1 \\ 1 \end{bmatrix}$$

where λ_i are the eigenvalues and e_i are the corresponding eigenvectors. Note that e_1 is the direction of the faster mode. Recall that the pulse response for $\mathbf{w}(k) = \mathbf{w}_0\delta(k)$ with zero initial state is exactly the same as the initial state response with $\mathbf{x}(0) = \mathbf{B}\mathbf{w}_0$ and no external disturbances. The possible location of the initial state $\mathbf{x}(0)$ corresponding to $\|\mathbf{w}_0\| \leq 1$ is given by the region inside the ellipse; $\mathbf{x}(0)^T(\mathbf{B}\mathbf{B}^T)^{-1}\mathbf{x}(0) \leq 1$ (see Figure 4.3). Thus, the worst-case state trajectory must start from a point on the ellipse and converge to the origin. From Figure 4.3, we see that, if the pulse disturbance tries to maximize the ℓ_2 norm of x_1 (Case: $\mathbf{C}_1$), the worst-case disturbance pushes $\mathbf{x}(0)$ close to the point $(x_1, x_2) = (0.1, 0)$, and the corresponding state trajectory is almost "horizontal" since the disturbance is chosen without caring for the size of x_2. Similarly, if the ℓ_2 norm of x_2 is of our interest (Case: $\mathbf{C}_2$), the worst-case trajectory becomes almost "vertical".

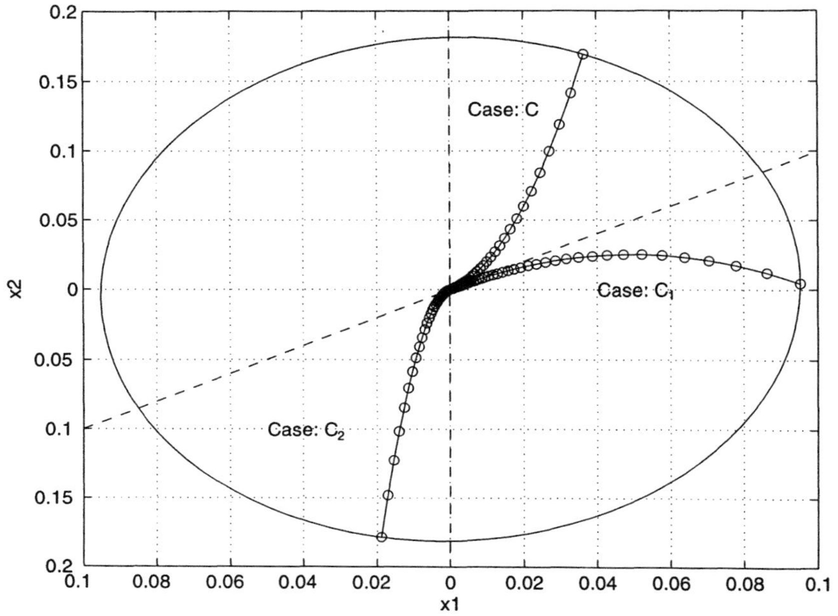

Figure 4.3 Υ_{pe} worst-case state trajectories.

4.7 ROBUST STABILITY AND PERFORMANCE ANALYSIS

4.7.1 Continuous-Time Systems

It is well known [126] that there are always four kinds of errors associated with a linear model: errors in model order, errors in disturbances, errors in nonlinearities, and errors in parameters. Furthermore, there is no control theory which guarantees stability or performance nonconservatively in the presence of all four of these error types. Below, we focus on parameter errors to motivate the analysis of uncertain systems.

Let a model of a linear system be described by

$$\begin{aligned} \dot{\mathbf{x}} &= \hat{\mathbf{A}}\mathbf{x} + \hat{\mathbf{K}}\mathbf{w} \\ \mathbf{y} &= \hat{\mathbf{M}}\mathbf{x} + \hat{\mathbf{H}}\mathbf{w}. \end{aligned} \tag{4.55}$$

Suppose that the system parameters $\hat{\mathbf{A}}$, $\hat{\mathbf{K}}$, $\hat{\mathbf{M}}$, $\hat{\mathbf{H}}$ are subject to the two types of uncertainties: scalar and matrix uncertainties, δ and $\boldsymbol{\Delta}$, respectively.

Example 4.7.1 The system matrix $\mathbf{A}$ may be given by

$$\mathbf{A} = \begin{bmatrix} 0 & 1 \\ -\delta_1 & -\delta_2 \end{bmatrix} = \begin{bmatrix} 0 & 1 \\ 0 & 0 \end{bmatrix} + \delta_1 \begin{bmatrix} 0 & 0 \\ -1 & 0 \end{bmatrix} + \delta_2 \begin{bmatrix} 0 & 0 \\ 0 & -1 \end{bmatrix}$$

where δ_1 and δ_2 are the uncertain parameters.

Example 4.7.2 Consider the vector second-order system described by $\hat{\mathbf{M}}\ddot{\mathbf{q}} + \hat{\mathbf{K}}\mathbf{q} = \hat{\mathbf{B}}\mathbf{u}$. Suppose that the stiffness matrix $\hat{\mathbf{K}}$ is subject to an uncertainty $\boldsymbol{\Delta}$ as in $\hat{\mathbf{K}} = \mathbf{K}_0 + \mathbf{L}\boldsymbol{\Delta}\mathbf{R}$.

The system can be put into the state space representation $\dot{\mathbf{x}} = \mathbf{A}\mathbf{x} + \mathbf{B}\mathbf{u}$ as follows:

$$\mathbf{A} = \begin{bmatrix} \mathbf{0} & \mathbf{I} \\ -\hat{\mathbf{M}}^{-1}\hat{\mathbf{K}} & \mathbf{0} \end{bmatrix}, \quad \mathbf{B} = \begin{bmatrix} \mathbf{0} \\ \hat{\mathbf{M}}^{-1}\hat{\mathbf{B}} \end{bmatrix}.$$

Note that the system matrix $\mathbf{A}$ depends linearly on the uncertainty $\boldsymbol{\Delta}$ and can be written as

$$\mathbf{A} = \begin{bmatrix} \mathbf{0} & \mathbf{I} \\ -\hat{\mathbf{M}}^{-1}\mathbf{K}_0 & \mathbf{0} \end{bmatrix} + \begin{bmatrix} \mathbf{0} \\ -\hat{\mathbf{M}}^{-1}\mathbf{L} \end{bmatrix} \boldsymbol{\Delta} \begin{bmatrix} \mathbf{R} & \mathbf{0} \end{bmatrix}.$$

Now let $\hat{\mathbf{A}}$, $\hat{\mathbf{K}}$, $\hat{\mathbf{M}}$, $\hat{\mathbf{H}}$ have the structure

$$\begin{aligned}
\hat{\mathbf{A}} &= \mathbf{A} + \sum_i \delta_i^A \mathbf{A}_i + \sum_i \mathbf{A}_i^L \boldsymbol{\Delta}_i^A \mathbf{A}_i^R = \mathbf{A} + \mathbf{A}_L \boldsymbol{\Delta}_A \mathbf{A}_R \\
\hat{\mathbf{K}} &= \mathbf{K} + \sum_i \delta_i^K \mathbf{K}_i + \sum_i \mathbf{K}_i^L \boldsymbol{\Delta}_i^K \mathbf{K}_i^R = \mathbf{K} + \mathbf{K}_L \boldsymbol{\Delta}_K \mathbf{K}_R \\
\hat{\mathbf{M}} &= \mathbf{M} + \sum_i \delta_i^M \mathbf{M}_i + \sum_i \mathbf{M}_i^L \boldsymbol{\Delta}_i^M \mathbf{M}_i^R = \mathbf{M} + \mathbf{M}_L \boldsymbol{\Delta}_M \mathbf{M}_R \\
\hat{\mathbf{H}} &= \mathbf{H} + \sum_i \delta_i^H \mathbf{H}_i + \sum_i \mathbf{H}_i^L \boldsymbol{\Delta}_i^H \mathbf{H}_i^R = \mathbf{H} + \mathbf{H}_L \boldsymbol{\Delta}_H \mathbf{H}_R
\end{aligned} \tag{4.56}$$

where $\mathbf{A}_L$, $\mathbf{A}_R$ and $\boldsymbol{\Delta}_A$ are defined by

$$\hat{\mathbf{A}} = \mathbf{A} + \begin{bmatrix} \mathbf{I} & \cdots & \mathbf{A}_1^L & \cdots \end{bmatrix} \begin{bmatrix} \delta_1^A \mathbf{I} & & & \\ & \ddots & & \\ & & \boldsymbol{\Delta}_1^A & \\ & & & \ddots \end{bmatrix} \begin{bmatrix} \mathbf{A}_1 \\ \vdots \\ \mathbf{A}_1^R \\ \vdots \end{bmatrix} = \mathbf{A} + \mathbf{A}_L \boldsymbol{\Delta}_A \mathbf{A}_R$$

and similarly for $\hat{\mathbf{K}}$, $\hat{\mathbf{M}}$, and $\hat{\mathbf{H}}$. It is then clear that, for the choices for $\mathbf{C}$ and $\mathbf{L}$ indicated below, we have the following description for the uncertain system:

$$\begin{aligned}
\dot{\mathbf{x}} &= \mathbf{A}\mathbf{x} + \mathbf{K}\mathbf{w} + \begin{bmatrix} \mathbf{A}_L & \mathbf{K}_L \end{bmatrix} \begin{bmatrix} \boldsymbol{\Delta}_A & \mathbf{0} \\ \mathbf{0} & \boldsymbol{\Delta}_K \end{bmatrix} \left(\begin{bmatrix} \mathbf{A}_R \\ \mathbf{0} \end{bmatrix} \mathbf{x} + \begin{bmatrix} \mathbf{0} \\ \mathbf{K}_R \end{bmatrix} \mathbf{w} \right) \\
&= \mathbf{A}\mathbf{x} + \mathbf{K}\mathbf{w} + \mathbf{B}\boldsymbol{\Delta}(\mathbf{C}\mathbf{x} + \mathbf{L}\mathbf{w}) \\
&= \mathbf{A}\mathbf{x} + \mathbf{K}\mathbf{w} + \mathbf{B}\boldsymbol{\omega}, \qquad \boldsymbol{\omega} = \boldsymbol{\Delta}\boldsymbol{\psi}, \qquad \boldsymbol{\psi} = \mathbf{C}\mathbf{x} + \mathbf{L}\mathbf{w}.
\end{aligned}$$

Similar development allows $\boldsymbol{\Delta}$ to contain uncertain parameters in all input, output matrices of (4.55). This motivates the model for a general class of uncertain systems given below.

We consider the following class of uncertain systems:

$$\begin{bmatrix} \dot{\mathbf{x}}(t) \\ \boldsymbol{\psi}(t) \\ \mathbf{y}(t) \end{bmatrix} = \begin{bmatrix} \mathbf{A} & \mathbf{B} & \mathbf{K} \\ \mathbf{C} & \mathbf{D} & \mathbf{L} \\ \mathbf{M} & \mathbf{N} & \mathbf{H} \end{bmatrix} \begin{bmatrix} \mathbf{x}(t) \\ \boldsymbol{\omega}(t) \\ \mathbf{w}(t) \end{bmatrix}, \quad \boldsymbol{\omega}(t) = \boldsymbol{\Delta}(t)\boldsymbol{\psi}(t), \tag{4.57}$$

where $\mathbf{x}$ is the state, $\mathbf{w}$ is the disturbance, $\mathbf{y}$ is the output of interest, $\boldsymbol{\psi}$ and $\boldsymbol{\omega}$ are the exogenous signals to describe the uncertainty $\boldsymbol{\Delta}$. The nominal system ($\boldsymbol{\Delta} \equiv \mathbf{0}$) is linear time-invariant, and the uncertainty $\boldsymbol{\Delta}$ is assumed to belong to the following set of norm-bounded time-varying structured uncertainties;

$$\mathcal{BU}_C \triangleq \{ \boldsymbol{\Delta} : \mathcal{R} \to \mathcal{R}^{m \times m}, \ \ \|\boldsymbol{\Delta}(t)\| \le 1, \ \ \boldsymbol{\Delta}(t) \in \mathcal{U} \},$$

where

$$\mathcal{U} \triangleq \{ \text{ block diag}(\delta_1 \mathbf{I}_{k_1} \cdots \delta_s \mathbf{I}_{k_s} \ \mathbf{\Delta}_1 \cdots \mathbf{\Delta}_f) : \ \delta_i \in \mathcal{R}, \ \ \mathbf{\Delta}_i \in \mathcal{R}^{k_{s+i} \times k_{s+i}} \ \}. \tag{4.58}$$

Note that the sizes of the (square) sub-blocks add up to m:

$$\sum_{i=1}^{s+f} k_i = m.$$

By considering the "square" uncertainty $\mathbf{\Delta}$, we have implicitly assumed that the dimensions of vector signals $\mathbf{w}$ and $\mathbf{y}$ are equal. This assumption can always be satisfied by adding rows or columns to appropriate system matrices in (4.57).

Recall that the nominal system ($\mathbf{\Delta} \equiv \mathbf{0}$) is (asymptotically) stable if $\mathbf{x}(t)$ approaches zero as $t \to \infty$ regardless of the initial state $\mathbf{x}(0)$. The first objective in this section is to extend this stability concept to the uncertain system (4.57).

Definition 4.7.1 The uncertain system (4.57) is said to be robustly stable if $\lim_{t\to\infty} \mathbf{x}(t) = \mathbf{0}$ regardless of the initial state $\mathbf{x}(0)$, and the uncertainty $\mathbf{\Delta} \in \mathcal{BU}_C$.

Note that the robust stability property of (4.57) is completely determined by matrices $\mathbf{A}$, $\mathbf{B}$, $\mathbf{C}$ and $\mathbf{D}$ and $\mathcal{BU}_C$. In this case, ignoring the other matrices, the uncertain system (4.57) can be described by

$$\dot{\mathbf{x}} = (\mathbf{A} + \mathbf{B}\mathbf{\Delta}(\mathbf{I} - \mathbf{D}\mathbf{\Delta})^{-1}\mathbf{C})\mathbf{x}.$$

We would like to find a necessary and sufficient condition for robust stability. However, the problem is difficult and no results are available in the literature (see [23, 89, 124, 142] for conditions in the case of dynamic uncertainty).

In the sequel, we shall develop a *sufficient* condition for robust stability. First, let us consider the simple special case where the uncertainty is unstructured, i.e. $\mathcal{U}$ in (4.58) is just a set of full block matrices ($s = 0$, $f = 1$ and $\mathcal{U} = \mathcal{R}^{m\times m}$).

Theorem 4.7.1 (Small Gain Theorem) *Let* $\mathbf{T}(s) \triangleq \mathbf{C}(s\mathbf{I}-\mathbf{A})^{-1}\mathbf{B}+\mathbf{D}$ *and suppose* $\|\mathbf{T}\|_\infty < 1$, *i.e. there exists* $\mathbf{P} > \mathbf{0}$ *such that*

$$\begin{bmatrix} \mathbf{PA} + \mathbf{A}^T\mathbf{P} & \mathbf{PB} \\ \mathbf{B}^T\mathbf{P} & -\mathbf{I} \end{bmatrix} + \begin{bmatrix} \mathbf{C}^T \\ \mathbf{D}^T \end{bmatrix} \begin{bmatrix} \mathbf{C} & \mathbf{D} \end{bmatrix} < \mathbf{0}. \tag{4.59}$$

Then the uncertain system (4.57) is robustly stable for all $\mathbf{\Delta} \in \mathcal{BU}_C$ *with* $\mathcal{U} \triangleq \mathcal{R}^{m\times m}$.

Proof. Let the uncertainty $\mathbf{\Delta} \in \mathcal{BU}_C$ be given and fixed (but unknown). Let $\mathbf{x}$, ω and ψ be any signals that satisfy (4.57) for some nonzero initial state $\mathbf{x}(0)$ with $\mathbf{w}(t) \equiv \mathbf{0}$. Then, from (4.59),

$$\begin{bmatrix} \mathbf{x}^T(t) & \omega^T(t) \end{bmatrix} \left(\begin{bmatrix} \mathbf{PA} + \mathbf{A}^T\mathbf{P} & \mathbf{PB} \\ \mathbf{B}^T\mathbf{P} & -\mathbf{I} \end{bmatrix} + \begin{bmatrix} \mathbf{C}^T \\ \mathbf{D}^T \end{bmatrix} \begin{bmatrix} \mathbf{C} & \mathbf{D} \end{bmatrix} \right) \begin{bmatrix} \mathbf{x}(t) \\ \omega(t) \end{bmatrix} < 0.$$

Using (4.57), we have

$$\mathbf{x}^T(t)\mathbf{P}\dot{\mathbf{x}}(t) + \dot{\mathbf{x}}^T(t)\mathbf{P}\mathbf{x}(t) - \omega^T(t)\omega(t) + \psi^T(t)\psi(t) < 0.$$

Note that, for $\mathbf{\Delta} \in \mathcal{BU}_C$, we have $\|\mathbf{\Delta}(t)\| \le 1$ and

$$\psi^T(t)\psi(t) - \omega^T(t)\omega(t) = \psi^T(t)(I - \mathbf{\Delta}^T(t)\mathbf{\Delta}(t))\psi(t) \ge 0.$$

Hence,

$$\mathbf{x}^T(t)\mathbf{P}\dot{\mathbf{x}}(t) + \dot{\mathbf{x}}^T(t)\mathbf{P}\mathbf{x}(t) = \frac{d}{dt}(\mathbf{x}^T(t)\mathbf{P}\mathbf{x}(t)) < 0.$$

Thus, the function $V(x) \triangleq \mathbf{x}^T\mathbf{P}\mathbf{x}$ satisfies $V(\mathbf{x}) > 0, \quad \forall\ \mathbf{x} \neq \mathbf{0}$ and $\dot{V}(\mathbf{x}) < 0$, which qualifies $V(\mathbf{x})$ as a Lyapunov function to prove stability of (4.57) for any given $\boldsymbol{\Delta} \in \mathcal{BU}_C$.

□

In the above proof, a quadratic Lyapunov function $V(\mathbf{x}) = \mathbf{x}^T\mathbf{P}\mathbf{x}$ is constructed to show robust stability of the uncertain system (4.57). Note that the Lyapunov function is independent of $\boldsymbol{\Delta}$, and works for all $\boldsymbol{\Delta} \in \mathcal{BU}$. Thus, the condition $\|\mathbf{T}\|_{\mathcal{H}_\infty} < 1$ implies the existence of a *single* quadratic Lyapunov function to prove stability of the uncertain system (4.57).

Definition 4.7.2 [79, 2] The uncertain system (4.57) is said to be quadratically stable if there exists a single quadratic Lyapunov function $V(\mathbf{x}) = \mathbf{x}^T\mathbf{P}\mathbf{x}$ to prove stability for all $\boldsymbol{\Delta} \in \mathcal{BU}_C$, i.e. if $(\mathbf{I} - \mathbf{D}\boldsymbol{\Delta})$ is invertible for any $\boldsymbol{\Delta} \in \mathcal{BU}_C$ and there exists $\mathbf{P} > \mathbf{0}$ such that

$$\mathbf{P}(\mathbf{A} + \mathbf{B}\boldsymbol{\Delta}(\mathbf{I} - \mathbf{D}\boldsymbol{\Delta})^{-1}\mathbf{C}) + (\mathbf{A} + \mathbf{B}\boldsymbol{\Delta}(\mathbf{I} - \mathbf{D}\boldsymbol{\Delta})^{-1}\mathbf{C})^T\mathbf{P} < \mathbf{0}, \quad \forall\ \boldsymbol{\Delta} \in \mathcal{BU}_C.$$

In general, every quadratically stable system is robustly stable, but the converse is not necessarily true. The gap is due to the requirement that the Lyapunov function be independent of the uncertainty. In view of Theorem 4.7.1 and its proof, the $\mathcal{H}_\infty$ norm condition $\|\mathbf{T}\|_{\mathcal{H}_\infty} < 1$ implies quadratic stability of (4.57). In fact, it is known [71] that the converse is also true, that is, if (4.57) is quadratically stable, then $\|\mathbf{T}\|_{\mathcal{H}_\infty} < 1$ holds. In summary, for the case of unstructured uncertainty, we have

$$\|\mathbf{T}\|_{\mathcal{H}_\infty} < 1 \quad \Leftrightarrow \text{ Quadratic Stability } \Rightarrow \text{ Robust Stability.}$$

Next, consider the *structured* uncertainty $\mathcal{BU}$ with $\mathcal{U}$ given by (4.58). In this case, in general, quadratic stability does not necessarily imply $\|\mathbf{T}\|_{\mathcal{H}_\infty} < 1$ [112], and we have

$$\|\mathbf{T}\|_{\mathcal{H}_\infty} < 1 \quad \Rightarrow \text{ Quadratic Stability } \Rightarrow \text{ Robust Stability.}$$

Thus, the Small Gain condition $\|\mathbf{T}\|_{\mathcal{H}_\infty} < 1$ guarantees robust stability of (4.57), but it may be very conservative. The conservatism due to the second "⇒" is by definition, and that due to the first "⇒" is (partly) caused by the fact that the available structure information on $\boldsymbol{\Delta}$ is totally ignored.

A standard approach [23, 115] to fill conservatism by taking the structure information into account, is to introduce the following subset of positive definite matrices:

$$\mathcal{S} \triangleq \{\text{ block diag}(\mathbf{S}_1 \cdots \mathbf{S}_s\ s_1\mathbf{I}_{k_{s+1}} \cdots s_f\mathbf{I}_{k_{s+f}}) : \ \mathbf{S}_i \in \mathcal{R}^{k_i \times k_i}, \quad s_i \in \mathcal{R}, \quad \mathbf{S}_i > \mathbf{0}, \quad s_i > 0\,\}.$$

Clearly, for each $\mathbf{S} \in \mathcal{S}$, we have

$$\boldsymbol{\Delta} = \mathbf{S}^{-1/2}\boldsymbol{\Delta}\mathbf{S}^{1/2}, \quad \forall\ \boldsymbol{\Delta} \in \mathcal{U}$$

where $\mathbf{S}^{1/2}$ denotes the positive definite square root of $\mathbf{S}$, and is introduced for later convenience. Hence, we can replace $\boldsymbol{\Delta}$ in (4.57) by $\mathbf{S}^{-1/2}\boldsymbol{\Delta}\mathbf{S}^{1/2}$ to obtain a new representation of the same uncertain system (4.57):

$$\begin{bmatrix} \dot{\mathbf{x}} \\ \hat{\psi} \end{bmatrix} = \begin{bmatrix} \hat{\mathbf{A}} & \hat{\mathbf{B}} \\ \hat{\mathbf{C}} & \hat{\mathbf{D}} \end{bmatrix} \begin{bmatrix} \mathbf{x} \\ \hat{\omega} \end{bmatrix}, \quad \hat{\omega} = \boldsymbol{\Delta}\hat{\psi} \tag{4.60}$$

where

$$\begin{bmatrix} \hat{\mathbf{A}} & \hat{\mathbf{B}} \\ \hat{\mathbf{C}} & \hat{\mathbf{D}} \end{bmatrix} \triangleq \begin{bmatrix} \mathbf{I} & \mathbf{0} \\ \mathbf{0} & \mathbf{S}^{1/2} \end{bmatrix} \begin{bmatrix} \mathbf{A} & \mathbf{B} \\ \mathbf{C} & \mathbf{D} \end{bmatrix} \begin{bmatrix} \mathbf{I} & \mathbf{0} \\ \mathbf{0} & \mathbf{S}^{-1/2} \end{bmatrix}. \tag{4.61}$$

From the Small Gain Theorem (Theorem 4.7.1), the uncertain system (4.57) is robustly stable if $\|\hat{\mathbf{T}}\|_{\mathcal{H}_\infty} < 1$ where

$$\hat{\mathbf{T}}(s) \triangleq \hat{\mathbf{C}}(s\mathbf{I} - \hat{\mathbf{A}})^{-1}\hat{\mathbf{B}} + \hat{\mathbf{D}}.$$

Note that $\hat{\mathbf{T}} = \mathbf{S}^{1/2}\mathbf{T}\mathbf{S}^{-1/2}$ and $\mathbf{S}$ could be any matrix such that $\mathbf{S} \in \mathcal{S}$. Thus, this freedom can be used to reduce the conservativeness. The following definition is due to [26].

Definition 4.7.3 The uncertain linear system (4.57) is said to be *$\mathcal{Q}$-stable* if there exists a matrix $\mathbf{S} \in \mathcal{S}$ such that $\|\mathbf{S}^{1/2}\mathbf{T}\mathbf{S}^{-1/2}\|_{\mathcal{H}_\infty} < 1$.

A characterization of $\mathcal{Q}$-stability can be obtained by replacing matrices $\mathbf{A}$, $\mathbf{B}$, $\mathbf{C}$ and $\mathbf{D}$ in (4.59) with $\hat{\mathbf{A}}$, $\hat{\mathbf{B}}$, $\hat{\mathbf{C}}$ and $\hat{\mathbf{D}}$ defined in (4.61):

$$\begin{bmatrix} \mathbf{PA} + \mathbf{A}^T\mathbf{P} & \mathbf{PBS}^{-1/2} \\ \mathbf{S}^{-1/2}\mathbf{B}^T\mathbf{P} & -\mathbf{I} \end{bmatrix} + \begin{bmatrix} \mathbf{C}^T\mathbf{S}^{1/2} \\ \mathbf{S}^{-1/2}\mathbf{D}^T\mathbf{S}^{1/2} \end{bmatrix} \begin{bmatrix} \mathbf{S}^{1/2}\mathbf{C} & \mathbf{S}^{1/2}\mathbf{D}\mathbf{S}^{-1/2} \end{bmatrix} < \mathbf{0}.$$

Using the congruent transformation with block diag $(\mathbf{I}, \mathbf{S}^{1/2})$, we have

$$\begin{bmatrix} \mathbf{PA} + \mathbf{A}^T\mathbf{P} & \mathbf{PB} \\ \mathbf{B}^T\mathbf{P} & -\mathbf{S} \end{bmatrix} + \begin{bmatrix} \mathbf{C}^T \\ \mathbf{D}^T \end{bmatrix} \mathbf{S} \begin{bmatrix} \mathbf{C} & \mathbf{D} \end{bmatrix} < \mathbf{0}. \tag{4.62}$$

Thus, the uncertain system (4.57) is $\mathcal{Q}$-stable if and only if there exist matrices $\mathbf{P} > \mathbf{0}$ and $\mathbf{S} \in \mathcal{S}$ satisfying (4.62).

Exercise 4.7.1 (i) Show that, for each $\mathbf{S} \in \mathcal{S}$, the following holds:

$$\mathbf{S} - \mathbf{\Delta}^T(t)\mathbf{S}\mathbf{\Delta}(t) \geq \mathbf{0}, \quad \forall\, \mathbf{\Delta} \in \mathcal{BU}_C.$$

(ii) Using the result of (i), prove that $\mathcal{Q}$-stability implies quadratic stability.

Now we can summarize the above discussion as follows:

$$\|\mathbf{T}\|_{\mathcal{H}_\infty} < 1 \quad \Rightarrow \quad \mathcal{Q}\text{-stability} \quad \Rightarrow \quad \text{Quadratic Stability} \quad \Rightarrow \quad \text{Robust Stability}$$

for the system (4.57) with structured uncertainty. Note that the notion of $\mathcal{Q}$-stability has been introduced to reduce the gap between $\|\mathbf{T}\|_{\mathcal{H}_\infty} < 1$ and quadratic stability. In general, $\mathcal{Q}$-stability is more conservative than quadratic stability [97, 110]. For the case of *dynamic time-varying* uncertainty $\mathbf{\Delta}$, $\mathcal{Q}$-stability is *equivalent* to robust stability [89, 124]. Quadratic Lyapunov matrix $\mathbf{P}$ associated with $\mathcal{Q}$-stability constitutes the basis for robust *performance* analysis in the following.

Our next objective is to define performance measures by certain "sizes" of the output signal $\mathbf{y}$ in response to certain classes of the disturbances $\mathbf{w}$, in the presence of the uncertainty $\mathbf{\Delta}$, and to provide (computable) characterizations of the performance measures. We shall define three robust performance measures. All three are given by the worst-case disturbance attenuation level in the presence of the uncertainty, but the measures of the "size" of the error signal and the classes of the disturbance are different.

The first one is a generalization of the (nominal) $\mathcal{H}_2$ norm performance measure for the uncertain system. Since the uncertain system is time-varying, the transfer matrix from $\mathbf{w}$ to $\mathbf{y}$ does not exist, and hence the $\mathcal{H}_2$ norm to extend the concept of the $\mathcal{H}_2$ norm to time-varying systems is to consider the energy ($\mathcal{L}_2$ norm) of the output signal $\mathbf{y}$ in response to the impulsive disturbance $\mathbf{w}$.

Robust $\mathcal{H}_2$ Performance: Let a Linear Quadratic (LQ) cost be given by

$$\hat{J}_{\mathcal{H}_2}(\mathbf{\Delta}, \mathbf{w}_0) \triangleq \|\mathbf{y}\|^2_{\mathcal{L}_2} = \int_0^\infty \mathbf{y}^T(t)\mathbf{y}(t)dt$$

where $\mathbf{y}$ is the output signal in the presence of the uncertainty $\mathbf{\Delta}$ when the system (4.57) is excited by an impulsive disturbance $\mathbf{w}(t) = \mathbf{w}_0\delta(t)$ with a zero initial state, i.e. $\mathbf{x}(0) = \mathbf{0}$. Define the robust $\mathcal{H}_2$ performance measure as the worst-case LQ cost:

$$J_{\mathcal{H}_2} \triangleq \sup_{\mathbf{w}_0,\ \mathbf{\Delta}} \{\hat{J}_{\mathcal{H}_2}(\mathbf{\Delta}, \mathbf{w}_0) : \ \|\mathbf{w}_0\| \leq 1, \ \ \mathbf{\Delta} \in \mathcal{BU}_C\}.$$

Note that $J_{\mathcal{H}_2}$ is the worst-case cost over all possible directions of the impulsive disturbance as well as all admissible perturbations. It is also possible to define a similar robust performance measure by taking the sum of the LQ costs for impulsive disturbances applied one at a time at each disturbance channel, instead of the worst-case $\mathbf{w}_0$.

Another measure we consider for the robust disturbance attenuation is the worst-case peak value of the output signal $\mathbf{y}$ in response to the disturbance $\mathbf{w}$ with its energy being below a specified (*a priori* known) level.

Robust $\mathcal{L}_\infty$ Performance: Consider the following cost function:

$$\hat{J}_{\mathcal{L}_\infty}(\mathbf{\Delta}, \mathbf{w}) \triangleq \|\mathbf{y}\|^2_{\mathcal{L}_\infty} = \sup_{t \geq 0} \mathbf{y}^T(t)\mathbf{y}(t)$$

where $\mathbf{y}$ is the output signal in the presence of disturbance $\mathbf{w}$ and the plant perturbation $\mathbf{\Delta}$ with a zero initial condition $\mathbf{x}(0) = \mathbf{0}$. The robust $\mathcal{L}_\infty$ performance measure is defined by the worst-case peak value of the error signal subject to a finite energy disturbance as follows:

$$J_{\mathcal{L}_\infty} \triangleq \sup_{\mathbf{w},\ \mathbf{\Delta}} \{\hat{J}_{\mathcal{L}_\infty}(\mathbf{\Delta}, \mathbf{w}) : \ \|\mathbf{w}\|_{\mathcal{L}_2} \leq 1, \ \ \mathbf{\Delta} \in \mathcal{BU}_C\}.$$

Thus the robust $\mathcal{L}_\infty$ performance measure is the worst-case energy ($\mathcal{L}_2$) to peak ($\mathcal{L}_\infty$) gain of the system subject to the uncertainty $\mathbf{\Delta}$.

The last robust performance measure considered here is the worst-case energy of the output signal $\mathbf{y}$ in response to the energy bounded disturbance $\mathbf{w}$ [26].

Robust $\mathcal{H}_\infty$ Performance: Consider

$$\hat{J}_{\mathcal{H}_\infty}(\mathbf{\Delta}, \mathbf{w}) \triangleq \int_0^\infty \mathbf{y}^T(t)\mathbf{y}(t)dt$$

where $\mathbf{y}$ is the output signal in response to the disturbance $\mathbf{w}$, in the presence of the uncertainty $\mathbf{\Delta}$, with the zero initial state $\mathbf{x}(0) = \mathbf{0}$. The robust $\mathcal{H}_\infty$ performance is defined by taking the worst case of $\hat{J}_{\mathcal{H}_\infty}(\boldsymbol{\Delta}, \mathbf{w})$ over all possible uncertainty $\boldsymbol{\Delta} \in \mathcal{BU}_C$ and the disturbance $\mathbf{w}$ with bounded energy:

$$J_{\mathcal{H}_\infty} \triangleq \sup_{\mathbf{w},\ \boldsymbol{\Delta}} \{\hat{J}_{\mathcal{H}_\infty}(\boldsymbol{\Delta}, \mathbf{w}) : \ \|\mathbf{w}\|_{\mathcal{L}_2} \le 1, \quad \boldsymbol{\Delta} \in \mathcal{BU}_C\}.$$

Hence, this performance measure is a robustified version of the $\mathcal{L}_2$ gain. Since the $\mathcal{L}_2$ gain of the nominal system is given by the $\mathcal{H}_\infty$ norm, we call this the robust $\mathcal{H}_\infty$ performance.

Now, we shall present characterizations of the robust performance measures defined above. We restrict our attention to the class of $\mathcal{Q}$-stable systems, by which the finiteness of each robust performance measure is guaranteed.

The following theorem provides an upper bound on the robust $\mathcal{H}_2$ performance [85, 56, 139].

Theorem 4.7.2 *The following statements are equivalent:*

(i) The uncertain system (4.57) is $\mathcal{Q}$-stable for all $\boldsymbol{\Delta} \in \mathcal{BU}_C$.

(ii) There exist $\mathbf{P} > \mathbf{0}$ and $\mathbf{S} \in \mathcal{S}$ such that

$$\begin{bmatrix} \mathbf{PA} + \mathbf{A}^T\mathbf{P} & \mathbf{PB} \\ \mathbf{B}^T\mathbf{P} & -\mathbf{S} \end{bmatrix} + \begin{bmatrix} \mathbf{C} & \mathbf{D} \\ \mathbf{M} & \mathbf{N} \end{bmatrix}^T \begin{bmatrix} \mathbf{S} & \mathbf{0} \\ \mathbf{0} & \mathbf{I} \end{bmatrix} \begin{bmatrix} \mathbf{C} & \mathbf{D} \\ \mathbf{M} & \mathbf{N} \end{bmatrix} < \mathbf{0}. \tag{4.63}$$

Suppose the above statements hold. Then the robust $\mathcal{H}_2$ performance measure $J_{\mathcal{H}_2}$ is finite if $\mathbf{L} = \mathbf{0}$ and $\mathbf{H} = \mathbf{0}$, in which case, it is bounded above by (recall that $\|\mathbf{M}\| = \bar{\sigma}[\mathbf{M}]$)

$$J_{\mathcal{H}_2} < \|\mathbf{K}^T\mathbf{PK}\|.$$

Proof. Recall that the system (4.57) is $\mathcal{Q}$-stable if and only if (4.62) holds. This implies the existence of $\bar{\mathbf{P}} > \mathbf{0}$, $\bar{\mathbf{S}} \in \mathcal{S}$ and a sufficiently small $\varepsilon > 0$ such that

$$\begin{bmatrix} \bar{\mathbf{P}}\mathbf{A} + \mathbf{A}^T\bar{\mathbf{P}} & \bar{\mathbf{P}}\mathbf{B} \\ \mathbf{B}^T\bar{\mathbf{P}} & -\bar{\mathbf{S}} \end{bmatrix} + \begin{bmatrix} \mathbf{C}^T \\ \mathbf{D}^T \end{bmatrix} \bar{\mathbf{S}} \begin{bmatrix} \mathbf{C} & \mathbf{D} \end{bmatrix} + \varepsilon \begin{bmatrix} \mathbf{M}^T \\ \mathbf{N}^T \end{bmatrix} \begin{bmatrix} \mathbf{M} & \mathbf{N} \end{bmatrix} < \mathbf{0}.$$

Dividing the left-hand side by ε, and defining $\mathbf{P} \triangleq \varepsilon^{-1}\bar{\mathbf{P}} > \mathbf{0}$ and $\mathbf{S} \triangleq \varepsilon^{-1}\bar{\mathbf{S}} \in \mathcal{S}$, we have (4.63). This proves the equivalence (i) $\Leftrightarrow$ (ii).

Now, consider the signals $\mathbf{x}$ and ω in response to the impulsive disturbance $\mathbf{w}(t) = \mathbf{w}_0\delta(t)$ with the zero initial state, and a (candidate) quadratic Lyapunov function $V(\mathbf{x}(t)) \triangleq \mathbf{x}^T(t)\mathbf{P}\mathbf{x}(t)$. Denote the left-hand side of (4.63) by Φ. Since $\Phi < \mathbf{0}$, for any given vector $\mathbf{v}(t) \triangleq [\ \mathbf{x}^T(t)\ \ \omega^T(t)\ ]^T \ne \mathbf{0}$, we have $\mathbf{v}^T(t)\Phi\mathbf{v}(t) < 0$, or equivalently,

$$\dot{V}(\mathbf{x}(t)) < \omega^T(t)\mathbf{S}\omega(t) - \psi^T(t)\mathbf{S}\psi(t) - \mathbf{y}^T(t)\mathbf{y}(t), \quad \forall\, t \ge 0,$$

where we note that $\mathbf{x}(0) = \mathbf{0}$ and $\mathbf{w}(t) = \mathbf{w}_0\delta(t)$ are equivalent to $\mathbf{x}(0) = \mathbf{K}\mathbf{w}_0$ and $\mathbf{w}(t) = \mathbf{0}$, $\forall\, t \ge 0$. Integrating both sides from $t = 0$ to ∞,

$$\mathbf{w}_0^T(\mathbf{K}^T\mathbf{PK})\mathbf{w}_0 > \int_0^\infty \|\mathbf{y}(t)\|^2 dt + \int_0^\infty (\psi^T(t)\mathbf{S}\psi(t) - \omega^T(t)\mathbf{S}\omega(t))dt,$$
$$\forall\, \boldsymbol{\Delta} \in \mathcal{BU}_C,$$

where we used the fact $\lim_{t\to\infty}\mathbf{x}(t) = \mathbf{0}$ due to the $\mathcal{Q}$-stability (which implies asymptotic stability). Noting that, for each $\mathbf{S} \in \mathcal{S}$,

$$\omega^T(t)\mathbf{S}\omega(t) - \psi^T(t)\mathbf{S}\psi(t) = \psi^T(t)(\mathbf{\Delta}^T(t)\mathbf{S}\mathbf{\Delta}(t) - \mathbf{S})\psi(t) \leq \mathbf{0}, \quad \forall\, \mathbf{\Delta} \in \mathcal{BU}_C, \quad t \geq 0, \tag{4.64}$$

and the worst-case disturbance intensity $\mathbf{w}_0$ is given by the eigenvector of $\mathbf{K}^T\mathbf{P}\mathbf{K}$ corresponding to the maximum eigenvalue, we see that

$$\|\mathbf{K}^T\mathbf{P}\mathbf{K}\| > \int_0^\infty \|\mathbf{y}(t)\|^2 dt, \quad \forall\, \mathbf{\Delta} \in \mathcal{BU}_C, \quad \|\mathbf{w}_0\| \leq 1.$$

This completes the proof. □

It can be seen in the above proof that the performance bound in Theorem 4.7.2 holds as long as the uncertainty set $\mathcal{BU}_C$ is defined such that

$$\int_0^\infty (\psi^T(t)\mathbf{S}\psi(t) - \omega^T(t)\mathbf{S}\omega(t))dt \geq 0, \quad \forall\, \mathbf{\Delta} \in \mathcal{BU}_C. \tag{4.65}$$

This simple observation allows us to investigate the conservativeness of the bound by searching for the "largest" uncertainty set $\mathcal{BU}_C$ such that (4.65) holds. Let us first consider the case where $\mathbf{\Delta}$ is a linear operator. In this case, defining $\hat{\omega} \triangleq \mathbf{S}^{1/2}\omega$, $\hat{\psi} \triangleq \mathbf{S}^{1/2}\psi$ and noting that $\hat{\omega} = \mathbf{S}^{1/2}\mathbf{\Delta}\psi = \mathbf{\Delta}\hat{\psi}$, the condition (4.65) becomes

$$\|\hat{\psi}\|_{\mathcal{L}_2} \geq \|\hat{\omega}\|_{\mathcal{L}_2}, \quad \forall\, \mathbf{\Delta} \in \mathcal{BU}_C.$$

Thus we see that the result of Theorem 4.7.2 holds for any linear dynamic (possibly noncausal) time-varying uncertainties with $\mathcal{L}_2$ gain less than or equal to 1. For the full block $\mathbf{\Delta}_i$ of the uncertainty, we can further relax the linearity assumption on the operator $\mathbf{\Delta}_i$ since, in this case, the scaling $\mathbf{S}$ is just a scalar and can be factored out in (4.65).

The following theorem provides a characterization of the robust $\mathcal{L}_\infty$ performance bound [4, 56]. As in the nominal performance case discussed in the previous section, the performance bound is given by the maximum singular value of a matrix associated with a controllability-Gramian-type Lyapunov matrix, which is the dual of the Lyapunov matrix used in the robust $\mathcal{H}_2$ performance characterization.

Theorem 4.7.3 *The following statements are equivalent.*

(i) The uncertain system (4.57) is $\mathcal{Q}$-stable for all $\mathbf{\Delta} \in \mathcal{BU}_C$.

(ii) There exist $\mathbf{Q} > \mathbf{0}$ and $\mathbf{S} \in \mathcal{S}$ such that

$$\begin{bmatrix} \mathbf{AQ} + \mathbf{QA}^T & \mathbf{QC}^T \\ \mathbf{CQ} & -\mathbf{S} \end{bmatrix} + \begin{bmatrix} \mathbf{B} & \mathbf{K} \\ \mathbf{D} & \mathbf{L} \end{bmatrix} \begin{bmatrix} \mathbf{S} & \mathbf{0} \\ \mathbf{0} & \mathbf{I} \end{bmatrix} \begin{bmatrix} \mathbf{B} & \mathbf{K} \\ \mathbf{D} & \mathbf{L} \end{bmatrix}^T < \mathbf{0}. \tag{4.66}$$

Suppose the above statements hold. Then the robust $\mathcal{L}_\infty$ performance measure $J_{\mathcal{L}_\infty}$ is finite if $\mathbf{N} = \mathbf{0}$ and $\mathbf{H} = \mathbf{0}$, in which case, it is bounded above by

$$J_{\mathcal{L}_\infty} < \|\mathbf{MQM}^T\|.$$

Proof. The equivalence (i) ⇔ (ii) can be verified by dualizing the result of Theorem 4.7.2. Now suppose statement (ii) holds. By the Schur complement formula, inequality (4.66) is equivalent to

$$\begin{bmatrix} \mathbf{AQ}+\mathbf{QA}^T & \mathbf{QC}^T & \mathbf{B} & \mathbf{K} \\ \mathbf{CQ} & -\mathbf{S} & \mathbf{D} & \mathbf{L} \\ \mathbf{B}^T & \mathbf{D}^T & -\mathbf{S}^{-1} & \mathbf{0} \\ \mathbf{K}^T & \mathbf{L}^T & \mathbf{0} & -\mathbf{I} \end{bmatrix} < \mathbf{0}.$$

After a congruent transformation involving $\mathbf{Q}^{-1}$, another use of the Schur complement formula yields

$$\boldsymbol{\Omega} \triangleq \begin{bmatrix} \mathbf{Q}^{-1}\mathbf{A}+\mathbf{A}^T\mathbf{Q}^{-1} & \mathbf{Q}^{-1}\mathbf{B} & \mathbf{Q}^{-1}\mathbf{K} \\ \mathbf{B}^T\mathbf{Q}^{-1} & -\mathbf{S}^{-1} & \mathbf{0} \\ \mathbf{K}^T\mathbf{Q}^{-1} & \mathbf{0} & -\mathbf{I} \end{bmatrix} + \begin{bmatrix} \mathbf{C}^T \\ \mathbf{D}^T \\ \mathbf{L}^T \end{bmatrix} \mathbf{S}^{-1} \begin{bmatrix} \mathbf{C} & \mathbf{D} & \mathbf{L} \end{bmatrix} < \mathbf{0}.$$

Now we can prove the result by a similar approach to the case of the robust $\mathcal{H}_2$ performance bound. Consider the signals $\mathbf{x}$ and $\boldsymbol{\omega}$ in response to some $\mathcal{L}_2$ disturbance $\mathbf{w}$ with the zero initial state, and a (candidate) Lyapunov function $V(\mathbf{x}(t)) \triangleq \mathbf{x}^T(t)\mathbf{Q}^{-1}\mathbf{x}(t)$. Then, for any vector $\mathbf{u}(t) \triangleq [\,\mathbf{x}^T(t)\;\; \boldsymbol{\omega}^T(t)\;\; \mathbf{w}^T(t)\,]^T \neq \mathbf{0}$, we have $\mathbf{u}^T(t)\boldsymbol{\Omega}\mathbf{u}(t) < 0$, or equivalently,

$$\dot{V}(\mathbf{x}(t)) < \|\mathbf{w}(t)\|^2 + \boldsymbol{\omega}^T(t)\mathbf{S}^{-1}\boldsymbol{\omega}(t) - \boldsymbol{\psi}^T(t)\mathbf{S}^{-1}\boldsymbol{\psi}(t), \quad \forall\, t \geq 0.$$

Integrating both sides from $t = 0$ to τ, we have

$$\mathbf{x}^T(\tau)\mathbf{Q}^{-1}\mathbf{x}(\tau) < \int_0^\tau \|\mathbf{w}(t)\|^2 dt + \int_0^\tau (\boldsymbol{\omega}^T(t)\mathbf{S}^{-1}\boldsymbol{\omega}(t) - \boldsymbol{\psi}^T(t)\mathbf{S}^{-1}\boldsymbol{\psi}(t))dt, \quad \forall\, \boldsymbol{\Delta} \in \mathcal{BU}_C, \quad \tau \geq 0.$$

It is easily verified that

$$\int_0^\tau (\boldsymbol{\omega}^T(t)\mathbf{S}^{-1}\boldsymbol{\omega}(t) - \boldsymbol{\psi}^T(t)\mathbf{S}^{-1}\boldsymbol{\psi}(t))dt \leq \mathbf{0}, \quad \forall\, \boldsymbol{\Delta} \in \mathcal{BU}_C, \quad \tau \geq 0 \tag{4.67}$$

and hence, for any (nonzero) finite energy disturbance $\mathbf{w}$ and any perturbation $\boldsymbol{\Delta} \in \mathcal{BU}_C$,

$$\mathbf{x}^T(t)\mathbf{Q}^{-1}\mathbf{x}(t) < \int_0^\tau \|\mathbf{w}(t)\|^2 dt \leq \|\mathbf{w}\|^2_{\mathcal{L}_2}, \quad \forall t \geq 0$$

or equivalently,

$$\begin{bmatrix} \|\mathbf{w}\|^2_{\mathcal{L}_2} & \mathbf{x}^T(t) \\ \mathbf{x}(t) & \mathbf{Q} \end{bmatrix} > \mathbf{0}, \quad \forall\, t \geq 0.$$

By a (nonsquare) congruent transformation,

$$\begin{bmatrix} 1 & \mathbf{0} \\ \mathbf{0} & \mathbf{M} \end{bmatrix}\begin{bmatrix} \|\mathbf{w}\|^2_{\mathcal{L}_2} & \mathbf{x}^T(t) \\ \mathbf{x}(t) & \mathbf{Q} \end{bmatrix}\begin{bmatrix} 1 & \mathbf{0} \\ \mathbf{0} & \mathbf{M}^T \end{bmatrix} = \begin{bmatrix} \|\mathbf{w}\|^2_{\mathcal{L}_2} & \mathbf{y}^T(t) \\ \mathbf{y}(t) & \mathbf{MQM}^T \end{bmatrix} \geq \mathbf{0}, \quad \forall\, t \geq 0$$

where the last inequality is strict if $\mathbf{M}$ has linearly independent rows. Using the Schur complement formula,

$$\frac{1}{\|\mathbf{w}\|^2_{\mathcal{L}_2}}\mathbf{y}(t)\mathbf{y}^T(t) \leq \mathbf{MQM}^T \leq \|\mathbf{MQM}^T\|\mathbf{I}, \quad \forall\, t \geq 0$$

or

$$\frac{\|\mathbf{y}(t)\|^2}{\|\mathbf{w}\|^2_{\mathcal{L}_2}} \leq \|\mathbf{MQM}^T\|, \quad \forall\, t \geq 0.$$

This completes the proof. □

As in the case of the robust $\mathcal{H}_2$ performance bound, we shall examine how conservative our performance bound is by searching for a larger class of uncertainties for which Theorem 4.7.3 is valid. From the proof of Theorem 4.7.3, our result holds for any uncertainty set $\mathcal{BU}_C$ such that (4.67) is satisfied. If $\mathbf{\Delta}$ is a linear operator, then following a similar procedure to that given for the case of the robust $\mathcal{H}_2$ performance, we see that (4.67) is equivalent to

$$\int_0^\tau (\|\hat{\omega}(t)\|^2 - \|\hat{\psi}(t)\|^2)dt \leq 0, \quad \forall\, \mathbf{\Delta} \in \mathcal{BU}_C, \quad \tau \geq 0, \tag{4.68}$$

where $\hat{\omega} = \mathbf{\Delta}\hat{\psi}$. According to Lemma 2.4 of [139], (4.68) holds if $\mathcal{BU}_C$ is the set of linear causal operators with $\mathcal{L}_2$ gain less than or equal to 1. Note that causality of the uncertainty came into play in contrast to the case of the robust $\mathcal{H}_2$ performance bound where Theorem 4.7.2 is valid for possibly noncausal uncertainties.

Finally, we state a characterization of the robust $\mathcal{H}_\infty$ performance bound [26, 96].

Theorem 4.7.4 *The following statements are equivalent:*

(i) The uncertain system (4.57) is $\mathcal{Q}$-stable for all $\mathbf{\Delta} \in \mathcal{BU}_C$.

(ii) There exist matrices $\mathbf{P} > \mathbf{0}$, $\mathbf{S} \in \mathcal{S}$ and a scalar $\gamma > 0$ such that

$$\begin{bmatrix} \mathbf{PA}+\mathbf{A}^T\mathbf{P} & \mathbf{PB} & \mathbf{PK} \\ \mathbf{B}^T\mathbf{P} & -\mathbf{S} & \mathbf{0} \\ \mathbf{K}^T\mathbf{P} & \mathbf{0} & -\gamma^2\mathbf{I} \end{bmatrix} + \begin{bmatrix} \mathbf{C}^T & \mathbf{M}^T \\ \mathbf{D}^T & \mathbf{N}^T \\ \mathbf{L}^T & \mathbf{H}^T \end{bmatrix} \begin{bmatrix} \mathbf{S} & \mathbf{0} \\ \mathbf{0} & \mathbf{I} \end{bmatrix} \begin{bmatrix} \mathbf{C} & \mathbf{D} & \mathbf{L} \\ \mathbf{M} & \mathbf{N} & \mathbf{H} \end{bmatrix} < \mathbf{0}. \tag{4.69}$$

Suppose the above statements hold. Then the robust $\mathcal{H}_\infty$ performance $J_{\mathcal{H}_\infty}$ is bounded above by

$$J_{\mathcal{H}_\infty} < \gamma.$$

Proof. Using Finsler's Theorem, there exists $\gamma > 0$ satisfying (4.69) if and only if there exist $\mathbf{P} > \mathbf{0}$ and $\mathbf{S} \in \mathcal{S}$ satisfying (4.63). Hence, the equivalence (i) $\Leftrightarrow$ (ii) follows from Theorem 4.7.2.

To prove the performance bound, consider the signals in (4.57) and let Ψ be the left-hand side of (4.69). Then, for any vector $\zeta(t) \triangleq [\ \mathbf{x}^T(t)\ \ \omega^T(t)\ \ \mathbf{w}^T(t)\ ]^T \neq \mathbf{0}$, we have $\zeta^T(t)\Psi\zeta(t) < 0$, or equivalently,

$$2\mathbf{x}^T(t)\mathbf{P}\dot{\mathbf{x}}(t) + \mathbf{y}^T(t)\mathbf{y}(t) - \gamma^2\mathbf{w}^T(t)\mathbf{w}(t) < \omega^T(t)\mathbf{S}\omega(t) - \psi^T(t)\mathbf{S}\psi(t) \leq 0,$$

where the last inequality holds since $\mathbf{S} \in \mathcal{S}$ and $\omega(t) = \mathbf{\Delta}(t)\psi(t)$ with $\mathbf{\Delta} \in \mathcal{BU}_C$. Thus,

$$\frac{d}{dt}(\mathbf{x}^T(t)\mathbf{P}\mathbf{x}(t)) + \mathbf{y}^T(t)\mathbf{y}(t) - \gamma^2\mathbf{w}^T(t)\mathbf{w}(t) < 0.$$

Integrating from $t = 0$ to ∞ with $\mathbf{x}(0) = \mathbf{0}$, we have

$$\int_0^\infty \mathbf{y}^T(t)\mathbf{y}(t)dt < \gamma^2 \int_0^\infty \mathbf{w}^T(t)\mathbf{w}(t)dt$$

where we used the fact that $\lim_{t\to\infty} \mathbf{x}(t) = \mathbf{0}$ due to $\mathcal{Q}$-stability. Hence,

$$\|\mathbf{y}\|_{\mathcal{L}_2} < \gamma\|\mathbf{w}\|_{\mathcal{L}_2} \leq \gamma.$$

This completes the proof. □

4.7.2 Discrete-Time Systems

Consider the uncertain discrete-time system

$$\begin{bmatrix} \mathbf{x}(k+1) \\ \psi(k) \\ \mathbf{y}(k) \end{bmatrix} = \begin{bmatrix} \mathbf{A} & \mathbf{B} & \mathbf{K} \\ \mathbf{C} & \mathbf{D} & \mathbf{L} \\ \mathbf{M} & \mathbf{N} & \mathbf{H} \end{bmatrix} \begin{bmatrix} \mathbf{x}(k) \\ \omega(k) \\ \mathbf{w}(k) \end{bmatrix}, \quad \omega(k) = \mathbf{\Delta}(k)\psi(k), \tag{4.70}$$

where $\mathbf{x}$ is the state, $\mathbf{w}$ is the disturbance, $\mathbf{y}$ is the output of interest, ψ and ω are the exogenous signals to describe the uncertainty $\mathbf{\Delta}$. Suppose that the uncertainty $\mathbf{\Delta}$ is known to belong to the following set of norm-bounded time-varying structured uncertainties:

$$\mathcal{BU}_D \triangleq \{ \mathbf{\Delta} : \mathcal{I} \to \mathcal{R}^{m\times m}, \quad \|\mathbf{\Delta}(k)\| \le 1, \quad \mathbf{\Delta}(k) \in \mathcal{U} \}$$

where $\mathcal{I}$ is the set of integers and $\mathcal{U}$ is defined in (4.58).

In this section, robust stability and performance of the uncertain system (4.70) are analyzed in a manner that is completely analogous to the continuous-time counterpart. Let us start by defining the concept for robust stability.

Definition 4.7.4 The uncertain discrete-time system (4.70) is said to be robustly stable if $\lim_{k\to\infty} \mathbf{x}(k) = \mathbf{0}$ regardless of the initial state $\mathbf{x}(0)$, and the uncertainty $\mathbf{\Delta} \in \mathcal{BU}_D$.

Definition 4.7.5 [79] The uncertain discrete-time system (4.70) is said to be quadratically stable if there exists a single quadratic Lyapunov function $V(\mathbf{x}) = \mathbf{x}^T\mathbf{P}\mathbf{x}$ to prove stability for all $\mathbf{\Delta} \in \mathcal{BU}_D$, i.e. if $(\mathbf{I} - \mathbf{D}\mathbf{\Delta})$ is invertible for any $\mathbf{\Delta} \in \mathcal{BU}_D$ and there exists $\mathbf{P} > \mathbf{0}$ such that

$$\mathbf{P} > (\mathbf{A} + \mathbf{B}\mathbf{\Delta}(\mathbf{I} - \mathbf{D}\mathbf{\Delta})^{-1}\mathbf{C})^T\mathbf{P}(\mathbf{A} + \mathbf{B}\mathbf{\Delta}(\mathbf{I} - \mathbf{D}\mathbf{\Delta})^{-1}\mathbf{C}), \quad \forall\, \mathbf{\Delta} \in \mathcal{BU}_D.$$

Definition 4.7.6 [26] The uncertain discrete-time system (4.70) is said to be $\mathcal{Q}$-stable if there exists a matrix $\mathbf{S} \in \mathcal{S}$ such that $\|\mathbf{S}^{1/2}\mathbf{T}\mathbf{S}^{-1/2}\|_{\mathcal{H}_\infty} < 1$ where $\mathbf{T}(z) \triangleq \mathbf{C}(z\mathbf{I}-\mathbf{A})^{-1}\mathbf{B}+\mathbf{D}$.

As in the continuous-time case, the above notions of stability can be ordered as follows:

$$\|\mathbf{T}\|_{\mathcal{H}_\infty} < 1 \quad \Rightarrow \quad \mathcal{Q}\text{-stability} \quad \Rightarrow \quad \text{Quadratic Stability} \quad \Rightarrow \quad \text{Robust Stability}.$$

Ideally, we would like to have a necessary and sufficient condition for robust stability, but this is difficult. The easiest (and hence most conservative) way to guarantee robust stability is the small gain condition $\|\mathbf{T}\|_{\mathcal{H}_\infty} < 1$. This condition and $\mathcal{Q}$-stability and quadratic stability are all equivalent if the uncertainty is unstructured. However, in general, there are gaps between these three. Conservatism in $\|\mathbf{T}\|_{\mathcal{H}_\infty} < 1$ can be reduced by taking the structure information on the uncertainty into account, which leads to the notion of $\mathcal{Q}$-stability or the scaled $\mathcal{H}_\infty$ norm condition. $\mathcal{Q}$-stability is still conservative, but has a nice (computable) characterization as follows [33, 96]. (A proof is left for the reader as an easy exercise.)

Theorem 4.7.5 *The following statements are equivalent:*

(i) The uncertain system (4.70) is $\mathcal{Q}$-stable.

(ii) There exist matrices $\mathbf{P} > \mathbf{0}$ and $\mathbf{S} \in \mathcal{S}$ such that

$$\begin{bmatrix} \mathbf{P} & \mathbf{0} \\ \mathbf{0} & \mathbf{S} \end{bmatrix} > \begin{bmatrix} \mathbf{A} & \mathbf{B} \\ \mathbf{C} & \mathbf{D} \end{bmatrix}^T \begin{bmatrix} \mathbf{P} & \mathbf{0} \\ \mathbf{0} & \mathbf{S} \end{bmatrix} \begin{bmatrix} \mathbf{A} & \mathbf{B} \\ \mathbf{C} & \mathbf{D} \end{bmatrix}.$$

Now, we shall define robust performance measures and provide characterizations based on the notion of $\mathcal{Q}$-stability.

Robust $\mathcal{H}_2$ Performance: Consider

$$\hat{F}_{\mathcal{H}_2}(\boldsymbol{\Delta}, \mathbf{w}_0) \triangleq \|\mathbf{y}\|_{\ell_2}^2 = \sum_{k=0}^{\infty} \mathbf{y}^T(k)\mathbf{y}(k)$$

where $\{ \mathbf{y}(k) \}_{k=0}^{\infty}$ is the output signal in response to

$$\mathbf{w}(k) = \begin{cases} \mathbf{w}_0 & (k = 0) \\ \mathbf{0} & (k \geq 1) \end{cases}, \qquad \mathbf{x}(0) = \mathbf{0}$$

in the presence of perturbation $\{ \boldsymbol{\Delta}(k) \}_{k=0}^{\infty}$. The robust $\mathcal{H}_2$ performance measure is defined by

$$F_{\mathcal{H}_2} \triangleq \sup_{\mathbf{w}_0,\ \boldsymbol{\Delta}} \{\hat{F}_{\mathcal{H}_2}(\boldsymbol{\Delta}, \mathbf{w}_0) : \ \|\mathbf{w}_0\| \leq 1, \quad \boldsymbol{\Delta} \in \mathcal{BU}_D\}.$$

Robust ℓ_∞ Performance: Consider

$$\hat{F}_{\ell_\infty}(\boldsymbol{\Delta}, \mathbf{w}) \triangleq \|\mathbf{y}\|_{\ell_\infty}^2 = \sup_{k \geq 0} \mathbf{y}^T(k)\mathbf{y}(k)$$

where $\{ \mathbf{y}(k) \}_{k=0}^{\infty}$ is the output signal for the disturbance $\{ \mathbf{w}(k) \}_{k=0}^{\infty}$ and the perturbation $\{ \boldsymbol{\Delta}(k) \}_{k=0}^{\infty}$, with a zero initial state $\mathbf{x}(0) = \mathbf{0}$. The robust ℓ_∞ performance measure is defined by

$$F_{\ell_\infty} \triangleq \sup_{\mathbf{w},\ \boldsymbol{\Delta}} \{\hat{F}_{\ell_\infty}(\boldsymbol{\Delta}, \mathbf{w}) : \ \|\mathbf{w}\|_{\ell_2} \leq 1, \quad \boldsymbol{\Delta} \in \mathcal{BU}_D\}.$$

Robust $\mathcal{H}_\infty$ Performance: Consider

$$\hat{F}_{\mathcal{H}_\infty}(\boldsymbol{\Delta}, \mathbf{w}) \triangleq \|\mathbf{y}\|_{\ell_2}^2 = \sum_{k=0}^{\infty} \mathbf{y}^T(k)\mathbf{y}(k)$$

where $\{ \mathbf{y}(k) \}_{k=0}^{\infty}$ is the output signal in response to the disturbance $\mathbf{w}$, with the zero initial state $\mathbf{x}(0) = \mathbf{0}$, in the presence of perturbation $\{ \boldsymbol{\Delta}(k) \}_{k=0}^{\infty}$. The robust $\mathcal{H}_\infty$ performance is defined by

$$F_{\mathcal{H}_\infty} \triangleq \sup_{\mathbf{w},\ \boldsymbol{\Delta}} \{\hat{F}_{\mathcal{H}_\infty}(\boldsymbol{\Delta}, \mathbf{w}) : \ \|\mathbf{w}\|_{\ell_2} \leq 1, \quad \boldsymbol{\Delta} \in \mathcal{BU}_D\}.$$

The above robust performance measures are completely analogous to the ones for the continuous-time case. We can obtain upper bounds for the performance measures by applying similar techniques to those given in the previous section. However, for the robust $\mathcal{H}_2$ performance, there is a difficulty associated with discrete-time systems, which results from the pulse disturbance as opposed to the (continuous-time) impulsive disturbance. To avoid this difficulty, let us introduce the following technical assumption.

Assumption $\quad \omega(0) = \mathbf{0}$.

This assumption holds if the uncertainty is modeled by a linear dynamic strictly proper system with the zero initial state.

Now we are ready to give characterizations for upper bounds on the robust performance measures for the uncertain discrete-time system (4.70) [56].

Theorem 4.7.6 *The following statements are equivalent.*

(i) The uncertain discrete-time system (4.70) is $\mathcal{Q}$-stable for all $\mathbf{\Delta} \in \mathcal{BU}_D$.

(ii) There exist $\mathbf{P} > \mathbf{0}$ and $\mathbf{S} \in \mathcal{S}$ such that

$$\begin{bmatrix} \mathbf{P} & \mathbf{0} \\ \mathbf{0} & \mathbf{S} \end{bmatrix} > \begin{bmatrix} \mathbf{A} & \mathbf{B} \\ \mathbf{C} & \mathbf{D} \end{bmatrix}^T \begin{bmatrix} \mathbf{P} & \mathbf{0} \\ \mathbf{0} & \mathbf{S} \end{bmatrix} \begin{bmatrix} \mathbf{A} & \mathbf{B} \\ \mathbf{C} & \mathbf{D} \end{bmatrix} + \begin{bmatrix} \mathbf{M}^T \\ \mathbf{N}^T \end{bmatrix} \begin{bmatrix} \mathbf{M} & \mathbf{N} \end{bmatrix}. \tag{4.71}$$

Suppose the above statements hold. Further we assume that $\mathbf{L} = \mathbf{0}$ and Assumption is satisfied. Then the robust $\mathcal{H}_2$ performance measure $F_{\mathcal{H}_2}$ is bounded above by

$$F_{\mathcal{H}_2} < \|\mathbf{K}^T\mathbf{P}\mathbf{K} + \mathbf{H}^T\mathbf{H}\|.$$

Proof. Using the discrete-time bounded real lemma (see e.g. [26, 96]), and following a similar procedure to the proof of Theorem 4.7.2, the equivalence of statements (i) and (ii) can be verified.

Now, let $\mathbf{x}$ and ω be the signals in response to the pulse disturbance. Then, multiplying (4.71) by $[\ \mathbf{x}^T(k)\ \ \omega^T(k)\ ]$ and its transpose from the left and right, respectively, we have

$$\mathbf{y}^T(k)\mathbf{y}(k) < \mathbf{x}^T(k)\mathbf{P}\mathbf{x}(k) - \mathbf{x}^T(k+1)\mathbf{P}\mathbf{x}(k+1) + \omega^T(k)\mathbf{S}\omega(k) - \psi^T(k)\mathbf{S}\psi(k), \quad \forall\, k \geq 1.$$

Taking the summation over $k = 1, 2, \ldots$, and adding $\|\mathbf{y}(0)\|^2$ to both sides, after some manipulations,

$$\begin{aligned} \|\mathbf{y}\|_{\ell_2}^2 \;&<\; \begin{bmatrix} \mathbf{x}^T(0) & \omega^T(0) & \mathbf{w}^T(0) \end{bmatrix} \begin{bmatrix} \mathbf{A}^T & \mathbf{C}^T & \mathbf{M}^T \\ \mathbf{B}^T & \mathbf{D}^T & \mathbf{N}^T \\ \mathbf{K}^T & \mathbf{L}^T & \mathbf{H}^T \end{bmatrix} \begin{bmatrix} \mathbf{P} & \mathbf{0} & \mathbf{0} \\ \mathbf{0} & \mathbf{S} & \mathbf{0} \\ \mathbf{0} & \mathbf{0} & \mathbf{I} \end{bmatrix} \\ &\quad \times \begin{bmatrix} \mathbf{A} & \mathbf{B} & \mathbf{K} \\ \mathbf{C} & \mathbf{D} & \mathbf{L} \\ \mathbf{M} & \mathbf{N} & \mathbf{H} \end{bmatrix} \begin{bmatrix} \mathbf{x}(0) \\ \omega(0) \\ \mathbf{w}(0) \end{bmatrix} \\ &\quad - \omega^T(0)\mathbf{S}\omega(0) + \sum_{k=0}^{\infty} \left(\omega^T(k)\mathbf{S}\omega(k) - \psi^T(k)\mathbf{S}\psi(k)\right). \end{aligned}$$

Note that, for $\mathbf{S} \in \mathcal{S}$ and $\mathbf{\Delta} \in \mathcal{BU}_D$, we have $\omega^T(k)\mathbf{S}\omega(k) - \psi^T(k)\mathbf{S}\psi(k) \leq 0$. Then the performance bound follows immediately with assumptions $\mathbf{x}(0) = \mathbf{0}$ and $\omega(0) = \mathbf{0}$, by taking the worst-case over the direction of the pulse disturbance. □

The result of Theorem 4.7.6 clearly shows analogy to a characterization of the $\mathcal{H}_2$ norm for the *nominal* system, based on the Lyapunov inequality. Indeed, the nominal case result can be recovered by letting the matrices associated with the uncertainty be zero. In this case, the upper bound for the $\mathcal{H}_2$ norm is tight [109]. However, the upper bound in Theorem 4.7.6 on the robust $\mathcal{H}_2$ performance may not be tight in general due to the conservativeness of $\mathcal{Q}$-stability. In view of the above proof, Theorem 4.7.6 holds for the larger class of uncertainties, i.e. linear noncausal time-varying uncertainties with ℓ_2 gain less than or equal to 1.

It is worth noting that the matrix block diag($\mathbf{P}$, $\mathbf{S}$) in Theorem 4.7.6 may be considered as a structured Lyapunov matrix to prove $\mathcal{Q}$-stability. In particular, it is the structured observability Gramian if we replace the inequality by equality, for the augmented system

$$\hat{\mathbf{A}} \triangleq \begin{bmatrix} \mathbf{A} & \mathbf{B} \\ \mathbf{C} & \mathbf{D} \end{bmatrix}, \quad \hat{\mathbf{C}} \triangleq \begin{bmatrix} \mathbf{M} & \mathbf{N} \end{bmatrix}.$$

The following theorem characterizes an upper bound on the robust ℓ_∞ performance [56]. The result exhibits a certain duality to the result of Theorem 4.7.6, and in this case, we have a structured controllability Gramian to describe the bound.

Theorem 4.7.7 *The following statements are equivalent.*

(i) The uncertain discrete-time system (4.70) is $\mathcal{Q}$-stable for all $\boldsymbol{\Delta} \in \mathcal{BU}_D$.

(ii) There exist $\mathbf{Q} > \mathbf{0}$ and $\mathbf{S} \in \mathcal{S}$ such that

$$\begin{bmatrix} \mathbf{Q} & \mathbf{0} \\ \mathbf{0} & \mathbf{S} \end{bmatrix} > \begin{bmatrix} \mathbf{A} & \mathbf{B} \\ \mathbf{C} & \mathbf{D} \end{bmatrix} \begin{bmatrix} \mathbf{Q} & \mathbf{0} \\ \mathbf{0} & \mathbf{S} \end{bmatrix} \begin{bmatrix} \mathbf{A} & \mathbf{B} \\ \mathbf{C} & \mathbf{D} \end{bmatrix}^T + \begin{bmatrix} \mathbf{K} \\ \mathbf{L} \end{bmatrix} \begin{bmatrix} \mathbf{K}^T & \mathbf{L}^T \end{bmatrix}. \tag{4.72}$$

Suppose the above statements hold. Then the robust ℓ_∞ performance measure F_{ℓ_∞} is finite if $\mathbf{N} = \mathbf{0}$, in which case, it is bounded above by

$$F_{\ell_\infty} < \|\mathbf{MQM}^T + \mathbf{HH}^T\|.$$

Proof. The equivalence (i) $\Leftrightarrow$ (ii) follows from Theorem 4.7.6 by dualizing the result. Note that (4.72) is equivalent to

$$\boldsymbol{\Omega}_D \triangleq \begin{bmatrix} \mathbf{Q}^{-1} & \mathbf{0} & \mathbf{0} \\ \mathbf{0} & \mathbf{S}^{-1} & \mathbf{0} \\ \mathbf{0} & \mathbf{0} & \mathbf{I} \end{bmatrix} - \begin{bmatrix} \mathbf{A}^T & \mathbf{C}^T \\ \mathbf{B}^T & \mathbf{D}^T \\ \mathbf{K}^T & \mathbf{L}^T \end{bmatrix} \begin{bmatrix} \mathbf{Q}^{-1} & \mathbf{0} \\ \mathbf{0} & \mathbf{S}^{-1} \end{bmatrix} \begin{bmatrix} \mathbf{A} & \mathbf{B} & \mathbf{K} \\ \mathbf{C} & \mathbf{D} & \mathbf{L} \end{bmatrix} > \mathbf{0},$$

by the Schur complement formula. Then, for $\boldsymbol{\zeta}(k) \triangleq [\ \mathbf{x}^T(k) \quad \boldsymbol{\omega}^T(k) \quad \mathbf{w}^T(k)\ ]^T \neq \mathbf{0}$, we have $\boldsymbol{\zeta}_k^T \boldsymbol{\Omega}_D \boldsymbol{\zeta}_k > 0$, or

$$\|\mathbf{w}(k)\|^2 + \mathbf{x}^T(k)\mathbf{Q}^{-1}\mathbf{x}(k) - \mathbf{x}^T(k+1)\mathbf{Q}^{-1}\mathbf{x}(k+1) - \boldsymbol{\psi}^T(k)\mathbf{S}^{-1}\boldsymbol{\psi}(k) + \boldsymbol{\omega}^T(k)\mathbf{S}^{-1}\boldsymbol{\omega}(k) > 0.$$

Taking the summation over $k = 0, 1, \ldots, N$,

$$\begin{aligned} \mathbf{x}^T(N+1)\mathbf{Q}^{-1}\mathbf{x}(N+1) \quad < \quad & \sum_{k=0}^{N} \|\mathbf{w}(k)\|^2 + \sum_{k=0}^{N} (\boldsymbol{\omega}^T(k)\mathbf{S}^{-1}\boldsymbol{\omega}(k) - \boldsymbol{\psi}^T(k)\mathbf{S}^{-1}\boldsymbol{\psi}(k)) \\ & + \mathbf{x}^T(0)\mathbf{Q}^{-1}\mathbf{x}(0). \end{aligned}$$

The rest of the proof is similar to the proof of Theorem 4.7.3. □

Finally, we give a characterization of robust $\mathcal{H}_\infty$ performance bound [26, 96]. The result can be proved by a similar procedure to the proof of Theorem 4.7.6, and hence is left for the reader as an exercise.

Theorem 4.7.8 *The following statements are equivalent:*

(i) The uncertain system (4.70) is $\mathcal{Q}$-stable for all $\boldsymbol{\Delta} \in \mathcal{BU}_D$.

(ii) There exist matrices $\mathbf{P} > \mathbf{0}$, $\mathbf{S} \in \mathcal{S}$ and a scalar $\gamma > 0$ such that

$$\begin{bmatrix} \mathbf{P} & \mathbf{0} & \mathbf{0} \\ \mathbf{0} & \mathbf{S} & \mathbf{0} \\ \mathbf{0} & \mathbf{0} & \gamma^2\mathbf{I} \end{bmatrix} > \begin{bmatrix} \mathbf{A} & \mathbf{B} & \mathbf{K} \\ \mathbf{C} & \mathbf{D} & \mathbf{L} \\ \mathbf{M} & \mathbf{N} & \mathbf{H} \end{bmatrix}^T \begin{bmatrix} \mathbf{P} & \mathbf{0} & \mathbf{0} \\ \mathbf{0} & \mathbf{S} & \mathbf{0} \\ \mathbf{0} & \mathbf{0} & \mathbf{I} \end{bmatrix} \begin{bmatrix} \mathbf{A} & \mathbf{B} & \mathbf{K} \\ \mathbf{C} & \mathbf{D} & \mathbf{L} \\ \mathbf{M} & \mathbf{N} & \mathbf{H} \end{bmatrix}. \tag{4.73}$$

Suppose the above statements hold. Then the robust $\mathcal{H}_\infty$ performance $F_{\mathcal{H}_\infty}$ is bounded above by

$$F_{\mathcal{H}_\infty} < \gamma.$$

CHAPTER 4 SUMMARY

This chapter presents both deterministic and stochastic interpretations of the covariance of the state or output of a dynamic system. The covariance matrix may be constructed by injecting impulses (pulses) into one input channel at a time in continuous (discrete) systems, and summing the second-order information (the outer product of the state with itself) over the number of input channels. It is shown that some problems that are nonlinear in the state space, remain linear in the state covariance equations. Hence, using second-order information (the covariance) permits a larger class of systems to be treated by linear methods.

It is of fundamental interest in this book that the system gains can be related to the second-order information (covariance) of the state. In later chapters, the ability to assign a specified covariance matrix will be interpreted in terms of the system gains that can be assigned by feedback. The system gains are defined to relate the norm of an output to the norm of an input. A basic motivation for defining these quantities is that many control design objectives can be given by the requirement that the error output signal be small regardless of the disturbance and/or command input. Thus, the system gains are measures for performance of control systems. A variety of nominal performance measures are treated in [19, 154, 163] while several robust performance measures can be found in [26, 56, 55, 139].

For further reading on stochastic processes in systems and control, see [10, 136, 77].

CHAPTER FIVE

Covariance Controllers

Covariance analysis is the workhorse of systems engineering. Almost every engineering discipline uses covariance analysis to assess errors and to evaluate performance. It was not until [52] that techniques were developed to use feedback control to *assign* a desired covariance. This chapter outlines and extends those procedures.

5.1 COVARIANCE CONTROL PROBLEM

Recall that the state covariance carries many system properties, and that the closed-loop stability is equivalent to the existence of a positive definite state covariance under controllability assumptions. This motivates the following *covariance control problem* [52]: Suppose a state covariance which satisfies certain closed-loop performance specifications is given. We wish to find all controllers which assign the given state covariance. We call such controllers *covariance controllers*. Note that not all positive definite matrices can be assigned as state covariances. A positive definite matrix $\mathbf{X}$ is *assignable* as a state covariance if it satisfies

$$\mathbf{A}_{c\ell}\mathbf{X} + \mathbf{X}\mathbf{A}_{c\ell}^T + \mathbf{B}_{c\ell}\mathbf{B}_{c\ell}^T = \mathbf{0} \quad \text{(continuous-time case)}$$

$$\mathbf{X} = \mathbf{A}_{c\ell}\mathbf{X}\mathbf{A}_{c\ell}^T + \mathbf{B}_{c\ell}\mathbf{B}_{c\ell}^T \quad \text{(discrete-time case)}$$

for some controller $\mathbf{G}$ (note that the matrices $\mathbf{A}_{c\ell}$ and $\mathbf{B}_{c\ell}$ are functions of $\mathbf{G}$). The solution to the covariance control problem consists of the following two parts: (1) necessary and sufficient conditions for a positive definite matrix to be assignable, and (2) an explicit formula for all controllers which assign a given state covariance to the closed-loop system. Throughout the chapter, we assume for simplicity that there is no correlation between the process- and measurement-noises, and we define the noise covariances $\mathbf{W}$ and $\mathbf{V}$ as

$$\begin{bmatrix} \mathbf{D}_p \\ \mathbf{D}_z \end{bmatrix} \begin{bmatrix} \mathbf{D}_p^T & \mathbf{D}_z^T \end{bmatrix} = \begin{bmatrix} \mathbf{W} & \mathbf{0} \\ \mathbf{0} & \mathbf{V} \end{bmatrix}. \tag{5.1}$$

We shall also assume that $(\mathbf{A}_p, \mathbf{D}_p)$ is controllable and $(\mathbf{A}_p, \mathbf{B}_p, \mathbf{M}_p)$ is a stabilizable and detectable triple. In the following section, we consider the covariance control problem for the linear time-invariant continuous-time system.

5.2 CONTINUOUS-TIME COVARIANCE CONTROLLERS

5.2.1 State Feedback

Consider the linear system given by (4.15) where we assume that all the states are available for feedback, i.e. $\mathbf{M}_p = \mathbf{I}$ and $\mathbf{D}_z = \mathbf{0}$. With a controller of constant state feedback gain $\mathbf{G}$, the closed-loop state covariance $\mathbf{X}$ satisfies the following Lyapunov equation:

$$(\mathbf{A}_p + \mathbf{B}_p\mathbf{G})\mathbf{X} + \mathbf{X}(\mathbf{A}_p + \mathbf{B}_p\mathbf{G})^T + \mathbf{W} = \mathbf{0}. \tag{5.2}$$

Theorem 5.2.1 *[52, 129] Let a positive definite matrix $\mathbf{X} \in \mathcal{R}^{n_p \times n_p}$ be given. Then the following statements are equivalent:*

(i) There exists a control gain $\mathbf{G}$ which assigns $\mathbf{X}$ as a state covariance.

(ii) $\mathbf{X}$ satisfies

$$(\mathbf{I} - \mathbf{B}_p\mathbf{B}_p^+)(\mathbf{A}_p\mathbf{X} + \mathbf{X}\mathbf{A}_p^T + \mathbf{W})(\mathbf{I} - \mathbf{B}_p\mathbf{B}_p^+) = \mathbf{0}.$$

In this case, all such control gains $\mathbf{G}$ are given by

$$\begin{aligned} \mathbf{G} &= -\frac{1}{2}\mathbf{B}_p^+(\mathbf{A}_p\mathbf{X} + \mathbf{X}\mathbf{A}_p^T + \mathbf{W})(2\mathbf{I} - \mathbf{B}_p\mathbf{B}_p^+)\mathbf{X}^{-1} \\ &+ \mathbf{B}_p^+\mathbf{S}_F\mathbf{B}_p\mathbf{B}_p^+\mathbf{X}^{-1} + (\mathbf{I} - \mathbf{B}_p^+\mathbf{B}_p)\mathbf{Z}_F \end{aligned} \tag{5.3}$$

where $\mathbf{Z}_F$ is arbitrary and $\mathbf{S}_F$ is an arbitrary skew-symmetric matrix.

Proof. Recall that a given matrix $\mathbf{X} > 0$ is assignable as a state covariance if and only if there exists a controller $\mathbf{G}$ satisfying (5.2). Rearranging (5.2), we have

$$\mathbf{B}_p\mathbf{G}\mathbf{X} + (\mathbf{B}_p\mathbf{G}\mathbf{X})^T + \mathbf{Q} = \mathbf{0} \tag{5.4}$$

$$\mathbf{Q} \triangleq \mathbf{A}_p\mathbf{X} + \mathbf{X}\mathbf{A}_p^T + \mathbf{W}.$$

Then the result directly follows by applying Theorem 2.3.9 to solve (5.4) for $\mathbf{GX}$. □

5.2.2 Covariance Construction for Assignability

The following result permits the construction of covariances assignable by state feedback (Theorem 5.2.1) in a finite number of steps. Let the SVD of $\mathbf{B}_p$ be given by

$$\mathbf{B}_p = \mathbf{U}\mathbf{\Sigma}\mathbf{V}^T = \begin{bmatrix} \mathbf{U}_1 & \mathbf{U}_2 \end{bmatrix} \begin{bmatrix} \mathbf{\Sigma}_1 & \mathbf{0} \\ \mathbf{0} & \mathbf{0} \end{bmatrix} \begin{bmatrix} \mathbf{V}_1^T \\ \mathbf{V}_2^T \end{bmatrix}$$

and define $\mathbf{A}_2, \mathbf{B}_2, \mathbf{W}_2, \mathbf{X}_2$ by

$$\mathbf{A}_2 = \mathbf{U}_2^T\mathbf{A}_p\mathbf{U}_2, \quad \mathbf{B}_2 = \mathbf{U}_2^T\mathbf{A}_p\mathbf{U}_1, \quad \mathbf{W}_2 = \mathbf{U}_2^T\mathbf{W}\mathbf{U}_2, \quad \mathbf{X}_2 = \mathbf{U}_2^T\mathbf{X}\mathbf{U}_2.$$

Lemma 5.2.1 *Given $\mathbf{A}_p, \mathbf{B}_p, \mathbf{W}$, the following statements are equivalent:*

(i) $\mathbf{X} > \mathbf{0}$ satisfies

$$(\mathbf{I} - \mathbf{B}_p\mathbf{B}_p^+)(\mathbf{A}_p\mathbf{X} + \mathbf{X}\mathbf{A}_p^T + \mathbf{W})(\mathbf{I} - \mathbf{B}_p\mathbf{B}_p^+) = \mathbf{0}. \tag{5.5}$$

(ii) There exist real matrices $\mathbf{X}_2 > \mathbf{0}$, $\mathbf{R} = \mathbf{R}^T > \mathbf{0}$, $\mathbf{S} = -\mathbf{S}^T$ *and* $\mathbf{Z}$ *such that*

$$\mathbf{X} = \mathbf{U}\begin{bmatrix} \mathbf{R} + \mathbf{G}_2\mathbf{X}_2\mathbf{G}_2{}^T & \mathbf{G}_2\mathbf{X}_2 \\ \mathbf{X}_2\mathbf{G}_2{}^T & \mathbf{X}_2 \end{bmatrix}\mathbf{U}^T \tag{5.6}$$

$$\begin{aligned}\mathbf{G}_2 &= -\frac{1}{2}\mathbf{B}_2^+(\mathbf{A}_2\mathbf{X}_2 + \mathbf{X}_2\mathbf{A}_2^T + \mathbf{W}_2)(2\mathbf{I} - \mathbf{B}_2\mathbf{B}_2^+)\mathbf{X}_2{}^{-1} \\ &+ \mathbf{B}_2^+\mathbf{S}\mathbf{B}_2\mathbf{B}_2^+\mathbf{X}_2{}^{-1} + (\mathbf{I} - \mathbf{B}_2^+\mathbf{B}_2)\mathbf{Z}\end{aligned} \tag{5.7}$$

where $\mathbf{X}_2$ *satisfies*

$$(\mathbf{I} - \mathbf{B}_2\mathbf{B}_2^+)(\mathbf{A}_2\mathbf{X}_2 + \mathbf{X}_2\mathbf{A}_2^T + \mathbf{W}_2)(\mathbf{I} - \mathbf{B}_2\mathbf{B}_2^+) = \mathbf{0}. \tag{5.8}$$

Proof. Suppose (i) holds. Define

$$\begin{bmatrix} \mathbf{Q} & \mathbf{L} \\ \mathbf{L}^T & \mathbf{X}_2 \end{bmatrix} \triangleq \begin{bmatrix} \mathbf{U}_1^T \\ \mathbf{U}_2^T \end{bmatrix}\mathbf{X}\begin{bmatrix} \mathbf{U}_1 & \mathbf{U}_2 \end{bmatrix}$$

where $\mathbf{X}_2 > \mathbf{0}$ follows from $\mathbf{X} > \mathbf{0}$. Note that $\mathbf{I} - \mathbf{B}_p\mathbf{B}_p^+ = \mathbf{U}_2\mathbf{U}_2^T$, and (5.5) yields

$$(\mathbf{A}_2 + \mathbf{B}_2\mathbf{G}_2)\mathbf{X}_2 + \mathbf{X}_2(\mathbf{A}_2 + \mathbf{B}_2\mathbf{G}_2)^T + \mathbf{W}_2 = \mathbf{0}, \tag{5.9}$$

for some $\mathbf{G}_2 = \mathbf{L}\mathbf{X}_2^{-1}$. Applying Theorem (5.2.1) to (5.9) yields (5.7)–(5.8), where $\mathbf{S} = -\mathbf{S}^T$ and $\mathbf{Z}$ is arbitrary. The existence of $\mathbf{R}$ follows from (5.6), from which

$$\mathbf{X} > \mathbf{0} \iff \mathbf{U}^T\mathbf{X}\mathbf{U} > \mathbf{0} \iff \begin{bmatrix} \mathbf{R} + \mathbf{G}_2\mathbf{X}_2\mathbf{G}_2^T & \mathbf{G}_2\mathbf{X}_2 \\ \mathbf{X}_2\mathbf{G}_2^T & \mathbf{X}_2 \end{bmatrix} > \mathbf{0}$$

or, using the Schur complement, $\{\mathbf{X} > \mathbf{0}\} \iff \{\mathbf{X}_2 > \mathbf{0}$ and $\mathbf{R} + \mathbf{G}_2\mathbf{X}_2\mathbf{G}_2^T - \mathbf{G}_2\mathbf{X}_2\mathbf{X}_2^{-1}\mathbf{X}_2\mathbf{G}_2^T = \mathbf{R} > \mathbf{0}\}$. This proves that (i)⇒(ii). To show the converse, suppose (ii) holds. Then (5.6) yields $\mathbf{X} > \mathbf{0}$, since $\mathbf{R} > \mathbf{0}$ and $\mathbf{X}_2 > \mathbf{0}$. From Theorem 5.2.1, $\mathbf{X}_2$ and $\mathbf{G}_2$ satisfy (5.9) for arbitrary $\mathbf{S} = -\mathbf{S}^T$ and $\mathbf{Z}$. Then by substitution $\mathbf{X}$ given by (5.6) satisfies (5.5). □

Note that if $\mathbf{X}_2 > \mathbf{0}$ can be found to satisfy (5.8), then (5.7) gives a set of $\mathbf{G}_2$ which can be placed into a matrix of the form (5.6) to produce an assignable $\mathbf{X}$, given an arbitrary $\mathbf{R} > \mathbf{0}$. So what is now needed is an $\mathbf{X}_2 > \mathbf{0}$ satisfying (5.8). Note that finding $\mathbf{X}_2$ to satisfy (5.8) is the same as the mathematical problem of finding $\mathbf{X}$ to satisfy (5.5). What has been accomplished is that (5.8) is a *smaller* matrix equation. Hence to solve (5.8) simply apply Theorem 5.2.1 again. This yields another problem of the same form, but of smaller size. This process can be repeated until the size of the problem is trivial to solve. This is the idea of the algorithm presented below.

Given matrices $\mathbf{A}_p$, $\mathbf{B}_p$, $\mathbf{W}$, define for $k = 1, 2, \ldots, q$, the set of matrices,

$$\begin{aligned}\mathbf{A}_{k+1} &\triangleq \mathbf{U}_{2k}^T\mathbf{A}_k\mathbf{U}_{2k} & \mathbf{A}_1 &\triangleq \mathbf{A}_p \\ \mathbf{B}_{k+1} &= \mathbf{U}_{2k}^T\mathbf{A}_k\mathbf{U}_{1k} & \mathbf{B}_1 &\triangleq \mathbf{B}_p \\ \mathbf{W}_{k+1} &= \mathbf{U}_{2k}^T\mathbf{W}_k\mathbf{U}_{2k} & \mathbf{W}_1 &\triangleq \mathbf{W}\end{aligned} \tag{5.10}$$

using the singular value decompositions,

$$\mathbf{B}_k = \mathbf{U}_k\boldsymbol{\Sigma}_k\mathbf{V}_k^T = \begin{bmatrix} \mathbf{U}_{1k} & \mathbf{U}_{2k} \end{bmatrix}\begin{bmatrix} \boldsymbol{\Sigma}_{1k} & \mathbf{0} \\ \mathbf{0} & \mathbf{0} \end{bmatrix}\begin{bmatrix} \mathbf{V}_{1k}^T \\ \mathbf{V}_{2k}^T \end{bmatrix} \tag{5.11}$$

where the terminal integer q is defined by the event

$$\mathbf{B}_q\mathbf{B}_q^T > \mathbf{0} \quad \text{or} \quad \mathbf{B}_q = \mathbf{0}.$$

To show that $q < \infty$, note that if $\mathbf{B}_k\mathbf{B}_k^T \not> \mathbf{0}$ and $\mathbf{B}_k \neq \mathbf{0}$ then $\mathbf{U}_{1k}$ and $\mathbf{U}_{2k}$ are still well-defined (neither is an empty matrix). Since $\mathbf{U}_{1k}$ is not empty, $\mathbf{U}_{2k}$ is not square and $\mathbf{U}_{2k}^T\mathbf{A}_k$ has fewer rows than $\mathbf{A}_k$. That is $\dim(\mathbf{A}_{k+1}) = \dim(\mathbf{B}_{k+1}) < \dim(\mathbf{A}_k)$ where $\dim(\cdot)$ denotes the row dimension of $(\cdot)$. Hence the dimension of the matrices reduces by at least one on each iteration, and thus, the number of iterations is $q < n$, where $\mathbf{A}_p$ is $n \times n$.

Recursive Algorithm for Assignable Covariances

STEP 1 Compute the sequences $\mathbf{A}_k, \mathbf{B}_k, \mathbf{W}_k, k = 1, 2, \ldots, q$ and $\mathbf{U}_k, k = 1, 2, \ldots, q-1$ from (5.10)–(5.11).

STEP 2 Initialize $k = q$ and choose $\mathbf{X}_q$ according to the following rule: If $\mathbf{B}_q\mathbf{B}_q^T > \mathbf{0}$, choose $\mathbf{X}_q$ as any positive definite matrix. If $\mathbf{B}_q = \mathbf{0}$, choose $\mathbf{X}_q$ as the unique positive definite solution to $\mathbf{A}_q\mathbf{X}_q + \mathbf{X}_q\mathbf{A}_q^T + \mathbf{W}_q = \mathbf{0}$.

STEP 3 Choose an arbitrary $\mathbf{Z}_k$ and an arbitrary $\mathbf{S}_k = -\mathbf{S}_k^T$, and compute

$$\begin{aligned} \mathbf{G}_k &= -\frac{1}{2}\mathbf{B}_k^+(\mathbf{A}_k\mathbf{X}_k + \mathbf{X}_k\mathbf{A}_k^T + \mathbf{W}_k)(2\mathbf{I} - \mathbf{B}_k\mathbf{B}_k^+)\mathbf{X}_k^{-1} \\ &\quad + \mathbf{B}_k^+\mathbf{S}_k\mathbf{B}_k\mathbf{B}_k^+\mathbf{X}_k^{-1} + (\mathbf{I} - \mathbf{B}_k^+\mathbf{B}_k)\mathbf{Z}_k. \end{aligned}$$

STEP 4 Choose an arbitrary $\mathbf{R}_k > \mathbf{0}$ and update $\mathbf{X}_k$

$$\mathbf{X}_{k-1} = \mathbf{U}_{k-1}\begin{bmatrix} \mathbf{R}_k + \mathbf{G}_k\mathbf{X}_k\mathbf{G}_k^T & \mathbf{G}_k\mathbf{X}_k \\ \mathbf{X}_k\mathbf{G}_k^T & \mathbf{X}_k \end{bmatrix}\mathbf{U}_{k-1}^T.$$

STEP 5 If $k = 2$ go to STEP 6. If $k > 2$, update k to $k-1$ and return to STEP 3.

STEP 6 Let $\mathbf{X} = \mathbf{X}_1$.

This algorithm is a natural consequence of Lemma 5.2.1, and the only thing that needs to be justified is the choice of $\mathbf{X}_q$ in STEP 2. At $k = q$ we need to solve

$$(\mathbf{I} - \mathbf{B}_q\mathbf{B}_q^+)(\mathbf{A}_q\mathbf{X}_q + \mathbf{X}_q\mathbf{A}_q^T + \mathbf{W}_q)(\mathbf{I} - \mathbf{B}_q\mathbf{B}_q^+) = \mathbf{0}$$

for $\mathbf{X}_q > \mathbf{0}$. If $\mathbf{B}_q\mathbf{B}_q^T > \mathbf{0}$ then $\mathbf{X}_q > \mathbf{0}$ is arbitrary. If $\mathbf{B}_q = \mathbf{0}$, and $\mathbf{A}_q$ is stable with $(\mathbf{A}_q, \mathbf{W}_q)$ controllable (see that $\mathbf{W}_q > \mathbf{0}$), then $\mathbf{X}_q > \mathbf{0}$ satisfying

$$\mathbf{A}_q\mathbf{X}_q + \mathbf{X}_q\mathbf{A}_q^T + \mathbf{W}_q = \mathbf{0} \tag{5.12}$$

is unique. If $\mathbf{B}_q = \mathbf{0}$ and $\mathbf{A}_q$ is not stable, then no $\mathbf{X} > \mathbf{0}$ solves (5.12). In this case the matrix pair $(\mathbf{A}, \mathbf{B})$ is not stabilizable. In fact, if $\mathbf{W} = \mathbf{B}\mathbf{B}^T$, then $\mathbf{W}_k = \mathbf{0}$ for all k, and the matrix pair $(\mathbf{A}, \mathbf{B})$ is stabilizable if and only if (5.12) has a positive definite solution $\mathbf{X}_q > \mathbf{0}$.

Example 5.2.1 Consider a fourth-order single-input system given in the controllable canonical form:

$$\mathbf{A}_p = \begin{bmatrix} 0 & 1 & 0 & 0 \\ 0 & 0 & 1 & 0 \\ 0 & 0 & 0 & 1 \\ -a_0 & -a_1 & -a_2 & -a_3 \end{bmatrix}, \quad \mathbf{B}_p = \begin{bmatrix} 0 \\ 0 \\ 0 \\ 1 \end{bmatrix}.$$

Let $\mathbf{W} = \mathbf{D}_p\mathbf{D}_p^T$ and $\mathbf{D}_p = \mathbf{B}_p$. In this case, $(\mathbf{A}_p + \mathbf{B}_p\mathbf{G}, \mathbf{D}_p)$ is controllable for any $\mathbf{G}$. We will apply the recursive algorithm to generate all assignable covariances (controllability Gramians).

In STEP 1 of the algorithm, using (5.10) and (5.11), we find $\mathbf{A}_1 = \mathbf{A}_p$, $\mathbf{B}_1 = \mathbf{B}_p$, $\mathbf{W}_1 = \mathbf{B}_p\mathbf{B}_p^T$ and

$$\mathbf{U}_1 = \left[\begin{array}{c|ccc} 0 & 1 & 0 & 0 \\ 0 & 0 & 1 & 0 \\ 0 & 0 & 0 & 1 \\ 1 & 0 & 0 & 0 \end{array}\right], \quad \mathbf{A}_2 = \begin{bmatrix} 0 & 1 & 0 \\ 0 & 0 & 1 \\ 0 & 0 & 0 \end{bmatrix}$$

$$\mathbf{B}_2 = \begin{bmatrix} 0 \\ 0 \\ 1 \end{bmatrix}, \quad \mathbf{U}_2 = \left[\begin{array}{c|cc} 0 & 1 & 0 \\ 0 & 0 & 1 \\ 1 & 0 & 0 \end{array}\right],$$

$$\mathbf{A}_3 = \begin{bmatrix} 0 & 1 \\ 0 & 0 \end{bmatrix}, \quad \mathbf{B}_3 = \begin{bmatrix} 0 \\ 1 \end{bmatrix}, \quad \mathbf{U}_3 = \left[\begin{array}{c|c} 0 & 1 \\ 1 & 0 \end{array}\right], \quad \mathbf{A}_4 = 0, \quad \mathbf{B}_4 = 1$$

and $\mathbf{W}_k = \mathbf{0}$ $(k = 2, 3, 4)$. Note that $q = 4$.

In STEP 2 of the algorithm, $\mathbf{B}_q\mathbf{B}_q^T > \mathbf{0}$ and therefore $\mathbf{X}_q$ can be chosen to be an arbitrary positive scalar; we denote it by $x_4 > 0$. For the computation of $\mathbf{G}_k$ in STEP 3 of the algorithm, the choices of parameters $\mathbf{S}_k = -\mathbf{S}_k^T$ and $\mathbf{Z}_k$ do not affect the value of $\mathbf{G}_k$ because of the special structure of $\mathbf{B}_k$. In STEP 4, the parameter $\mathbf{R}_k$ is a positive scalar and we denote it by r_k.

Iterating STEPS 3 and 4 for $k = 4, 3, 2$, we have

$$\mathbf{G}_4 = 0, \qquad \mathbf{X}_3 = \begin{bmatrix} x_4 & 0 \\ 0 & r_4 \end{bmatrix},$$

$$\mathbf{G}_3 = -\begin{bmatrix} r_4/x_4 \\ 0 \end{bmatrix}^T, \qquad \mathbf{X}_2 = \begin{bmatrix} x_4 & 0 & -r_4 \\ 0 & r_4 & 0 \\ -r_4 & 0 & r_5 \end{bmatrix},$$

$$\mathbf{G}_2 = -\begin{bmatrix} 0 \\ r_3/r_4 + r_4/x_4 \\ 0 \end{bmatrix}^T, \qquad \mathbf{X}_1 = \begin{bmatrix} x_4 & 0 & -r_4 & 0 \\ 0 & r_4 & 0 & -r_5 \\ -r_4 & 0 & r_5 & 0 \\ 0 & -r_5 & 0 & r_6 \end{bmatrix}$$

where

$$r_5 \triangleq r_3 + \frac{r_4^2}{x_4}, \qquad r_6 \triangleq r_2 + \frac{r_5^2}{r_4}.$$

Thus, the set of all assignable covariances is parametrized by $\mathbf{X} = \mathbf{X}_1$ in terms of positive scalars r_2, r_3, r_4 and x_4.

Note that $\mathbf{X}$ has the signature Hankel structure. In fact, as shown in the next subsection, when an nth order single-input system is in the controllable canonical form, the set of assignable controllability Gramians coincides with the set of positive definite signature Hankel matrices. This fourth-order example can be generalized to give a parametrization of $n \times n$ positive definite signature Hankel matrices in terms of n positive scalars.

Finally, using (5.3), the unique state feedback gain **G**, that assigns a given **X**, is found to be

$$\mathbf{G} = \begin{bmatrix} a_0 - p_2 p_4 \\ a_1 - p_1(p_3 + p_4) \\ a_2 - p_2 - p_3 - p_4 \\ a_3 - p_1 \end{bmatrix}^T, \quad \begin{array}{l} p_1 \triangleq 1/(2r_2), \\ p_2 \triangleq r_2/r_3, \\ p_3 \triangleq r_3/r_4, \\ p_4 \triangleq r_4/x_4. \end{array}$$

Since $p_k > 0$ $(k = 1, \ldots, 4)$ if and only if $r_k > 0$ $(k = 2, 3, 4)$ and $x_4 > 0$, the above formula for **G** provides a parametrization of all stabilizing state feedback gains in terms of positive scalars p_k $(k = 1, \ldots, 4)$. From the analysis point of view, we see that the set of all stable fourth-order polynomials[1] is parametrized by

$$\phi(\lambda) = \lambda^4 + p_1\lambda^3 + (p_2 + p_3 + p_4)\lambda^2 + p_1(p_3 + p_4)\lambda + p_2 p_4$$

where $p_k > 0$ $(k = 1, \ldots, 4)$.

5.2.3 State Feedback for Single-Input Systems

The above algorithm collapses to a closed form analytical solution for single-input systems. For the system described by

$$\dot{\mathbf{x}} = \mathbf{A}\mathbf{x} + \mathbf{B}(u + w) \tag{5.13}$$

suppose $(\mathbf{A}, \mathbf{B})$ is in phase variable form, and

$$\mathbf{A} = \left[\begin{array}{c|c} \mathbf{0} & \mathbf{I}_{n_x} \\ \hline \multicolumn{2}{c}{-\mathbf{a}^T} \end{array}\right], \quad \mathbf{B} = \begin{bmatrix} 0 \\ 1 \end{bmatrix} \tag{5.14}$$

$$\mathbf{a}^T = \begin{bmatrix} a_1 & a_2 & \cdots & a_n \end{bmatrix}, \quad u = \mathbf{G}\mathbf{x}$$

$$\mathbf{G} = \begin{bmatrix} g_1 & g_2 & \cdots & g_n \end{bmatrix} \tag{5.15}$$

where the characteristic equation for **A** is

$$\lambda^n + a_n\lambda^{n-1} + \cdots a_2\lambda + a_1 = 0. \tag{5.16}$$

and the characteristic equation of the closed-loop system is the same as (5.16) with a_i replaced by $a_i - g_i$, $i = 1, \ldots, n_x$. Define $\sigma_i \triangleq \mathbf{X}_{ii}$ where **X** satisfies

$$\mathbf{0} = \mathbf{X}(\mathbf{A} + \mathbf{B}\mathbf{G})^T + (\mathbf{A} + \mathbf{B}\mathbf{G})\mathbf{X} + \mathbf{B}\mathbf{B}^T. \tag{5.17}$$

Hence σ_i is the impulse-to-energy gain between the input and the ith state variable, or σ_i is the variance of the ith state variable if the input w is zero mean white noise. There are n values of a_i and n values of σ_i. We seek to show the one-to-one relationship between them.

Many engineering problems, such as antenna or telescope pointing, have performance requirements naturally stated in terms of inequalities on the variances, σ_i. Hence, it might be of interest to explicitly relate stability and performance by expressing the allowed variance values σ_i in terms of the coefficients a_i, which contain all stability information. Note that $\mathbf{x}^T\mathbf{X}^{-1}\mathbf{x}$ is a Lyapunov function for (5.13).

The motivation for this section was influenced by two known results:

[1] We say that a polynomial is *stable* if all its roots have negative real parts.

I) The work of [87] proved the Hurwitz–Routh test [113, 54] from Lyapunov stability theory, by using a transformation of (5.13) which yields a diagonal Lyapunov matrix. Positive definiteness of this Lyapunov matrix is shown to be equivalent to the Hurwitz–Routh test. Other studies connecting Lyapunov to the Hurwitz–Routh test include [80, 122, 13, 31, 146, 123, 149, 68, 106, 100, 99, 101, 87, 66].

II) The covariance control theory in section 5.2.1 gives the necessary and sufficient conditions under which a given state covariance matrix $\mathbf{X}$ may be assigned by a feedback controller.

These two results (I, II) are combined as follows. The set of all $\mathbf{X}$ that may be "assigned" by a state feedback controller for the controllable system $\{\dot{\mathbf{x}} = \mathbf{A}\mathbf{x} + \mathbf{B}(u + w), u = \mathbf{G}\mathbf{x}\}$ has special structure (to be shown), and $\mathbf{A} + \mathbf{B}\mathbf{G}$ is asymptotically stable, if and only if the "assigned" $\mathbf{X}$ is positive definite. Hence, we exploit this special structure of $\mathbf{X}$ and use $\mathbf{x}^T(t)\mathbf{X}^{-1}\mathbf{x}(t)$ as a Lyapunov function, in lieu of the diagonal choice of a Lyapunov matrix used by [87]. The potential advantages are the new explicit relationships between the *physical* performance entities σ_i, $i = 1, \ldots, n$ and stability (the coefficients $a_i, i = 1, \ldots, n$). Such information may also be useful in model reduction [131].

Let the even and odd coefficients of (5.16) be collected in vectors $\mathbf{a}_e$, $\mathbf{a}_o$, respectively,

$$\mathbf{a}_o \triangleq \begin{bmatrix} a_1 \\ a_3 \\ a_5 \\ \vdots \end{bmatrix}, \quad \mathbf{a}_e \triangleq \begin{bmatrix} a_2 \\ a_4 \\ a_6 \\ \vdots \end{bmatrix} \tag{5.18}$$

and define two Hankel matrices composed of the σ_i

$$\mathbf{X}_o \triangleq \begin{bmatrix} \sigma_1 & -\sigma_2 & \sigma_3 & -\sigma_4 & \sigma_5 & -\sigma_6 & \\ -\sigma_2 & \sigma_3 & -\sigma_4 & \sigma_5 & -\sigma_6 & \sigma_7 & \\ \sigma_3 & -\sigma_4 & \sigma_5 & -\sigma_6 & \sigma_7 & -\sigma_8 & \\ -\sigma_4 & \sigma_5 & -\sigma_6 & \sigma_7 & -\sigma_8 & \sigma_9 & \\ \sigma_5 & -\sigma_6 & \sigma_7 & -\sigma_8 & \sigma_9 & -\sigma_{10} & \\ -\sigma_6 & \sigma_7 & -\sigma_8 & \sigma_9 & -\sigma_{10} & \sigma_{11} & \\ & & & & & & \ddots \\ & & & & & & \text{etc.} \end{bmatrix} \tag{5.19}$$

$$\mathbf{X}_e \triangleq \begin{bmatrix} \sigma_2 & -\sigma_3 & \sigma_4 & -\sigma_5 & \sigma_6 & -\sigma_7 & \\ -\sigma_3 & \sigma_4 & -\sigma_5 & \sigma_6 & -\sigma_7 & \sigma_8 & \\ \sigma_4 & -\sigma_5 & \sigma_6 & -\sigma_7 & \sigma_8 & -\sigma_9 & \\ -\sigma_5 & \sigma_6 & -\sigma_7 & \sigma_8 & -\sigma_9 & \sigma_{10} & \\ \sigma_6 & -\sigma_7 & \sigma_8 & -\sigma_9 & \sigma_{10} & -\sigma_{11} & \\ -\sigma_7 & \sigma_8 & -\sigma_9 & \sigma_{10} & -\sigma_{11} & \sigma_{12} & \\ & & & & & & \ddots \\ & & & & & & \text{etc.} \end{bmatrix} \tag{5.20}$$

where $\mathbf{a}_o \in \mathcal{R}^{n_o}$, $\mathbf{a}_e \in \mathcal{R}^{n_e}$. When n = even, $n_o = n_e = n/2$. When n = odd, $n_o = n_e + 1, n_e = (n-1)/2$. Also $\mathbf{X}_o \in \mathcal{R}^{n_o \times n_o}$, $\mathbf{X}_e \in \mathcal{R}^{n_e \times n_e}$. Define the phrase "stable polynomial" to mean that all roots of the polynomial lie in the left half plane.

Theorem 5.2.2 *For the linear time-invariant system (5.13), let σ_i, $i = 1, \ldots, n$ denote the state variances, and let a_i, $i = 1, \ldots, n$ denote the coefficients of the characteristic polynomial (5.16). Then the following statements are equivalent:*

(i) The matrices $\mathbf{X}_o, \mathbf{X}_e$ in (5.19) and (5.20) are positive definite.
(ii) The set of $\mathbf{a}_i$ corresponds to a stable polynomial.
(iii) For $n = even$

$$\mathbf{a}_o = \mathbf{X}_o^{-1}\mathbf{X}_e\mathbf{1}, \quad \mathbf{a}_e = \frac{1}{2}\mathbf{X}_e^{-1}\mathbf{1}, \quad \mathbf{1} \triangleq \begin{bmatrix} \mathbf{0} \\ 1 \end{bmatrix} \tag{5.21}$$

and for $n = odd$,

$$\mathbf{a}_o = \frac{1}{2}\mathbf{X}_o^{-1}\mathbf{1}, \quad \mathbf{a}_e = \mathbf{X}_e^{-1}\begin{bmatrix} \mathbf{0} & \mathbf{I}_e \end{bmatrix}\mathbf{X}_o\mathbf{1}, \tag{5.22}$$

where $\mathbf{X}_e, \mathbf{X}_o$ are any positive definite matrices having structure and $\mathbf{a}_o$, $\mathbf{a}_e$ are defined by (5.18).

Theorem 5.2.2 gives the explicit (unique) dependence of the coefficients of the characteristic polynomial on the assignable variances of each state variable of (5.13). Table 5.1 lists the results of Theorem 5.2.2 for $n = 2, 3, 4, 5, 6$. Note that the stability conditions for $n = 5$ include all the conditions for $n = 4$ plus one additional condition $\Delta_5 > 0$ (denoted "$+\Delta_5 > 0$" in Table 5.1).

Corollary 5.2.1 *Define the positive scalar η by*

$$\begin{aligned} \eta &\triangleq \frac{1}{2}\mathbf{1}^T\mathbf{X}_e^{-1}\mathbf{1}, \quad n = \text{even} \\ &= \frac{1}{2}\mathbf{1}^T\mathbf{X}_o^{-1}\mathbf{1}, \quad n = \text{odd}. \end{aligned}$$

All poles of the system (5.13) lie between the vertical lines in the complex plane $-\eta + j\omega$ and $0 + j\omega$ and $(0 \le \omega \le \infty)$. Furthermore, the average of all pole locations is $-\eta/n$.

Proof. The sum of all poles is

$$\sum_{i=1}^{n} \lambda_k = tr[\mathbf{A}] = -a_n \tag{5.23}$$

where a_n is given by (5.21). Hence, $a_n = \eta$ and $(1/n)\sum_{i=1}^{n} \lambda_i = (1/n)tr[\mathbf{A}] = (-1/n)a_n = (-1/n)\eta$. Now consider (5.17), with $\mathbf{G} = \mathbf{0}$ and

$$\ell_i^* \mathbf{A} = \lambda_i \ell_i^*.$$

Then from (5.17)

$$\ell_i^* \left[\mathbf{X}\mathbf{A}^T + \mathbf{A}\mathbf{X} + \mathbf{B}\mathbf{B}^T\right] \ell_i = 0$$

yields

$$\lambda_i + \lambda_i^* = -\frac{\ell_i^* \mathbf{B}\mathbf{B}^T \ell_i}{\ell_i^* \mathbf{X} \ell_i} \ge -\bar{\lambda}(\mathbf{X}^{-1}\mathbf{B}\mathbf{B}^*),$$

Table 5.1 The set of Hurwitz coefficients a_i

	$n=2$	$n=3$	$n=4$	$n=5$	$n=6$
Stability conditions $X > 0$	$\sigma_1 > 0,\ \sigma_2 > 0$	$+\Delta_3 > 0$	$+\Delta_4 > 0$	$+\Delta_5 > 0$	$+\Delta_6 > 0$
a_1	$\frac{1}{\sigma_1}[\sigma_2]$	$\frac{1}{2\Delta_3}[\sigma_2]$	$\frac{1}{\Delta_3}[\Delta_4]$	$\frac{1}{2\Delta_5}[\Delta_4]$	$\frac{1}{\Delta_5}[\sigma_4(\sigma_3\sigma_5 - \sigma_4^2) + \sigma_5(\sigma_3\sigma_4 - \sigma_2\sigma_5) + \sigma_6\Delta_4]$
a_2	$\frac{1}{2\sigma_2}$	$\frac{1}{\sigma_2}[\sigma_3]$	$\frac{1}{2\Delta_4}[\sigma_3]$	$\frac{1}{\Delta_4}[\sigma_3\sigma_5 - \sigma_4^2]$	$\frac{1}{2\Delta_6}[\sigma_3\sigma_5 - \sigma_4^2]$
a_3		$\frac{1}{2\Delta_3}[\sigma_1]$	$\frac{\sigma_1\sigma_4 - \sigma_2\sigma_3}{\Delta_3}$	$\frac{1}{2\Delta_5}[\sigma_1\sigma_4 - \sigma_2\sigma_3]$	$\frac{1}{\Delta_5}[\sigma_1(\sigma_2\sigma_5 - \sigma_3\sigma_4) - \sigma_5(\sigma_1\sigma_5 - \sigma_3^2) + \sigma_6(\sigma_1\sigma_4 - \sigma_2\sigma_3)]$
a_4			$\frac{1}{2\Delta_4}[\sigma_2]$	$\frac{1}{\Delta_4}[\sigma_2\sigma_5 - \sigma_3\sigma_4]$	$\frac{1}{2\Delta_6}[\sigma_2\sigma_5 - \sigma_3\sigma_4]$
a_5				$\frac{1}{\Delta_5}\Delta_3$	$\frac{1}{\Delta_5}[\sigma_4\Delta_4 + \sigma_6\Delta_3 + \sigma_5(\sigma_2\sigma_3 - \sigma_1\sigma_4)]$
a_6					$\frac{1}{2\Delta_6}[\Delta_4]$

$$\Delta_3 \triangleq \sigma_1\sigma_3 - \sigma_2^2,$$
$$\Delta_4 \triangleq \sigma_2\sigma_4 - \sigma_3^2,$$
$$\Delta_5 \triangleq \sigma_3\Delta_4 + \sigma_5\Delta_3 + \sigma_4(\sigma_2\sigma_3 - \sigma_1\sigma_4),$$
$$\Delta_6 \triangleq \sigma_4(\sigma_3\sigma_5 - \sigma_4^2) + \sigma_5(\sigma_3\sigma_4 - \sigma_2\sigma_5) + \sigma_6\Delta_4.$$

and $\bar{\lambda}$ denotes maximum eigenvalue. For any nonsingular $\mathbf{T}$

$$\bar{\lambda}(\mathbf{X}^{-1}\mathbf{B}\mathbf{B}^{*}) = \bar{\lambda}\left[\mathbf{T}^{-1}(\mathbf{X}^{-1}\mathbf{B}\mathbf{B}^{T})\mathbf{T}\right].$$

For a certain choice of $\mathbf{T}$ (to be discussed), it is possible to show that

$$\bar{\lambda}\left[\mathbf{T}^{-1}(\mathbf{X}^{-1}\mathbf{B}\mathbf{B}^{T})\mathbf{T}\right] = \eta. \qquad \square$$

Now consider $\mathbf{G} \neq \mathbf{0}$ in (5.17). From Theorem 5.2.1, the set of all covariances assignable by $\mathbf{G}$ satisfy

$$(\mathbf{I} - \mathbf{B}\mathbf{B}^{+})(\mathbf{X}\mathbf{A}^{T} + \mathbf{A}\mathbf{X})(\mathbf{I} - \mathbf{B}\mathbf{B}^{+}) = \mathbf{0}. \tag{5.24}$$

Lemma 5.2.2 *Let* $(\mathbf{A}, \mathbf{B})$ *have structure (5.14). Then the following statements are equivalent.*

(i) A matrix $\mathbf{X}$ *has the "signature Hankel" structure*

$$\mathbf{X} = \begin{bmatrix} \sigma_1 & 0 & -\sigma_2 & 0 & \sigma_3 & 0 & -\sigma_4 & & \\ 0 & \sigma_2 & 0 & -\sigma_3 & 0 & \sigma_4 & 0 & & \\ -\sigma_2 & 0 & \sigma_3 & 0 & -\sigma_4 & 0 & \sigma_5 & & \\ 0 & -\sigma_3 & 0 & \sigma_4 & 0 & -\sigma_5 & 0 & & \\ \sigma_3 & 0 & -\sigma_4 & 0 & \sigma_5 & 0 & -\sigma_6 & & \\ 0 & \sigma_4 & 0 & -\sigma_5 & 0 & \sigma_6 & 0 & & \\ -\sigma_4 & 0 & \sigma_5 & 0 & -\sigma_6 & 0 & \sigma_7 & & \\ & & & & & & & \ddots & \\ & & & & & & & & \text{etc.} \end{bmatrix} \tag{5.25}$$

(ii) $\mathbf{X}$ *satisfies (5.24).*

Lemma 5.2.3 *Equation (5.13) describes an asymptotically stable system, if and only if matrix* $\mathbf{X}$ *in (5.17) is a positive definite matrix of "signature Hankel" structure (5.25). Furthermore, a suitable Lyapunov function is*

$$\mathcal{V}(\mathbf{x}(t)) = \mathbf{x}^{T}(t)\mathbf{X}^{-1}\mathbf{x}(t), \tag{5.26}$$

and its time rate change is (for n *= even),*

$$\dot{\mathcal{V}}(\mathbf{x}(t)) = -4(\mathbf{x}_e^{T}\mathbf{a}_e)^2 \tag{5.27}$$

where

$$\mathbf{x}_e^{T} \triangleq \begin{bmatrix} x_2 & x_4 & x_6 & x_8 & \cdots \end{bmatrix}, \quad \mathbf{a}_e^{T} \triangleq \begin{bmatrix} a_2 & a_4 & a_6 & a_8 & \cdots \end{bmatrix} \tag{5.28}$$

and, for n *= odd,*

$$\dot{\mathcal{V}}(\mathbf{x}(t)) = -4(\mathbf{x}_o^{T}\mathbf{a}_o)^2, \tag{5.29}$$

where

$$\mathbf{x}_o^{T} \triangleq \begin{bmatrix} x_1 & x_3 & x_5 & x_7 & \cdots \end{bmatrix}, \quad \mathbf{a}_o^{T} \triangleq \begin{bmatrix} a_1 & a_3 & a_5 & a_7 & \cdots \end{bmatrix}. \tag{5.30}$$

Furthermore, the "time constant" of the system, defined by τ where

$$\tau^{-1} \triangleq \sup_{\mathbf{x}} \frac{\dot{\mathcal{V}}(\mathbf{x}(t))}{\mathcal{V}(\mathbf{x}(t))} \tag{5.31}$$

is

$$\tau = \begin{cases} (\mathbf{a}_e^T \mathbf{X}_e \mathbf{a}_e)^{-1} & (n = \text{even}) \\ (\mathbf{a}_o^T \mathbf{X}_o \mathbf{a}_o)^{-1} & (n = \text{odd}). \end{cases} \tag{5.32}$$

Remark 1: Note from (5.17) that (5.26) is a Lyapunov function suitable for proving stability, since $\mathbf{X} > \mathbf{0}$ is a necessary and sufficient condition for the stability of $\mathbf{A}$, given the controllability of $(\mathbf{A}, \mathbf{B})$ assumed in (5.17).

Remark 2: The $\mathcal{V}$ in (5.26) is negative semidefinite and zero only at the origin $\mathbf{x} = \mathbf{0}$. Hence, $\mathbf{X} > \mathbf{0}$ is necessary and sufficient for stability. Relationships (5.21) and (5.22) have been exploited to obtain expressions (5.27) and (5.29).

Remark 3: Note that the Lyapunov matrix (the inverse of the covariance matrix) has been completely specified in terms of only the n variances σ_i, $i = 1, \ldots, n$. Hence, the study of positive definiteness of $\mathbf{X}$ (and the stability of $\mathbf{A}$) involves only n positive numbers. We should not be surprised, therefore, to later find a one-to-one correspondence between the two sets of parameters (a_i and σ_i).

Proof of Lemma 5.2.2 From Lemma 5.2.2,

$$\mathbf{I} - \mathbf{B}\mathbf{B}^+ = \begin{bmatrix} \mathbf{I}_{n-1} & \mathbf{0} \\ \mathbf{0} & 0 \end{bmatrix}.$$

Hence (5.24) is equivalent to the $(n-1) \times (n-1)$ block of $\mathbf{X}\mathbf{A}^T + \mathbf{A}\mathbf{X}$ equal to zero,

$$(\mathbf{X}\mathbf{A}^T + \mathbf{A}\mathbf{X})_{11} = \mathbf{0}_{(n-1)\times(n-1)} \tag{5.33}$$

or equivalently,

$$\begin{bmatrix} \mathbf{0} & \mathbf{I}_{n-1} \end{bmatrix} \mathbf{X} \begin{bmatrix} \mathbf{I}_{n-1} \\ \mathbf{0} \end{bmatrix} + \begin{bmatrix} \mathbf{I}_{n-1} & \mathbf{0} \end{bmatrix} \mathbf{X} \begin{bmatrix} \mathbf{0} \\ \mathbf{I}_{n-1} \end{bmatrix} = \mathbf{0} \tag{5.34}$$

or equivalently,

$$\mathbf{X}_{i+1,j} = -\mathbf{X}_{i,j+1}, \quad i, j = 1, \ldots, n-1. \tag{5.35}$$

Note from (5.35) and symmetry that $\mathbf{X}_{\alpha\beta} = 0$ when $\alpha + \beta = \text{odd}$. By defining $\sigma_i \triangleq \mathbf{X}_{ii}$ and using symmetry ($\mathbf{X}_{\alpha\beta} = \mathbf{X}_{\beta\alpha}$), (5.35) is equivalent to (5.25).

Consider the unitary coordinate transformation of (5.13) defined by $\mathbf{T} = [\mathbf{T}_o \quad \mathbf{T}_e]$,

$$\mathbf{T}_o \triangleq \begin{bmatrix} 1 & 0 & 0 & 0 & \\ 0 & 0 & 0 & 0 & \\ 0 & 1 & 0 & 0 & \\ 0 & 0 & 0 & 0 & \\ 0 & 0 & 1 & 0 & \\ 0 & 0 & 0 & 0 & \\ 0 & 0 & 0 & 1 & \\ & & & & \ddots \end{bmatrix}_{n \times n_o} \qquad \mathbf{T}_e \triangleq \begin{bmatrix} 0 & 0 & 0 & \\ 1 & 0 & 0 & \\ 0 & 0 & 0 & \\ 0 & 1 & 0 & \\ 0 & 0 & 0 & \\ 0 & 0 & 1 & \\ 0 & 0 & 0 & \\ & & & \ddots \end{bmatrix}_{n \times n_e} \tag{5.36}$$

where

$$n_o \triangleq \frac{\bar{n}}{2}, \qquad \bar{n} \triangleq \begin{cases} n & \text{if } n \text{ is even} \\ n+1 & \text{if } n \text{ is odd} \end{cases}$$

$$n_e \triangleq \frac{\underline{n}}{2}, \qquad \underline{n} \triangleq \begin{cases} n & \text{if } n \text{ is even} \\ n-1 & \text{if } n \text{ is odd} \end{cases}$$

where $\mathbf{x}_o^T \mathbf{x}_e(t) = 0$ since $\mathbf{T}_o \mathbf{T}_e = \mathbf{0}$.

The transformed state vector has these properties

$$\mathbf{T}\mathbf{x} = \begin{bmatrix} \mathbf{x}_o \\ \mathbf{x}_e \end{bmatrix}, \quad \mathbf{x}_o = \begin{bmatrix} x_1 \\ x_3 \\ x_5 \\ \vdots \\ x_{\text{odd}} \end{bmatrix}, \quad \mathbf{x}_e = \begin{bmatrix} x_2 \\ x_4 \\ x_6 \\ \vdots \\ x_{\text{even}} \end{bmatrix}. \tag{5.37}$$

We shall refer to $\mathbf{x}_o \in \mathcal{R}^{n_o}$ and $\mathbf{x}_e \in \mathcal{R}^{n_e}$ as the "odd" and "even" state vectors. Likewise, we shall refer to "odd" and "even" coefficients $\mathbf{a}_o$, $\mathbf{a}_e$ in (5.18).

Define the "signature Toeplitz" matrices

$$\mathbf{\Lambda}_{oo} \in \mathcal{R}^{\frac{n+1}{2} \times n}, \ \mathbf{\Lambda}_{eo} \in \mathcal{R}^{\frac{n-1}{2} \times n},$$

$$\mathbf{\Lambda}_{eo} = \begin{bmatrix} 0 & \cdot & \cdot & \cdot & & \cdot & \cdot & 0 & \cdots & 0 \\ \vdots & \ddots & \ddots & \ddots & & \ddots & \ddots & & & \vdots \\ \vdots & & \ddots & \tilde{\mathbf{a}}_2 & \tilde{\mathbf{a}}_4 & \tilde{\mathbf{a}}_6 & \cdots & \tilde{\mathbf{a}}_{n-1} & \tilde{\mathbf{1}} & \vdots \\ \vdots & & & \ddots & \ddots & \ddots & \ddots & & \ddots & \ddots \ 0 \\ 0 & & \cdot^{\cdot^{\cdot}} & & 0 & \cdot & \cdot & \cdot & \cdot & \cdot \end{bmatrix} \tag{5.38}$$

$$\mathbf{\Lambda}_{oo} = \begin{bmatrix} \cdot & \cdot & \cdot & \cdot & 0 & \cdots & 0 \\ 0 & \ddots & \ddots & \ddots & \ddots & \ddots & \vdots \\ \vdots & \ddots & \tilde{\mathbf{a}}_1 & \tilde{\mathbf{a}}_3 & \tilde{\mathbf{a}}_5 \ \cdots \ \tilde{\mathbf{a}}_n & \ddots & \vdots \\ \vdots & & \ddots & \ddots & \ddots \ \ddots & \ddots & 0 \\ 0 & & \cdots & 0 & \cdot \ \cdot \ \cdot & & \cdot \end{bmatrix} \tag{5.39}$$

and the two $n/2 \times n$ "signature Toeplitz" matrices $\mathbf{\Lambda}_{oe}$, $\mathbf{\Lambda}_{ee}$, composed, respectively, of odd and even coefficients,

$$\mathbf{\Lambda}_{oe} = \begin{bmatrix} \cdot & \cdot & \cdot & \cdot & & \cdot & \cdot & 0 & \cdots & 0 \\ 0 & \ddots & \ddots & \ddots & & \ddots & \ddots & & \ddots & \vdots \\ \vdots & \ddots & \tilde{\mathbf{a}}_1 & \tilde{\mathbf{a}}_3 & \tilde{\mathbf{a}}_5 & \cdots & & \tilde{\mathbf{a}}_{n-1} & \tilde{\mathbf{1}} & \vdots \\ \vdots & & \ddots & \ddots & \ddots & \ddots & & \ddots & \ddots & 0 \\ 0 & & \cdots & 0 & \cdot & \cdot & \cdot & & \cdot & \cdot \end{bmatrix} \tag{5.40}$$

$$\mathbf{\Lambda}_{ee} = \begin{bmatrix} 0 & \cdot & \cdot & \cdot & \cdot & 0 & \cdots & 0 \\ \vdots & \ddots & \ddots & \ddots & \ddots & \ddots & \ddots & \vdots \\ \vdots & & \ddots & \tilde{\mathbf{a}}_2 & \tilde{\mathbf{a}}_4 & \tilde{\mathbf{a}}_6 \cdots & \tilde{\mathbf{a}}_n & \vdots \\ \vdots & & \ddots & \ddots & \ddots & \ddots & \ddots & 0 \\ 0 & \cdots & & 0 & \cdot & \cdot & \cdot & \cdot \end{bmatrix} \tag{5.41}$$

where the diagonal dotted lines denote the elements of a vector (such as $\tilde{\mathbf{a}}_i$) arranged diagonally in the matrix, and where

$$\tilde{\mathbf{a}}_i \triangleq \bar{\mathbf{1}} a_i, \qquad i = \begin{cases} 1, 5, 9, 13, 17, \cdots \\ 2, 6, 10, 14, 18, \cdots \end{cases} \; i \leq n \tag{5.42}$$

$$\tilde{\mathbf{a}}_j \triangleq -\bar{\mathbf{1}} a_j, \qquad j = \begin{cases} 3, 7, 11, 15, \cdots \\ 4, 8, 12, 16, \cdots \end{cases} \; j \leq n \tag{5.43}$$

where

$$\bar{\mathbf{1}}^T \triangleq [1, -1, 1, -1, 1, -1, \cdots] \tag{5.44}$$

$$\tilde{\mathbf{1}} \triangleq (-1)^{n_e} \bar{\mathbf{1}}. \tag{5.45}$$

Theorem 5.2.3 *Let $a_i, i - 1, \ldots, n$ represent any coefficients associated with a stable polynomial. Then the variances of all state variables are given uniquely by, for n = even,*

$$\sigma = \frac{1}{2}\mathbf{\Lambda}_e^{-1}\mathbf{1}, \qquad \mathbf{\Lambda}_e \triangleq \begin{bmatrix} \mathbf{\Lambda}_{oe} \\ \mathbf{\Lambda}_{ee} \end{bmatrix} \tag{5.46}$$

and, for n = odd,

$$\sigma = \frac{1}{2}\mathbf{\Lambda}_o^{-1}\mathbf{1}, \qquad \mathbf{\Lambda}_o \triangleq \begin{bmatrix} \mathbf{\Lambda}_{eo} \\ \mathbf{\Lambda}_{oo} \end{bmatrix}. \tag{5.47}$$

In Theorem 5.2.2, it is clear that the n a_i may be determined from n σ_i by inverting a square matrix of size no greater than $(n+1)/2$. In (5.46) and (5.47), it appears that a full $n \times n$ matrix must be inverted to find the σ_i in terms of given a_i. However, the matrices $\mathbf{\Lambda}_e, \mathbf{\Lambda}_o$ have special structure, and only the last column of $\mathbf{\Lambda}_o^{-1}$ or $\mathbf{\Lambda}_e^{-1}$ need be computed.

Tables 5.2 (a) and (b) list the results of Theorem 5.2.3 for $n = 2, 3, 4, 5, 6$. Note that the stability condition ($\mathbf{X} > \mathbf{0}$) is the same as for Table 5.1, but when expressed in terms of the coefficients, yield the same test as the Hurwitz–Routh stability test.

The Lyapunov equation (5.17) expressed in the transformed coordinates ($\mathbf{T}^T\mathbf{A}\mathbf{T}$, $\mathbf{T}^T\mathbf{B}^T$, $\mathbf{T}^T\mathbf{X}\mathbf{T}$) yields the Lyapunov equation

$$\mathbf{T}^T\mathbf{A}\mathbf{T}\mathbf{T}^T\mathbf{X}\mathbf{T} + \mathbf{T}^T\mathbf{X}\mathbf{T}\mathbf{T}^T\mathbf{A}^T\mathbf{T} + \mathbf{T}^T\mathbf{B}\mathbf{B}^T\mathbf{T} = \mathbf{0} \tag{5.48}$$

where

$$\mathbf{T}^T\mathbf{X}\mathbf{T} = \begin{bmatrix} \mathbf{X}_o & \mathbf{0} \\ \mathbf{0} & \mathbf{X}_e \end{bmatrix}, \qquad \text{since } \mathbf{T}_o^T\mathbf{X}\mathbf{T}_e = \mathbf{0}. \tag{5.49}$$

For n = even, the 1-2 block partition of (5.48) yields

$$\mathbf{X}_e + \mathbf{X}_o \left[\begin{array}{c|c} \mathbf{0} & -\mathbf{a}_o \\ \mathbf{I}_o & \end{array}\right] = \mathbf{0}$$

leading to (5.21) and

$$\mathbf{X}_e\begin{bmatrix}\mathbf{I}_o\\ \mathbf{0}\end{bmatrix}+\mathbf{X}_o\begin{bmatrix}\mathbf{0}\\ \mathbf{I}_o\end{bmatrix} = \mathbf{0}$$

which holds by virtue of (5.19). In a similar way, the 2-2 block of (5.48) yields $\mathbf{a}_e$ in (5.22). For n = odd, the 1-1 block of (5.48) and the 1-2 block of (5.48) yield (5.22).

To prove stability, note that $\mathbf{A}, \mathbf{B}$ is controllable, hence stability is equivalent to $\mathbf{X} > \mathbf{0}$, or $\mathbf{T}^T\mathbf{X}\mathbf{T} > \mathbf{0}$, or $\mathbf{X}_o > \mathbf{0}$, $\mathbf{X}_e > \mathbf{0}$. □

Proof of Theorem 5.2.3. Consider n = even and verify (5.46). Let both equations in (5.21) be arranged in one equation as follows:

$$\begin{bmatrix}\mathbf{X}_o\mathbf{a}_o - \mathbf{X}_e\mathbf{1}\\ \mathbf{X}_3\mathbf{a}_e\end{bmatrix} = \begin{bmatrix}\mathbf{0}\\ \frac{1}{2}\mathbf{1}\end{bmatrix}. \tag{5.50}$$

This equation can be rearranged to collect the n elements σ_i, $i = 1, \ldots, n$ contained in $\mathbf{X}_o$, $\mathbf{X}_e$ into a single vector $\boldsymbol{\sigma}$:

$$\boldsymbol{\sigma}^T = \begin{bmatrix}\sigma_1 & \sigma_2 & \sigma_3\cdots & \sigma_n\end{bmatrix},$$

so that the left-hand side of (5.50) becomes equivalent to $\boldsymbol{\Lambda}_e\boldsymbol{\sigma}$, where $\boldsymbol{\Lambda}_e$ is given by (5.46), (5.40) and (5.41). For example, for $n = 8$, (5.50) is rearranged as follows:

$$\boldsymbol{\Lambda}_e\boldsymbol{\sigma} = \left[\begin{array}{cccc|cccc}
a_1 & -a_3 & a_5 & -a_7 & 1 & 0 & 0 & 0\\
0 & -a_1 & a_3 & -a_5 & a_7 & -1 & 0 & 0\\
0 & 0 & a_1 & -a_3 & a_5 & -a_7 & 1 & 0\\
0 & 0 & 0 & -a_1 & a_3 & -a_5 & a_7 & -1\\ \hline
0 & a_2 & -a_4 & a_6 & -a_8 & 0 & 0 & 0\\
0 & 0 & -a_2 & a_4 & -a_6 & a_8 & 0 & 0\\
0 & 0 & 0 & a_2 & -a_4 & a_6 & -a_8 & 0\\
0 & 0 & 0 & 0 & -a_2 & a_4 & -a_6 & a_8
\end{array}\right]\begin{bmatrix}\sigma_1\\ \sigma_2\\ \sigma_3\\ \sigma_4\\ \sigma_5\\ \sigma_6\\ \sigma_7\\ \sigma_8\end{bmatrix} = \left[\begin{array}{c}0\\0\\0\\0\\ \hline 0\\0\\0\\ \frac{1}{2}\end{array}\right]$$

which is defined by the notation in (5.40), and where

$$\boldsymbol{\Lambda} = \left[\begin{array}{c}\boldsymbol{\Lambda}_{oe}\\ \hline \boldsymbol{\Lambda}_{ee}\end{array}\right] = \left[\begin{array}{c|c}\boldsymbol{\Lambda}_{oe_1} & \boldsymbol{\Lambda}_{oe_2}\\ \hline \boldsymbol{\Lambda}_{ee_1} & \boldsymbol{\Lambda}_{ee_2}\end{array}\right]. \tag{5.51}$$

Now consider (5.47). Rearrange (5.47) to get

$$\begin{bmatrix}\mathbf{X}_e\mathbf{a}_e - \begin{bmatrix}\mathbf{0} & \mathbf{I}_e\end{bmatrix}\mathbf{X}_o\mathbf{1}\\ \mathbf{X}_o\mathbf{a}_o\end{bmatrix} = \begin{bmatrix}\mathbf{0}\\ \frac{1}{2}\mathbf{1}\end{bmatrix}. \tag{5.52}$$

Similarly, since $\mathbf{X}_o$, $\mathbf{X}_e$ contain only n positive numbers σ_i, $i = 1, \cdots, n$, each of the terms $\mathbf{X}_e\mathbf{a}_e$, $\mathbf{X}_o\mathbf{a}_o$, $[\mathbf{0} \;\; \mathbf{I}_e]\,\mathbf{X}_o\mathbf{1}$, can be rearranged as some matrix times the n-vector $\boldsymbol{\sigma}$. Hence, by straight forward construction (5.52) leads to $\boldsymbol{\Lambda}_o\boldsymbol{\sigma} = (1/2)\mathbf{1}$ with the $\boldsymbol{\Lambda}_o$ given by

$$\boldsymbol{\Lambda}_o = \left[\begin{array}{c}\boldsymbol{\Lambda}_{eo}\\ \hline \boldsymbol{\Lambda}_{oo}\end{array}\right] = \left[\begin{array}{ccccccc}
0 & a_2 & -a_4 & a_6 & -1 & 0 & 0\\
0 & 0 & -a_2 & a_4 & -a_6 & 1 & 0\\
0 & 0 & 0 & a_2 & -a_4 & a_6 & -1\\ \hline
a_1 & -a_3 & a_5 & -a_7 & 0 & 0 & 0\\
0 & -a_1 & a_3 & -a_5 & a_7 & 0 & 0\\
0 & 0 & a_1 & -a_3 & a_5 & -a_7 & 0\\
0 & 0 & 0 & -a_1 & a_3 & -a_5 & a_7
\end{array}\right] \tag{5.53}$$

leading to the notation (5.38). This completes the Proof of Theorem 5.2.3. □

Table 5.2 (a) Assignable variances

	$n=2$	$n=3$	$n=4$	$n=5$
Stability conditions $X>0$	$\sigma_1>0,\ \sigma_2>0$	$a_1>0$ $a_3>0$ $\Omega_3>0$	$a_1>0$ $a_4>0$ $a_3a_4-a_2>0,\ \Omega_4>0$	$a_1>0, a_5>0, a_4a_5-a_3>0, \Omega_5>0$ $a_3(a_4a_5-a_3)-a_5(a_2a_5-a_1)>0$
σ_1	$\frac{1}{2a_1a_2}$	$\frac{a_3}{2\Omega_3 a_1}$	$\frac{a_3a_4-a_2}{2\Omega_4 a_1}$	$\frac{1}{2a_1\Omega_5}[a_3(a_4a_5-a_3)-a_5(a_2a_5-a_1)]$
σ_2	$\frac{1}{2a_2}$	$\frac{1}{2\Omega_3}$	$\frac{a_4}{2\Omega_4}$	$\frac{a_4a_5-a_3}{2\Omega_5}$
σ_3		$\frac{a_2}{2\Omega_3}$	$\frac{a_2}{2\Omega_4}$	$\frac{a_2a_5-a_1}{2\Omega_5}$
σ_4			$\frac{a_2a_3-a_1a_4}{2\Omega_4}$	$\frac{a_2a_3-a_1a_4}{2\Omega_5}$
σ_5				$\frac{1}{2\Omega_5}[a_4(a_2a_3-a_1a_4)-a_2(a_2a_5-a_1)]$
σ_6				

$$\Omega_3 \triangleq a_2a_3-a_1,$$
$$\Omega_4 \triangleq a_2(a_3a_4-a_2)-a_1a_4^2,$$
$$\Omega_5 \triangleq -a_5[a_2(a_2a_5-a_1)+a_4(a_1a_4-a_2a_3)]-a_3(a_2a_3-a_1a_4)+a_1(a_2a_5-a_1).$$

Table 5.2 (b) Assignable variances

	$n = 6$
σ_1	$\frac{1}{2\Omega_6 a_1}[a_3(a_4^2 - a_2 a_6) + a_2(a_2 - a_4 a_5) + a_6[a_3(a_3 a_6 - a_4 a_5) + a_5(a_2 a_5 - a_1 a_6) + (a_1 a_4 - a_2 a_3)]]$
σ_2	$\frac{1}{2\Omega_6}[a_4^2 - a_2 a_6 + a_6(a_3 a_6 - a_4 a_5)]$
σ_3	$\frac{1}{2\Omega_6}[a_2 a_4 + a_6(a_1 a_6 - a_2 a_5)]$
σ_4	$\frac{1}{2\Omega_6}[a_2^2 + a_6(a_1 a_4 - a_2 a_3)]$
σ_5	$\frac{1}{2\Omega_6}[a_1(a_4^2 - a_2 a_6) + a_2(a_2 a_5 - a_3 a_4)]$
σ_6	$\frac{1}{2\Omega_6}[a_1[a_4(a_4 a_5 - a_3 a_6) + a_6(a_1 a_6 - a_2 a_5)] + a_2[a_3(a_3 a_6 - a_4 a_5) + a_1(a_4 - a_5 a_6) + a_2(a_5^2 - a_3)]]$

$$\Omega_6 \triangleq a_1 a_4(-a_4^2 + 2a_2 a_6) + a_1 a_6(a_1 a_6^2 + a_4^2 a_5 - a_2 a_5 a_6 - a_3 a_4 a_6)$$
$$+a_2(a_3 a_4^2 + a_2^2 - a_2 a_4 a_5 - a_2 a_3 a_6) + a_2 a_6(a_3^2 a_6 + a_2 a_5^2 + a_1 a_4 - a_2 a_3 - a_1 a_5 a_6 - a_3 a_4 a_5).$$

Theorems 5.2.2 and 5.2.3 have two uses. When $\mathbf{G} = \mathbf{0}$ these theorems characterize the one-to-one relationship between the coefficients of the characteristic polynomial and the variances of each state variable, due to a unit intensity white noise input. If $\mathbf{G} \neq \mathbf{0}$ then $\mathbf{a}_{cl}^T = \mathbf{a}_{ol}^T = -\mathbf{G}$ where $\mathbf{a}_{cl}^T$ and $\mathbf{a}_{ol}^T$ contain the characteristic polynomial coefficients of the closed-loop and open-loop systems, respectively, and $\mathbf{G}$ is the state feedback gain. Obviously, then,

$$\mathbf{G} = \mathbf{a}_{ol}^T - \mathbf{a}_{cl}^T$$

where $\mathbf{a}_{cl}^T$ is given by Theorem 5.2.2 expressed as a function of variances σ_i.

5.2.4 Output Feedback without Measurement Noise

In this section, we generalize the result of the previous section for the case where not all the states are available for feedback. However, we assume that a linear combination of the states can be measured *without noises*, i.e. $\mathbf{D}_z = \mathbf{0}$ (or equivalently, $\mathbf{V} = \mathbf{0}$). In this case, the covariance equation is given by

$$(\mathbf{A} + \mathbf{BGM})\mathbf{X} + \mathbf{X}(\mathbf{A} + \mathbf{BGM})^T + \mathbf{DD}^T = \mathbf{0} \tag{5.54}$$

where the matrices $\mathbf{A}$, $\mathbf{B}$, $\mathbf{M}$, $\mathbf{D}$ and $\mathbf{G}$ are the augmented matrices defined in (4.15). Accordingly, we shall partition the state covariance $\mathbf{X}$ and define the Schur complement of $\mathbf{X}$ as follows:

$$\mathbf{X} = \begin{bmatrix} \mathbf{X}_p & \mathbf{X}_{pc} \\ \mathbf{X}_{pc} & \mathbf{X}_c \end{bmatrix}, \quad \bar{\mathbf{X}}_p \triangleq \mathbf{X}_p - \mathbf{X}_{pc}\mathbf{X}_c^{-1}\mathbf{X}_{pc}^T.$$

Theorem 5.2.4 *[157] Let a positive definite matrix $\mathbf{X} \in \mathcal{R}^{(n_p+n_c)\times(n_p+n_c)}$ be given. Then the following statements are equivalent:*

(i) There exists a controller $\mathbf{G}$ which assigns $\mathbf{X}$ as a state covariance.

(ii) $\mathbf{X}$ satisfies

$$(\mathbf{I} - \mathbf{B}_p\mathbf{B}_p^+)(\mathbf{A}_p\mathbf{X}_p + \mathbf{X}_p\mathbf{A}_p^T + \mathbf{W})(\mathbf{I} - \mathbf{B}_p\mathbf{B}_p^+) = \mathbf{0} \tag{5.55}$$

$$(\mathbf{I} - \mathbf{M}_p^+\mathbf{M}_p)\bar{\mathbf{X}}_p^{-1}(\mathbf{A}_p\bar{\mathbf{X}}_p + \bar{\mathbf{X}}_p\mathbf{A}_p^T + \mathbf{W})\bar{\mathbf{X}}_p^{-1}(\mathbf{I} - \mathbf{M}_p^+\mathbf{M}_p) = \mathbf{0} \tag{5.56}$$

$$(\mathbf{I} - \Gamma\Gamma^+)(\mathbf{I} - \mathbf{M}^+\mathbf{M})\mathbf{X}^{-1}\mathbf{Q} = \mathbf{0}$$

where

$$\Gamma \triangleq (\mathbf{I} - \mathbf{M}^+\mathbf{M})\mathbf{X}^{-1}\mathbf{BB}^+.$$

In this case, all such controllers $\mathbf{G}$ are given by

$$\mathbf{G} = -\frac{1}{2}\mathbf{B}^+(\mathbf{Q} + \mathbf{S})\mathbf{X}^{-1}\mathbf{M}^+ + \mathbf{Z}_F - \mathbf{B}^+\mathbf{B}\mathbf{Z}_F\mathbf{MM}^+$$

where $\mathbf{Z}_F$ is arbitrary and

$$\mathbf{Q} \triangleq \mathbf{AX} + \mathbf{XA}^T + \mathbf{DD}^T \tag{5.57}$$

$$\mathbf{S} \triangleq [\Theta^+\Phi + (\mathbf{I} - \Theta^+\Theta)\mathbf{S}_F](\mathbf{I} - \Theta^+\Theta) - (\Theta^+\Phi)^T \tag{5.58}$$

$$\Theta \triangleq \begin{bmatrix} \mathbf{I} - \mathbf{BB}^+ \\ (\mathbf{I} - \mathbf{M}^+\mathbf{M})\mathbf{X}^{-1} \end{bmatrix}, \quad \Phi \triangleq \begin{bmatrix} -(\mathbf{I} - \mathbf{BB}^+) \\ (\mathbf{I} - \mathbf{M}^+\mathbf{M})\mathbf{X}^{-1} \end{bmatrix}\mathbf{Q}$$

where $\mathbf{S}_F$ is an arbitrary skew-symmetric matrix.

Proof. A given $\mathbf{X} > \mathbf{0}$ is assignable if and only if there exists a $\mathbf{G}$ satisfying (5.54), that is,

$$\mathbf{BGMX} + (\mathbf{BGMX})^T + \mathbf{Q} = \mathbf{0}.$$

Then the result directly follows from Theorem 2.3.7. In this case, the solvability conditions are given by

$$(\mathbf{I} - \mathbf{BB}^+)\mathbf{Q}(\mathbf{I} - \mathbf{BB}^+) = \mathbf{0} \tag{5.59}$$

$$(\mathbf{I} - \mathbf{M}^+\mathbf{M})\mathbf{X}^{-1}\mathbf{Q}\mathbf{X}^{-1}(\mathbf{I} - \mathbf{M}^+\mathbf{M}) = \mathbf{0} \tag{5.60}$$

$$(\mathbf{I} - \Gamma\Gamma^+)(\mathbf{I} - \mathbf{M}^+\mathbf{M})\mathbf{X}^{-1}\mathbf{Q} = \mathbf{0}.$$

Substituting the partitioned matrices defined in (4.15) into (5.59) and (5.60) yields (5.55) and (5.56), respectively. □

In Theorem 5.2.4, the controller order n_c is fixed by the dimension of the prespecified state covariance $\mathbf{X} \in \mathcal{R}^{(n_p+n_c)\times(n_p+n_c)}$. The static output feedback case ($n_c = 0$) can be deduced by replacing the augmented matrices $\mathbf{A}$, $\mathbf{B}$, etc. with the original matrices $\mathbf{A}_p$, $\mathbf{B}_p$, etc., and setting $\mathbf{X}_p = \bar{\mathbf{X}}_p = \mathbf{X}$.

5.2.5 Static Output Feedback with Noisy Measurements

This section treats the case where the measurements are *completely* contaminated by noises, i.e. $\mathbf{D}_z\mathbf{D}_z^T \triangleq \mathbf{V} > \mathbf{0}$. As in the state feedback case, we consider the covariance controller of static (constant) feedback gain $\mathbf{G}$. In this case, the covariance equation is given by

$$(\mathbf{A}_p + \mathbf{B}_p\mathbf{G}\mathbf{M}_p)\mathbf{X} + \mathbf{X}(\mathbf{A}_p + \mathbf{B}_p\mathbf{G}\mathbf{M}_p)^T + (\mathbf{D}_p + \mathbf{B}_p\mathbf{G}\mathbf{D}_z)(\mathbf{D}_p + \mathbf{B}_p\mathbf{G}\mathbf{D}_z)^T = \mathbf{0} \tag{5.61}$$

or equivalently, using the definition (5.1),

$$(\mathbf{A}_p + \mathbf{B}_p\mathbf{G}\mathbf{M}_p)\mathbf{X} + \mathbf{X}(\mathbf{A}_p + \mathbf{B}_p\mathbf{G}\mathbf{M}_p)^T + \mathbf{W} + \mathbf{B}_p\mathbf{G}\mathbf{V}\mathbf{G}^T\mathbf{B}_p^T = \mathbf{0}. \tag{5.62}$$

Theorem 5.2.5 *Let a positive definite matrix $\mathbf{X} \in \mathcal{R}^{n_p\times n_p}$ be given. Then the following statements are equivalent:*

(i) There exists a control gain $\mathbf{G}$ which assigns $\mathbf{X}$ as a state covariance.
(ii) $\mathbf{X}$ satisfies

$$(\mathbf{I} - \mathbf{B}_p\mathbf{B}_p^+)(\mathbf{A}_p\mathbf{X} + \mathbf{X}\mathbf{A}_p^T + \mathbf{W})(\mathbf{I} - \mathbf{B}_p\mathbf{B}_p^+) = \mathbf{0}$$

$$\mathbf{A}_p\mathbf{X} + \mathbf{X}\mathbf{A}_p^T - \mathbf{X}\mathbf{M}_p^T\mathbf{V}^{-1}\mathbf{M}_p\mathbf{X} + \mathbf{W} + \mathbf{L}_p\mathbf{L}_p^T = \mathbf{0}$$

for some matrix $\mathbf{L}_p \in \mathcal{R}^{n_p\times n_z}$.

In this case, all such control gains $\mathbf{G}$ are given by

$$\mathbf{G} = \mathbf{B}_p^+(\mathbf{L}_p\mathbf{U}_p\mathbf{V}^{-\frac{1}{2}} - \mathbf{X}\mathbf{M}_p^T\mathbf{V}^{-1}) + (\mathbf{I} - \mathbf{B}_p^+\mathbf{B}_p)\mathbf{Z}_F \tag{5.63}$$

where $\mathbf{Z}_F$ is arbitrary and

$$\mathbf{U}_p = \mathbf{V}_L\begin{bmatrix} \mathbf{I} & \mathbf{0} \\ \mathbf{0} & \mathbf{U}_F \end{bmatrix}\mathbf{V}_R^T$$

$$(\mathbf{I} - \mathbf{B}_p\mathbf{B}_p^+)\mathbf{L}_p = \mathbf{U}_L\begin{bmatrix} \Sigma_L & \mathbf{0} \\ \mathbf{0} & \mathbf{0} \end{bmatrix}\mathbf{V}_L^T \qquad (SVD)$$

$$(\mathbf{I} - \mathbf{B}_p\mathbf{B}_p^+)\mathbf{X}\mathbf{M}_p^T\mathbf{V}^{-\frac{1}{2}} = \mathbf{U}_L\begin{bmatrix} \Sigma_L & \mathbf{0} \\ \mathbf{0} & \mathbf{0} \end{bmatrix}\mathbf{V}_R^T \qquad (SVD)$$

where $\mathbf{U}_F$ is an arbitrary orthogonal matrix.

Proof. We shall solve the equation (5.62) for $\mathbf{G}$. After expanding, completing the square with respect to $\mathbf{B}_p\mathbf{G}$ yields

$$(\mathbf{B}_p\mathbf{G} + \mathbf{X}\mathbf{M}_p^T\mathbf{V}^{-1})\mathbf{V}(\mathbf{B}_p\mathbf{G} + \mathbf{X}\mathbf{M}_p^T\mathbf{V}^{-1})^T = \mathbf{Q}$$

$$\mathbf{Q} \triangleq \mathbf{X}\mathbf{M}_p^T\mathbf{V}^{-1}\mathbf{M}_p\mathbf{X} - \mathbf{A}_p\mathbf{X} - \mathbf{X}\mathbf{A}_p^T - \mathbf{W}.$$

Applying Theorem 2.3.9, the above equation is solvable for $\mathbf{G}$ if and only if

$$\mathbf{Q} \geq 0, \quad \text{rank}(\mathbf{Q}) \leq n_z$$

$$(\mathbf{I} - \mathbf{B}_p\mathbf{B}_p^+)(\mathbf{Q} - \mathbf{X}\mathbf{M}_p^T\mathbf{V}^{-1}\mathbf{M}_p\mathbf{X})(\mathbf{I} - \mathbf{B}_p\mathbf{B}_p^+) = \mathbf{0}$$

or equivalently, there exists $\mathbf{L}_p \in \mathcal{R}^{n_p \times n_z}$ such that

$$\mathbf{Q} = \mathbf{L}_p\mathbf{L}_p^T$$

$$(\mathbf{I} - \mathbf{B}_p\mathbf{B}_p^+)(\mathbf{A}_p\mathbf{X} + \mathbf{X}\mathbf{A}_p^T + \mathbf{W})(\mathbf{I} - \mathbf{B}_p\mathbf{B}_p^+) = \mathbf{0}.$$

In this case, all solutions $\mathbf{G}$ are given by (5.63). □

5.2.6 Dynamic Output Feedback with Noisy Measurements

In this section, we impose the same assumption ($\mathbf{V} > \mathbf{0}$), as in Section 5.2.5, but we consider the dynamic output feedback covariance controller. In this case, the state covariance satisfies

$$(\mathbf{A} + \mathbf{B}\mathbf{G}\mathbf{M})\mathbf{X} + \mathbf{X}(\mathbf{A} + \mathbf{B}\mathbf{G}\mathbf{M})^T + (\mathbf{D} + \mathbf{B}\mathbf{G}\mathbf{E})(\mathbf{D} + \mathbf{B}\mathbf{G}\mathbf{E})^T = \mathbf{0}. \tag{5.64}$$

The above Lyapunov equation has the same structure as in (5.61) for the static output feedback case. However, due to the singularity of the matrix

$$\mathbf{E}\mathbf{E}^T = \begin{bmatrix} \mathbf{D}_z\mathbf{D}_z^T & \mathbf{0} \\ \mathbf{0} & \mathbf{0} \end{bmatrix} = \begin{bmatrix} \mathbf{V} & \mathbf{0} \\ \mathbf{0} & \mathbf{0} \end{bmatrix}, \quad \mathbf{V} > \mathbf{0},$$

the derivation of the covariance controller is much more involved.

Theorem 5.2.6 *Let a positive definite matrix $\mathbf{X} \in \mathcal{R}^{(n_p+n_c)\times(n_p+n_c)}$ be given. Then the following statements are equivalent:*

(i) There exists a controller which assigns $\mathbf{X}$ as a state covariance.

(ii) $\mathbf{X}$ satisfies

$$(\mathbf{I} - \mathbf{B}_p\mathbf{B}_p^+)(\mathbf{A}_p\mathbf{X}_p + \mathbf{X}_p\mathbf{A}_p^T + \mathbf{W})(\mathbf{I} - \mathbf{B}_p\mathbf{B}_p^+) = \mathbf{0}$$

$$\mathbf{A}_p\bar{\mathbf{X}}_p + \bar{\mathbf{X}}_p\mathbf{A}_p^T - \bar{\mathbf{X}}_p\mathbf{M}_p^T\mathbf{V}^{-1}\mathbf{M}_p\bar{\mathbf{X}}_p + \mathbf{W} + \mathbf{L}_p\mathbf{L}_p^T = \mathbf{0}$$

$$\bar{\mathbf{Q}} = (\mathbf{B}\mathbf{P}_K)(\mathbf{B}\mathbf{P}_K)^+\bar{\mathbf{Q}}(\bar{\mathbf{M}}\mathbf{P}_C\mathbf{P}_N)^+(\bar{\mathbf{M}}\mathbf{P}_C\mathbf{P}_N) \tag{5.65}$$

for some $\mathbf{L}_p \in \mathcal{R}^{n_p \times n_z}$ where

$$\mathbf{X} \triangleq \begin{bmatrix} \mathbf{X}_p & \mathbf{X}_{pc} \\ \mathbf{X}_{pc}^T & \mathbf{X}_c \end{bmatrix}, \quad \bar{\mathbf{X}}_p \triangleq \mathbf{X}_p - \mathbf{X}_{pc}\mathbf{X}_c^{-1}\mathbf{X}_{pc}^T$$

$$\bar{\mathbf{M}} \triangleq [\ \mathbf{M}_p \quad \mathbf{0}\ ], \quad \mathbf{C} \triangleq [\ \mathbf{0} \quad \mathbf{I}_{n_c}\ ] \tag{5.66}$$
$$\bar{\mathbf{Q}} \triangleq (\mathbf{I} - \mathbf{B}\mathbf{K}^+\mathbf{P}_C\mathbf{X}^{-1})(\mathbf{A}\mathbf{X} + \mathbf{X}\mathbf{A}^T + \mathbf{D}\mathbf{D}^T)\mathbf{X}^{-1}\mathbf{P}_C\mathbf{P}_N$$
$$\mathbf{P}_C \triangleq \mathbf{I} - \mathbf{C}^+\mathbf{C}, \quad \mathbf{P}_N \triangleq \mathbf{I} - \mathbf{N}\mathbf{N}^+, \quad \mathbf{P}_K \triangleq \mathbf{I} - \mathbf{K}^+\mathbf{K}$$
$$\mathbf{K} \triangleq \mathbf{P}_C\mathbf{X}^{-1}\mathbf{B}, \quad \mathbf{N} \triangleq \mathbf{K}\mathbf{B}^+.$$

In this case, all such controllers are given by

$$\begin{aligned}\begin{bmatrix} \mathbf{C}_c \\ \mathbf{A}_c \end{bmatrix} &= -\frac{1}{2}\mathbf{B}^+[\hat{\mathbf{Q}} + \Theta^+\Phi(\mathbf{I} - \Theta^+\Theta) - (\Theta^+\Phi)^T]\mathbf{X}^{-1}\mathbf{C}^+ \\ &\quad + \mathbf{B}^+(\mathbf{I} - \Theta^+\Theta)\mathbf{S}_F(\mathbf{I} - \Theta^+\Theta)\mathbf{X}^{-1}\mathbf{C}^+ \\ &\quad + (\mathbf{I} - \mathbf{B}\mathbf{B}^+)\mathbf{Z}_{F1}\end{aligned}$$

$$\begin{aligned}\begin{bmatrix} \mathbf{D}_c \\ \mathbf{B}_c \end{bmatrix} &= \mathbf{K}^+(\mathbf{L}\mathbf{U}\mathbf{V}^{-\frac{1}{2}} - \mathbf{P}_C\bar{\mathbf{M}}^T\mathbf{V}^{-1}) \\ &\quad - \mathbf{P}_K(\mathbf{B}\mathbf{P}_K)^+\bar{\mathbf{Q}}(\bar{\mathbf{M}}\mathbf{P}_C\mathbf{P}_N)^+ \\ &\quad - \mathbf{P}_K\mathbf{Z}_{F2} + \mathbf{P}_K(\mathbf{B}\mathbf{P}_K)^+(\mathbf{B}\mathbf{P}_K)\mathbf{Z}_{F2}(\bar{\mathbf{M}}P_C\mathbf{P}_N)(\bar{\mathbf{M}}\mathbf{P}_C\mathbf{P}_N)^+\end{aligned} \tag{5.67}$$

where

$$\hat{\mathbf{Q}} \triangleq (\mathbf{A} + \mathbf{B}\mathbf{G}_1\bar{\mathbf{M}})\mathbf{X} + \mathbf{X}(\mathbf{A} + \mathbf{B}\mathbf{G}_1\bar{\mathbf{M}})^T + \mathbf{D}\mathbf{D}^T + \mathbf{B}\mathbf{G}_1\mathbf{V}\mathbf{G}_1^T\mathbf{B}^T$$
$$\Theta \triangleq \begin{bmatrix} \mathbf{I} - \mathbf{B}\mathbf{B}^+ \\ \mathbf{P}_C\mathbf{X}^{-1} \end{bmatrix}, \quad \Phi \triangleq \begin{bmatrix} -(\mathbf{I} - \mathbf{B}\mathbf{B}^+) \\ \mathbf{P}_C\mathbf{X}^{-1} \end{bmatrix}\hat{\mathbf{Q}}$$
$$\mathbf{L} \triangleq \begin{bmatrix} \mathbf{L}_p \\ \mathbf{0} \end{bmatrix}, \quad \mathbf{G}_1 \triangleq \begin{bmatrix} \mathbf{D}_c \\ \mathbf{B}_c \end{bmatrix}$$
$$\mathbf{U} \triangleq \mathbf{V}_L\begin{bmatrix} \mathbf{I} & \mathbf{0} \\ \mathbf{0} & \mathbf{U}_F \end{bmatrix}\mathbf{V}_R^T \tag{5.68}$$
$$\mathbf{P}_N\mathbf{L} \triangleq \mathbf{U}_L\begin{bmatrix} \Sigma_L & \mathbf{0} \\ \mathbf{0} & \mathbf{0} \end{bmatrix}\mathbf{V}_L^T \qquad (SVD)$$
$$\mathbf{P}_N\mathbf{P}_C\bar{\mathbf{M}}^T\mathbf{V}^{-\frac{1}{2}} = \mathbf{U}_L\begin{bmatrix} \Sigma_L & \mathbf{0} \\ \mathbf{0} & \mathbf{0} \end{bmatrix}\mathbf{V}_R^T \qquad (SVD)$$

where $\mathbf{S}_F$ is an arbitrary skew-symmetric matrix and $\mathbf{U}_F$ is an arbitrary orthogonal matrix and $\mathbf{Z}_{F1}$, $\mathbf{Z}_{F2}$ are arbitrary.

Proof. We shall solve (5.64) for controller $\mathbf{G}$. Defining $\bar{\mathbf{M}}$ and $\mathbf{C}$ as in (5.66) and

$$\mathbf{Q} \triangleq \mathbf{A}\mathbf{X} + \mathbf{X}\mathbf{A}^T + \mathbf{D}\mathbf{D}^T, \quad \mathbf{G}_1 \triangleq \begin{bmatrix} \mathbf{D}_c \\ \mathbf{B}_c \end{bmatrix}, \quad \mathbf{G}_2 \triangleq \begin{bmatrix} \mathbf{C}_c \\ \mathbf{A}_c \end{bmatrix},$$

(5.64) can be written

$$\mathbf{B}\mathbf{G}_1\mathbf{V}\mathbf{G}_1^T\mathbf{B}^T + (\mathbf{B}\mathbf{G}_1\bar{\mathbf{M}}\mathbf{X})^T + (\mathbf{B}\mathbf{G}_1\bar{\mathbf{M}}\mathbf{X})^T + \mathbf{B}\mathbf{G}_2\mathbf{C}\mathbf{X} + (\mathbf{B}\mathbf{G}_2\mathbf{C}\mathbf{X})^T + \mathbf{Q} = \mathbf{0}.$$

Completing the square with respect to $\mathbf{B}\mathbf{G}_1$ yields

$$\begin{aligned}\mathbf{0} &= (\mathbf{B}\mathbf{G}_1 + \mathbf{X}\bar{\mathbf{M}}^T\mathbf{V}^{-1})\mathbf{V}(\mathbf{B}\mathbf{G}_1 + \mathbf{X}\bar{\mathbf{M}}^T\mathbf{V}^{-1})^T + \mathbf{B}\mathbf{G}_2\mathbf{C}\mathbf{X} + (\mathbf{B}\mathbf{G}_2\mathbf{C}\mathbf{X})^T \\ &\quad + \mathbf{Q} - \mathbf{X}\bar{\mathbf{M}}^T\mathbf{V}^{-1}\bar{\mathbf{M}}\mathbf{X}\end{aligned}$$

or equivalently, using $\hat{\mathbf{Q}}$ defined as above,

$$\mathbf{B}\mathbf{G}_2\mathbf{C}\mathbf{X} + (\mathbf{B}\mathbf{G}_2\mathbf{C}\mathbf{X})^T + \hat{\mathbf{Q}} = \mathbf{0}.$$

From Theorem 2.3.7, the above equation is solvable for $\mathbf{G}_2$ if and only if

$$(\mathbf{I} - \mathbf{B}\mathbf{B}^+)\hat{\mathbf{Q}}(\mathbf{I} - \mathbf{B}\mathbf{B}^+) = \mathbf{0} \tag{5.69}$$

$$(\mathbf{I} - \mathbf{C}^+\mathbf{C})\mathbf{X}^{-1}\hat{\mathbf{Q}}\mathbf{X}^{-1}(\mathbf{I} - \mathbf{C}^+\mathbf{C}) = \mathbf{0} \tag{5.70}$$

$$(\mathbf{I} - \mathbf{N}\mathbf{N}^+)(\mathbf{I} - \mathbf{C}^+\mathbf{C})\mathbf{X}^{-1}\hat{\mathbf{Q}} = \mathbf{0} \tag{5.71}$$

hold, in which case, all solutions $\mathbf{G}_2$ are given by (5.67). Using the property $(\mathbf{I} - \mathbf{B}\mathbf{B}^+)\mathbf{B} = \mathbf{0}$, (5.69) is equivalent to

$$(\mathbf{I} - \mathbf{B}\mathbf{B}^+)\mathbf{Q}(\mathbf{I} - \mathbf{B}\mathbf{B}^+) = \mathbf{0}. \tag{5.72}$$

Note that

$$\mathbf{P}_N\mathbf{N} = \mathbf{0} \;\Rightarrow\; \mathbf{P}_N\mathbf{N}\mathbf{B} = (\mathbf{I} - \mathbf{N}\mathbf{N}^+)(\mathbf{I} - \mathbf{C}^+\mathbf{C})\mathbf{X}^{-1}\mathbf{B} = \mathbf{0}.$$

Hence (5.71) is equivalent to

$$\mathbf{P}_N\mathbf{P}_C\mathbf{X}^{-1}(\mathbf{Q} + \mathbf{X}\bar{\mathbf{M}}^T\mathbf{G}_1^T\mathbf{B}^T) = \mathbf{0}. \tag{5.73}$$

Note that (5.70) holds if and only if there exists $\mathbf{L} \in \mathcal{R}^{(n_p+n_c)\times n_z}$ such that

$$\mathbf{P}_C\mathbf{X}^{-1}(\mathbf{Q} - \mathbf{X}\bar{\mathbf{M}}^T\mathbf{V}^{-1}\bar{\mathbf{M}}\mathbf{X})\mathbf{X}^{-1}\mathbf{P}_C + \mathbf{L}\mathbf{L}^T = \mathbf{0} \tag{5.74}$$

$$\mathbf{L}\mathbf{L}^T = \mathbf{P}_C\mathbf{X}^{-1}(\mathbf{B}\mathbf{G}_1 + \mathbf{X}\bar{\mathbf{M}}^T\mathbf{V}^{-1})\mathbf{V}(\mathbf{B}\mathbf{G}_1 + \mathbf{X}\bar{\mathbf{M}}^T\mathbf{V}^{-1})^T\mathbf{X}^{-1}\mathbf{P}_C. \tag{5.75}$$

Now we claim that, for any given $\mathbf{L}$ satisfying (5.74), (5.75) is always solvable for $\mathbf{G}_1$ provided (5.72) holds. This can be shown as follows.
First note that (5.75) can be written

$$\mathbf{L}\mathbf{L}^T = (\mathbf{K}\mathbf{G}_1 + \mathbf{P}_C\bar{\mathbf{M}}^T\mathbf{V}^{-1})\mathbf{V}(\mathbf{K}\mathbf{G}_1 + \mathbf{P}_C\bar{\mathbf{M}}^T\mathbf{V}^{-1})^T. \tag{5.76}$$

Note that (5.76) is solvable for $\mathbf{G}_1$ if and only if

$$\mathbf{L}\mathbf{L}^T = (\mathbf{N}\hat{\mathbf{G}}_1 + \mathbf{P}_C\bar{\mathbf{M}}^T\mathbf{V}^{-1})\mathbf{V}(\mathbf{N}\hat{\mathbf{G}}_1 + \mathbf{P}_C\bar{\mathbf{M}}^T\mathbf{V}^{-1})^T \tag{5.77}$$

is solvable for $\hat{\mathbf{G}}_1$ since, if $\mathbf{G}_1$ exists, then $\hat{\mathbf{G}}_1 \triangleq \mathbf{B}\mathbf{G}_1$ is a solution to (5.77), and if $\hat{\mathbf{G}}_1$ exists, then $\mathbf{G}_1 \triangleq \mathbf{B}^+\hat{\mathbf{G}}_1$ is a solution to (5.76). Hence, from Theorem 2.3.9, there exists $\mathbf{G}_1$ satisfying (5.76) if and only if

$$\mathbf{P}_N(\mathbf{L}\mathbf{L}^T - \mathbf{P}_C\bar{\mathbf{M}}^T\mathbf{V}^{-1}\bar{\mathbf{M}}\mathbf{P}_C)\mathbf{P}_N = \mathbf{0} \tag{5.78}$$

holds, in which case, all such $\mathbf{G}_1$ are given by

$$\mathbf{G}_1 = \mathbf{K}^+(\mathbf{L}\mathbf{U}\mathbf{V}^{-\frac{1}{2}} - \mathbf{P}_C\bar{\mathbf{M}}^T\mathbf{V}^{-1}) + \mathbf{P}_K\mathbf{Z}_K \tag{5.79}$$

where $\mathbf{Z}_K$ is arbitrary and $\mathbf{U}$ is given in (5.68). Using (5.74), the existence condition (5.78) is equivalent to

$$\mathbf{P}_N\mathbf{P}_C\mathbf{X}^{-1}\mathbf{Q}\mathbf{X}^{-1}\mathbf{P}_C\mathbf{P}_N = \mathbf{0}. \tag{5.80}$$

Since $\mathbf{Q}$ satisfies (5.72), there exists $\mathbf{Z}_Q$ such that

$$\mathbf{Q} = \mathbf{Z}_Q - (\mathbf{I} - \mathbf{B}\mathbf{B}^+)\mathbf{Z}_Q(\mathbf{I} - \mathbf{B}\mathbf{B}^+).$$

Noting that $\mathbf{P}_N\mathbf{P}_C\mathbf{X}^{-1}\mathbf{B}\mathbf{B}^+ = \mathbf{P}_N\mathbf{N} = \mathbf{0}$, it is easy to see that (5.80) holds. Thus (5.75) is solvable for $\mathbf{G}_1$, and all solutions are given by (5.79). Finally, we need to find the existence condition and all matrices $\mathbf{U}_F$ and $\mathbf{Z}_K$ in (5.79) such that (5.73) holds. Substituting (5.79) into (5.73) yields

$$\mathbf{Q}\mathbf{X}^{-1}\mathbf{P}_C\mathbf{P}_N + \mathbf{B}\mathbf{K}^+(\mathbf{L}\mathbf{U}\mathbf{V}^{-\frac{1}{2}} - \mathbf{P}_C\bar{\mathbf{M}}^T\mathbf{V}^{-1})\bar{\mathbf{M}}\mathbf{P}_C\mathbf{P}_N + \mathbf{B}\mathbf{P}_K\mathbf{Z}_K\bar{\mathbf{M}}\mathbf{P}_C\mathbf{P}_N = \mathbf{0}.$$

Using the definition of $\mathbf{U}$ in (5.68), the second term in the above equation is

$$\begin{aligned}
&\mathbf{B}\mathbf{K}^+(\mathbf{L}\mathbf{V}_L\begin{bmatrix} \mathbf{I} & \mathbf{0} \\ \mathbf{0} & \mathbf{U}_F \end{bmatrix}\mathbf{V}_R^T\mathbf{V}^{-\frac{1}{2}} - \mathbf{P}_C\bar{\mathbf{M}}^T\mathbf{V}^{-1})\bar{\mathbf{M}}\mathbf{P}_C\mathbf{P}_N \\
&= \mathbf{B}\mathbf{K}^+(\mathbf{L}\mathbf{V}_L\begin{bmatrix} \mathbf{I} & \mathbf{0} \\ \mathbf{0} & \mathbf{U}_F \end{bmatrix}\mathbf{V}_R^T\mathbf{V}_R\begin{bmatrix} \Sigma_L & \mathbf{0} \\ \mathbf{0} & \mathbf{0} \end{bmatrix}\mathbf{U}_L) - \mathbf{B}\mathbf{K}^+\mathbf{P}_C\bar{\mathbf{M}}^T\mathbf{V}^{-1}\bar{\mathbf{M}}\mathbf{P}_C\mathbf{P}_N \\
&= \mathbf{B}\mathbf{K}^+(\mathbf{L}\mathbf{L}^T - \mathbf{P}_C\bar{\mathbf{M}}\mathbf{V}^{-1}\bar{\mathbf{M}}\mathbf{P}_C)\mathbf{P}_N \\
&= -\mathbf{B}\mathbf{K}^+\mathbf{P}_C\mathbf{X}^{-1}\mathbf{Q}\mathbf{X}^{-1}\mathbf{P}_C\mathbf{P}_N
\end{aligned}$$

where we used (5.74) in the last equality. Hence we have

$$(\mathbf{I} - \mathbf{B}\mathbf{K}^+\mathbf{P}_C\mathbf{X}^{-1})\mathbf{Q}\mathbf{X}^{-1}\mathbf{P}_C\mathbf{P}_N = -\mathbf{B}\mathbf{P}_K\mathbf{Z}_K\bar{\mathbf{M}}\mathbf{P}_C\mathbf{P}_N$$

or equivalently,

$$\bar{\mathbf{Q}} = -\mathbf{B}\mathbf{P}_K\mathbf{Z}_K\bar{\mathbf{M}}\mathbf{P}_C\mathbf{P}_N.$$

There exists $\mathbf{Z}_K$ solving the above equation if and only if

$$\bar{\mathbf{Q}} = (\mathbf{B}\mathbf{P}_K)(\mathbf{B}\mathbf{P}_K)^+\bar{\mathbf{Q}}(\bar{\mathbf{M}}\mathbf{P}_C\mathbf{P}_N)^+(\bar{\mathbf{M}}\mathbf{P}_C\mathbf{P}_N)$$

holds, in which case, all such $\mathbf{Z}_K$ are

$$\mathbf{Z}_K = -(\mathbf{B}\mathbf{P}_K)^+\bar{\mathbf{Q}}(\bar{\mathbf{M}}\mathbf{P}_C\mathbf{P}_N)^+ + \mathbf{Z}_{F2} - (\mathbf{B}\mathbf{P}_K)^+(\mathbf{B}\mathbf{P}_K)\mathbf{Z}_{F2}(\bar{\mathbf{M}}\mathbf{P}_C\mathbf{P}_N)(\bar{\mathbf{M}}\mathbf{P}_C\mathbf{P}_N)^+ \quad (5.81)$$

where $\mathbf{Z}_{F2}$ is arbitrary. Substituting (5.81) into (5.79) yields (5.67). Finally, the first two conditions in statement (ii) can be obtained by substituting the augmented matrices into (5.72) and (5.74). □

5.2.7 Structure of Covariance Controllers

In this section, we shall consider full-order ($n_c = n_p$) dynamic controllers. The closed-loop state covariance

$$\mathbf{X} = \begin{bmatrix} \mathbf{X}_p & \mathbf{X}_{pc} \\ \mathbf{X}_{pc}^T & \mathbf{X}_c \end{bmatrix} \quad (5.82)$$

has dimensions $2n_p \times 2n_p$. For simplicity, we assume that the square matrix $\mathbf{X}_{pc}$ is invertible. Under this assumption, the following theorem shows that the third assignability condition (5.65) in Theorem 5.2.6 becomes redundant, and the structure of all full-order covariance controllers is shown to be observer-based.

Theorem 5.2.7 *Let a positive definite matrix* $\mathbf{X} \in \mathcal{R}^{2n_p \times 2n_p}$ *be given. Then the following statements are equivalent:*

(i) There exists a full-order dynamic controller which assigns $\mathbf{X}$ *as a state covariance.*
(ii) $\mathbf{X}$ *satisfies*

$$(\mathbf{I}-\mathbf{B}_p\mathbf{B}_p^+)(\mathbf{A}_p\mathbf{X}_p+\mathbf{X}_p\mathbf{A}_p^T+\mathbf{W})(\mathbf{I}-\mathbf{B}_p\mathbf{B}_p^+)=\mathbf{0}, \tag{5.83}$$

$$\mathbf{A}_p\bar{\mathbf{X}}_p+\bar{\mathbf{X}}_p\mathbf{A}_p^T-\bar{\mathbf{X}}_p\mathbf{M}_p\mathbf{V}^{-1}\mathbf{M}_p\bar{\mathbf{X}}_p+\mathbf{W}+\mathbf{L}_p\mathbf{L}_p^T=\mathbf{0} \tag{5.84}$$

for some $\mathbf{L}_p\in\mathcal{R}^{n_p\times n_z}$ *where*

$$\bar{\mathbf{X}}_p\triangleq\mathbf{X}_p-\mathbf{X}_{pc}\mathbf{X}_c^{-1}\mathbf{X}_{pc}^T. \tag{5.85}$$

If these conditions hold, all controllers which assign $\mathbf{X}$ *are given by*

$$\begin{aligned}\mathbf{A}_c &= \mathbf{X}_c\mathbf{X}_{pc}^{-1}(\mathbf{A}_p+\mathbf{B}_p\mathbf{D}_c\mathbf{M}_p+\mathbf{L}_p\mathbf{V}^{\frac{1}{2}}\mathbf{B}_c^T\mathbf{X}_{pc}^{-1})\mathbf{X}_{pc}\mathbf{X}_c^{-1}\\ &\quad -\mathbf{B}_c\mathbf{M}_p\mathbf{X}_{pc}\mathbf{X}_c^{-1}+\mathbf{X}_c\mathbf{X}_{pc}^{-1}\mathbf{B}_p\mathbf{C}_c \qquad (5.86)\\ \mathbf{B}_c &= \mathbf{X}_c\mathbf{X}_{pc}^{-1}(\bar{\mathbf{X}}_p\mathbf{M}_p^T\mathbf{V}^{-1}+\mathbf{B}_p\mathbf{D}_c-\mathbf{L}_p\mathbf{V}^{-\frac{1}{2}}) \qquad (5.87)\\ \mathbf{C}_c &= \frac{1}{2}\mathbf{B}_p^+\hat{\mathbf{Q}}_p(2\mathbf{I}-\mathbf{B}_p\mathbf{B}_p^+)\mathbf{X}_{pc}^{-T}\\ &\quad +\mathbf{B}_p^+\mathbf{S}_F\mathbf{B}_p\mathbf{B}_p^+\mathbf{X}_{pc}^{-T}+(\mathbf{I}-\mathbf{B}_p^+\mathbf{B}_p)\mathbf{Z}_F \qquad (5.88)\\ \mathbf{D}_c &= \textit{arbitrary} \qquad (5.89)\end{aligned}$$

where $\mathbf{Z}_F$ *is arbitrary and* $\mathbf{S}_F$ *is an arbitrary skew-symmetric matrix and*

$$\hat{\mathbf{Q}}_p\triangleq(\mathbf{A}_p+\mathbf{B}_p\mathbf{D}_c\mathbf{M}_p)\mathbf{X}_p+\mathbf{X}_p(\mathbf{A}_p+\mathbf{B}_p\mathbf{D}_c\mathbf{M}_p)^T+\mathbf{W}+\mathbf{B}_p\mathbf{D}_c\mathbf{V}\mathbf{D}_c^T\mathbf{B}_p^T. \tag{5.90}$$

Proof. Consider the covariance equation

$$\mathbf{Q}\triangleq(\mathbf{A}+\mathbf{B}\mathbf{G}\mathbf{M})\mathbf{X}+\mathbf{X}(\mathbf{A}+\mathbf{B}\mathbf{G}\mathbf{M})^T+(\mathbf{D}+\mathbf{B}\mathbf{G}\mathbf{E})(\mathbf{D}+\mathbf{B}\mathbf{G}\mathbf{E})^T=\mathbf{0}.$$

Computing each partitioned block of $\mathbf{Q}$, we have

$$\mathbf{Q}_{11}\triangleq\hat{\mathbf{Q}}_p+\mathbf{B}_p\mathbf{C}_c\mathbf{X}_{pc}^T+\mathbf{X}_{pc}\mathbf{C}_c^T\mathbf{B}_p^T=\mathbf{0} \tag{5.91}$$

$$\mathbf{Q}_{12}\triangleq(\mathbf{A}_p+\mathbf{B}_p\mathbf{D}_c\mathbf{M}_p)\mathbf{X}_{pc}+\mathbf{X}_{pc}\mathbf{A}_c^T+(\bar{\mathbf{X}}_p\mathbf{M}_p^T+\mathbf{B}_p\mathbf{D}_c\mathbf{V})\mathbf{B}_c^T+\mathbf{B}_p\mathbf{C}_c\mathbf{X}_c=\mathbf{0} \tag{5.92}$$

$$\mathbf{Q}_{22}\triangleq\mathbf{A}_c\mathbf{X}_c+\mathbf{X}_c\mathbf{A}_c^T+\mathbf{B}_c\mathbf{M}_p\mathbf{X}_{pc}+\mathbf{X}_{pc}^T\mathbf{M}_p^T\mathbf{B}_c^T+\mathbf{B}_c\mathbf{V}\mathbf{B}_c=\mathbf{0} \tag{5.93}$$

where $\hat{\mathbf{Q}}_p$ is defined by (5.90).
Necessity: Suppose a given matrix $\mathbf{X}>\mathbf{0}$ is assignable as a state covariance. Then there exists a controller $(\mathbf{A}_c,\mathbf{B}_c,\mathbf{C}_c,\mathbf{D}_c)$ satisfying (5.91)–(5.93). Pre- and post-multiplying (5.91) by $(\mathbf{I}-\mathbf{B}_p\mathbf{B}_p^+)$ immediately yields (5.83). Let

$$\mathbf{P}_x\triangleq\begin{bmatrix}\mathbf{I} & -\mathbf{X}_{pc}\mathbf{X}_c^{-1}\end{bmatrix},\quad \bar{\mathbf{Q}}\triangleq\mathbf{P}_x\mathbf{Q}\mathbf{P}_x^T. \tag{5.94}$$

Using (5.91)–(5.93), (5.94) yields

$$\begin{aligned}\bar{\mathbf{Q}} &= \bar{\mathbf{Q}}_p+(\mathbf{B}_p\mathbf{D}_c-\mathbf{X}_{pc}\mathbf{X}_c^{-1})\mathbf{M}_p\bar{\mathbf{X}}_p+\bar{\mathbf{X}}_p\mathbf{M}_p^T(\mathbf{B}_p\mathbf{D}_c-\mathbf{X}_{pc}\mathbf{X}_c^{-1}\mathbf{B}_c)^T\\ &\quad +(\mathbf{B}_p\mathbf{D}_c-\mathbf{X}_{pc}\mathbf{X}_c^{-1}\mathbf{B}_c)\mathbf{V}(\mathbf{B}_p\mathbf{D}_c-\mathbf{X}_{pc}\mathbf{X}_c^{-1}\mathbf{B}_c)^T=\mathbf{0} \qquad (5.95)\end{aligned}$$

where

$$\bar{\mathbf{Q}}_p \triangleq \mathbf{A}_p\bar{\mathbf{X}}_p + \bar{\mathbf{X}}_p\mathbf{A}_p^T + \mathbf{W}. \tag{5.96}$$

Completing the square, (5.95) is equivalent to

$$\bar{\mathbf{Q}}_p - \bar{\mathbf{X}}_p\mathbf{M}_p^T\mathbf{V}^{-1}\mathbf{M}_p\bar{\mathbf{X}}_p + \mathbf{L}_p\mathbf{L}_p^T = \mathbf{0} \tag{5.97}$$

where

$$\mathbf{L}_p \triangleq (\mathbf{B}_p\mathbf{D}_c - \mathbf{X}_{pc}\mathbf{X}_c^{-1}\mathbf{B}_c + \bar{\mathbf{X}}_p\mathbf{M}_p^T\mathbf{V}^{-1})\mathbf{V}^{\frac{1}{2}}. \tag{5.98}$$

Thus the condition (5.84) is necessary.

Sufficiency: Suppose the assignability conditions (5.83) and (5.84) are satisfied for a matrix $\mathbf{X} > \mathbf{0}$. Sufficiency will be shown by constructing all controller matrices satisfying (5.91)–(5.93). Applying Theorem 2.3.9 to the equation (5.91), the existence of $\mathbf{C}_c$ satisfying $\mathbf{Q}_{11} = 0$ is guaranteed (for any choice of $\mathbf{D}_c$) by (5.83) and invertibility of $\mathbf{X}_{pc}$, and all such matrices $\mathbf{C}_c$ are given by

$$\begin{aligned}\mathbf{C}_c \quad &= \quad -\frac{1}{2}\mathbf{B}_p^+\hat{\mathbf{Q}}_p(2\mathbf{I} - \mathbf{B}_p\mathbf{B}_p^+)\mathbf{X}_{pc}^{-T} \\ &\quad +\mathbf{B}_p^+\mathbf{S}_F\mathbf{B}_p\mathbf{B}_p^+\mathbf{X}_{pc}^{-T} + (\mathbf{I} - \mathbf{B}_p^+\mathbf{B}_p)\mathbf{Z}_F\end{aligned} \tag{5.99}$$

where $\mathbf{Z}_F$ is arbitrary and $\mathbf{S}_F$ is an arbitrary skew-symmetric matrix. Now, instead of solving (5.92) and (5.93), we consider the following equivalent equation:

$$\begin{bmatrix} \mathbf{I} & -\mathbf{X}_{pc}\mathbf{X}_c^{-1} \\ \mathbf{0} & \mathbf{I} \end{bmatrix}\begin{bmatrix} \mathbf{Q}_{12} \\ \mathbf{Q}_{22} \end{bmatrix} = \mathbf{0}. \tag{5.100}$$

The first row block of (5.100) gives

$$(\mathbf{A}_p + \mathbf{B}_p\mathbf{D}_c\mathbf{M}_p - \mathbf{X}_{pc}\mathbf{X}_c^{-1}\mathbf{B}_c\mathbf{M}_p)\mathbf{X}_{pc} + \mathbf{B}_p\mathbf{C}_c\mathbf{X}_c - \mathbf{X}_{pc}\mathbf{X}_c^{-1}\mathbf{A}_c\mathbf{X}_c + \mathbf{L}_p\mathbf{V}^{\frac{1}{2}}\mathbf{B}_c^T = \mathbf{0}.$$

Solving for $\mathbf{A}_c$, we have

$$\begin{aligned}\mathbf{A}_c \quad &= \quad \mathbf{X}_c\mathbf{X}_{pc}^{-1}(\mathbf{A}_p + \mathbf{B}_p\mathbf{D}_c\mathbf{M}_p + \mathbf{L}_p\mathbf{V}^{\frac{1}{2}}\mathbf{B}_c^T\mathbf{X}_{pc}^{-1})\mathbf{X}_{pc}\mathbf{X}_c^{-1} \\ &\quad -\mathbf{B}_c\mathbf{M}_p\mathbf{X}_{pc}\mathbf{X}_c^{-1} + \mathbf{X}_c\mathbf{X}_{pc}^{-1}\mathbf{B}_p\mathbf{C}_c.\end{aligned} \tag{5.101}$$

Finally, we need to show the existence of $\mathbf{B}_c$ and $\mathbf{D}_c$ satisfying the second row block of (5.100), or $\mathbf{Q}_{22} = \mathbf{0}$, with $\mathbf{A}_c$ and $\mathbf{C}_c$ given by (5.101) and (5.99), respectively. Recall that, if any solution $\mathbf{B}_c$ exists, then it must satisfy (5.98) for some $\mathbf{L}_p$ where $\mathbf{L}_p$ satisfies (5.97). Solving (5.98) for $\mathbf{B}_c$, yields (5.87)

$$\mathbf{B}_c = \mathbf{X}_c\mathbf{X}_{pc}^{-1}(\bar{\mathbf{X}}_p\mathbf{M}_p^T\mathbf{V}^{-1} + \mathbf{B}_p\mathbf{D}_c - \mathbf{L}_p\mathbf{V}^{-\frac{1}{2}}). \tag{5.102}$$

Now we claim that, given any matrix $\mathbf{D}_c$, the controller matrices $\mathbf{A}_c$, $\mathbf{B}_c$ and $\mathbf{C}_c$ given by (5.101), (5.102) and (5.99) satisfy $\mathbf{Q}_{22} = \mathbf{0}$. This can be verified as follows. Substituting $\mathbf{A}_c$ and $\mathbf{B}_c$ into (5.93), then using (5.91) to eliminate $\mathbf{C}_c$, we have

$$\mathbf{Q}_{22} = -\mathbf{X}_c\mathbf{X}_{pc}^{-1}(\bar{\mathbf{Q}}_p - \bar{\mathbf{X}}_p\mathbf{M}_p^T\mathbf{V}^{-1}\mathbf{M}_p\bar{\mathbf{X}}_p^{-T} + \mathbf{L}_p\mathbf{L}_p^T)\mathbf{X}_{pc}^{-T}\mathbf{X}_c = \mathbf{0}$$

where the last equality holds due to (5.97). This completes the proof. □

The following corollary simplifies Theorem 5.2.7.

Corollary 5.2.2 *Assume that $(\mathbf{A}_p, \mathbf{D}_p)$ is controllable and $(\mathbf{A}_p, \mathbf{B}_p, \mathbf{M}_p)$ is a stabilizable and detectable triple, where from (5.1), $\mathbf{D}_p\mathbf{D}_p^T = \mathbf{W}$. Let a positive definite matrix $\mathbf{X}_p \in \mathcal{R}^{n_p \times n_p}$ be given. Then the following statements are equivalent:*

(i) There exists a full-order ($n_c = n_p$) controller which assigns $\mathbf{X}_p$ as a plant state covariance.

(ii) $\mathbf{X}_p$ satisfies

$$(\mathbf{I} - \mathbf{B}_p\mathbf{B}_p^+)(\mathbf{X}_p\mathbf{A}_p^T + \mathbf{A}_p\mathbf{X}_p + \mathbf{W})(\mathbf{I} - \mathbf{B}_p\mathbf{B}_p^+) = \mathbf{0} \tag{5.103}$$

$$\mathbf{X}_p > \mathbf{P}$$

where $\mathbf{P} > \mathbf{0}$ is the solution to

$$\mathbf{P}\mathbf{A}_p^T + \mathbf{A}_p\mathbf{P} - \mathbf{P}\mathbf{M}_p^T\mathbf{V}^{-1}\mathbf{M}_p\mathbf{P} + \mathbf{W} = \mathbf{0} \tag{5.104}$$

which stabilizes

$$\mathbf{A}_p - \mathbf{P}\mathbf{M}_p^T\mathbf{V}^{-1}\mathbf{M}_p. \tag{5.105}$$

Proof. From Theorem 5.2.7 statement (i) is equivalent to (5.83) and the existence of $\mathbf{X}_{pc}$, $\mathbf{X}_c > \mathbf{0}$, and $\mathbf{L}_p$ satisfying (5.84) and (5.85). To prove that (i) implies (ii), suppose (5.83) and (5.84) hold. We need the following lemma.

Lemma 5.2.4 *Let matrices $\mathbf{A}$, $\mathbf{Q}$, and $\mathbf{W}_i$ ($i = 1, 2$) be given, where $\mathbf{Q} \geq \mathbf{0}$. Suppose $\mathbf{X}_i$ ($i = 1, 2$) satisfy*

$$\mathbf{0} \;=\; \mathbf{X}_i\mathbf{A}^T + \mathbf{A}\mathbf{X}_i - \mathbf{X}_i\mathbf{Q}\mathbf{X}_i + \mathbf{W}_i, \quad i = 1, 2. \tag{5.106}$$

Then, $\mathbf{X}_2 \geq \mathbf{X}_1$ if $\mathbf{W}_2 \geq \mathbf{W}_1$ and $\mathbf{A} - \mathbf{X}_2\mathbf{Q}$ is stable.

Proof. Subtract equations (5.106) to get, for $\tilde{\mathbf{X}} = \mathbf{X}_2 - \mathbf{X}_1$,

$$\begin{aligned}\mathbf{0} &= \tilde{\mathbf{X}}\mathbf{A}^T + \mathbf{A}\tilde{\mathbf{X}} + \mathbf{W}_2 - \mathbf{W}_1 - \mathbf{X}_2\mathbf{Q}\mathbf{X}_2 + \mathbf{X}_1\mathbf{Q}\mathbf{X}_1 \\ &= \tilde{\mathbf{X}}(\mathbf{A}^T - \mathbf{Q}\mathbf{X}_2) + (\mathbf{A} - \mathbf{X}_2\mathbf{Q})\tilde{\mathbf{X}} + \tilde{\mathbf{Q}}\end{aligned}$$

where

$$\tilde{\mathbf{Q}} = \tilde{\mathbf{X}}\mathbf{Q}\tilde{\mathbf{X}} + \mathbf{W}_2 - \mathbf{W}_1 \geq \mathbf{0}.$$

Since $\mathbf{A} - \mathbf{X}_2\mathbf{Q}$ is stable,

$$\tilde{\mathbf{X}} = \int_0^\infty e^{(\mathbf{A}-\mathbf{X}_2\mathbf{Q})t}\tilde{\mathbf{Q}}_e^{(\mathbf{A}-\mathbf{X}_2\mathbf{Q})^T t}\, dt \;\geq \mathbf{0}.$$

This proves Lemma 5.2.4. □

From Lemma 5.2.4, note that the stabilizing solution $\mathbf{P}$ to (5.104) and $\bar{\mathbf{X}}_p$ in (5.84) satisfy $\bar{\mathbf{X}}_p \geq \mathbf{P}$ for any choice of $\mathbf{L}_p$ in (5.84). Since $\mathbf{X}_{pc}$ is of full rank (assumed in section 5.2.7)

$$\mathbf{X}_p = \bar{\mathbf{X}}_p + \mathbf{X}_{pc}\mathbf{X}_c^{-1}\mathbf{X}_{pc}^T > \bar{\mathbf{X}}_p.$$

Hence, $\mathbf{X}_p > \mathbf{P}$, and (i) implies (ii).

To prove the converse, assume that (ii) holds. We will construct $\mathbf{X}_{pc}$, $\mathbf{X}_c > \mathbf{0}$, and $\mathbf{L}_p$ satisfying (5.84) and (5.85). Choose $\mathbf{L}_p = \mathbf{0}$. Then $\bar{\mathbf{X}}_p = \mathbf{P}$, and hence $\mathbf{X}_p - \bar{\mathbf{X}}_p > \mathbf{0}$. Now we can choose

$$\mathbf{X}_{pc} = \mathbf{X}_c = (\mathbf{X}_p - \bar{\mathbf{X}}_p).$$

Clearly this choice satisfies (5.85). This completes the proof of Corollary 5.2.2. □

In Theorem 5.2.5, an assignable state covariance can be constructed by solving the linear equation (5.103) for $\mathbf{X}_p > \mathbf{P}$ and the Riccati equation (5.84) for $\bar{\mathbf{X}}_p > \mathbf{P}$. In this case, the parameter $\mathbf{L}_p$ must be chosen such that $\bar{\mathbf{X}}_p < \mathbf{X}_p$ since, from (5.85), we have

$$\mathbf{X}_p - \bar{\mathbf{X}}_p = \mathbf{X}_{pc}\mathbf{X}_c^{-1}\mathbf{X}_{pc}^T > \mathbf{0}$$

where the right-hand side is positive definite since $\mathbf{X}_c > \mathbf{0}$ and $\mathbf{X}_{pc}$ is square nonsingular. If $\bar{\mathbf{X}}_p < \mathbf{X}_p$ holds, then we can find $\mathbf{X}_{pc}$ and $\mathbf{X}_c > 0$ satisfying the above equality. Thus we have an assignable state covariance as in (5.82) where the positive definiteness of $\mathbf{X}$ is guaranteed by $\mathbf{X}_c > \mathbf{0}$ and $\bar{\mathbf{X}}_p > \mathbf{0}$. Note that the choice of $\mathbf{X}_{pc}$ and $\mathbf{X}_c$ for given $\mathbf{X}_p$ and $\bar{\mathbf{X}}_p$ is immaterial and contributes only to the coordinate transformation on the controller states. If we choose

$$\mathbf{X}_{pc} = \mathbf{X}_c = (\mathbf{X}_p - \bar{\mathbf{X}}_p) \triangleq \bar{\mathbf{X}}_c,$$

and consider the strictly proper controller ($\mathbf{D}_c = \mathbf{0}$), then the controller formulae (5.86)–(5.88) become

$$\mathbf{A}_c = \mathbf{A}_p - \mathbf{B}_c\mathbf{M}_p + \mathbf{B}_p\mathbf{C}_c + \mathbf{L}_p\mathbf{V}^{\frac{1}{2}}\mathbf{B}_c^T\bar{\mathbf{X}}_c^{-1} \quad (5.107)$$

$$\mathbf{B}_c = \bar{\mathbf{X}}_p\mathbf{M}_p^T\mathbf{V}^{-1} - \mathbf{L}_p\mathbf{V}^{-\frac{1}{2}} \quad (5.108)$$

$$\begin{aligned}\mathbf{C}_c = & -\frac{1}{2}\mathbf{B}_p^+(\mathbf{A}_p\mathbf{X}_p + \mathbf{X}_p\mathbf{A}_p^T + \mathbf{W})(2\mathbf{I} - \mathbf{B}_p\mathbf{B}_p^+)\bar{\mathbf{X}}_c^{-1} \\ & + \mathbf{B}_p^+\mathbf{S}_F\mathbf{B}_p\mathbf{B}_p^+\bar{\mathbf{X}}_c^{-1} + (\mathbf{I} - \mathbf{B}_p^+\mathbf{B}_p)\mathbf{Z}_F \end{aligned} \quad (5.109)$$

where $\mathbf{Z}_F$ is arbitrary and $\mathbf{S}_F$ is an arbitrary skew-symmetric matrix. Note that the controller has the following observer-based structure;

$$\dot{\mathbf{x}}_c = \mathbf{A}_p\mathbf{x}_c + \mathbf{B}_p\mathbf{u} + \mathbf{B}_c(\mathbf{z} - \mathbf{M}_p\mathbf{x}_c) + \mathbf{L}_p\mathbf{V}^{\frac{1}{2}}\mathbf{B}_c^T\bar{\mathbf{X}}_c^{-1}\mathbf{x}_c,$$

$$\mathbf{u} = \mathbf{C}_c\mathbf{x}_c.$$

If we call the estimator part of the covariance controller obtained by choosing $\mathbf{L}_p = \mathbf{0}$ the "central estimator", it is apparent that the central estimator is the Kalman filter. Recall that the Riccati solution $\bar{\mathbf{X}}_p$ for the Kalman filter ($\mathbf{L}_p = \mathbf{0}$) (we shall denote this solution by $\mathbf{P}$) has a physical significance; the estimation error covariance [1]. In fact, for nonzero choices of $\mathbf{L}_p$, the Riccati solution $\bar{\mathbf{X}}_p$ can still be considered as the estimation error covariance since

$$\begin{aligned}\mathcal{E}[(\mathbf{x}_p - \mathbf{x}_c)(\mathbf{x}_p - \mathbf{x}_c)^T] &= \mathcal{E}[\mathbf{x}_p\mathbf{x}_p^T] - \mathcal{E}[\mathbf{x}_p\mathbf{x}_c^T] - \mathcal{E}[\mathbf{x}_c\mathbf{x}_p^T] + \mathcal{E}[\mathbf{x}_c\mathbf{x}_c^T] \\ &= \mathbf{X}_p - \bar{\mathbf{X}}_c - \bar{\mathbf{X}}_c + \bar{\mathbf{X}}_c = \bar{\mathbf{X}}_p.\end{aligned}$$

Since $\mathbf{P}$ satisfies $\mathbf{P} \leq \bar{\mathbf{X}}_p$ for any choice of $\mathbf{L}_p$, the Kalman filter optimizes not only the scalar objective $tr(\bar{\mathbf{X}}_p)$ as in the standard LQG theorem but also the matrix-valued, or multiobjective function $\bar{\mathbf{X}}_p$ (in the sense $\mathbf{P} \leq \bar{\mathbf{X}}_p$ over all $\mathbf{L}_p$). On the other hand, nonzero choices of the free matrix $\mathbf{L}_p$ may improve some other performances such as robustness.

For the estimated-state feedback part of the covariance controller, we see that the feedback gain $\mathbf{C}_c$ given by (5.109) is identical to the state feedback covariance controller $\mathbf{G}$ in (5.3) if we replace $\bar{\mathbf{X}}_c$ by $\mathbf{X}_p$. This difference can be interpreted as the compensation for the estimation error due to noisy measurements by subtracting the error covariance $\bar{\mathbf{X}}_p$ from $\mathbf{X}_p$ to obtain $\bar{\mathbf{X}}_c$. Also note that the set of all assignable plant state covariances with full-order controllers is a subset of that with state feedback with the additional constraint $\bar{\mathbf{X}}_p < \mathbf{X}_p$. This fact makes sense physically since the estimation error covariance $\bar{\mathbf{X}}_p$ is zero if all the states are available without noise.

Finally, we see that the separation principle does not hold for the covariance controller, i.e. the state estimator and the estimated-state feedback gain cannot be designed separately since the determination of the estimator parameters $\mathbf{A}_c$ and $\mathbf{B}_c$ involves closed-loop information $\mathbf{X}_p$, and the computation of the estimated-state feedback gain $\mathbf{C}_c$ requires the estimator information $\bar{\mathbf{X}}_p$. Nevertheless, the plant state covariance $\mathbf{X}_p$ and the estimation error covariance $\bar{\mathbf{X}}_p$ to be assigned can be specified by solving the linear equation (5.83) and the Riccati equation (5.84) with a simple constraint $\mathbf{0} < \bar{\mathbf{X}}_p < \mathbf{X}_p$.

5.3 DISCRETE-TIME COVARIANCE CONTROLLERS

5.3.1 State Feedback

Consider the linear system given by (4.31), where we assume that all the states are available for feedback, i.e. $\mathbf{M}_p = \mathbf{I}$ and $\mathbf{D}_z = \mathbf{0}$. With a controller of constant state feedback gain $\mathbf{G}$, the closed-loop state covariance $\mathbf{X}$ satisfies the following Lyapunov equation.

$$\mathbf{X} = (\mathbf{A}_p + \mathbf{B}_p\mathbf{G})\mathbf{X}(\mathbf{A}_p + \mathbf{B}_p\mathbf{G})^T + \mathbf{W}. \tag{5.110}$$

Theorem 5.3.1 *Let a positive definite matrix $\mathbf{X} \in \mathcal{R}^{n_p \times n_p}$ be given. Then the following statements are equivalent:*

(i) There exists a state feedback gain $\mathbf{G}$ which assigns $\mathbf{X}$ as a state covariance.

(ii) $\mathbf{X}$ satisfies

$$(\mathbf{I} - \mathbf{B}_p\mathbf{B}_p^+)(\mathbf{X} - \mathbf{A}_p\mathbf{X}\mathbf{A}_p^T - \mathbf{W})(\mathbf{I} - \mathbf{B}_p\mathbf{B}_p^+) = \mathbf{0}$$
$$\mathbf{X} \geq \mathbf{W}.$$

In this case, all such control gains $\mathbf{G}$ are given by

$$\mathbf{G} = \mathbf{B}_p^+(\mathbf{L}_p\mathbf{U}_p\mathbf{X}^{-\frac{1}{2}} - \mathbf{A}_p) + (\mathbf{I} - \mathbf{B}_p^+\mathbf{B}_p)\mathbf{Z}_F$$

where $\mathbf{Z}_F$ is arbitrary and

$$\mathbf{L}_p \triangleq (\mathbf{X} - \mathbf{W})^{\frac{1}{2}}$$
$$\mathbf{U}_p \triangleq \mathbf{V}_L \begin{bmatrix} \mathbf{I} & \mathbf{0} \\ \mathbf{0} & \mathbf{U}_F \end{bmatrix} \mathbf{V}_R^T$$
$$(\mathbf{I} - \mathbf{B}_p\mathbf{B}_p^+)\mathbf{L} = \mathbf{U}_L \begin{bmatrix} \Sigma_L & \mathbf{0} \\ \mathbf{0} & \mathbf{0} \end{bmatrix} \mathbf{V}_L^T \quad (SVD)$$
$$(\mathbf{I} - \mathbf{B}_p\mathbf{B}_p^+)\mathbf{A}_p\mathbf{X}^{\frac{1}{2}} = \mathbf{U}_L \begin{bmatrix} \Sigma_L & \mathbf{0} \\ \mathbf{0} & \mathbf{0} \end{bmatrix} \mathbf{V}_R^T \quad (SVD)$$

where $\mathbf{U}_F$ is an arbitrary orthogonal matrix.

Proof. Rearranging (5.110), we have

$$(\mathbf{B}_p\mathbf{G}+\mathbf{A}_p)\mathbf{X}(\mathbf{B}_p\mathbf{G}+\mathbf{A}_p)^T = \mathbf{X}-\mathbf{W}.$$

Then the result directly follows from Theorem 2.3.9. Note that the rank condition

$$\text{rank}(\mathbf{X}-\mathbf{W}) \le n_p$$

is redundant since the dimension of $\mathbf{X}-\mathbf{W}$ is n_p. □

5.3.2 Output Feedback

In this section, we generalize the result of the previous section for the case where not all the states are available for feedback. As opposed to the case for continuous-time systems, we can allow general structure for the measurement noise for the discrete-time plant, that is, $\mathbf{V} \triangleq \mathbf{D}_z\mathbf{D}_z^T$ can have arbitrary rank (of course it should be less than or equal to n_z by the dimension constraint). However, we assume that there is no redundant measured output, or equivalently, $\mathbf{M}_p\mathbf{M}_p^T > \mathbf{0}$. Practically speaking, this assumption is reasonable since we don't want to add a costly sensor to obtain redundant information, and also from the theoretical point of view, this assumption can be removed by increasing the complexity of the result. Consider the dynamic output feedback controller of fixed-order given in (4.31). The closed-loop state covariance satisfies

$$\mathbf{X} = (\mathbf{A}+\mathbf{BGM})\mathbf{X}(\mathbf{A}+\mathbf{BGM})^T + (\mathbf{D}+\mathbf{BGE})(\mathbf{D}+\mathbf{BGE})^T. \tag{5.111}$$

where $\mathbf{X}_0 = \mathbf{0},\ \mathbf{W}_0 = \mathbf{I}$ in (4.36).

Theorem 5.3.2 *Let a positive definite matrix* $\mathbf{X} \in \mathcal{R}^{(n_p+n_c)\times(n_p+n_c)}$ *be given. Then the following statements are equivalent:*

(i) There exists a controller $\mathbf{G}$ *which assigns* $\mathbf{X}$ *as a state covariance.*
(ii) $\mathbf{X}$ *satisfies*

$$(\mathbf{I}-\mathbf{BB}^+)(\mathbf{X}-\mathbf{AXA}^T-\mathbf{DD}^T)(\mathbf{I}-\mathbf{BB}^+) = \mathbf{0}$$
$$\mathbf{X} = \mathbf{AXA}^T - \mathbf{AXM}^T(\mathbf{MXM}^T+\mathbf{EE}^T)^{-1}\mathbf{MXA}^T + \mathbf{DD}^T + \mathbf{LL}^T$$

for some $\mathbf{L} \in \mathcal{R}^{(n_p+n_c)\times(n_z+n_c)}$.

In this case, all such controllers $\mathbf{G}$ *are given by*

$$\mathbf{G} = \mathbf{B}^+(\mathbf{LUR}^{-\frac{1}{2}} - \mathbf{AXM}^T\mathbf{R}^{-1}) + (\mathbf{I}-\mathbf{B}^+\mathbf{B})\mathbf{Z}_F \tag{5.112}$$

where $\mathbf{Z}_F$ *is arbitrary and*

$$\mathbf{R} \triangleq \mathbf{MXM}^T + \mathbf{EE}^T$$
$$\mathbf{U} \triangleq \mathbf{V}_L\begin{bmatrix}\mathbf{I} & \mathbf{0}\\ \mathbf{0} & \mathbf{U}_F\end{bmatrix}\mathbf{V}_R^T$$
$$(\mathbf{I}-\mathbf{BB}^+)\mathbf{L} = \mathbf{U}_L\begin{bmatrix}\Sigma_L & \mathbf{0}\\ \mathbf{0} & \mathbf{0}\end{bmatrix}\mathbf{V}_L^T \qquad (SVD)$$
$$(\mathbf{I}-\mathbf{BB}^+)\mathbf{AXM}^T\mathbf{R}^{-\frac{1}{2}} = \mathbf{U}_L\begin{bmatrix}\Sigma_L & \mathbf{0}\\ \mathbf{0} & \mathbf{0}\end{bmatrix}\mathbf{V}_R^T \qquad (SVD)$$

where $\mathbf{U}_F$ *is an arbitrary orthogonal matrix.*

Proof. Noting that $\mathbf{E}\mathbf{D}^T = 0$, since we assume that there is no correlation between the process and measurement noises ($\mathbf{D}_p\mathbf{D}_z^T = \mathbf{0}$), the covariance equation (5.111) can be expanded as

$$\mathbf{X} = \mathbf{A}\mathbf{X}\mathbf{A}^T + \mathbf{D}\mathbf{D}^T + \mathbf{B}\mathbf{G}\mathbf{M}\mathbf{X}\mathbf{A}^T + (\mathbf{B}\mathbf{G}\mathbf{M}\mathbf{X}\mathbf{A}^T)^T + \mathbf{B}\mathbf{G}\mathbf{R}\mathbf{G}^T\mathbf{B}^T$$

where

$$\mathbf{R} \triangleq \mathbf{M}\mathbf{X}\mathbf{M}^T + \mathbf{E}\mathbf{E}^T.$$

Since $\mathbf{X} > \mathbf{0}$ and there is no redundant sensor ($\mathbf{M}\mathbf{M}^T > \mathbf{0}$), we have $\mathbf{R} > \mathbf{0}$ for any matrix $\mathbf{E}$. Hence, we can complete the square with respect to $\mathbf{BG}$ as follows:

$$(\mathbf{BG} + \mathbf{A}\mathbf{X}\mathbf{M}^T\mathbf{R}^{-1})\mathbf{R}(\mathbf{BG} + \mathbf{A}\mathbf{X}\mathbf{M}^T\mathbf{R}^{-1})^T = \mathbf{Q}$$

$$\mathbf{Q} \triangleq \mathbf{X} - \mathbf{A}\mathbf{X}\mathbf{A}^T + \mathbf{A}\mathbf{X}\mathbf{M}^T\mathbf{R}^{-1}\mathbf{M}\mathbf{X}\mathbf{A}^T - \mathbf{D}\mathbf{D}^T.$$

Using Theorem 2.3.9, the above equation is solvable for $\mathbf{G}$ if and only if

$$\mathbf{Q} \geq \mathbf{0}, \quad \text{rank}(\mathbf{Q}) \leq n_p + n_c \tag{5.113}$$

$$(\mathbf{I} - \mathbf{B}\mathbf{B}^+)(\mathbf{Q} - \mathbf{A}\mathbf{X}\mathbf{M}^T\mathbf{R}^{-1}\mathbf{M}\mathbf{X}\mathbf{A}^T)(\mathbf{I} - \mathbf{B}\mathbf{B}^+) = \mathbf{0}$$

hold, in which case, all solutions $\mathbf{G}$ are given by (5.112), where $\mathbf{L} \in \mathcal{R}^{(n_p+n_c)\times(n_c+n_z)}$ is such that $\mathbf{Q} = \mathbf{L}\mathbf{L}^T$. Note that the existence of such a matrix $\mathbf{L}$ is equivalent to the condition (5.113). □

5.3.3 Plant Covariance Assignment

In the previous section, we have considered a problem of assigning a matrix $\mathbf{X} > \mathbf{0}$ as a closed-loop state covariance. For a dynamic controller of order n_c, the closed-loop state covariance $\mathbf{X}$ can be partitioned as:

$$\mathbf{X} = \begin{bmatrix} \mathbf{X}_p & \mathbf{X}_{pc} \\ \mathbf{X}_{pc}^T & \mathbf{X}_c \end{bmatrix} \tag{5.114}$$

where $\mathbf{X}_p$ is the plant state covariance, $\mathbf{X}_c$ is the controller state covariance, and $\mathbf{X}_{pc}$ is the correlation between the plant and controller states. Since $\mathbf{X}_c$ and $\mathbf{X}_{pc}$ are dependent upon the choice of the controller coordinate, there is less motivation to assign a prespecified matrix as a controller state covariance or a correlation (except for the case with observer-based control where these matrices are related to the estimation error covariance). Moreover, an important output performance can be completely specified by the plant state covariance $\mathbf{X}_p$. These facts motivate us to ask: (1) When is a given matrix $\mathbf{X}_p$ assignable as a plant state covariance? (2) If $\mathbf{X}_p$ is assignable, what are the controllers which assign this $\mathbf{X}_p$? Note that we can find all controllers which assign a given $\mathbf{X}_p$ using the controller formula given in (5.112) if we can find all assignable *closed-loop* state covariances $\mathbf{X}$ whose 11-block is a specified assignable plant state covariance $\mathbf{X}_p$. The following theorem answers these questions.

Theorem 5.3.3 *Let a positive definite matrix $\mathbf{X}_p \in \mathcal{R}^{n_p\times n_p}$ be given. Then the following statements are equivalent:*

(i) There exists a controller (of some unspecified order) which assigns $\mathbf{X}_p$ as a plant state covariance.

(ii) $\mathbf{X}_p$ *satisfies*

$$(\mathbf{I}-\mathbf{B}_p\mathbf{B}_p^+)(\mathbf{X}_p-\mathbf{A}_p\mathbf{X}_p\mathbf{A}_p^T-\mathbf{W})(\mathbf{I}-\mathbf{B}_p\mathbf{B}_p^+)=\mathbf{0},$$

$$\mathbf{X}_p \geq \mathbf{P}$$

where $\mathbf{P} > \mathbf{0}$ *is the solution to*

$$\mathbf{P}=\mathbf{A}_p\mathbf{P}\mathbf{A}_p^T-\mathbf{A}_p\mathbf{P}\mathbf{M}_p^T(\mathbf{M}_p\mathbf{P}\mathbf{M}_p^T+\mathbf{V})^{-1}\mathbf{M}_p\mathbf{P}\mathbf{A}_p^T+\mathbf{W}$$

which stabilizes $\mathbf{A}_p-\mathbf{P}\mathbf{M}_p{}^T\mathbf{V}^{-1}\mathbf{M}_p$.

Suppose $\mathbf{X}_p$ *is assignable as a plant state covariance. Then all assignable closed-loop state covariances* $\mathbf{X} > \mathbf{0}$ *whose 11-block is* $\mathbf{X}_p$ *can be constructed as follows. Choose an arbitrary matrix* $\mathbf{L}_p \in \mathcal{R}^{n_p\times n_z}$ *such that the stabilizing solution* $\bar{\mathbf{X}}_p > \mathbf{0}$ *to*

$$\bar{\mathbf{X}}_p=\mathbf{A}_p\bar{\mathbf{X}}_p\mathbf{A}_p^T-\mathbf{A}_p\bar{\mathbf{X}}_p\mathbf{M}_p^T(\mathbf{M}_p\bar{\mathbf{X}}_p\mathbf{M}_p^T+\mathbf{V})^{-1}\mathbf{M}_p\bar{\mathbf{X}}_p\mathbf{A}_p^T+\mathbf{W}+\mathbf{L}_p\mathbf{L}_p^T$$

satisfies

$$\bar{\mathbf{X}}_p \leq \mathbf{X}_p.$$

Then let $\mathbf{X}_c$ *and* $\mathbf{X}_{pc}$ *be any matrix factor such that*

$$\mathbf{X}_{pc}\mathbf{X}_c^{-1}\mathbf{X}_{pc}^T=\mathbf{X}_p-\bar{\mathbf{X}}_p,\quad \mathbf{X}_c > \mathbf{0}$$

and form a matrix

$$\mathbf{X}=\begin{bmatrix}\mathbf{X}_p & \mathbf{X}_{pc}\\ \mathbf{X}_{pc}^T & \mathbf{X}_c\end{bmatrix}$$

which is assignable as a closed-loop state covariance.

Proof. From Theorem 5.3.2, a given positive definite matrix $\mathbf{X} \in \mathcal{R}^{(n_p+n_c)\times(n_p+n_c)}$ is assignable as a closed-loop state covariance if and only if it satisfies

$$(\mathbf{I}-\mathbf{B}\mathbf{B}^+)(\mathbf{X}-\mathbf{A}\mathbf{X}\mathbf{A}^T-\mathbf{D}\mathbf{D}^T)(\mathbf{I}-\mathbf{B}\mathbf{B}^+)=\mathbf{0},$$

$$\mathbf{X}=\mathbf{A}\mathbf{X}\mathbf{A}^T-\mathbf{A}\mathbf{X}\mathbf{M}^T(\mathbf{M}\mathbf{X}\mathbf{M}^T+\mathbf{E}\mathbf{E}^T)^{-1}\mathbf{M}\mathbf{X}\mathbf{A}^T+\mathbf{D}\mathbf{D}^T+\mathbf{L}\mathbf{L}^T$$

for some $\mathbf{L} \in \mathcal{R}^{(n_p+n_c)\times(n_z+n_c)}$. Substituting the augmented matrices defined in (4.32), (5.114) and

$$\mathbf{L} \triangleq \begin{bmatrix}\mathbf{L}_1\\ \mathbf{L}_2\end{bmatrix},\quad \mathbf{L}_1 \in \mathcal{R}^{n_p\times(n_c+n_z)},\quad \mathbf{L}_2 \in \mathcal{R}^{n_c\times(n_c+n_z)},$$

we have

$$(\mathbf{I}-\mathbf{B}_p\mathbf{B}_p^+)(\mathbf{X}_p-\mathbf{A}_p\mathbf{X}_p\mathbf{A}_p^T-\mathbf{W})(\mathbf{I}-\mathbf{B}_p\mathbf{B}_p^+)=\mathbf{0} \tag{5.115}$$

$$\bar{\mathbf{X}}_p=\mathbf{A}_p\bar{\mathbf{X}}_p\mathbf{A}_p^T-\mathbf{A}_p\bar{\mathbf{X}}_p\mathbf{M}_p^T(\mathbf{M}_p\bar{\mathbf{X}}_p\mathbf{M}_p^T+\mathbf{V})^{-1}\mathbf{M}_p\bar{\mathbf{X}}_p\mathbf{A}_p^T+\mathbf{W}+\mathbf{L}_1\mathbf{L}_1^T-\mathbf{X}_{pc}\mathbf{X}_c^{-1}\mathbf{X}_{pc}^T$$

$$\mathbf{X}_{pc}=\mathbf{L}_1\mathbf{L}_2^T,\quad \mathbf{X}_c=\mathbf{L}_2\mathbf{L}_2^T$$

where

$$\bar{\mathbf{X}}_p \triangleq \mathbf{X}_p-\mathbf{X}_{pc}\mathbf{X}_c^{-1}\mathbf{X}_{pc}^T. \tag{5.116}$$

Consider the Singular Value Decomposition (SVD) of $\mathbf{L}_2$,

$$\mathbf{L}_2=\mathbf{U}\begin{bmatrix}\Sigma & \mathbf{0}\end{bmatrix}\mathbf{V}^T,$$

$$\begin{bmatrix} \bar{\mathbf{L}}_1 & \bar{\mathbf{L}}_2 \end{bmatrix} \triangleq \mathbf{L}_1\mathbf{V}, \quad \bar{\mathbf{L}}_1 \in \mathcal{R}^{n_p \times n_c}, \quad \bar{\mathbf{L}}_2 \in \mathcal{R}^{n_p \times n_z}.$$

In the above SVD of $\mathbf{L}_2$, we used the fact that $\mathbf{L}_2$ has full row rank since $\mathbf{X}_c = \mathbf{L}_2\mathbf{L}_2^T > \mathbf{0}$. Using these definitions, we have

$$\bar{\mathbf{X}}_p = \mathbf{A}_p\bar{\mathbf{X}}_p\mathbf{A}_p^T - \mathbf{A}_p\bar{\mathbf{X}}_p\mathbf{M}_p^T(\mathbf{M}_p\bar{\mathbf{X}}_p\mathbf{M}_p^T + \mathbf{V})^{-1}\mathbf{M}_p\bar{\mathbf{X}}_p\mathbf{A}_p^T + \mathbf{W} + \bar{\mathbf{L}}_2\bar{\mathbf{L}}_2^T, \tag{5.117}$$

$$\mathbf{X}_{pc} = \bar{\mathbf{L}}_1\Sigma\mathbf{U}^T, \quad \mathbf{X}_c = \mathbf{U}\Sigma^2\mathbf{U}^T. \tag{5.118}$$

Therefore, $\mathbf{X} > \mathbf{0}$ is assignable if and only if (5.115)–(5.118) hold for some $\bar{\mathbf{L}}_1$, $\bar{\mathbf{L}}_2$, $\mathbf{U}$ and Σ where $\mathbf{U}$ is an orthogonal matrix and Σ is a diagonal matrix with positive diagonal elements. Note that (5.118) is always satisfied by some $\bar{\mathbf{L}}_1$, $\mathbf{U}$ and Σ for any given $\mathbf{X}_{pc}$ and $\mathbf{X}_c > 0$. Thus we need only (5.115) and (5.117) for $\mathbf{X} > \mathbf{0}$ to be assignable.

Now, a given matrix $\mathbf{X}_p > 0$ is assignable as a plant state covariance if and only if it satisfies (5.115) and there exists $\bar{\mathbf{L}}_2$ such that the positive definite solution $\bar{\mathbf{X}}_p$ to (5.117) satisfies

$$\bar{\mathbf{X}}_p \leq \mathbf{X}_p,$$

which is necessary and sufficient for the existence of $\mathbf{X}_{pc}$ and $\mathbf{X}_c > \mathbf{0}$ such that (5.116) holds. Conceptually, such a matrix $\bar{\mathbf{L}}_2$ exists if and only if $\mathbf{X}_p$ is "larger" than the "smallest" $\bar{\mathbf{X}}_p$ satisfying (5.117) for some $\bar{\mathbf{L}}_2$. Let the solution $\bar{\mathbf{X}}_p$ for the choice $\bar{\mathbf{L}}_2 = \mathbf{0}$ be denoted by $\mathbf{P}$. From the monotonicity property of the stabilizing solution $\bar{\mathbf{X}}_p \geq \mathbf{0}$ to the Riccati equation (5.117) with respect to the forcing term $\bar{\mathbf{L}}_2\bar{\mathbf{L}}_2^T$, the "smallest" $\bar{\mathbf{X}}_p$ is given by $\mathbf{P}$. The term "smallest" means that $\mathbf{P} \leq \bar{\mathbf{X}}_p$ holds for any choice of $\bar{\mathbf{L}}_2$. Thus, if $\mathbf{X}_p$ does not satisfy $\mathbf{P} \leq \mathbf{X}_p$, then there is no $\bar{\mathbf{L}}_2$ such that $\bar{\mathbf{X}}_p \leq \mathbf{X}_p$. Conversely, if $\mathbf{X}_p$ satisfies $\mathbf{P} \leq \mathbf{X}_p$, then a choice $\bar{\mathbf{L}}_2 = \mathbf{0}$ will surely yield $\bar{\mathbf{X}}_p \leq \mathbf{X}_p$ (since $\bar{\mathbf{X}}_p = \mathbf{P}$ in this case). This completes the proof. □

For a given assignable plant state covariance $\mathbf{X}_p$, all controllers which assign this $\mathbf{X}_p$ can be obtained by first constructing an assignable closed-loop state covariance $\mathbf{X}$ by solving the Riccati equation, then computing a controller using the explicit formula given in Theorem 5.3.2. In this case, the controller order n_c is determined when we factor the matrix $\mathbf{X}_p - \bar{\mathbf{X}}_p$ to find $\mathbf{X}_{pc}$ and $\mathbf{X}_c$, and is minimal if the factorization is such that $\mathbf{X}_{pc}$ has full column rank, in which case, we have

$$n_c = \text{rank}(\mathbf{X}_p - \bar{\mathbf{X}}_p).$$

Hence, to assign a given plant state covariance $\mathbf{X}_p$, we can search for a lower-order controller ($n_c < n_p$) by choosing the parameter $\mathbf{L}_p$ to reduce the rank of $\mathbf{X}_p - \bar{\mathbf{X}}_p$.

Finally, the plant covariance assignment formulation considered in this section can also be applied to the full-order covariance controllers for continuous-time systems (see Section 5.2.7). However, the formulation for more general (the controller order $n_c < n_p$) continuous-time covariance controllers is much more difficult due to the third assignability condition (5.65) in Theorem 5.2.6. The reader may want to prove a result parallel to Theorem 5.2.7 for the discrete-time system.

5.4 MINIMAL ENERGY COVARIANCE CONTROL

Covariance control theory provides a parametrization of all controllers which assign a particular state covariance matrix to the closed-loop system. Such controllers are not unique.

In this section we obtain the analytic expressions for the covariance controllers which minimize the required control effort. Both continuous-time and discrete-time systems are discussed.

5.4.1 Continuous-Time Output Feedback

We seek the continuous-time covariance controller to minimize the required control effort

$$\|\mathbf{u}\| \triangleq \left[\lim_{t\to\infty} \mathcal{E}\mathbf{u}^T(t)\mathbf{R}\mathbf{u}(t)\right]^{1/2} \tag{5.119}$$

subject to

$$\lim_{t\to\infty} \mathcal{E}[\mathbf{x}(t)\mathbf{x}^T(t)] = \mathbf{X} \tag{5.120}$$

where $\mathbf{R} > \mathbf{0}$ is a given weighting matrix and $\mathbf{X} > \mathbf{0}$ is a given assignable covariance. An equivalent deterministic formulation of the control effort (5.119) can be provided as in Chapter 4. We have seen in Section 5.2 that all covariance controllers which assign $\mathbf{X}$ to the closed-loop system (4.15) can be parametrized in the following form (Theorem 5.2.1):

$$\mathbf{G} = \mathbf{G}_1 + \mathbf{G}_2\mathbf{S}_F\mathbf{G}_3 \tag{5.121}$$

where $\mathbf{S}_F$ is an arbitrary skew-symmetric matrix and the $\mathbf{G}_1$, $\mathbf{G}_2$ and $\mathbf{G}_3$ are known matrices which depend on the plant parameters and the state covariance $\mathbf{X}$. For simplicity we have assumed that the input matrix $\mathbf{B}_p$ has full column rank and the measurement matrix $\mathbf{M}_p$ has full row rank. Hence $\mathbf{B}^+\mathbf{B} = \mathbf{I}$ and $\mathbf{M}\mathbf{M}^+ = \mathbf{I}$ and in this case $\mathbf{Z}_F - \mathbf{B}^+\mathbf{B}\mathbf{Z}_F\mathbf{M}\mathbf{M}^+ = \mathbf{0}$ in the parametrization of all controllers provided in Theorem 5.2.4.

The following result provides the solution of the continuous-time minimum effort covariance control.

Theorem 5.4.1 *The continuous-time covariance controller* $\mathbf{G}$, *which solves the minimum effort covariance control problem, is provided by any choice of the skew-symmetric matrix* $\mathbf{S}_F$ *which solves the following generalized Sylvester equation*

$$\mathbf{K}_1\mathbf{S}_F\mathbf{K}_2 + \mathbf{K}_2\mathbf{S}_F\mathbf{K}_1 + \mathbf{K}_3 = \mathbf{0} \tag{5.122}$$

where

$$\begin{aligned}
\mathbf{K}_1 &= \mathbf{G}_2^T\mathbf{I}_1^T\mathbf{R}\mathbf{I}_1\mathbf{G}_2 = \mathbf{K}_1^T \\
\mathbf{K}_2 &= \mathbf{G}_3\mathbf{M}\mathbf{X}\mathbf{M}^T\mathbf{G}_3^T = \mathbf{K}_2^T \\
\mathbf{K}_3 &= \mathbf{G}_2^T\mathbf{I}_1^T\mathbf{R}\mathbf{I}_1\mathbf{G}_1\mathbf{M}\mathbf{X}\mathbf{M}^T\mathbf{G}_3^T - \mathbf{G}_3\mathbf{M}\mathbf{X}\mathbf{M}^T\mathbf{G}_1^T\mathbf{I}_1^T\mathbf{R}\mathbf{I}_1\mathbf{G}_2 = -\mathbf{K}_3^T
\end{aligned} \tag{5.123}$$

and $\mathbf{I}_1 = [\mathbf{I}\ \ \mathbf{0}]$.

Proof. The control effort $\|\mathbf{u}\|$ can be computed as follows:

$$\begin{aligned}
\|\mathbf{u}\|^2 &= \mathcal{E}_\infty\left[(\mathbf{C}_c\mathbf{x}_c + \mathbf{D}_c\mathbf{M}_p\mathbf{x}_p)^T\mathbf{R}(\mathbf{C}_c\mathbf{x}_c + \mathbf{D}_c\mathbf{M}_p\mathbf{x}_p)\right] \\
&= \mathcal{E}_\infty\left[(\mathbf{I}_1\mathbf{G}\mathbf{M}\mathbf{x})^T\mathbf{R}(\mathbf{I}_1\mathbf{G}\mathbf{M}\mathbf{x})\right] = tr(\mathbf{I}_1\mathbf{G}\mathbf{M}\mathbf{X}\mathbf{M}^T\mathbf{G}^T\mathbf{I}_1^T\mathbf{R}).
\end{aligned}$$

Substituting the expression $\mathbf{G} = \mathbf{G}_1 + \mathbf{G}_2\mathbf{S}_F\mathbf{G}_3$ we obtain

$$\begin{aligned}\|\mathbf{u}\|^2 &= tr\left[\mathbf{I}_1(\mathbf{G}_1 + \mathbf{G}_2\mathbf{S}_F\mathbf{G}_3)\mathbf{MXM}^T(\mathbf{G}_1 + \mathbf{G}_2\mathbf{S}_F\mathbf{G}_3)^T\mathbf{I}_1^T\mathbf{R}\right] \\ &= tr(\mathbf{I}_1\mathbf{G}_1\mathbf{MXM}^T\mathbf{G}_1^T\mathbf{I}_1^T\mathbf{R}) \\ &\quad + 2\,tr(\mathbf{I}_1\mathbf{G}_2\mathbf{S}_F\mathbf{G}_3\mathbf{MXM}^T\mathbf{G}_1^T\mathbf{I}_1^T\mathbf{R}) \\ &\quad + tr(\mathbf{I}_1\mathbf{G}_2\mathbf{S}_F\mathbf{G}_3\mathbf{MXM}^T\mathbf{G}_3^T\mathbf{S}_F^T\mathbf{G}_2^T\mathbf{I}_1^T\mathbf{R}).\end{aligned}$$

Since we are looking for a skew-symmetric solution, we substitute $\mathbf{S}_F = (\mathbf{S}_{F1} - \mathbf{S}_{F1}^T)/2$ and we minimize with respect to $\mathbf{S}_{F1}$. The minimum control effort solution is obtained by differentiating the above expression with respect to $\mathbf{S}_{F1}$. This differentiation provides the following condition for a global minimizer

$$\mathbf{G}_2^T\mathbf{I}_1^T\mathbf{R}\mathbf{I}_1\mathbf{G}_2(\mathbf{S}_{F1} - \mathbf{S}_{F1}^T)\mathbf{G}_3\mathbf{MXM}^T\mathbf{G}_3^T \tag{5.124}$$

$$+\mathbf{G}_3\mathbf{MXM}^T\mathbf{G}_3^T(\mathbf{S}_{F1} - \mathbf{S}_{F1}^T)\mathbf{G}_3\mathbf{MXM}^T\mathbf{G}_3^T \tag{5.125}$$

$$+2\mathbf{G}_2^T\mathbf{I}_1^T\mathbf{R}\mathbf{I}_1\mathbf{G}_1\mathbf{MXM}^T\mathbf{G}_3^T - 2\mathbf{G}_3\mathbf{MXM}^T\mathbf{G}_1^T\mathbf{I}_1^T\mathbf{R}\mathbf{I}_1\mathbf{G}_2 = \mathbf{0}. \tag{5.126}$$

Substituting $(\mathbf{S}_{F1} - \mathbf{S}_{F1}^T)/2 = \mathbf{S}_F$ provides the condition (5.122). □

Next we provide an explicit expression for the skew-symmetric solutions of the equation (5.122) using a Kronecker matrix algebra approach. (Note that the equation (5.122) always has a skew-symmetric solution.) To begin we need the following lemma [86]:

Lemma 5.4.1 *Let $\mathbf{S}_F \in \mathcal{R}^{k\times k}$ be a skew-symmetric matrix, and let*

$$\mathbf{s} = [s_{12}, s_{13}, \ldots, s_{1k}, s_{23}, \ldots, s_{2k}, \ldots s_{(k-1)k}]^T \tag{5.127}$$

where $\mathbf{S} = [s_{ij}]$. Then there exists a $k^2 \times k(k-1)/2$ matrix $\mathbf{\Delta}$ such that

$$vec\,\mathbf{S} = \mathbf{\Delta s} \tag{5.128}$$

where the vec operator stacks the columns of a matrix one underneath the other.

Note that the columns of $\mathbf{\Delta}$ form a basis for the vector space of $k \times k$ skew-symmetric matrices. The choices of the skew-symmetric matrices $\mathbf{S}_F$ which provide a solution to the minimum effort covariance control problem are parametrized in the following result.

Theorem 5.4.2 *The minimum effort continuous-time covariance controller $\mathbf{G}$ is provided by the following choices of the skew-symmetric matrix $\mathbf{S}_F$ in the parametrization (5.121)*

$$vec\,\mathbf{S}_F = -\mathbf{\Delta}(\mathbf{K\Delta})^+ vec\,(\mathbf{K}_3) + \mathbf{\Delta}[\mathbf{I} - (\mathbf{K\Delta})^+\mathbf{K\Delta}]\mathbf{q} \tag{5.129}$$

where $\mathbf{K} \triangleq \mathbf{K}_2^T \otimes \mathbf{K}_1 + \mathbf{K}_1^T \otimes \mathbf{K}_2$, $\mathbf{q}$ is an arbitrary $(n_p+n_c)(n_p+n_c-1)/2$-dimensional vector, $\mathbf{K}_1$, $\mathbf{K}_2$ and $\mathbf{K}_3$ are defined in (5.123), and $\mathbf{\Delta}$ is defined by (5.128). The Kronecker product $\otimes$ is defined in Appendix A.

The optimal choice (5.129) is obtained by solving equation (5.122) using the vec operator. The static state feedback case is provided by setting $\mathbf{M}_p = \mathbf{I}$ and $n_c = 0$.

Example 5.4.1 Consider the following continuous-time model of a missile attitude regulator

$$\dot{\mathbf{x}} = \begin{bmatrix} 0 & 0 & 0 & 0 \\ 10 & -1 & 0 & -1 \\ 0 & 1 & 0 & 2 \\ 1 & 0 & 0 & -1 \end{bmatrix} \mathbf{x} + \begin{bmatrix} 1 & 0 \\ 0.5 & 0.4 \\ 0.8 & 1 \\ 0 & 0 \end{bmatrix} \mathbf{u} + \begin{bmatrix} 0 \\ 1 \\ 1 \\ 0 \end{bmatrix} \mathbf{w}$$

where $\mathbf{w}(t)$ is a white noise process with intensity $\mathbf{W} = 10$. Suppose that the following state feedback control gain has been designed to satisfy output variance requirements:

$$\mathbf{G}_o = \begin{bmatrix} -40.8760 & -66.5402 & -78.3062 & -100.6755 \\ 27.4402 & 47.1713 & 56.7027 & 73.5772 \end{bmatrix}.$$

This controller assigns the following covariance matrix to the closed-loop system

$$\mathbf{X} = \begin{bmatrix} 0.5968 & -0.3167 & -0.1201 & 0.0604 \\ -0.3167 & 0.6068 & -0.3172 & 0.0324 \\ -0.1201 & -0.3172 & 0.4848 & -0.1298 \\ 0.0604 & 0.0324 & -0.1298 & 0.0604 \end{bmatrix}$$

and the required control effort (for unit weight matrix $\mathbf{R} = \mathbf{I}$) is

$$\|\mathbf{u}\|^2 = 535.6851.$$

We wish to redesign the controller $\mathbf{G}_o$ preserving the original closed-loop state covariance matrix $\mathbf{X}$ such that $\|\mathbf{u}\|^2$ is minimized. Theorem 5.4.1 provides the following state feedback gain as the solution to this problem

$$\mathbf{G} = \begin{bmatrix} -14.5586 & -24.4277 & -18.5667 & -21.1357 \\ -11.1921 & -8.1730 & -16.3240 & -15.1148 \end{bmatrix}.$$

This controller assigns the same covariance matrix $\mathbf{X}$ to the closed-loop system but the required control effort is

$$\|\mathbf{u}\|^2 = 109.6756.$$

Note that the controller $\mathbf{G}_o$ corresponds to the choice $\mathbf{S} = \mathbf{0}$ in the covariance controller parametrization (5.121) but the minimum effort controller $\mathbf{G}$ corresponds to the following optimal choice provided by the expression (5.129)

$$\mathbf{S}_F = \begin{bmatrix} 0 & -0.8465 & -2.1162 & 0 \\ 0.8465 & 0 & -0.3809 & 0 \\ 2.1162 & 0.3809 & 0 & 0 \\ 0 & 0 & 0 & 0 \end{bmatrix}.$$

5.4.2 Discrete-Time Output Feedback

In this section we seek the discrete-time dynamic output feedback controller with measurement noise to minimize the control effort

$$\|\mathbf{u}\| \triangleq \lim_{t\to\infty} \mathcal{E}[\mathbf{u}(k)^T \mathbf{R} \mathbf{u}(k)]^{1/2} \tag{5.130}$$

subject to

$$\lim_{t\to\infty} \mathcal{E}_\infty[\mathbf{x}(k)\mathbf{x}(k)^T] = \mathbf{X} \tag{5.131}$$

where $\mathbf{R} > \mathbf{0}$ is a given weighting matrix and $\mathbf{X} > \mathbf{0}$ is a given assignable covariance. According to the results in section 5.3, the set of all covariance controllers which assign $\mathbf{X}$ to the closed-loop system is parametrized by (5.112):

$$\mathbf{G} = \mathbf{B}^{+}(\mathbf{L}\mathbf{V}_L \begin{bmatrix} \mathbf{I}_k & \mathbf{0} \\ \mathbf{0} & \mathbf{U}_F \end{bmatrix} \mathbf{V}_R^T \mathbf{R}^{-1/2} - \mathbf{A}\mathbf{X}\mathbf{M}^T\mathbf{R}^{-1}) \tag{5.132}$$

where $\mathbf{U}_F$ is an arbitrary orthogonal matrix. We have assumed that the input matrix $\mathbf{B}_p$ has full column rank. The discrete-time minimum effort covariance controller is provided by the following result.

Theorem 5.4.3 *The discrete-time covariance controller with measurement noise* $\mathbf{G}$*, which solves the minimum effort covariance control problem, is provided by the following choice of the orthogonal matrix* $\mathbf{U}_F$ *in the parametrization (5.132):*

$$\mathbf{U}_F = \mathbf{U}_1\mathbf{U}_2^T \tag{5.133}$$

where $\mathbf{U}_1$ *and* $\mathbf{U}_2$ *are defined from the singular value decomposition*

$$\mathbf{\Phi}_2^T\mathbf{R}\mathbf{\Theta}_2 = \mathbf{U}_1\mathbf{\Lambda}\mathbf{U}_2^T \tag{5.134}$$

and the matrices $\mathbf{\Phi}_2$ *and* $\mathbf{\Theta}_2$ *are defined by the following expressions*

$$\begin{aligned} [\mathbf{\Theta}_1\ \mathbf{\Theta}_2] &\triangleq \mathbf{I}_1\mathbf{B}^{+}\mathbf{L}\mathbf{V}_L \\ [\mathbf{\Phi}_1\ \mathbf{\Phi}_2] &\triangleq \mathbf{I}_1\mathbf{B}^{+}\mathbf{A}\mathbf{X}\mathbf{M}^T\mathbf{V}_R \end{aligned} \tag{5.135}$$

where $\mathbf{I}_1 = [\mathbf{I}\ \ \mathbf{0}]$. *The value of the minimum control effort is*

$$\|\mathbf{u}\|_{\min} = [\|\mathbf{\Theta}_1 - \mathbf{\Phi}_1\|^2 + \|\mathbf{\Theta}_2\|^2 + \|\mathbf{\Phi}_2\|^2 - 2(\lambda_1 + \ldots + \lambda_k)]^{1/2} \tag{5.136}$$

where $\mathbf{\Lambda} = \mathrm{diag}(\lambda_1, \ldots, \lambda_k)$.

For the proof we need the following lemma (see [121], [50]).

Lemma 5.4.2 *Let* $\mathbf{N} \in \mathcal{R}^{n\times n}$ *be a given matrix and let* $\mathbf{N} = \mathbf{U}_1\mathbf{\Sigma}\mathbf{U}_2$ *be a singular value decomposition of* $\mathbf{N}$. *Then the optimization problem*

$$\textit{maximize}\ \ tr\,(\mathbf{N}\mathbf{U}), \qquad \mathbf{U}\mathbf{U}^T = \mathbf{I} \tag{5.137}$$

has the solution $\mathbf{U} = \mathbf{U}_1\mathbf{U}_2^T$, *and the value of the maximum is* $\sigma_1 + \ldots + \sigma_n$, *where* σ_i *are the singular values of* $\mathbf{N}$.

We now prove Theorem 5.4.3 using the above lemma.

Proof. The control effort can be computed as follows:

$$\|\mathbf{u}\|^2 = \mathcal{E}_\infty[\mathbf{u}(k)^T\mathbf{R}\mathbf{u}(k)]^{1/2} = tr\left[\mathbf{I}_1\mathbf{G}(\mathbf{M}\mathbf{X}\mathbf{M}^T + \mathbf{E}\mathbf{E}^T)\mathbf{G}^T\mathbf{I}_1^T\mathbf{R}\right]. \tag{5.138}$$

By substituting the expression (5.132) for the covariance controller $\mathbf{G}$ we obtain after simplifications

$$\|\mathbf{u}\|^2 = \|\mathbf{I}_1\mathbf{B}^{+}\mathbf{L}\mathbf{V}_L \begin{bmatrix} \mathbf{I}_k & \mathbf{0} \\ \mathbf{0} & \mathbf{U}_F \end{bmatrix} - \mathbf{I}_1\mathbf{B}^{+}\mathbf{A}\mathbf{X}\mathbf{M}^T\mathbf{R}^{-1/2}\mathbf{V}_R\|^2. \tag{5.139}$$

The definitions (5.135) result in the following expression:

$$\|\mathbf{u}\|^2 = \| [\boldsymbol{\Theta}_1 \ \boldsymbol{\Theta}_2\mathbf{U}_F] - [\boldsymbol{\Phi}_1 \ \boldsymbol{\Theta}_2\mathbf{U}_F] \|^2 = \|\boldsymbol{\Theta}_1 - \boldsymbol{\Phi}_1\|^2 + \|\boldsymbol{\Theta}_2\mathbf{U}_F - \boldsymbol{\Phi}_2\|^2. \quad (5.140)$$

Hence, the discrete-time minimum effort covariance control problem is provided by the solution of the minimum norm problem

$$\textit{minimize} \ \|\boldsymbol{\Theta}_2\mathbf{U}_F - \boldsymbol{\Phi}_2\|^2, \quad \mathbf{U}_F\mathbf{U}_F{}^T = \mathbf{I}. \quad (5.141)$$

Expanding this expression we obtain

$$\|\boldsymbol{\Theta}_2\mathbf{U}_F - \boldsymbol{\Phi}_2\|^2 = \|\boldsymbol{\Theta}_2\|^2 + \|\boldsymbol{\Phi}_2\|^2 - 2tr(\boldsymbol{\Phi}_2^T\mathbf{R}\boldsymbol{\Theta}_2\mathbf{U}_F) \quad (5.142)$$

where the orthogonality of $\mathbf{U}_F$ has been used. Hence, our problem requires the solution of the maximization problem

$$\textit{maximize} \ tr(\boldsymbol{\Phi}_2^T\mathbf{R}\boldsymbol{\Theta}_2\mathbf{U}_F), \quad \mathbf{U}_F \textit{ is orthogonal.} \quad (5.143)$$

By applying Lemma 5.4.2 we obtain the optimal choice of the orthogonal matrix $\mathbf{U}_F$ and the value of the minimum control effort. □

Example 5.4.2 Consider the following discretized state space model of a simply supported beam with 4 states, 2 inputs and 2 outputs

$$\begin{aligned}\mathbf{x}(k+1) \ = \ & \begin{bmatrix} 0.8778 & 0.4782 & 0 & 0 \\ -0.4782 & 0.8730 & 0 & 0 \\ 0 & 0 & -0.4075 & 0.2251 \\ 0 & 0 & -3.6011 & -0.4165 \end{bmatrix}\mathbf{x}(k) \\ & + \begin{bmatrix} 0.0718 & -0.1222 \\ 0.2811 & -0.4782 \\ 0.0837 & 0.1759 \\ 0.2141 & 0.4501 \end{bmatrix}\mathbf{u}(k) \\ & + \begin{bmatrix} 0.0214 & 0.0055 & 0 & 0 \\ -0.0055 & 0.0214 & 0 & 0 \\ 0 & 0 & 0.0102 & 0.0039 \\ 0 & 0 & -0.0629 & 0.0101 \end{bmatrix}\mathbf{w}(k)\end{aligned}$$

where $\mathbf{w}(k)$ is a white noise process with intensity $\mathbf{W} = 100\mathbf{I}$. Also, consider the following state feedback control that has been designed to satisfy covariance control requirements

$$\mathbf{G}_o = \begin{bmatrix} -0.7604 & -3.4667 & 9.5586 & 0.2955 \\ -0.0010 & 1.2221 & 3.3200 & -0.2642 \end{bmatrix}.$$

The corresponding control effort (for a unit weight matrix $\mathbf{R} = \mathbf{I}$) is

$$\|\mathbf{u}\|^2 = 3.3546.$$

Theorem 5.4.3 provides a controller that assigns the same closed-loop state covariance matrix to the closed-loop system with minimum control effort. This minimum effort covariance controller is

$$\mathbf{G} = \begin{bmatrix} -0.0645 & -0.6566 & 1.6037 & -0.0619 \\ -0.0219 & 0.6336 & 1.7486 & -0.1774 \end{bmatrix}$$

and the minimum required control effort is

$$\|\mathbf{u}\|^2 = 0.3549.$$

The orthogonal matrix $\mathbf{U}_F$ in the covariance controller parametrization (5.133) that corresponds to the minimum effort covariance controller is

$$\mathbf{U}_F = \begin{bmatrix} 0.4704 & 0.8824 \\ -0.8824 & 0.4704 \end{bmatrix}.$$

5.5 FINITE WORDLENGTH COVARIANCE CONTROL

This section introduces a control design which takes into account the state quantization errors for controllers synthesized in finite precision fixed-point machines, using either synchronized or skewed sampling. In addition, the problem of controller complexity for fixed point arithmetic is addressed to answer the question: "What memory (total length of all words) is required in the control computer to guarantee a specified RMS performance bound on each output of the plant?" The tools of covariance control are used to solve these problems.

The integration of the signal processing and control disciplines has received considerable attention in recent years. No longer is it considered wise to separate the design of controllers (with the assumption of infinite precision implementation), and the signal processing concerns (the synthesis of the given controller). It is well known [152], [36] that there exists an optimal realization of a given controller, so that synthesis in these optimal coordinates will minimize the noise gain from state round-off effects. (A controller that is designed without regard to controller synthesis, can then be implemented in an optimal realization for this given controller.) It is also known that such controllers are not optimal overall. That is, the design and synthesis problems are not independent problems.

This chapter makes some improvements in digital control theory. First, we parametrize all state covariances that are assignable to the closed-loop system in the presence of quantization error in the control computer and in the A/D and D/A devices. Secondly, we characterize all dynamic controllers which assign these covariances to the closed-loop system. To this end, desired performance objectives expressed in terms of the covariance matrix can be traded with controller complexity (controller order and available wordlength). All of these results are also derived for problems with skewed sampling. We shall refer to "skewing" as the asynchronous sampling of the measurement and control. The sampling instants occur at the same rate, but are "skewed" by amount δ seconds. Typically, δ is chosen as the computer duty cycle (time required by the control computer to compute a new control given a new measurement). In this way, computational delays are accommodated in the model and control design process.

5.6 SYNCHRONOUS SAMPLING

Consider a controller which is synthesized in a digital computer with synchronous sampling and fixed-point arithmetic. It is well known [152], [36] that the effects of the quantization error in the control computer depend on the realization of the controller. To this end, we shall study the control design problem in a transformed set of controller parameters $\mathbf{A}_c =$

$\mathbf{T}_c^{-1}\tilde{\mathbf{A}}_c\mathbf{T}_c$, $\mathbf{B}_c = \mathbf{T}_c^{-1}\tilde{\mathbf{B}}_c$, $\mathbf{C}_c = \tilde{\mathbf{C}}_c\mathbf{T}_c$, $\mathbf{D}_c = \tilde{\mathbf{D}}_c$ and write the controller dynamics

$$\begin{aligned} \mathbf{x}_c(k+1) &= \mathbf{T}_c^{-1}\tilde{\mathbf{A}}_c\mathbf{T}_c(\mathbf{x}_c(k)+\mathbf{e}_x(k)) + \mathbf{T}_c^{-1}\tilde{\mathbf{B}}_c(\mathbf{z}(k)+\mathbf{e}_z(k)) \\ \mathbf{u}(k) &= \tilde{\mathbf{C}}_c\mathbf{T}_c\,(\mathbf{x}_c(k)+\mathbf{e}_x(k)) + \tilde{\mathbf{D}}_c\,(\mathbf{z}(k)+\mathbf{e}_z(k)) \end{aligned} \tag{5.144}$$

where $\mathbf{e}_x(k)$ is the quantization error introduced by the controller state computation $\mathbf{x}_c(k)$ in the control computer (with wordlength β_x, and $\mathbf{e}_z(k)$ is the quantization error introduced by the A/D converter (with wordlength β_z). The plant is described by

$$\begin{aligned} \mathbf{x}_p(k+1) &= \mathbf{A}_p\mathbf{x}_p(k) + \mathbf{B}_p(\mathbf{u}(k)+\mathbf{e}_u(k)) + \mathbf{D}_p\mathbf{w}_p(k) \\ \mathbf{y}_p(k) &= \mathbf{C}_p\mathbf{x}_p(k) \\ \mathbf{z}(k) &= \mathbf{M}_p\mathbf{x}_p(k) + \mathbf{v}(k) \end{aligned} \tag{5.145}$$

where $\mathbf{e}_u(k)$ is the quantization error introduced by the D/A converter (with wordlength β_i). Under sufficient excitation conditions we can approximate the quantization errors $\mathbf{e}_x(k)$, $\mathbf{e}_z(k)$ and $\mathbf{e}_u(k)$ as zero-mean white noise processes $\mathbf{W}_x = \text{diag}[\ldots q_i \ldots]$, $\mathbf{W}_z = q_z\mathbf{I}$ and $\mathbf{W}_u = q_u\mathbf{I}$, respectively, where $q_i = (1/12)\,2^{-2\beta_i}$, $q_z = (1/12)\,2^{-2\beta_z}$, $q_u = (1/12)\,2^{-2\beta_u}$, and β_i is the length of the fractional part of the word storing the ith controller state variable.

Define the matrix

$$\mathbf{W} = \begin{bmatrix} q_u\mathbf{I} & \mathbf{0} & \mathbf{0} & \mathbf{0} \\ \mathbf{0} & \mathbf{W}_p & \mathbf{0} & \mathbf{0} \\ \mathbf{0} & \mathbf{0} & \mathbf{W}_z+\mathbf{V} & \mathbf{0} \\ \mathbf{0} & \mathbf{0} & \mathbf{0} & \mathbf{T}_c\mathbf{W}_x\mathbf{T}_c^T \end{bmatrix} \tag{5.146}$$

where $\mathbf{W}_p$ and $\mathbf{V}$ are the covariances of the plant noise $\mathbf{w}_p(k)$ and the measurement noise $\mathbf{v}(k)$, respectively.

$$\begin{aligned} &\mathbf{A} = \begin{bmatrix} \mathbf{A}_p & \mathbf{0} \\ \mathbf{0} & \mathbf{0} \end{bmatrix}, \quad \mathbf{B} = \begin{bmatrix} \mathbf{B}_p & \mathbf{0} \\ \mathbf{0} & \mathbf{I} \end{bmatrix}, \quad \mathbf{C} = \begin{bmatrix} \mathbf{C}_p & \mathbf{0} \\ \mathbf{0} & \mathbf{0} \end{bmatrix} \\ &\mathbf{M} = \begin{bmatrix} \mathbf{M}_p & \mathbf{0} \\ \mathbf{0} & \mathbf{I} \end{bmatrix}, \quad \mathbf{D} = \begin{bmatrix} \mathbf{B}_p & \mathbf{D}_p & \mathbf{0} & \mathbf{0} \\ \mathbf{0} & \mathbf{I} & \mathbf{0} & \mathbf{0} \end{bmatrix}, \quad \mathbf{H} = \begin{bmatrix} \mathbf{0} & \mathbf{0} \\ \mathbf{I} & \mathbf{0} \end{bmatrix} \\ &\mathbf{J} = \begin{bmatrix} \mathbf{0} & \mathbf{0} & \mathbf{I} & \mathbf{0} \\ \mathbf{0} & \mathbf{0} & \mathbf{0} & \mathbf{I} \end{bmatrix}, \quad \mathbf{T} = \begin{bmatrix} \mathbf{I} & \mathbf{0} \\ \mathbf{0} & \mathbf{T}_c \end{bmatrix} \\ &\tilde{\mathbf{G}} = \begin{bmatrix} \mathbf{D}_c & \tilde{\mathbf{C}}_c \\ \tilde{\mathbf{B}}_c & \tilde{\mathbf{A}}_c \end{bmatrix}, \quad \mathbf{G} = \mathbf{T}^{-1}\tilde{\mathbf{G}}\mathbf{T} = \begin{bmatrix} \mathbf{D}_c & \mathbf{C}_c \\ \mathbf{B}_c & \mathbf{A}_c \end{bmatrix}. \end{aligned} \tag{5.147}$$

Then the closed-loop system is described by

$$\begin{aligned} \mathbf{x}(k+1) &= (\mathbf{A}+\mathbf{BGM})\mathbf{x}(k) + (\mathbf{D}+\mathbf{BGJ})\mathbf{w}(k) \\ \mathbf{y}(k) &= (\mathbf{C}+\mathbf{HGM})\mathbf{x}(k) + \mathbf{HGJw}(k) \end{aligned} \tag{5.148}$$

where $\mathbf{x} = \begin{bmatrix}\mathbf{x}_p^T & \mathbf{x}_c^T\end{bmatrix}^T$, $\mathbf{w} = \begin{bmatrix}\mathbf{e}_u^T & \mathbf{w}_p^T & \mathbf{e}_z^T+\mathbf{v}^T & \mathbf{e}_x^T\end{bmatrix}^T$ and $\mathbf{y} = \begin{bmatrix}\mathbf{y}_p^T & \mathbf{u}^T\end{bmatrix}^T$. The state and output covariances satisfy

$$\begin{aligned} \mathbf{X} &= (\mathbf{A}+\mathbf{BGM})\mathbf{X}(\mathbf{A}+\mathbf{BGM})^T + (\mathbf{D}+\mathbf{BGJ})\mathbf{W}(\mathbf{D}+\mathbf{BGJ})^T \\ \mathbf{Y} &= (\mathbf{C}+\mathbf{HGM})\mathbf{X}(\mathbf{C}+\mathbf{HGM})^T + \mathbf{HGJWJ}^T\mathbf{G}^T\mathbf{H}^T. \end{aligned} \tag{5.149}$$

We seek to obtain necessary and sufficient conditions for assignability of a covariance matrix $\mathbf{X}$, as well as a parametrization of all covariance controllers which assign a particular covariance.

5.7 SKEWED SAMPLING

In this section we present the finite wordlength control problem for the case where the computational time delay δ in the control computer is taken into account. With quantization errors, the dynamics of the plant for skewed sampling of $\mathbf{z}$ are described by

$$\begin{aligned} \mathbf{x}_p(k+1) &= \mathbf{A}_p\mathbf{x}_p(k) + \mathbf{B}_p\left(\mathbf{u}(k) + \mathbf{e}_u(k)\right) + \mathbf{D}_p\mathbf{w}_p(k) \\ \mathbf{z}_\delta(k+1) &= \mathbf{M}_\delta\mathbf{x}_p(k) + \mathbf{H}_\delta\left(\mathbf{u}(k) + \mathbf{e}_u(k)\right) + \mathbf{v}(k) \end{aligned} \tag{5.150}$$

where $\mathbf{z}_\delta(k+1)$ is a measurement occurring δ seconds *before* $\mathbf{x}_p(k+1)$. The matrix $\mathbf{H}_\delta$ is a leakage term (allowing the plant inputs to appear in the output) that goes to zero as the skewing δ goes to zero. The controller with quantization error is described by

$$\begin{aligned} \mathbf{x}_c(k+1) &= \mathbf{A}_c\left(\mathbf{x}_c(k) + \mathbf{e}_x(k)\right) + \mathbf{B}_c\left(\mathbf{z}_\delta(k+1) + \mathbf{e}_z(k+1)\right) \\ \mathbf{u}(k) &= \mathbf{C}_c\left(\mathbf{x}_c(k) + \mathbf{e}_x(k)\right) + \mathbf{D}_c\left(\mathbf{z}_\delta(k+1) + \mathbf{e}_z(k+1)\right). \end{aligned} \tag{5.151}$$

Hence the closed-loop system is described by

$$\begin{bmatrix} \mathbf{x}_p(k+1) \\ \mathbf{u}(k+1) \\ \mathbf{x}_c(k+1) \end{bmatrix} = \begin{bmatrix} \mathbf{A}_p & \mathbf{B}_p & \mathbf{0} \\ (\mathbf{D}_c + \mathbf{C}_c\mathbf{B}_c)\mathbf{M}_\delta & (\mathbf{D}_c + \mathbf{C}_c\mathbf{B}_c)\mathbf{H}_\delta & \mathbf{C}_c\mathbf{A}_c \\ \mathbf{B}_c\mathbf{M}_\delta & \mathbf{B}_c\mathbf{H}_\delta & \mathbf{A}_c \end{bmatrix} \begin{bmatrix} \mathbf{x}_p(k) \\ \mathbf{u}(k) \\ \mathbf{x}_c(k) \end{bmatrix}$$

$$\begin{bmatrix} \mathbf{B}_p & \mathbf{D}_p & \mathbf{0} & \mathbf{0} \\ (\mathbf{D}_c + \mathbf{C}_c\mathbf{B}_c)\mathbf{H}_\delta & \mathbf{0} & \mathbf{0} & \mathbf{0} \\ \mathbf{B}_c\mathbf{H}_\delta & \mathbf{0} & \mathbf{0} & \mathbf{0} \end{bmatrix} \begin{bmatrix} \mathbf{e}_u \\ \mathbf{w}_p \\ \mathbf{e}_z + \mathbf{v} \\ \mathbf{e}_x \end{bmatrix} \tag{5.152}$$

where the extra state variable $\mathbf{u}(k)$ is required due to the skew delay δ. In the presence of controller state, input and output quantization errors the closed-loop skewed system can be described by

$$\begin{aligned} \mathbf{x}(k+1) &= (\mathbf{A} + \mathbf{B}\bar{\mathbf{G}}\mathbf{M})\mathbf{x}(k) + (\mathbf{D} + \mathbf{B}\bar{\mathbf{G}}\mathbf{J})\mathbf{w}(k) \\ \mathbf{y}(k) &= (\mathbf{C} + \mathbf{H}\bar{\mathbf{G}}\mathbf{M})\mathbf{x}(k) + \mathbf{H}\bar{\mathbf{G}}\mathbf{J}\mathbf{w}(k) \end{aligned} \tag{5.153}$$

where $\mathbf{x} = \begin{bmatrix} \mathbf{x}_p^T & \mathbf{u}^T & \mathbf{x}_c^T \end{bmatrix}^T$, $\mathbf{w} = \begin{bmatrix} \mathbf{e}_u^T & \mathbf{w}_p^T & \mathbf{e}_z^T + \mathbf{v}^T & \mathbf{e}_x^T \end{bmatrix}^T$ and $\mathbf{y} = \begin{bmatrix} \mathbf{y}_p^T & \mathbf{u}^T \end{bmatrix}^T$, and

$$\mathbf{A} = \begin{bmatrix} \mathbf{A}_s & \mathbf{0} \\ \mathbf{0} & \mathbf{0} \end{bmatrix}, \quad \mathbf{A}_s = \begin{bmatrix} \mathbf{A}_p & \mathbf{B}_p \\ \mathbf{0} & \mathbf{0} \end{bmatrix}, \quad \mathbf{B} = \begin{bmatrix} \mathbf{B}_s & \mathbf{0} \\ \mathbf{0} & \mathbf{I} \end{bmatrix}, \quad \mathbf{B}_s = \begin{bmatrix} \mathbf{0} \\ \mathbf{I} \end{bmatrix}$$

$$\mathbf{D} = \begin{bmatrix} \mathbf{D}_s & \mathbf{0} \\ \mathbf{0} & \mathbf{0} \end{bmatrix}, \quad \mathbf{D}_s = \begin{bmatrix} \mathbf{B}_p & \mathbf{D}_p \\ \mathbf{0} & \mathbf{0} \end{bmatrix}, \quad \mathbf{M} = \begin{bmatrix} \mathbf{M}_s & \mathbf{0} \\ \mathbf{0} & \mathbf{I} \end{bmatrix}, \quad \mathbf{M}_s = \begin{bmatrix} \mathbf{M}_\delta & \mathbf{H}_\delta \end{bmatrix}$$

$$\mathbf{C} = \begin{bmatrix} \mathbf{C}_p & \mathbf{0} & \mathbf{0} \\ \mathbf{0} & \mathbf{I} & \mathbf{0} \end{bmatrix}, \quad \mathbf{J} = \begin{bmatrix} \mathbf{H}_\delta & \mathbf{0} & \mathbf{0} & \mathbf{0} \\ \mathbf{0} & \mathbf{0} & \mathbf{0} & \mathbf{0} \end{bmatrix}, \quad \mathbf{H} = \begin{bmatrix} \mathbf{0} & \mathbf{0} \\ \mathbf{I} & \mathbf{0} \end{bmatrix} \tag{5.154}$$

and

$$\bar{\mathbf{G}} = \begin{bmatrix} \mathbf{D}_c + \mathbf{C}_c\mathbf{B}_c & \mathbf{C}_c\mathbf{A}_c \\ \mathbf{B}_c & \mathbf{A}_c \end{bmatrix} \tag{5.155}$$

The state covariance of system (5.153) satisfies the same equation as (5.149), except $\overline{\mathbf{G}}$ replaces $\mathbf{G}$ in (5.149).

Now suppose that $\overline{\mathbf{G}}$ is found to provide the closed-loop system the desired response $\mathbf{x}(k)$. The question is whether or not there exists a controller that yields this response. The following result provides a constraint on $\overline{\mathbf{G}}$ which allows the synthesis of a finite wordlength controller to give the desired response.

Lemma 5.7.1 *Given a matrix $\overline{\mathbf{G}}$, there exist controller matrices $\mathbf{A}_c$, $\mathbf{B}_c$, $\mathbf{C}_c$ and $\mathbf{D}_c$ satisfying (5.155) if and only if*

$$\overline{\mathbf{G}}_{12}(\mathbf{I} - \overline{\mathbf{G}}_{22}^{+}\overline{\mathbf{G}}_{22}) = \mathbf{0} \tag{5.156}$$

in which case all such controller matrices are given by

$$\mathbf{A}_c = \overline{\mathbf{G}}_{22},\quad \mathbf{B}_c = \overline{\mathbf{G}}_{21},\quad \mathbf{C}_c = \overline{\mathbf{G}}_{12}\overline{\mathbf{G}}_{22}^{+} + \mathbf{Z}(\mathbf{I} - \overline{\mathbf{G}}_{22}\overline{\mathbf{G}}_{22}^{+}),\quad \mathbf{D}_c = \overline{\mathbf{G}}_{11} - \mathbf{C}_c\overline{\mathbf{G}}_{21} \tag{5.157}$$

where $\mathbf{Z}$ is arbitrary.

Note that the controller satisfying (5.156) is unique if the controller matrix $\mathbf{A}_c = \overline{\mathbf{G}}_{22}$ is invertible.

5.8 COVARIANCE ASSIGNMENT

In this section we shall describe the necessary and sufficient conditions for a covariance matrix to be assignable (satisfies (5.149)) and parametrize all finite wordlength controllers $\mathbf{G}$ which assign this covariance to the closed-loop system.

Theorem 5.8.1 *The set of all covariance matrices $\mathbf{X} > \mathbf{0}$ that satisfy the finite wordlength covariance equation (5.149), for some $\mathbf{G}$, are parametrized by the following conditions:*

$$(\mathbf{I} - \mathbf{B}\mathbf{B}^{+})(\mathbf{X} - \mathbf{A}\mathbf{X}\mathbf{A}^{T} - \mathbf{D}\mathbf{W}\mathbf{D}^{T})(\mathbf{I} - \mathbf{B}\mathbf{B}^{+}) = \mathbf{0} \tag{5.158}$$

$$\begin{aligned}\mathbf{X} \;=\; & \mathbf{A}\mathbf{X}\mathbf{A}^{T} + \mathbf{D}\mathbf{W}\mathbf{D}^{T} \\ & -(\mathbf{A}\mathbf{X}\mathbf{M}^{T} + \mathbf{D}\mathbf{W}\mathbf{J}^{T})(\mathbf{M}\mathbf{X}\mathbf{M}^{T} + \mathbf{J}\mathbf{W}\mathbf{J}^{T})^{-1}(\mathbf{A}\mathbf{X}\mathbf{M}^{T} + \mathbf{D}\mathbf{W}\mathbf{J}^{T})^{T} + \mathbf{L}\mathbf{L}^{T}\end{aligned} \tag{5.159}$$

where $\mathbf{L}$ is some $(n_p + n_u + n_c) \times (n_z + n_c)$ matrix. For any such $\mathbf{X}$, the set of all matrices $\mathbf{G}$ satisfying the covariance equation (5.149) is given by

$$\mathbf{G} = \mathbf{B}^{+}(\mathbf{L}\mathbf{U}\mathbf{R}^{-1/2} - (\mathbf{A}\mathbf{X}\mathbf{M}^{T} + \mathbf{D}\mathbf{W}\mathbf{J}^{T})\mathbf{R}^{-1}) + \mathbf{Z}_F - \mathbf{B}^{+}\mathbf{B}\mathbf{Z}_F \tag{5.160}$$

where $\mathbf{Z}_F$ is an arbitrary matrix and

$$\mathbf{U} = \mathbf{V}_L \begin{bmatrix} \mathbf{I} & \mathbf{0} \\ \mathbf{0} & \mathbf{U}_F \end{bmatrix} \mathbf{V}_R{}^{T},\quad \mathbf{U}_F\mathbf{U}_F^{T} = \mathbf{I} \tag{5.161}$$

where $\mathbf{V}_L$ and $\mathbf{V}_M$ are obtained from the singular value decompositions

$$\begin{aligned}(\mathbf{I} - \mathbf{B}\mathbf{B}^{+})\mathbf{L} &= \mathbf{U}_L\Sigma_L\mathbf{V}_L^{T} \\ (\mathbf{I} - \mathbf{B}\mathbf{B}^{+})(\mathbf{A}\mathbf{X}\mathbf{M}^{T} + \mathbf{D}\mathbf{W}\mathbf{J}^{T})\mathbf{R}^{-\frac{1}{2}} &= \mathbf{U}_L\Sigma_M\mathbf{V}_M^{T},\quad \Sigma_M = [\Sigma_L\ \ \mathbf{0}]\end{aligned} \tag{5.162}$$

and $\mathbf{L}$ is defined by (5.159), and $\mathbf{R}$ is defined by

$$\mathbf{R} \triangleq \mathbf{M}\mathbf{X}\mathbf{M}^{T} + \mathbf{J}\mathbf{W}\mathbf{J}^{T}. \tag{5.163}$$

Proof. The proof follows the proof of Theorem 5.3.2 without the assumption $\mathbf{ED}^T = \mathbf{0}$ made there. Rearranging equation (5.149) we obtain

$$\mathbf{X} - \mathbf{AXA}^T - \mathbf{DWD}^T =$$

$$\mathbf{BGMXA}^T + \mathbf{AXMG}^T\mathbf{B}^T + \mathbf{BGFG}^T\mathbf{B}^T + \mathbf{DWJ}^T\mathbf{G}^T\mathbf{B}^T + \mathbf{BGJWD}^T$$

where $\mathbf{R} \triangleq \mathbf{MXM}^T + \mathbf{JWJ}^T > \mathbf{0}$. Adding to both sides the term

$$(\mathbf{AXM}^T + \mathbf{DWJ}^T)\mathbf{R}^{-1}(\mathbf{AXM}^T + \mathbf{DWJ}^T)^T$$

and completing the square provides

$$\mathbf{P} = (\mathbf{BG} + \mathbf{SR}^{-1})\mathbf{R}(\mathbf{BG} + \mathbf{SR}^{-1})^T$$

where $\mathbf{S} = \mathbf{AXM}^T + \mathbf{DWJ}^T$. Using Theorem 2.3.9 we obtain the parametrization of all assignable covariances (5.159) and the set of all finite wordlength covariance controllers (5.160). □

CHAPTER 5 SUMMARY

Covariance analysis is a cornerstone of systems theory. Engineers have often evaluated performance and performed error analysis using variances and covariances because they have physical meaning and can be computed from signals. This is the first book chapter to provide a general method to *control* covariances, rather than merely analyzing them after a system is developed. The significance of these results go beyond control theory, since they also allow new criteria for design by passive systems. For example, since the L_2 to L_∞ gain of a linear system is the maximum singular value of the output covariance matrix, one can choose physical parameters (springs mass, stiffness, material properties) to achieve a given covariance upper bound or a given L_2 to L_∞ gain. Peak stresses in aircraft structures can be limited by such designs. This is a new direction in structure design.

This chapter provides the necessary and sufficient conditions for a given covariance matrix to be assignable by output feedback control. The set of all covariances that can be assigned to a given linear system is parametrized. The set of all controllers that assign a given covariance matrix is given by an explicit formula with free parameters. A choice of the free parameters is given to minimize control energy while assigning a covariance. The necessary conditions for the existence of $\mathbf{G}$ to solve (5.2) were perhaps first presented in [92]. Since then many other papers on covariance control have appeared; interested readers are referred to [14, 30, 53, 60, 94, 95, 129, 130, 132, 133, 134, 135, 147, 155, 156, 157]. This chapter also describes the covariance assignment problem for discrete-time systems subject to skewed sampling between measurement and control, and subject to finite precision computing in the A/D, D/A devices and in the controller state noise with variance related to the wordlength. The covariance equation for the problem has similar structure to other problems in this book, and the same theorems for control design apply.

CHAPTER SIX

Covariance Upper Bound Controllers

The motivation for this chapter is that upper bounds on performances might be more useful than the more difficult task of assigning exact performance. Mathematically, this allows the use of inequality constraints in lieu of equality constraints on covariance matrices.

6.1 COVARIANCE BOUNDING CONTROL PROBLEM

Consider the linear time-invariant systems given in Sections 4.2 and 4.5 for the continuous-time and discrete-time cases, respectively. As we have seen in Chapter 4, the output covariance is closely related to many performance properties, such as the $\mathcal{H}_2$ norm, the $\mathcal{L}_2$ to $\mathcal{L}_\infty$ gain and the variance of each output signal. In many cases, the "smaller" the output covariance, the better the system performance. This provides motivation to consider a performance specification given in terms of a bound on the output covariance. In this chapter, consider the following covariance bounding control problem:

> Determine if there exists a controller which stabilizes the system and yields an output covariance bounded above by a given matrix $\mho$. Find all such controllers when one exists.

In Chapter 5, we have formulated and solved the covariance control problem, where a given matrix $\mathbf{X}$ is assigned to the closed-loop system as the state covariance. This chapter is concerned with the design of a controller which guarantees an upper bound on the state covariance. Note also that our interest here is the *output* covariance which is more general than the state covariance in the sense that the state covariance is a special case ($\mathbf{C}_{c\ell} = \mathbf{I}$) of the output covariance.

Consider the following Lyapunov inequalities:

$$\mathbf{A}_{c\ell}\mathbf{X} + \mathbf{X}\mathbf{A}_{c\ell}^T + \mathbf{B}_{c\ell}\mathbf{B}_{c\ell}^T < \mathbf{0} \quad \text{(continuous-time case)}$$

$$\mathbf{X} > \mathbf{A}_{c\ell}\mathbf{X}\mathbf{A}_{c\ell}^T + \mathbf{B}_{c\ell}\mathbf{B}_{c\ell}^T \quad \text{(discrete-time case).}$$

It is easy to verify that any matrix $\mathbf{X} > \mathbf{0}$ satisfying each of the above inequalities is an upper bound on the state covariance when the system is excited by white noise $\mathbf{w}$ with covariance

$\mathbf{I}$, i.e.

$$\mathbf{X} > \lim_{t\to\infty} \mathcal{E}\,[\mathbf{x}(t)\mathbf{x}^T(t)] \quad \text{(continuous-time case)}$$
$$\mathbf{X} > \lim_{k\to\infty} \mathcal{E}\,[\mathbf{x}(k)\mathbf{x}^T(k)] \quad \text{(discrete-time case).}$$

Thus, each inequality defines a set of covariance upper bounds. Moreover, the upper bound is tight, i.e. for any $\varepsilon > 0$, one can find a matrix $\mathbf{X} > \mathbf{0}$ satisfying the above inequality such that

$$\|\mathbf{X} - \lim_{\alpha\to\infty} \mathcal{E}\,[\mathbf{x}(\alpha)\mathbf{x}^T(\alpha)]\| < \varepsilon,$$

where $\alpha = t\ (k)$ for continuous-time (discrete-time) systems. From the above discussion, it can be seen that an inequality constraint on the output covariance matrix can be equivalently written in terms of the above Lyapunov inequality as follows.

Lemma 6.1.1 *Let a symmetric matrix $\mathcal{U}$ be given. Consider the linear time-invariant continuous-time system (4.14) where* $\mathbf{w}$ *is the stochastic white noise process with intensity* $\mathbf{I}$*. Suppose the system is strictly proper, i.e.* $\mathbf{D}_{c\ell} = \mathbf{0}$*. Then the following statements are equivalent:*

(i) The system is asymptotically stable and the output covariance is bounded above by $\mathcal{U}$*;*

$$\lim_{t\to\infty} \mathcal{E}\,[\mathbf{y}(t)\mathbf{y}^T(t)] < \mathcal{U}.$$

(ii) There exists a matrix $\mathbf{X} > \mathbf{0}$ *such that*

$$\mathbf{C}_{c\ell}\mathbf{X}\mathbf{C}_{c\ell}^T < \mathcal{U}, \quad \mathbf{A}_{c\ell}\mathbf{X} + \mathbf{X}\mathbf{A}_{c\ell}^T + \mathbf{B}_{c\ell}\mathbf{B}_{c\ell}^T < \mathbf{0}. \tag{6.1}$$

The assumption $\mathbf{D}_{c\ell} = \mathbf{0}$ is necessary for the output covariance to be finite for continuous-time systems. Note that the strict properness of the system is not required for discrete-time systems since the output covariance is finite even if $\mathbf{D}_{c\ell} \neq \mathbf{0}$, provided the system is asymptotically stable. Thus, we have the following lemma for the discrete-time case.

Lemma 6.1.2 *Let a symmetric matrix $\mathcal{U}$ be given. Consider the linear time-invariant discrete-time system (4.31) where* $\mathbf{w}$ *is the stochastic white noise process with covariance* $\mathbf{I}$*. Then the following statements are equivalent:*

(i) The system is asymptotically stable and the output covariance is bounded above by $\mathcal{U}$*,*

$$\lim_{k\to\infty} \mathcal{E}\,[\mathbf{y}(k)\mathbf{y}^T(k)] < \mathcal{U}.$$

(ii) There exists a matrix $\mathbf{X} > \mathbf{0}$ *such that*

$$\mathbf{C}_{c\ell}\mathbf{X}\mathbf{C}_{c\ell}^T + \mathbf{D}_{c\ell}\mathbf{D}_{c\ell}^T < \mathcal{U}, \quad \mathbf{X} > \mathbf{A}_{c\ell}\mathbf{X}\mathbf{A}_{c\ell}^T + \mathbf{B}_{c\ell}\mathbf{B}_{c\ell}^T. \tag{6.2}$$

Now, using Lemmas 6.1.1 and 6.1.2, the covariance bounding control problem can be converted to an algebraic problem of finding matrices $\mathbf{X}$ and $\mathbf{G}$ satisfying the matrix inequalities (6.1) or (6.2). In the following, we shall assume that there is no redundant actuator ($\mathbf{B}_p^T\mathbf{B}_p > \mathbf{0}$) nor redundant sensor ($\mathbf{M}_p\mathbf{M}_p^T > \mathbf{0}$). This assumption reflects a reasonable practical situation, and can easily be removed at the expense of more complicated controller formulae.

6.2 CONTINUOUS-TIME CASE

This section provides a solution to the covariance bounding control problem for the continuous-time case. Recall that the output covariance is finite only if the closed-loop system is strictly proper. In order to assure $\mathbf{D}_{c\ell} = \mathbf{0}$, in (4.15) we require $\mathbf{D}_y = \mathbf{0}$ and either $\mathbf{B}_y = \mathbf{0}$ or $\mathbf{D}_z = \mathbf{0}$, when designing proper (possibly $\mathbf{D}_c \neq \mathbf{0}$) controllers. In the subsequent sections, we shall assume $\mathbf{D}_y = \mathbf{0}$ and $\mathbf{B}_y = \mathbf{0}$ (no penalty on the control input) and allow possibly nonzero $\mathbf{D}_z$ (which accounts for the measurement noise). The case where $\mathbf{B}_y \neq \mathbf{0}$ with strictly proper controllers will be treated in Section 6.2.4.

6.2.1 State Feedback

In this section, we assume that all the state variables are available for feedback without measurement noises ($\mathbf{M}_p = \mathbf{I}$, $\mathbf{D}_z = \mathbf{0}$). In this case, the Lyapunov inequality associated with the covariance bounding control problem is the following:

$$(\mathbf{A}_p + \mathbf{B}_p\mathbf{G})\mathbf{X} + \mathbf{X}(\mathbf{A}_p + \mathbf{B}_p\mathbf{G})^T + \mathbf{D}_p\mathbf{D}_p^T < \mathbf{0}. \tag{6.3}$$

Theorem 6.2.1 *Let a symmetric matrix $\mho$ be given and consider the system (4.14) with $\mathbf{D}_y = \mathbf{0}$, $\mathbf{B}_y = \mathbf{0}$, $\mathbf{M}_p = \mathbf{I}$ and $\mathbf{D}_z = \mathbf{0}$. Then the following statements are equivalent:*

(i) There exists a stabilizing state feedback gain $\mathbf{G}$ such that

$$\lim_{t\to\infty} \mathcal{E}\,[\mathbf{y}(t)\mathbf{y}^T(t)] < \mho.$$

(ii) There exists a matrix $\mathbf{X} > \mathbf{0}$ such that

$$\mathbf{C}_p\mathbf{X}\mathbf{C}_p^T < \mho, \quad \mathbf{B}_p^{\perp}(\mathbf{A}_p\mathbf{X} + \mathbf{X}\mathbf{A}_p^T + \mathbf{D}_p\mathbf{D}_p^T)\mathbf{B}_p^{\perp T} < \mathbf{0}.$$

(iii) There exist a scalar $\gamma > 0$ and a matrix $\mathbf{Q} > \mathbf{0}$ such that a positive definite solution $\mathbf{P} > \mathbf{0}$ to

$$\mathbf{P}\mathbf{A}_p + \mathbf{A}_p^T\mathbf{P} + \mathbf{P}(\frac{1}{\gamma^2}\mathbf{D}_p\mathbf{D}_p^T - \mathbf{B}_p\mathbf{B}_p^T)\mathbf{P} + \mathbf{Q} = \mathbf{0} \tag{6.4}$$

satisfies

$$\gamma^2\mathbf{C}_p\mathbf{P}^{-1}\mathbf{C}_p^T < \mho.$$

In this case, all such state feedback gains are given by

$$\mathbf{G} = -\mathbf{B}_p^T\mathbf{P} + \mathbf{L}\mathbf{Q}^{1/2} \tag{6.5}$$

where the matrices $\mathbf{P}$ and $\mathbf{Q}$ are those in (iii) and $\mathbf{L}$ is an arbitrary matrix such that $\|\mathbf{L}\| < 1$.

Proof. From Lemma 6.1.1, statement (i) holds if and only if there exist matrices $\mathbf{G}$ and $\mathbf{X} > \mathbf{0}$ satisfying (6.3) and $\mathbf{C}_p\mathbf{X}\mathbf{C}_p^T < \mho$. Note that a given matrix pair $(\mathbf{X}, \mathbf{G})$ satisfies the inequality in (6.3) if and only if there exists a scalar $\gamma > 0$ such that

$$(\mathbf{A}_p + \mathbf{B}_p\mathbf{G})\mathbf{X} + \mathbf{X}(\mathbf{A}_p + \mathbf{B}_p\mathbf{G})^T + \frac{1}{\gamma^2}\mathbf{X}\mathbf{G}^T\mathbf{G}\mathbf{X} + \mathbf{D}_p\mathbf{D}_p^T < \mathbf{0}.$$

Completing the square with respect to $\mathbf{G}$, we have

$$(\frac{1}{\gamma}\mathbf{X}\mathbf{G}^T + \gamma\mathbf{B}_p)(\frac{1}{\gamma}\mathbf{X}\mathbf{G}^T + \gamma\mathbf{B}_p)^T < \gamma^2\mathbf{B}_p\mathbf{B}_p^T - \mathbf{D}_p\mathbf{D}_p^T - \mathbf{A}_p\mathbf{X} - \mathbf{X}\mathbf{A}_p^T =: \mathbf{\Phi}.$$

From Corollary 2.3.6 there exists a matrix $\mathbf{G}$ satisfying the above inequality if and only if $\mathbf{\Phi} > \mathbf{0}$, in which case, all such $\mathbf{G}$ are given by

$$\mathbf{G} = -\gamma^2\mathbf{B}_p^T\mathbf{X}^{-1} + \gamma\mathbf{L}\mathbf{\Phi}^{1/2}\mathbf{X}^{-1}, \quad \|\mathbf{L}\| < 1.$$

Defining $\mathbf{P} \triangleq \gamma^2\mathbf{X}^{-1}$ and $\mathbf{Q} \triangleq \gamma^2\mathbf{X}^{-1}\mathbf{\Phi}\mathbf{X}^{-1}$, we have statement (iii). Finally, the equivalence of $\mathbf{\Phi} > \mathbf{0}$ and the second inequality in statement (ii) follows immediately from Finsler's Theorem (Theorem 2.3.10). □

There are several ways to interpret Theorem 6.2.1. In view of the covariance control problem considered in Chapter 5, we can define the set of *assignable covariance bounds* by any matrices $\mathbf{X} > \mathbf{0}$ satisfying (6.3) for some controller $\mathbf{G}$. Then such a set is characterized by the Linear Matrix Inequalities (LMIs) in statement (ii). Since the set is convex, an assignable covariance bound can be computed by solving a convex feasibility problem[1]. Once we find such an $\mathbf{X}$, all controllers which assign $\mathbf{X}$ to the closed-loop system as a covariance upper bound can be obtained by finding γ such that $\mathbf{\Phi} > \mathbf{0}$. Note that $\mathbf{\Phi} > \mathbf{0}$ can be achieved by sufficiently large $\gamma > 0$. Then $\mathbf{P}$ is given by $\mathbf{P} = \gamma^2\mathbf{X}^{-1}$ and $\mathbf{Q}$ is determined by the Riccati equation in statement (iii). Alternatively, one may find an assignable covariance bound by solving the Riccati equation in statement (iii) by choosing parameters $\gamma > 0$ and $\mathbf{Q} > \mathbf{0}$. Note that, for any $\mathbf{Q} > \mathbf{0}$, there exists a positive definite solution $\mathbf{P}$ to the Riccati equation if γ is sufficiently large, provided $(\mathbf{A}_p, \mathbf{B}_p)$ is stabilizable.

Clearly, there are many controllers which achieve a given covariance bound. Such freedoms are captured by γ, $\mathbf{Q}$ and $\mathbf{L}$. We shall investigate the physical significance of these parameters. Note that, for any vector $\mathbf{x} \in \mathcal{R}^{n_p}$, we have

$$\mathbf{x}^T(\mathbf{P}\mathbf{A}_p + \mathbf{A}_p^T\mathbf{P} + \mathbf{P}(\frac{1}{\gamma^2}\mathbf{D}_p\mathbf{D}_p^T - \mathbf{B}_p\mathbf{B}_p^T)\mathbf{P} + \mathbf{Q})\mathbf{x} = 0$$

or equivalently,

$$\mathbf{x}^T[(\mathbf{P}\mathbf{B}_p + \mathbf{G}^T)(\mathbf{B}_p^T\mathbf{P} + \mathbf{G}) - \mathbf{G}^T\mathbf{G} + \mathbf{Q})\mathbf{x} + \gamma^2\mathbf{w}^T\mathbf{w}$$

$$-(\frac{1}{\gamma}\mathbf{x}^T\mathbf{P}\mathbf{D}_p - \gamma\mathbf{w}^T)(\frac{1}{\gamma}\mathbf{D}_p^T\mathbf{P}\mathbf{x} - \gamma\mathbf{w}) - 2\mathbf{x}^T\mathbf{P}(\mathbf{A}_p\mathbf{x} + \mathbf{B}_p\mathbf{u} + \mathbf{D}_p\mathbf{w}) = 0$$

where $\mathbf{w} \in \mathcal{R}^{n_w}$ and $\mathbf{G} \in \mathcal{R}^{n_u \times n_p}$ are arbitrary. Now considering any trajectory of the system (4.15) with the state feedback given in Theorem 6.2.1, we have

$$\begin{aligned}\|\mathbf{u}(t)\|^2 &= \|\mathbf{L}\mathbf{Q}^{1/2}\mathbf{x}(t)\|^2 - \|\mathbf{Q}^{1/2}\mathbf{x}(t)\|^2 + \gamma^2\|\mathbf{w}(t)\|^2 \\ &\quad -\gamma^2\|\mathbf{w}(t) - \frac{1}{\gamma^2}\mathbf{D}_p^T\mathbf{P}\mathbf{x}(t)\|^2 - \frac{d}{dt}\left(\mathbf{x}(t)^T\mathbf{P}\mathbf{x}(t)\right).\end{aligned}$$

Integrating the both sides from $t = 0$ to ∞, we have

$$\begin{aligned}\int_0^\infty \|\mathbf{u}(t)\|^2 dt &= \gamma^2\int_0^\infty \|\mathbf{w}(t)\|^2 dt + \int_0^\infty (\|\mathbf{L}\mathbf{Q}^{1/2}\mathbf{x}(t)\|^2 - \|\mathbf{Q}^{1/2}\mathbf{x}(t)\|^2)dt \\ &\quad -\gamma^2\int_0^\infty \|\mathbf{w}(t) - \frac{1}{\gamma^2}\mathbf{D}_p^T\mathbf{P}\mathbf{x}(t)\|^2 dt + \mathbf{x}(0)^T\mathbf{P}\mathbf{x}(0)\end{aligned}$$

where we used the stability property $\lim_{t\to\infty}\mathbf{x}(t) = \mathbf{0}$. From the above observation, we have the following.

[1] See Chapters 10 and 11.

Theorem 6.2.2 *Let* $\mathbf{G}$ *be a controller generated by Theorem 6.2.1. Then the following statements hold:*

(i) The $\mathcal{L}_2$ *gain from the disturbance to the control input is bounded above by* γ*, i.e.*

$$\int_0^\infty \|\mathbf{u}(t)\|^2 dt \leq \gamma^2 \int_0^\infty \|\mathbf{w}(t)\|^2 dt$$

for all $\mathbf{L}$ *such that* $\|\mathbf{L}\| < 1$.

(ii) If we choose the parameter $\mathbf{L} = \mathbf{0}$*, then the* $\mathcal{L}_2$ *gain from the disturbance to* $\begin{bmatrix} \mathbf{u} \\ \mathbf{Q}^{1/2}\mathbf{x} \end{bmatrix}$ *is bounded above by* γ*, i.e.*

$$\int_0^\infty (\mathbf{x}^T(t)\mathbf{Q}\mathbf{x}(t) + \mathbf{u}(t)^T\mathbf{u}(t))dt \leq \gamma^2 \int_0^\infty \|\mathbf{w}(t)\|^2 dt.$$

If we let the output covariance bound be $\mathcal{U} = \sigma\mathbf{I}$ and take the limit $\sigma \to \infty$, then the performance specification in Theorem 6.2.1 can be effectively removed, in which case, Theorem 6.2.1 provides a parametrization of all stabilizing (static) state feedback gains. Note that the choice of $\mathbf{D}_p$ is immaterial since the closed-loop stability specification does not involve $\mathbf{D}_p$. Thus, letting $\mathbf{D}_p = \mathbf{0}$, we have the following.

Corollary 6.2.1 *[62] Consider a system* $\dot{\mathbf{x}} = \mathbf{A}\mathbf{x} + \mathbf{B}\mathbf{u}$*. Then the following statements are equivalent:*

(i) There exists a stabilizing state feedback controller $\mathbf{u} = \mathbf{G}\mathbf{x}$.

(ii) There exists a matrix $\mathbf{X} > \mathbf{0}$ *such that*

$$\mathbf{B}^\perp(\mathbf{A}\mathbf{X} + \mathbf{X}\mathbf{A}^T)\mathbf{B}^{\perp T} < \mathbf{0}.$$

(iii) There exists a matrix $\mathbf{Q} > \mathbf{0}$ *such that the Riccati equation*

$$\mathbf{P}\mathbf{A} + \mathbf{A}^T\mathbf{P} - \mathbf{P}\mathbf{B}\mathbf{B}^T\mathbf{P} + \mathbf{Q} = \mathbf{0} \tag{6.6}$$

has a solution $\mathbf{P} > \mathbf{0}$.

In this case, all such state feedback gains are given by

$$\mathbf{G} = -\mathbf{B}^T\mathbf{P} + \mathbf{L}\mathbf{Q}^{1/2} \tag{6.7}$$

where the matrices $\mathbf{P}$ *and* $\mathbf{Q}$ *are the ones in statement (iii) and* $\mathbf{L}$ *is an arbitrary matrix such that* $\|\mathbf{L}\| < 1$.

Note that, for each choice of the matrix $\mathbf{Q} > \mathbf{0}$, there exists a positive definite matrix $\mathbf{P}$ satisfying the Riccati equation (6.6) if and only if $(\mathbf{A}, \mathbf{B})$ is stabilizable, and in fact such a solution is unique. Hence, Corollary 6.2.1 parametrizes the set of all stabilizing state feedback gains in terms of an arbitrary positive definite matrix $\mathbf{Q}$ and arbitrary matrix $\mathbf{L}$ with the norm bound $\|\mathbf{L}\| < 1$. Clearly, a choice of the freedom $\mathbf{L} = \mathbf{0}$ yields the LQ optimal state feedback gain with respect to the cost function

$$J \triangleq \int_0^\infty (\mathbf{x}^T(t)\mathbf{Q}\mathbf{x}(t) + \mathbf{u}^T(t)\mathbf{u}(t))dt, \quad \mathbf{x}(0) = \mathbf{x}_0.$$

Indeed, for a stabilizing state feedback gain $\mathbf{G}$ generated by (6.7),

$$\begin{aligned}\mathbf{x}^T\mathbf{Q}\mathbf{x}+\mathbf{u}^T\mathbf{u} &= \mathbf{x}^T(\mathbf{P}\mathbf{B}\mathbf{B}^T\mathbf{P}-\mathbf{A}^T\mathbf{P}-\mathbf{P}\mathbf{A}+\mathbf{G}^T\mathbf{G})\mathbf{x}\\ &= \mathbf{x}^T[(\mathbf{G}+\mathbf{B}^T\mathbf{P})^T\mathbf{R}(\mathbf{G}+\mathbf{B}^T\mathbf{P})\\ &\quad -\mathbf{P}(\mathbf{A}+\mathbf{B}\mathbf{G})-(\mathbf{A}+\mathbf{B}\mathbf{G})^T\mathbf{P}]\mathbf{x}\\ &= \mathbf{x}^T\mathbf{Q}^{1/2}\mathbf{L}^T\mathbf{L}\mathbf{Q}^{1/2}\mathbf{x}-\frac{d}{dt}(\mathbf{x}^T\mathbf{P}\mathbf{x}).\end{aligned}$$

Integrating the both sides from $t = 0$ to ∞, and using the stability property $\lim_{t\to\infty}(\mathbf{x}^T(t)\mathbf{P}\mathbf{x}(t)) = 0$, we have

$$J = \mathbf{x}_0^T\mathbf{P}\mathbf{x}_0 + \int_0^\infty \|\mathbf{L}\mathbf{Q}^{1/2}\mathbf{x}(t)\|^2 dt.$$

Thus, the choice $\mathbf{L} = \mathbf{0}$ provides the optimal controller which minimizes the quadratic cost function J.

Example 6.2.1 Consider the double integrator system given by the following state space realization:

$$\mathbf{A}_p = \begin{bmatrix} 1 & 0 \\ 0 & 0 \end{bmatrix}, \quad \mathbf{B}_p = \begin{bmatrix} 0 \\ 1 \end{bmatrix}, \quad \mathbf{C}_p = \begin{bmatrix} 1 & 0 \end{bmatrix}, \quad \mathbf{D}_p = \begin{bmatrix} 1 \\ 1 \end{bmatrix}.$$

We will design a covariance upper bound controller with covariance bound $\mho = 4$ using the result in Theorem 6.2.1. Choosing the parameters $\gamma = 2$ and $\mathbf{Q} = \mathbf{I}_2$, one can obtain a positive definite solution $\mathbf{P}$ to the Riccati equation in (6.4) as follows:

$$\mathbf{P} = \begin{bmatrix} 6.854 & 6.997 \\ 6.997 & 8.794 \end{bmatrix}.$$

In this case, we have $\gamma^2\mathbf{C}_p\mathbf{P}^{-1}\mathbf{C}_p^T = 3.110 < \mho$ and thus the prescribed performance bound is achieved.

Note that

$$\mathbf{X} := \gamma^2\mathbf{P}^{-1} = \begin{bmatrix} 3.110 & -2.475 \\ -2.475 & 2.424 \end{bmatrix}$$

satisfies the conditions in statement (ii) of Theorem 6.2.1:

$$\mathbf{C}_p\mathbf{X}\mathbf{C}_p^T = 3.1102 < \mho, \quad \mathbf{B}_p^\perp(\mathbf{A}_p\mathbf{X}+\mathbf{X}\mathbf{A}_p^T+\mathbf{D}_p\mathbf{D}_p^T)\mathbf{B}_p^{\perp T} = -3.950 < 0.$$

Now, a class of state feedback gains that achieves the performance bound is given by (6.5) with $\mathbf{L}$ being the free parameter. Choosing $\mathbf{L} = \mathbf{0}$, we have

$$\mathbf{G} = -\mathbf{B}_p^T\mathbf{P} = -\begin{bmatrix} 6.997 & 8.794 \end{bmatrix}.$$

With this controller, the state covariance $\mathbf{X}_{cov} := \lim_{t\to\infty}\mathcal{E}[\mathbf{x}(t)\mathbf{x}^T(t)]$, which is the solution to the following Lyapunov equation:

$$(\mathbf{A}_p+\mathbf{B}_p\mathbf{G})\mathbf{X}_{cov}+\mathbf{X}_{cov}(\mathbf{A}_p+\mathbf{B}_p\mathbf{G})^T+\mathbf{D}_p\mathbf{D}_p^T = \mathbf{0},$$

is given by

$$\mathbf{X}_{cov} = \begin{bmatrix} 0.836 & -0.500 \\ -0.500 & 0.4547 \end{bmatrix}.$$

Note that $\mathbf{X}$ computed above is an upper bound on the state covariance $\mathbf{X}_{cov}$; since $\lambda(\mathbf{X} - \mathbf{X}_{cov}) = 4.102,\ 0.141$ where λ denotes the eigenvalue, we have $\mathbf{X} > \mathbf{X}_{cov}$. Finally, the actual output covariance is given by

$$\lim_{t\to\infty} \mathcal{E}[\mathbf{y}(t)\mathbf{y}^T(t)] = \mathbf{C}_p\mathbf{X}_{cov}\mathbf{C}_p^T = 0.8363 < \mho,$$

confirming that the control design objective is achieved.

6.2.2 Static Output Feedback

In this section, we consider the covariance bounding control problem with static output feedback. The Lyapunov inequality considered here is then given by

$$(\mathbf{A}_p + \mathbf{B}_p\mathbf{G}\mathbf{M}_p)\mathbf{X} + \mathbf{X}(\mathbf{A}_p + \mathbf{B}_p\mathbf{G}\mathbf{M}_p)^T + (\mathbf{D}_p + \mathbf{B}_p\mathbf{G}\mathbf{D}_z)(\mathbf{D}_p + \mathbf{B}_p\mathbf{G}\mathbf{D}_z)^T < \mathbf{0}. \quad (6.8)$$

In the following, we shall assume

$$\begin{bmatrix} \mathbf{D}_p \\ \mathbf{D}_z \end{bmatrix} \mathbf{D}_z^T = \begin{bmatrix} \mathbf{0} \\ \mathbf{V} \end{bmatrix}, \quad \mathbf{V} > \mathbf{0}, \quad (6.9)$$

i.e. there is no correlation between the process and measurement noises, and the measured outputs are fully contaminated by noises.

Theorem 6.2.3 *Let a symmetric matrix $\mho$ be given and consider the system (4.15) with $\mathbf{D}_y = \mathbf{0}$ and $\mathbf{B}_y = \mathbf{0}$. Suppose the assumption (6.9) holds. Then the following statements are equivalent.*

(i) There exists a stabilizing static output feedback gain G such that

$$\lim_{t\to\infty} \mathcal{E}\,[\mathbf{y}(t)\mathbf{y}^T(t)] < \mho.$$

(ii) There exists a matrix $\mathbf{X} > \mathbf{0}$ such that

$$\mathbf{C}_p\mathbf{X}\mathbf{C}_p^T < \mho$$

$$\mathbf{B}_p^{\perp}(\mathbf{A}_p\mathbf{X} + \mathbf{X}\mathbf{A}_p^T + \mathbf{D}_p\mathbf{D}_p^T)\mathbf{B}_p^{\perp T} < \mathbf{0}$$

$$\mathbf{A}_p\mathbf{X} + \mathbf{X}\mathbf{A}_p^T - \mathbf{X}\mathbf{M}_p^T\mathbf{V}^{-1}\mathbf{M}_p\mathbf{X} + \mathbf{D}_p\mathbf{D}_p^T < \mathbf{0}.$$

In this case, all such gains are given by

$$\mathbf{G} = -(\mathbf{B}_p^T\mathbf{Q}\mathbf{B}_p)^{-1}\mathbf{B}_p^T\mathbf{Q}\mathbf{X}\mathbf{M}_p^T\mathbf{V}^{-1} + (\mathbf{B}_p^T\mathbf{Q}\mathbf{B}_p)^{-1/2}\mathbf{L}\mathbf{S}^{1/2}$$

where $\mathbf{L}$ is an arbitrary matrix such that $\|\mathbf{L}\| < 1$ and

$$\mathbf{Q} \triangleq (\mathbf{X}\mathbf{M}_p^T\mathbf{V}^{-1}\mathbf{M}_p\mathbf{X} - \mathbf{A}_p\mathbf{X} - \mathbf{X}\mathbf{A}_p^T - \mathbf{D}_p\mathbf{D}_p^T)^{-1} > \mathbf{0}$$

$$\mathbf{S} \triangleq \mathbf{V}^{-1} - \mathbf{V}^{-1}\mathbf{M}_p\mathbf{X}[\mathbf{Q} - \mathbf{Q}\mathbf{B}_p(\mathbf{B}_p^T\mathbf{Q}\mathbf{B}_p)^{-1}\mathbf{B}_p^T\mathbf{Q}]\mathbf{X}\mathbf{M}_p^T\mathbf{V}^{-1} > \mathbf{0}.$$

Proof. Consider the Lyapunov inequality (6.8), which after expanding, completing the square with respect to $\mathbf{G}$ yields

$$(\mathbf{B}_p\mathbf{G} + \mathbf{X}\mathbf{M}_p^T\mathbf{V}^{-1})\mathbf{V}(\mathbf{B}_p\mathbf{G} + \mathbf{X}\mathbf{M}_p^T\mathbf{V}^{-1})^T < \mathbf{Q}^{-1}.$$

Then the result follows directly from Corollary 2.3.6. □

As in the state feedback case, we can obtain a parametrization of all stabilizing static output feedback gains by removing the performance constraint. In particular, a variety of necessary and sufficient conditions for stabilizability via static output feedback can be derived as follows.

Corollary 6.2.2 *[55] Consider a system* $\dot{\mathbf{x}} = \mathbf{A}\mathbf{x} + \mathbf{B}\mathbf{u}$, $\mathbf{z} = \mathbf{M}\mathbf{x}$. *Then the following statements are equivalent:*

(i) There exists a static output feedback controller $\mathbf{u} = \mathbf{G}\mathbf{z}$ *which stabilizes the system.*
(ii) There exists a matrix $\mathbf{X} > \mathbf{0}$ *such that*

$$\mathbf{B}^{\perp}(\mathbf{A}\mathbf{X} + \mathbf{X}\mathbf{A}^T)\mathbf{B}^{\perp T} < \mathbf{0} \tag{6.10}$$

$$\mathbf{A}\mathbf{X} + \mathbf{X}\mathbf{A}^T - \mathbf{X}\mathbf{M}^T\mathbf{M}\mathbf{X} < \mathbf{0}.$$

(iii) There exists a matrix pair $(\mathbf{X}, \mathbf{Y})$ *such that*

$$\mathbf{X} = \mathbf{Y}^{-1} > \mathbf{0}$$

$$\mathbf{B}^{\perp}(\mathbf{A}\mathbf{X} + \mathbf{X}\mathbf{A}^T)\mathbf{B}^{\perp T} < \mathbf{0}$$

$$\mathbf{M}^{T\perp}(\mathbf{Y}\mathbf{A} + \mathbf{A}^T\mathbf{Y})\mathbf{M}^{T\perp T} < \mathbf{0}.$$

(iv) There exist matrices $\mathbf{Q} > \mathbf{0}$ *and* $\mathbf{W} > \mathbf{0}$ *such that the solutions* $\mathbf{X} > \mathbf{0}$ *and* $\mathbf{Y} > \mathbf{0}$ *to*

$$\mathbf{A}\mathbf{X} + \mathbf{X}\mathbf{A}^T - \mathbf{X}\mathbf{M}^T\mathbf{M}\mathbf{X} + \mathbf{W} = \mathbf{0}$$

$$\mathbf{Y}\mathbf{A} + \mathbf{A}^T\mathbf{Y} - \mathbf{Y}\mathbf{B}\mathbf{B}^T\mathbf{Y} + \mathbf{Q} = \mathbf{0}$$

exist and satisfy $\mathbf{X}\mathbf{Y} = \rho\mathbf{I}$ *for some* $\rho > 0$.
(v) There exist $\mathbf{K}$, $\mathbf{F}$ *and* $\mathbf{X} > \mathbf{0}$ *such that*

$$(\mathbf{A} + \mathbf{B}\mathbf{K})\mathbf{X} + \mathbf{X}(\mathbf{A} + \mathbf{B}\mathbf{K})^T < \mathbf{0}$$

$$(\mathbf{A} + \mathbf{F}\mathbf{M})\mathbf{X} + \mathbf{X}(\mathbf{A} + \mathbf{F}\mathbf{M})^T < \mathbf{0}.$$

Proof. Letting $\mathfrak{U}$ be arbitrarily large, $\mathbf{D}_p$ be zero and $\mathbf{V}$ be identity in Theorem 6.2.3, we immediately have (i) $\Leftrightarrow$ (ii). In the above, if we let $\mathbf{V} = \varepsilon\mathbf{I}$ with a sufficiently small $\varepsilon > 0$ instead of $\mathbf{V} = \mathbf{I}$, we see the equivalence (i) $\Leftrightarrow$ (iii) by Finsler's Theorem 2.3.10. To prove (ii) $\Leftrightarrow$ (iv), note that (6.10) holds if and only if

$$\mathbf{A}\mathbf{X} + \mathbf{X}\mathbf{A}^T - \rho\mathbf{B}\mathbf{B}^T < \mathbf{0}$$

holds for some $\rho > 0$ by Finsler's Theorem, or equivalently,

$$\mathbf{X}^{-1}\mathbf{A} + \mathbf{A}^T\mathbf{X}^{-1} - \rho\mathbf{X}^{-1}\mathbf{B}\mathbf{B}^T\mathbf{X}^{-1} < \mathbf{0}.$$

Multiplying both sides by ρ and defining $\mathbf{Y} \triangleq \rho\mathbf{X}^{-1}$, (iv) can be verified. Finally, (iii) $\Leftrightarrow$ (v) follows from Corollary 6.2.1 and some algebraic manipulations. □

6.2.3 Reduced-Order Dynamic Output Feedback

This section obtains all dynamic output feedback controllers, of order less than or equal to the plant, which solve the covariance bounding control problem. The Lyapunov inequality associated with the closed-loop state covariance bounds is given by

$$(\mathbf{A}+\mathbf{BGM})\mathbf{X}+\mathbf{X}(\mathbf{A}+\mathbf{BGM})^T+(\mathbf{D}+\mathbf{BGE})(\mathbf{D}+\mathbf{BGE})^T<\mathbf{0}. \tag{6.11}$$

Note that the above inequality has exactly the same structure as (6.8) for the static output feedback case. However, $\mathbf{EE}^T$ is not positive definite (invertible) even if $\mathbf{V}\triangleq\mathbf{D}_z\mathbf{D}_z^T>\mathbf{0}$, and hence the result in the previous section (Theorem 6.2.3) cannot be applied directly to the dynamic output feedback case. The following result is derived *without any assumptions on matrices* **D** *and* **E**, and thus can be specialized to the previous result.

Theorem 6.2.4 *Let a symmetric matrix* $\mho$ *be given and consider the system (4.15) with* $\mathbf{D}_y=\mathbf{0}$ *and* $\mathbf{B}_y=\mathbf{0}$. *Then the following statements are equivalent:*

(i) There exists a dynamic output feedback controller of order n_c *which stabilizes the system and yields*

$$\lim_{t\to\infty}\mathcal{E}\,[\mathbf{y}(t)\mathbf{y}^T(t)]<\mho.$$

(ii) There exists a positive definite matrix $\mathbf{X}\in\mathcal{R}^{(n_p+n_c)\times(n_p+n_c)}$ *such that*

$$\mathbf{B}^{\perp}(\mathbf{AX}+\mathbf{XA}^T+\mathbf{DD}^T)\mathbf{B}^{\perp T}<\mathbf{0},\quad \mathbf{CXC}^T<\mho$$

$$\begin{bmatrix}\mathbf{M}^T\\ \mathbf{E}^T\end{bmatrix}^{\perp}\begin{bmatrix}\mathbf{X}^{-1}\mathbf{A}+\mathbf{A}^T\mathbf{X}^{-1} & \mathbf{X}^{-1}\mathbf{D}\\ \mathbf{D}^T\mathbf{X}^{-1} & -\mathbf{I}\end{bmatrix}\begin{bmatrix}\mathbf{M}^T\\ \mathbf{E}^T\end{bmatrix}^{\perp T}<\mathbf{0}.$$

In this case, all such controllers are given by

$$\mathbf{G}=-\mathbf{R}^{-1}\mathbf{\Gamma}^T\mathbf{\Phi}\mathbf{\Lambda}^T(\mathbf{\Lambda}\mathbf{\Phi}\mathbf{\Lambda}^T)^{-1}+\mathbf{S}^{1/2}\mathbf{L}(\mathbf{\Lambda}\mathbf{\Phi}\mathbf{\Lambda}^T)^{-1/2}$$

where **L** *is an arbitrary matrix such that* $\|\mathbf{L}\|<1$ *and* **R** *is an arbitrary positive definite matrix such that*

$$\mathbf{\Phi}\triangleq(\mathbf{\Gamma}\mathbf{R}^{-1}\mathbf{\Gamma}^T-\mathbf{\Theta})^{-1}>\mathbf{0}$$

and

$$\mathbf{S}\triangleq\mathbf{R}^{-1}-\mathbf{R}^{-1}\mathbf{\Gamma}^T[\mathbf{\Phi}-\mathbf{\Phi}\mathbf{\Lambda}^T(\mathbf{\Lambda}\mathbf{\Phi}\mathbf{\Lambda}^T)^{-1}\mathbf{\Lambda}\mathbf{\Phi}]\mathbf{\Gamma}\mathbf{R}^{-1}$$

$$\mathbf{\Theta}\triangleq\begin{bmatrix}\mathbf{AX}+\mathbf{XA}^T & \mathbf{D}\\ \mathbf{D}^T & -\mathbf{I}\end{bmatrix},\quad \mathbf{\Gamma}\triangleq\begin{bmatrix}\mathbf{B}\\ \mathbf{0}\end{bmatrix},\quad \mathbf{\Lambda}\triangleq\begin{bmatrix}\mathbf{MX} & \mathbf{E}\end{bmatrix}.$$

Proof. Using the Schur complement formula, (6.11) can be equivalently written

$$\begin{bmatrix}(\mathbf{A}+\mathbf{BGM})\mathbf{X}+\mathbf{X}(\mathbf{A}+\mathbf{BGM})^T & \mathbf{D}+\mathbf{BGE}\\ (\mathbf{D}+\mathbf{BGE})^T & -\mathbf{I}\end{bmatrix}<\mathbf{0}.$$

By the definitions for $\mathbf{\Theta}$, $\mathbf{\Gamma}$ and $\mathbf{\Lambda}$ given in Theorem 6.2.4, the above inequality can be rewritten

$$\mathbf{\Gamma G\Lambda}+(\mathbf{\Gamma G\Lambda})^T+\mathbf{\Theta}<\mathbf{0}.$$

Then the result follows from Theorem 2.3.12 where we note that

$$\mathbf{\Gamma}^{\perp}=\begin{bmatrix}\mathbf{B}\\ \mathbf{0}\end{bmatrix}^{\perp}=\begin{bmatrix}\mathbf{B}^{\perp} & \mathbf{0}\\ \mathbf{0} & \mathbf{I}\end{bmatrix}$$

$$\Lambda^{T\perp} = \begin{bmatrix} \mathbf{XM}^T \\ \mathbf{E}^T \end{bmatrix}^{\perp} = \begin{bmatrix} \mathbf{M}^T \\ \mathbf{E}^T \end{bmatrix}^{\perp} \begin{bmatrix} \mathbf{X}^{-1} & \mathbf{0} \\ \mathbf{0} & \mathbf{I} \end{bmatrix}.$$

This completes the proof. □

Exercise 6.2.1 Specialize Theorem 6.2.4 for the static output feedback case with the assumption (6.9). Verify your result by comparing with Theorem 6.2.3.

Let the set of assignable state covariance bounds be defined by any matrices $\mathbf{X} > \mathbf{0}$ satisfying the matrix inequalities in statement (ii) of Theorem 6.2.4. Note that the controller order n_c which assigns a given state covariance bound $\mathbf{X}$ to the closed-loop system is fixed by the dimension of $\mathbf{X}$.

Recall that matrices $\mathbf{A}$, $\mathbf{B}$, etc. are the augmented matrices defined in (4.15). Utilizing the structure of these matrices, we have another characterization of all assignable state covariance bounds as follows.

Corollary 6.2.3 *Let a matrix* $\mathbf{X} > \mathbf{0}$ *be given. Then the following statements are equivalent:*

(i) $\mathbf{X}$ *is an assignable state covariance bound, i.e.* $\mathbf{X}$ *satisfies the matrix inequalities in statement (ii) of Theorem 6.2.4.*

(ii) $\mathbf{X}$ *satisfies*

$$\mathbf{B}_p^{\perp}(\mathbf{A}_p\mathbf{X}_p + \mathbf{X}_p\mathbf{A}_p^T + \mathbf{D}_p\mathbf{D}_p^T)\mathbf{B}_p^{\perp T} < \mathbf{0}, \quad \mathbf{C}_p\mathbf{X}_p\mathbf{C}_p^T < \mathcal{U} \tag{6.12}$$

$$\begin{bmatrix} \mathbf{M}_p^T \\ \mathbf{D}_z^T \end{bmatrix}^{\perp} \begin{bmatrix} \mathbf{Y}_p\mathbf{A}_p + \mathbf{A}_p^T\mathbf{Y}_p & \mathbf{Y}_p\mathbf{D}_p \\ \mathbf{D}_p^T\mathbf{Y}_p & -\mathbf{I} \end{bmatrix} \begin{bmatrix} \mathbf{M}_p^T \\ \mathbf{D}_z^T \end{bmatrix}^{\perp T} < \mathbf{0} \tag{6.13}$$

where $\mathbf{X}_p > \mathbf{0}$ *and* $\mathbf{Y}_p > \mathbf{0}$ *are defined by*

$$\mathbf{X} = \begin{bmatrix} \mathbf{X}_p & \mathbf{X}_{pc} \\ \mathbf{X}_{pc}^T & \mathbf{X}_c \end{bmatrix}, \quad \mathbf{Y}_p \triangleq (\mathbf{X}_p - \mathbf{X}_{pc}\mathbf{X}_c^{-1}\mathbf{X}_{pc}^T)^{-1}. \tag{6.14}$$

Proof. The result follows by noting that

$$\mathbf{B}^{\perp} = \begin{bmatrix} \mathbf{B}_p & \mathbf{0} \\ \mathbf{0} & \mathbf{I} \end{bmatrix}^{\perp} = [\ \mathbf{B}_p^{\perp} \ \ \mathbf{0}\ ],$$

$$\begin{bmatrix} \mathbf{M}^T \\ \mathbf{E}^T \end{bmatrix}^{\perp} = \begin{bmatrix} \mathbf{M}_p^T & \mathbf{0} \\ \mathbf{0} & \mathbf{I} \\ \mathbf{D}_z^T & \mathbf{0} \end{bmatrix}^{\perp} = \begin{bmatrix} \begin{bmatrix} \mathbf{M}_p^T \\ \mathbf{D}_z^T \end{bmatrix}^{\perp} & \mathbf{0} \end{bmatrix} \begin{bmatrix} \mathbf{I} & \mathbf{0} & \mathbf{0} \\ \mathbf{0} & \mathbf{0} & \mathbf{I} \\ \hline \mathbf{0} & \mathbf{I} & \mathbf{0} \end{bmatrix},$$

and noting the matrix inversion lemma

$$\mathbf{X}^{-1} = \begin{bmatrix} \mathbf{Y}_p & * \\ * & * \end{bmatrix}$$

where $*$ denotes appropriate partitioned blocks. □

Corollary 6.2.3 provides a way to construct an assignable state covariance bound. Specifically, we first find matrices $\mathbf{X}_p > \mathbf{0}$ and $\mathbf{Y}_p > \mathbf{0}$ satisfying (6.12) and (6.13), then determine $\mathbf{X}_{pc}$ and $\mathbf{X}_c > \mathbf{0}$ such that

$$\mathbf{X}_{pc}\mathbf{X}_c^{-1}\mathbf{X}_{pc}^T = \mathbf{X}_p - \mathbf{Y}_p^{-1} \tag{6.15}$$

to construct $\mathbf{X}$ as in (6.14). Note that $\mathbf{X}_p$ and $\mathbf{Y}_p$ must satisfy $\mathbf{X}_p \geq \mathbf{Y}_p^{-1}$ in order for $\mathbf{X}_{pc}$ and $\mathbf{X}_c > \mathbf{0}$ satisfying (6.15) to exist. Note that the constraint $\mathbf{X}_p \geq \mathbf{Y}_p^{-1} > \mathbf{0}$ is convex since it is equivalent to the following LMI by the Schur complement formula:

$$\begin{bmatrix} \mathbf{X}_p & \mathbf{I} \\ \mathbf{I} & \mathbf{Y}_p \end{bmatrix} \geq 0. \tag{6.16}$$

Clearly, the resulting controller order n_c is minimal for a given pair $(\mathbf{X}_p, \mathbf{Y}_p)$ if we choose $\mathbf{X}_{pc}$ to be full column rank. In this case, we have

$$n_c = \text{rank}(\mathbf{X}_p - \mathbf{Y}_p^{-1}). \tag{6.17}$$

Example 6.2.2 Consider the double integrator plant with the following state space realization:

$$\begin{bmatrix} \mathbf{A}_p & \mathbf{D}_p & \mathbf{B}_p \\ \mathbf{C}_p & \mathbf{D}_y & \mathbf{B}_y \\ \mathbf{M}_p & \mathbf{D}_z & * \end{bmatrix} = \left[\begin{array}{cc|cc|c} 0 & 1 & 0 & 0 & 0 \\ 0 & 0 & 1 & 0 & 1 \\ \hline 1 & 0 & 0 & 0 & 0 \\ \hline 1 & 0 & 0 & 1 & * \end{array}\right].$$

We will design a first-order dynamic controller that achieves the output covariance upper bound $\mho = 1.5$. Note that the plant is of order 2, and hence this is a reduced (low) order control design. To design such a controller, we need to find matrices $\mathbf{X}_p$ and $\mathbf{Y}_p$ satisfying (6.12), (6.13), (6.16) and (6.17) with $n_c = 1$. Note that the size of $\mathbf{X}_p - \mathbf{Y}_p^{-1}$ in (6.17) is 2×2, and thus we need to reduce its rank by one. This rank reduction occurs when the matrix in (6.16) is singular, i.e. matrices $\mathbf{X}_p$ and $\mathbf{Y}_p$ satisfy constraint (6.16) on the boundary. This observation leads us to formulate a problem of minimizing $tr(\mathbf{X}_p + \mathbf{Y}_p)$ subject to constraints (6.12), (6.13), (6.16). This is a minimization problem with a linear objective function and LMI constraints, and hence can be solved using a commercial software package (e.g. MATLAB). In this way, we have obtained

$$\mathbf{X}_p = \begin{bmatrix} 1.500 & -0.000 \\ -0.000 & 13.078 \end{bmatrix}, \quad \mathbf{Y}_p = \begin{bmatrix} 1.414 & -1.000 \\ -1.000 & 1.414 \end{bmatrix}.$$

Note that, for this particular example, $\text{rank}(\mathbf{X}_p - \mathbf{Y}_p^{-1}) = 1$ is achieved, although the above minimization formulation did not guarantee satisfaction of this rank condition. Using the singular value decomposition

$$\mathbf{X}_p - \mathbf{Y}_p^{-1} = \mathbf{U}\mathbf{\Sigma}\mathbf{U}^T = \begin{bmatrix} 0.085 & -0.996 \\ -0.996 & -0.085 \end{bmatrix} \begin{bmatrix} 11.749 & 0 \\ 0 & 0.0000 \end{bmatrix} \begin{bmatrix} 0.085 & -0.996 \\ -0.996 & -0.085 \end{bmatrix}^T,$$

the matrices $\mathbf{X}_{pc}$ and $\mathbf{X}_c$ satisfying (6.15) are found to be the first column of $\mathbf{U}$ and $1/11.749$, respectively, to give

$$\mathbf{X} = \begin{bmatrix} \mathbf{X}_p & \mathbf{X}_{pc} \\ \mathbf{X}_{pc}^T & \mathbf{X}_c \end{bmatrix} = \begin{bmatrix} 1.500 & -0.000 & 0.0854 \\ -0.000 & 13.078 & -0.996 \\ 0.085 & -0.996 & 0.085 \end{bmatrix}.$$

It can be verified that this $\mathbf{X}$ satisfies the conditions in statement (ii) of Theorem 6.2.4. Now, using the controller formula given in Theorem 6.2.4 with $\mathbf{L} = \mathbf{0}$ and $\mathbf{R} = \varepsilon\mathbf{I}$ ($\varepsilon > 0$ sufficiently small), we have a controller that achieves the performance bound $\mho$:

$$\mathbf{G} = \begin{bmatrix} -17.493 & 154.083 \\ 1.409 & -13.077 \end{bmatrix},$$

or in the transfer function form,

$$\mathbf{C}_c(s\mathbf{I} - \mathbf{A}_c)^{-1}\mathbf{B}_c + \mathbf{D}_c = -17.493 \times \frac{s + 0.667}{s + 13.077}$$

which is a first-order controller. The resulting closed-loop output covariance is found to be

$$\lim_{t\to\infty} \mathcal{E}[\mathbf{y}(t)\mathbf{y}^T(t)] = 1.49996 < 1.5 = \mho.$$

To see the effect of the parameter $\mho$ on the closed-loop performance, the above design process is repeated for $\mho = 1.5,\ 2,\ 5$ to obtain two additional first-order controllers. The impulse responses ($w_1(t) = \delta(t)$, $w_2(t) \equiv 0$) of the closed-loop system for the three controllers thus obtained are plotted in Figure 6.1.

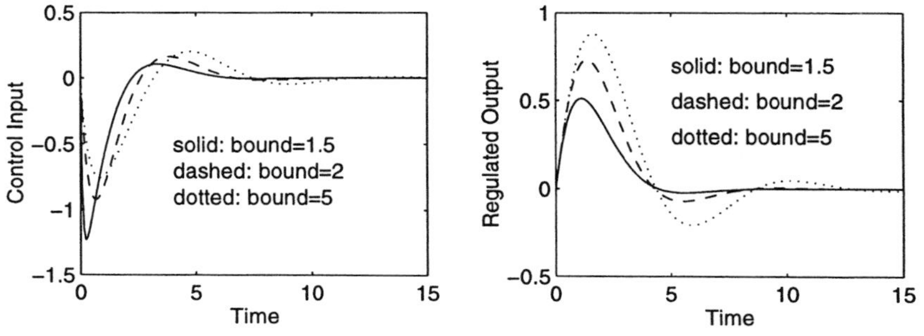

Figure 6.1 Closed-loop impulse responses.

We see from Figure 6.1 that the output performance (e.g. small peak and fast settling) is improved by reducing the value of the performance bound $\mho$, at the expense of more control effort.

6.2.4 Full-Order Dynamic Output Feedback

This section derives a full-order ($n_c = n_p$) dynamic output feedback controller which solves the covariance bounding control problem. In the following, we will assume for simplicity that

$$\mathbf{D}_y = \mathbf{0}, \quad \mathbf{D}_z \begin{bmatrix} \mathbf{D}_p^T & \mathbf{D}_z^T \end{bmatrix} = \begin{bmatrix} \mathbf{0} & \mathbf{I} \end{bmatrix}, \quad \mathbf{B}_y^T \begin{bmatrix} \mathbf{C}_p & \mathbf{B}_y \end{bmatrix} = \begin{bmatrix} \mathbf{0} & \mathbf{I} \end{bmatrix} \tag{6.18}$$

hold.

For the closed-loop output covariance to be bounded, it is necessary that the closed-loop transfer matrix from w to y is strictly proper:

$$\mathbf{D}_{c\ell} = \mathbf{D}_y + \mathbf{B}_y\mathbf{D}_c\mathbf{D}_z = \mathbf{0}$$

where $\mathbf{D}_c$ is the high frequency gain of the controller. With assumptions (6.18), this condition requires that the controller be strictly proper, i.e. $\mathbf{D}_c = \mathbf{0}$.

Recall from Lemma 6.1.1 that the closed-loop system is stable and the output covariance is bounded above by $\mho$ if and only if there exists a symmetric positive definite matrix $\mathbf{X} > \mathbf{0}$ such that

$$\mathbf{C}_{c\ell}\mathbf{X}\mathbf{C}_{c\ell}^T < \mho, \quad \mathbf{A}_{c\ell}\mathbf{X} + \mathbf{X}\mathbf{A}_{c\ell}^T + \mathbf{B}_{c\ell}\mathbf{B}_{c\ell}^T < \mathbf{0}. \tag{6.19}$$

In the above inequalities, we need to consider matrix $\mathbf{X}$ of the following structure only:

$$\mathbf{X} = \begin{bmatrix} \mathbf{X}_p & \mathbf{Z} \\ \mathbf{Z} & \mathbf{Z} \end{bmatrix}.$$

This is because, if there exist a controller $(\mathbf{A}_c, \mathbf{B}_c, \mathbf{C}_c, \mathbf{D}_c)$ and $\mathbf{X}$ satisfying (6.19), then there always exists a coordinate transformation matrix $\mathbf{T}_c$ such that the controller realization $(\mathbf{T}_c^{-1}\mathbf{A}_c\mathbf{T}_c, \mathbf{T}_c^{-1}\mathbf{B}_c, \mathbf{C}_c\mathbf{T}_c, \mathbf{D}_c)$ solves (6.19) for some positive definite matrix $\hat{\mathbf{X}}$ of the above structure. Verification of this simple fact is left for the reader as a straightforward exercise.

Consider the coordinate transformation for the closed-loop states by

$$\mathbf{T} \triangleq \begin{bmatrix} \mathbf{I} & \mathbf{0} \\ \mathbf{I} & -\mathbf{I} \end{bmatrix}. \tag{6.20}$$

This transformation is possible since the controller is of full order, and the resulting new closed-loop states consist of the plant state $\mathbf{x}_p$ and the difference $\mathbf{x}_p - \mathbf{x}_c$, where the latter may be interpreted as the estimation error if the controller were to dynamically estimate the plant state. The following identity is useful:

$$\begin{bmatrix} \mathbf{T} & \mathbf{0} \\ \mathbf{0} & \mathbf{I} \end{bmatrix} \begin{bmatrix} \mathbf{A}_{c\ell} & \mathbf{B}_{c\ell} \\ \mathbf{C}_{c\ell} & \mathbf{D}_{c\ell} \end{bmatrix} \begin{bmatrix} \mathbf{X} & \mathbf{0} \\ \mathbf{0} & \mathbf{I} \end{bmatrix} \begin{bmatrix} \mathbf{T}^T & \mathbf{0} \\ \mathbf{0} & \mathbf{I} \end{bmatrix} =$$

$$\left[\begin{array}{cc|c} \mathbf{A}_p\mathbf{X}_p + \mathbf{B}_p\mathbf{C}_c\mathbf{Z} & \mathbf{A}_p\mathbf{Y}_p^{-1} & \mathbf{D}_p \\ (\mathbf{A}_p - \mathbf{B}_c\mathbf{M}_p)\mathbf{X}_p + (\mathbf{B}_p\mathbf{C}_c - \mathbf{A}_c)\mathbf{Z} & (\mathbf{A}_p - \mathbf{B}_c\mathbf{M}_p)\mathbf{Y}_p^{-1} & \mathbf{D}_p - \mathbf{B}_c\mathbf{D}_z \\ \hline \mathbf{C}_p\mathbf{X}_p + \mathbf{B}_y\mathbf{C}_c\mathbf{Z} & \mathbf{C}_p\mathbf{Y}_p^{-1} & \mathbf{D}_y \end{array}\right] \tag{6.21}$$

where $\mathbf{Y}_p \triangleq (\mathbf{X}_p - \mathbf{Z})^{-1}$. Using this identity, it can readily be verified that the congruent transformation by $\mathbf{T}$ for the Lyapunov inequality in (6.19) yields

$$\mathbf{\Psi} = \begin{bmatrix} \mathbf{\Psi}_{11} & \mathbf{\Psi}_{12} \\ \mathbf{\Psi}_{12}^T & \mathbf{\Psi}_{22} \end{bmatrix} \triangleq \mathbf{T}(\mathbf{A}_{c\ell}\mathbf{X} + \mathbf{X}\mathbf{A}_{c\ell}^T + \mathbf{B}_{c\ell}\mathbf{B}_{c\ell}^T)\mathbf{T}^T < \mathbf{0}$$

where

$$\mathbf{\Psi}_{11} = \mathbf{A}_p\mathbf{X}_p + \mathbf{X}_p\mathbf{A}_p^T + \mathbf{B}_p\mathbf{C}_c\mathbf{Z} + \mathbf{Z}\mathbf{C}_c^T\mathbf{B}_p^T + \mathbf{D}_p\mathbf{D}_p^T \tag{6.22}$$

$$\mathbf{\Psi}_{12}^T = \mathbf{Y}_p^{-1}\mathbf{A}_p^T + (\mathbf{A}_p - \mathbf{B}_c\mathbf{M}_p)\mathbf{X}_p + (\mathbf{B}_p\mathbf{C}_c - \mathbf{A}_c)\mathbf{Z} + (\mathbf{D}_p - \mathbf{B}_c\mathbf{D}_z)\mathbf{D}_p^T \tag{6.23}$$

$$\mathbf{\Psi}_{22} = (\mathbf{A}_p - \mathbf{B}_c\mathbf{M}_p)\mathbf{Y}_p^{-1} + \mathbf{Y}_p^{-1}(\mathbf{A}_p - \mathbf{B}_c\mathbf{M}_p)^T + (\mathbf{D}_p - \mathbf{B}_c\mathbf{D}_z)(\mathbf{D}_p - \mathbf{B}_c\mathbf{D}_z)^T. \tag{6.24}$$

Note that the first inequality in (6.19) and $\mathbf{X} > \mathbf{0}$ are equivalent to

$$\begin{bmatrix} \mathbf{I} & \mathbf{0} \\ \mathbf{0} & \mathbf{T} \end{bmatrix} \begin{bmatrix} \mho & \mathbf{C}_{c\ell}\mathbf{X} \\ \mathbf{X}\mathbf{C}_{c\ell}^T & \mathbf{X} \end{bmatrix} \begin{bmatrix} \mathbf{I} & \mathbf{0} \\ \mathbf{0} & \mathbf{T}^T \end{bmatrix} > \mathbf{0}$$

or equivalently, with a congruent transformation involving $\mathbf{Y}_p$,

$$\begin{bmatrix} \mho & \mathbf{C}_p\mathbf{X}_p + \mathbf{B}_y\mathbf{C}_c\mathbf{Z} & \mathbf{C}_p \\ \mathbf{X}_p\mathbf{C}_p^T + \mathbf{Z}\mathbf{C}_c^T\mathbf{B}_y^T & \mathbf{X}_p & \mathbf{I} \\ \mathbf{C}_p^T & \mathbf{I} & \mathbf{Y}_p \end{bmatrix} > \mathbf{0}. \tag{6.25}$$

Our objective is to find $(\mathbf{A}_c, \mathbf{B}_c, \mathbf{C}_c, \mathbf{D}_c)$ and $(\mathbf{X}_p, \mathbf{Y}_p)$ such that $\mathbf{\Psi} < \mathbf{0}$ and (6.25) hold. To do this, we will eliminate some of the controller parameters and reduce the control design problem to a convex LMI problem (see Chapters 10 and 11 for the details of numerical solution procedures for LMI problems).

We eliminate $\mathbf{A}_c$ first. Note that $\mathbf{A}_c$ appears only in inequality $\boldsymbol{\Psi} < \mathbf{0}$ (not in (6.25)). If $\boldsymbol{\Psi} < \mathbf{0}$ holds for some $\mathbf{A}_c$, then $\boldsymbol{\Psi}_{11} < \mathbf{0}$ and $\boldsymbol{\Psi}_{22} < \mathbf{0}$ hold. Conversely, if $\boldsymbol{\Psi}_{11} < \mathbf{0}$ and $\boldsymbol{\Psi}_{22} < \mathbf{0}$ hold, then there exists $\mathbf{A}_c$ such that $\boldsymbol{\Psi} < \mathbf{0}$ and one such choice is given by solving $\boldsymbol{\Psi}_{12} = 0$ as follows:

$$\mathbf{A}_c = \mathbf{Y}_p^{-1}\mathbf{A}_p^T\mathbf{Z}^{-1} + (\mathbf{A}_p - \mathbf{B}_c\mathbf{M}_p)\mathbf{X}_p\mathbf{Z}^{-1} + \mathbf{B}_p\mathbf{C}_c + (\mathbf{D}_p - \mathbf{B}_c\mathbf{D}_z)\mathbf{D}_p^T\mathbf{Z}^{-1}. \tag{6.26}$$

Thus $\boldsymbol{\Psi} < \mathbf{0}$ has been replaced by $\boldsymbol{\Psi}_{11} < \mathbf{0}$ and $\boldsymbol{\Psi}_{22} < \mathbf{0}$ by eliminating $\mathbf{A}_c$. Next, we eliminate $\mathbf{B}_c$ which appears only in $\boldsymbol{\Psi}_{22} < \mathbf{0}$. Completing the square with respect to $\mathbf{B}_c$, we have

$$(\mathbf{B}_c - \mathbf{Y}_p^{-1}\mathbf{M}_p^T)(\mathbf{B}_c^T - \mathbf{M}_p\mathbf{Y}_p^{-1}) + \mathbf{A}_p\mathbf{Y}_p^{-1} + \mathbf{Y}_p^{-1}\mathbf{A}_p^T - \mathbf{Y}_p^{-1}\mathbf{M}_p^T\mathbf{M}_p\mathbf{Y}_p^{-1} + \mathbf{D}_p\mathbf{D}_p^T < \mathbf{0}.$$

This inequality holds for some $\mathbf{B}_c$ if and only if

$$\mathbf{A}_p\mathbf{Y}_p^{-1} + \mathbf{Y}_p^{-1}\mathbf{A}_p^T - \mathbf{Y}_p^{-1}\mathbf{M}_p^T\mathbf{M}_p\mathbf{Y}_p^{-1} + \mathbf{D}_p\mathbf{D}_p^T < \mathbf{0} \tag{6.27}$$

holds, in which case a choice of $\mathbf{B}_c$ is given by

$$\mathbf{B}_c = \mathbf{Y}_p^{-1}\mathbf{M}_p^T. \tag{6.28}$$

Thus we have shown that there exists a stabilizing controller achieving the covariance upper bound $\mho$ if and only if there exist matrices $\mathbf{X}_p > \mathbf{0}$, $\mathbf{Y}_p > \mathbf{0}$ and $\mathbf{C}_c$ such that $\boldsymbol{\Psi}_{11} < \mathbf{0}$, (6.27) and (6.25) hold, where $\boldsymbol{\Psi}_{11}$ is defined in (6.22) and $\mathbf{Z} \triangleq \mathbf{X}_p - \mathbf{Y}_p^{-1}$. Converting these inequalities to LMIs via change of variables, we have the following.

Theorem 6.2.5 *Let a symmetric matrix $\mho$ be given and consider the system (4.15). Suppose the assumptions in (6.18) hold. Then the following statements are equivalent:*

(i) There exists a dynamic output feedback controller which stabilizes the system and yields

$$\lim_{t\to\infty} \mathcal{E}\,[\mathbf{y}(t)\mathbf{y}^T(t)] < \mho.$$

(ii) There exist matrices $\mathbf{X}_p$, $\mathbf{Y}_p$ and $\mathbf{K}$ such that

$$\mathbf{A}_p\mathbf{X}_p + \mathbf{X}_p\mathbf{A}_p^T + \mathbf{B}_p\mathbf{K} + \mathbf{K}^T\mathbf{B}_p^T + \mathbf{D}_p\mathbf{D}_p^T < \mathbf{0} \tag{6.29}$$

$$\begin{bmatrix} \mathbf{Y}_p\mathbf{A}_p + \mathbf{A}_p^T\mathbf{Y}_p - \mathbf{M}_p^T\mathbf{M}_p & \mathbf{Y}_p\mathbf{D}_p \\ \mathbf{D}_p^T\mathbf{Y}_p & -\mathbf{I} \end{bmatrix} < \mathbf{0} \tag{6.30}$$

$$\begin{bmatrix} \mho & \mathbf{C}_p\mathbf{X}_p + \mathbf{B}_y\mathbf{K} & \mathbf{C}_p \\ \mathbf{X}_p\mathbf{C}_p^T + \mathbf{K}^T\mathbf{B}_y^T & \mathbf{X}_p & \mathbf{I} \\ \mathbf{C}_p^T & \mathbf{I} & \mathbf{Y}_p \end{bmatrix} > \mathbf{0}. \tag{6.31}$$

In this case, one such controller is given by

$$\begin{aligned} \mathbf{A}_c &= \mathbf{A}_p + \mathbf{B}_p\mathbf{C}_c - \mathbf{B}_c\mathbf{M}_p - \mathbf{Y}_p^{-1}\boldsymbol{\Omega}(\mathbf{I} - \mathbf{X}_p\mathbf{Y}_p)^{-1} \\ \mathbf{B}_c &= \mathbf{Y}_p^{-1}\mathbf{M}_p^T \\ \mathbf{C}_c &= \mathbf{K}(\mathbf{X}_p - \mathbf{Y}_p^{-1})^{-1} \\ \mathbf{D}_c &= \mathbf{0} \end{aligned}$$

where

$$\boldsymbol{\Omega} \triangleq \mathbf{Y}_p\mathbf{A}_p + \mathbf{A}_p^T\mathbf{Y}_p + \mathbf{Y}_p\mathbf{D}_p\mathbf{D}_p^T\mathbf{Y}_p - \mathbf{M}_p^T\mathbf{M}_p.$$

Proof. Inequalities (6.29) and (6.31) follow from $\mathbf{\Psi}_{11} < \mathbf{0}$ and (6.25) by letting $\mathbf{K} \triangleq \mathbf{C}_c\mathbf{Z}$. Inequality (6.30) follows from (6.27) using the Schur complement formula and the congruent transformation. Controller formulae can be derived using (6.26), (6.28) and $\mathbf{K} \triangleq \mathbf{C}_c\mathbf{Z}$. □

The above result can be specialized to the standard LQG (Linear Quadratic Gaussian) result. The LQG problem is to minimize $tr(\mathbf{\mho})$ subject to the conditions in statement (i) of Theorem 6.2.5. The solution to this problem can be obtained as follows.

In view of inequality (6.31), if $\mathbf{Y}_p$ is larger (in the sense of positive definite matrices), then the performance bound $tr(\mathbf{\mho}) < \gamma$ is more likely to be satisfied. On the other hand, the "largest" $\mathbf{Y}_p$ satisfying (6.30) is given by the inverse[2] of the stabilizing solution $\mathbf{Q}$ to

$$\mathbf{A}_p\mathbf{Q} + \mathbf{Q}\mathbf{A}_p^T - \mathbf{Q}\mathbf{M}_p^T\mathbf{M}_p\mathbf{Q} + \mathbf{D}_p\mathbf{D}_p^T = 0. \tag{6.32}$$

Therefore we choose $\mathbf{Y}_p$ in (6.30) and (6.31) to be $\mathbf{Q}^{-1}$.

Using the Schur complement formula, (6.31) is equivalent to

$$\begin{bmatrix} \mathbf{\mho} - \mathbf{C}_p\mathbf{Q}\mathbf{C}_p^T & (\mathbf{C}_p + \mathbf{B}_y\mathbf{C}_c)\mathbf{Z} \\ \mathbf{Z}(\mathbf{C}_p + \mathbf{B}_y\mathbf{C}_c)^T & \mathbf{Z} \end{bmatrix} > \mathbf{0}$$

or

$$\mathbf{\mho} > \mathbf{C}_p\mathbf{Q}\mathbf{C}_p^T + (\mathbf{C}_p + \mathbf{B}_y\mathbf{C}_c)\mathbf{Z}(\mathbf{C}_p + \mathbf{B}_y\mathbf{C}_c)^T \tag{6.33}$$

where we used $\mathbf{K} = \mathbf{C}_c\mathbf{Z}$. Also using (6.32), inequality (6.29) can be written

$$(\mathbf{A}_p + \mathbf{B}_p\mathbf{C}_c)\mathbf{Z} + \mathbf{Z}(\mathbf{A}_p + \mathbf{B}_p\mathbf{C}_c)^T + \mathbf{Q}\mathbf{M}_p^T\mathbf{M}_p\mathbf{Q} < \mathbf{0}. \tag{6.34}$$

Now, it can be verified using the dual characterization of the $\mathcal{H}_2$ norm that, for given γ and $\mathbf{C}_c$ such that $\mathbf{A}_p + \mathbf{B}_p\mathbf{C}_c$ is asymptotically stable, there exist $\mathbf{Z}$ and $\mathbf{\mho}$ satisfying (6.33), (6.34) and $tr(\mathbf{\mho}) < \gamma$ if and only if there exists $\mathbf{P} > \mathbf{0}$ such that

$$\mathbf{P}(\mathbf{A}_p + \mathbf{B}_p\mathbf{C}_c) + (\mathbf{A}_p + \mathbf{B}_p\mathbf{C}_c)^T\mathbf{P} + (\mathbf{C}_p + \mathbf{B}_y\mathbf{C}_c)^T(\mathbf{C}_p + \mathbf{B}_y\mathbf{C}_c) < \mathbf{0} \tag{6.35}$$

$$tr(\mathbf{M}_p\mathbf{Q}\mathbf{P}\mathbf{Q}\mathbf{M}_p^T) < \gamma - tr(\mathbf{C}_p\mathbf{Q}\mathbf{C}_p^T). \tag{6.36}$$

Completing the square with respect to $\mathbf{C}_c$ in (6.35),

$$(\mathbf{P}\mathbf{B}_p + \mathbf{C}_c^T)(\mathbf{B}_p^T\mathbf{P} + \mathbf{C}_c) + \mathbf{P}\mathbf{A}_p + \mathbf{A}_p^T\mathbf{P} - \mathbf{P}\mathbf{B}_p\mathbf{B}_p^T\mathbf{P} + \mathbf{C}_p^T\mathbf{C}_p < \mathbf{0}. \tag{6.37}$$

In view of (6.36), smaller $\mathbf{P}$ yields smaller γ. From the monotonicity property of the Riccati equation solution [107], the smallest $\mathbf{P}$ such that (6.37) is given by the stabilizing solution to

$$\mathbf{P}\mathbf{A}_p + \mathbf{A}_p^T\mathbf{P} - \mathbf{P}\mathbf{B}_p\mathbf{B}_p^T\mathbf{P} + \mathbf{C}_p^T\mathbf{C}_p = \mathbf{0}$$

with

$$\mathbf{C}_c = -\mathbf{B}_p^T\mathbf{P}.$$

In summary, we have the following.

[2] The inverse of $\mathbf{Q}$ exists if $(\mathbf{A}, \mathbf{D}_p, \mathbf{M}_p)$ is a controllable, detectable triple. This assumption can be relaxed by avoiding the use of $\mathbf{Q}^{-1}$ in the above discussion.

Theorem 6.2.6 *Consider the system given in (4.15). Suppose the assumptions in (6.18) hold. Then the controller that stabilizes (4.15) and solves*

$$\gamma_{\text{opt}} \triangleq \min\ tr(\lim_{t\to\infty} \mathcal{E}[\mathbf{y}(t)\mathbf{y}^T(t)])$$

is given by

$$\begin{aligned}\mathbf{A}_c &= \mathbf{A}_p + \mathbf{B}_p\mathbf{C}_c - \mathbf{B}_c\mathbf{M}_p \\ \mathbf{B}_c &= \mathbf{Q}\mathbf{M}_p^T \\ \mathbf{C}_c &= -\mathbf{B}_p^T\mathbf{P} \\ \mathbf{D}_c &= \mathbf{0}\end{aligned}$$

where $\mathbf{P}$ *and* $\mathbf{Q}$ *are the stabilizing solutions to*

$$\begin{aligned}\mathbf{P}\mathbf{A}_p + \mathbf{A}_p^T\mathbf{P} - \mathbf{P}\mathbf{B}_p\mathbf{B}_p^T\mathbf{P} + \mathbf{C}_p^T\mathbf{C}_p &= \mathbf{0} \\ \mathbf{A}_p\mathbf{Q} + \mathbf{Q}\mathbf{A}_p^T - \mathbf{Q}\mathbf{M}_p^T\mathbf{M}_p\mathbf{Q} + \mathbf{D}_p\mathbf{D}_p^T &= \mathbf{0}.\end{aligned}$$

Furthermore, the minimum value of the cost function is given by

$$\gamma_{\text{opt}} = tr(\mathbf{M}_p\mathbf{Q}\mathbf{P}\mathbf{Q}\mathbf{M}_p^T) + tr(\mathbf{C}_p\mathbf{Q}\mathbf{C}_p^T).$$

6.3 DISCRETE-TIME CASE

This section presents a solution to the covariance bounding control problem for the discrete-time case. In the previous section, we have made the closed-loop system strictly proper ($\mathbf{D}_{c\ell} = \mathbf{0}$) since the output covariance is not defined for proper ($\mathbf{D}_{c\ell} \neq \mathbf{0}$) systems. For discrete-time systems, however, the output covariance is finite even if the system is not strictly proper, provided the system is stable. In the following, we do not assume $\mathbf{D}_y = \mathbf{0}$. It is desirable to allow possibly nonzero $\mathbf{B}_y$, which corresponds to a control input covariance constraint. However, $\mathbf{B}_y \neq \mathbf{0}$ introduces a technical difficulty due to the fact that $\mathbf{C}_{c\ell}$ becomes dependent upon the controller parameter $\mathbf{G}$. For this reason, we shall treat the simple case $\mathbf{B}_y = \mathbf{0}$ first, and remove this assumption later.

6.3.1 State Feedback

In this section, we consider the (static) state feedback case where $\mathbf{M}_p = \mathbf{I}$ and $\mathbf{D}_z = \mathbf{0}$. The Lyapunov inequality which characterizes a set of state covariance bounds is given by

$$\mathbf{X} > (\mathbf{A}_p + \mathbf{B}_p\mathbf{G})\mathbf{X}(\mathbf{A}_p + \mathbf{B}_p\mathbf{G})^T + \mathbf{D}_p\mathbf{D}_p^T.$$

Theorem 6.3.1 *Let a symmetric matrix* $\boldsymbol{\mho}$ *be given and consider the system (4.32) with* $\mathbf{B}_y = \mathbf{0}$, $\mathbf{M}_p = \mathbf{I}$ *and* $\mathbf{D}_z = \mathbf{0}$. *Then the following statements are equivalent:*

(i) There exists a stabilizing state feedback gain $\mathbf{G}$ *such that*

$$\lim_{k\to\infty} \mathcal{E}\left[\mathbf{y}(k)\mathbf{y}^T(k)\right] < \boldsymbol{\mho}.$$

(ii) There exists a matrix $\mathbf{X}$ *such that*

$$\mathbf{D}_p\mathbf{D}_p^T < \mathbf{X}, \quad \mathbf{C}_p\mathbf{X}\mathbf{C}_p^T + \mathbf{D}_y\mathbf{D}_y^T < \mathcal{U}$$

$$\mathbf{B}_p^{\perp}(\mathbf{X} - \mathbf{A}_p\mathbf{X}\mathbf{A}_p^T - \mathbf{D}_p\mathbf{D}_p^T)\mathbf{B}_p^{\perp T} > \mathbf{0}.$$

In this case, all such state feedback gains are given by

$$\mathbf{G} = -(\mathbf{B}_p^T\mathbf{P}\mathbf{B}_p)^{-1}\mathbf{B}_p^T\mathbf{P}\mathbf{A}_p + (\mathbf{B}_p^T\mathbf{P}\mathbf{B}_p)^{-1/2}\mathbf{L}\mathbf{S}^{1/2}$$

where $\mathbf{L}$ *is an arbitrary matrix such that* $\|\mathbf{L}\| < 1$ *and*

$$\mathbf{P} \triangleq (\mathbf{X} - \mathbf{D}_p\mathbf{D}_p^T)^{-1} > \mathbf{0}$$

$$\mathbf{S} \triangleq \mathbf{X}^{-1} - \mathbf{A}_p^T\mathbf{P}\mathbf{A}_p + \mathbf{A}_p^T\mathbf{P}\mathbf{B}_p(\mathbf{B}_p^T\mathbf{P}\mathbf{B}_p)^{-1}\mathbf{B}_p^T\mathbf{P}\mathbf{A}_p > \mathbf{0}.$$

Proof. The result directly follows by solving

$$(\mathbf{B}_p\mathbf{G} + \mathbf{A}_p)\mathbf{X}(\mathbf{B}_p\mathbf{G} + \mathbf{A}_p)^T < \mathbf{X} - \mathbf{D}_p\mathbf{D}_p^T$$

for $\mathbf{G}$ using Corollary 2.3.6. □

As a simple consequence, we have the following.

Corollary 6.3.1 *Let a discrete-time system* $\mathbf{x}(k+1) = \mathbf{A}\mathbf{x}(k) + \mathbf{B}\mathbf{u}(k)$ *be given. Then the following statements are equivalent:*

(i) There exists a (static) state feedback controller $\mathbf{u}(k) = \mathbf{G}\mathbf{x}(k)$ *which stabilizes the system.*

(ii) There exist matrices $\mathbf{K}$ *and* $\mathbf{X} > \mathbf{0}$ *such that*

$$\mathbf{X} > (\mathbf{A} + \mathbf{B}\mathbf{K})\mathbf{X}(\mathbf{A} + \mathbf{B}\mathbf{K})^T.$$

(iii) There exists a matrix $\mathbf{X} > \mathbf{0}$ *such that*

$$\mathbf{B}^{\perp}(\mathbf{X} - \mathbf{A}\mathbf{X}\mathbf{A}^T)\mathbf{B}^{\perp T} > \mathbf{0}.$$

(iv) There exists a matrix $\mathbf{Y} > \mathbf{0}$ *such that*

$$\mathbf{Y} > \mathbf{A}^T\mathbf{Y}\mathbf{A} - \mathbf{A}^T\mathbf{Y}\mathbf{B}(\mathbf{B}^T\mathbf{Y}\mathbf{B})^{-1}\mathbf{B}^T\mathbf{Y}\mathbf{A}. \tag{6.38}$$

Proof. The equivalence of (i) and (ii) is a direct consequence of Lyapunov's stability theory for discrete-time systems. (i) ⇔ (iii) follows from Theorem 6.3.1 by letting the performance bound $\mathcal{U}$ be arbitrarily large and specializing to the case $\mathbf{D}_p = \mathbf{0}$ without loss of generality. Finally, (iii) ⇔ (iv) can be proved as follows. Note that a matrix $\mathbf{Y} > \mathbf{0}$ satisfies (6.38) if and only if there exists $\mathbf{R} > \mathbf{0}$ such that

$$\mathbf{Y} > \mathbf{A}^T\mathbf{Y}\mathbf{A} - \mathbf{A}^T\mathbf{Y}\mathbf{B}(\mathbf{B}^T\mathbf{Y}\mathbf{B} + \mathbf{R})^{-1}\mathbf{B}^T\mathbf{Y}\mathbf{A}.$$

Using the matrix inversion lemma, we have

$$\mathbf{Y} > \mathbf{A}^T(\mathbf{Y}^{-1} + \mathbf{B}\mathbf{R}^{-1}\mathbf{B}^T)^{-1}\mathbf{A},$$

or equivalently, using the Schur complement formula,

$$\mathbf{Y}^{-1} + \mathbf{B}\mathbf{R}^{-1}\mathbf{B}^T > \mathbf{A}\mathbf{Y}^{-1}\mathbf{A}^T.$$

From Finsler's Theorem (Corollary 2.3.5), there exists $\mathbf{R} > \mathbf{0}$ satisfying the above inequality if and only if

$$\mathbf{B}^{\perp}(\mathbf{Y}^{-1} - \mathbf{A}\mathbf{Y}^{-1}\mathbf{A}^T)\mathbf{B}^{\perp T} > \mathbf{0}$$

holds. Thus letting $\mathbf{X} \triangleq \mathbf{Y}^{-1}$, we have (iii). □

6.3.2 Static Output Feedback

This section considers the covariance bounding control problem for the static output feedback case. The Lyapunov inequality in this case is given by

$$\mathbf{X} > (\mathbf{A}_p + \mathbf{B}_p\mathbf{G}\mathbf{M}_p)\mathbf{X}(\mathbf{A}_p + \mathbf{B}_p\mathbf{G}\mathbf{M}_p)^T + (\mathbf{D}_p + \mathbf{B}_p\mathbf{G}\mathbf{D}_z)(\mathbf{D}_p + \mathbf{B}_p\mathbf{G}\mathbf{D}_z)^T. \quad (6.39)$$

As in the continuous-time case, we shall assume orthogonality between the process and measurement noises ($\mathbf{D}_p\mathbf{D}_z^T = \mathbf{0}$) for simplicity. However, we do not assume that the measured outputs are fully contaminated by noises ($\mathbf{V} \triangleq \mathbf{D}_z\mathbf{D}_z^T > \mathbf{0}$). In other words, we allow possibly singular $\mathbf{V} \geq \mathbf{0}$ since such a generality does not introduce an additional complexity in the following derivation.

Theorem 6.3.2 *Let a symmetric matrix* $\boldsymbol{\mho}$ *be given and consider the system (4.31) with* $\mathbf{B}_y = \mathbf{0}$ *and* $\mathbf{D}_p\mathbf{D}_z^T = \mathbf{0}$. *Then the following statements are equivalent:*

(i) There exists a stabilizing static output feedback gain $\mathbf{G}$ *such that*

$$\lim_{k\to\infty} \mathcal{E}\left[\mathbf{y}(k)\mathbf{y}^T(k)\right] < \boldsymbol{\mho}.$$

(ii) There exists a matrix $\mathbf{X} > \mathbf{0}$ *such that*

$$\mathbf{B}_p^{\perp}(\mathbf{X} - \mathbf{A}_p\mathbf{X}\mathbf{A}_p^T - \mathbf{D}_p\mathbf{D}_p^T)\mathbf{B}_p^{\perp T} > \mathbf{0}, \quad \mathbf{C}_p\mathbf{X}\mathbf{C}_p^T + \mathbf{D}_y\mathbf{D}_y^T < \boldsymbol{\mho}$$

$$\mathbf{X} > \mathbf{A}_p\mathbf{X}\mathbf{A}_p^T - \mathbf{A}_p\mathbf{X}\mathbf{M}_p^T(\mathbf{M}_p\mathbf{X}\mathbf{M}_p^T + \mathbf{V})^{-1}\mathbf{M}_p\mathbf{X}\mathbf{A}_p^T + \mathbf{D}_p\mathbf{D}_p^T$$

where $\mathbf{V} \triangleq \mathbf{D}_z\mathbf{D}_z^T$.

In this case, all such static output feedback gains are given by

$$\mathbf{G} = -(\mathbf{B}_p^T\mathbf{P}\mathbf{B}_p)^{-1}\mathbf{B}_p^T\mathbf{P}\mathbf{A}_p\mathbf{X}\mathbf{M}_p^T(\mathbf{M}_p\mathbf{X}\mathbf{M}_p^T + \mathbf{V})^{-1} + (\mathbf{B}_p^T\mathbf{P}\mathbf{B}_p)^{-1/2}\mathbf{L}\mathbf{S}^{1/2}$$

where $\mathbf{L}$ *is an arbitrary matrix such that* $\|\mathbf{L}\| < 1$ *and*

$$\mathbf{P} \triangleq (\mathbf{X} - \mathbf{A}_p\mathbf{X}\mathbf{A}_p^T + \boldsymbol{\Gamma}\mathbf{R}\boldsymbol{\Gamma}^T - \mathbf{D}_p\mathbf{D}_p^T)^{-1} > \mathbf{0}$$

$$\mathbf{S} \triangleq \mathbf{R}^{-1} - \boldsymbol{\Gamma}^T[\mathbf{P} - \mathbf{P}\mathbf{B}_p(\mathbf{B}_p^T\mathbf{P}\mathbf{B}_p)^{-1}\mathbf{B}_p^T\mathbf{P}]\boldsymbol{\Gamma} > \mathbf{0}$$

$$\mathbf{R} \triangleq \mathbf{M}_p\mathbf{X}\mathbf{M}_p^T + \mathbf{V}, \quad \boldsymbol{\Gamma} \triangleq \mathbf{A}_p\mathbf{X}\mathbf{M}_p^T\mathbf{R}^{-1}.$$

Proof. After expanding each term in (6.39), completing the square with respect to $\mathbf{G}$ yields

$$(\mathbf{B}_p\mathbf{G}+\boldsymbol{\Gamma})\mathbf{R}(\mathbf{B}_p\mathbf{G}+\boldsymbol{\Gamma})^T < \mathbf{P}^{-1}.$$

Then the result follows by applying Corollary 2.3.6. □

Corollary 6.3.2 *Let a discrete-time system*

$$\mathbf{x}(k+1)=\mathbf{A}\mathbf{x}(k)+\mathbf{B}\mathbf{u}(k), \quad \mathbf{z}(k)=\mathbf{M}\mathbf{x}(k)$$

be given. Then the following statements are equivalent:

(i) There exists a static output feedback controller $\mathbf{u}(k)=\mathbf{G}\mathbf{z}(k)$ *which stabilizes the system.*

(ii) There exists a matrix $\mathbf{X}>\mathbf{0}$ *such that*

$$\mathbf{B}^{\perp}(\mathbf{X}-\mathbf{A}\mathbf{X}\mathbf{A}^T)\mathbf{B}^{\perp T}>\mathbf{0}$$

$$\mathbf{X}>\mathbf{A}\mathbf{X}\mathbf{A}^T-\mathbf{A}\mathbf{X}\mathbf{M}^T(\mathbf{M}\mathbf{X}\mathbf{M}^T)^{-1}\mathbf{M}\mathbf{X}\mathbf{A}^T.$$

(iii) There exists a matrix pair $(\mathbf{X},\mathbf{Y})$ *such that*

$$\mathbf{X}=\mathbf{Y}^{-1}>\mathbf{0}$$

$$\mathbf{B}^{\perp}(\mathbf{X}-\mathbf{A}\mathbf{X}\mathbf{A}^T)\mathbf{B}^{\perp T}>\mathbf{0}$$

$$\mathbf{M}^{T\perp}(\mathbf{Y}-\mathbf{A}^T\mathbf{Y}\mathbf{A})\mathbf{M}^{T\perp T}>\mathbf{0}.$$

(iv) There exist a matrix pair $(\mathbf{X},\mathbf{Y})$ *and a scalar* $\mu>0$ *such that*

$$\mathbf{X}=\mathbf{Y}^{-1}>\mathbf{0}$$

$$\mathbf{X}>\mathbf{A}\mathbf{X}\mathbf{A}^T-\mu\mathbf{B}\mathbf{B}^T$$

$$\mathbf{Y}>\mathbf{A}^T\mathbf{Y}\mathbf{A}-\mu\mathbf{M}^T\mathbf{M}.$$

(v) There exist matrices $\mathbf{K}$, $\mathbf{F}$ *and* $\mathbf{X}>\mathbf{0}$ *such that*

$$\mathbf{X}>(\mathbf{A}+\mathbf{B}\mathbf{K})\mathbf{X}(\mathbf{A}+\mathbf{B}\mathbf{K})^T$$

$$\mathbf{X}>(\mathbf{A}+\mathbf{F}\mathbf{M})\mathbf{X}(\mathbf{A}+\mathbf{F}\mathbf{M})^T.$$

(vi) There exists a coordinate transformation matrix $\mathbf{T}$ *such that the transformed system*

$$\hat{\mathbf{A}}\triangleq\mathbf{T}^{-1}\mathbf{A}\mathbf{T}, \quad \hat{\mathbf{B}}\triangleq\mathbf{T}^{-1}\mathbf{B}, \quad \hat{\mathbf{M}}\triangleq\mathbf{M}\mathbf{T}$$

satisfies

$$\|\hat{\mathbf{A}}\mathbf{x}\|<\|\mathbf{x}\|, \quad \forall\,\mathbf{x}\neq\mathbf{0} \text{ such that } \hat{\mathbf{M}}\mathbf{x}=\mathbf{0}$$

$$\|\hat{\mathbf{A}}^T\mathbf{x}\|<\|\mathbf{x}\|, \quad \forall\,\mathbf{x}\neq\mathbf{0} \text{ such that } \hat{\mathbf{B}}^T\mathbf{x}=\mathbf{0}.$$

Proof. The equivalence (i) $\Leftrightarrow$ (ii) follows from Theorem 6.3.2 by letting $\mathcal{U}$ be arbitrarily large and $\mathbf{D}_p = \mathbf{0}$ and $\mathbf{V} = \mathbf{0}$. (ii) $\Leftrightarrow$ (iii) and (iii) $\Leftrightarrow$ (v) are immediate from Corollary 6.3.1 and its proof. (iii) $\Leftrightarrow$ (iv) is a direct consequence of Finsler's Theorem. Finally, we shall prove (iii) $\Leftrightarrow$ (vi). Choosing the coordinate transformation matrix $\mathbf{T} \triangleq \mathbf{X}^{1/2}$ and noting that $\hat{\mathbf{B}}^{\perp} = \mathbf{B}^{\perp}\mathbf{T}$, it can be shown that (iii) is equivalent to the existence of $\mathbf{T}$ such that

$$\hat{\mathbf{B}}^{\perp}(\mathbf{I} - \hat{\mathbf{A}}\hat{\mathbf{A}}^T)\hat{\mathbf{B}}^{\perp T} > \mathbf{0}$$

$$\hat{\mathbf{M}}^{T\perp}(\mathbf{I} - \hat{\mathbf{A}}^T\hat{\mathbf{A}})\hat{\mathbf{M}}^{T\perp T} > \mathbf{0}.$$

Then the result simply follows once we notice that, for given matrices $\mathbf{Q}$ and $\mathbf{D}$, the following holds;

$$\mathbf{D}^{\perp}\mathbf{Q}\mathbf{D}^{\perp T} > \mathbf{0} \quad \Leftrightarrow \quad \mathbf{x}^T\mathbf{Q}\mathbf{x} > \mathbf{0}, \;\; \forall\, \mathbf{x} \neq \mathbf{0} \text{ such that } \mathbf{D}^T\mathbf{x} = \mathbf{0}.$$

This completes the proof. □

Corollary 6.3.2 provides several characterizations of conditions for stabilizability via static output feedback in terms of matrix inequalities. Although none of them are immediately verifiable, they may be useful for developing computational algorithms or algebraically verifiable tests to determine if a given system is stabilizable via static output feedback. For this purpose, we shall discuss each of the above conditions. Condition (iii) can be viewed as the existence of a matrix pair $(\mathbf{X}, \mathbf{Y})$ such that $\mathbf{X}$ and $\mathbf{Y}$ belong to convex sets defined by LMIs, and satisfies the nonconvex coupling condition $\mathbf{X} = \mathbf{Y}^{-1} > \mathbf{0}$. Convexity of the sets may be useful for developing computational algorithms.[3] Condition (iv) is given in terms of Lyapunov inequalities with a *negative* forcing term. Condition (ii) shows that the set of Lyapunov matrices is given by the intersection of the two sets defined by an LMI and a Riccati-like inequality. Condition (v) exhibits a certain "separation" property of the static output feedback stabilization problem. In particular, each of the inequalities in (v) corresponds to the state feedback or state estimation problem. What makes the whole problem difficult is the requirement of a single quadratic Lyapunov function (i.e. Lyapunov matrix $\mathbf{X}$) to prove stability of both problems. Finally, condition (vi) says that the system is stabilizable via static output feedback if and only if there exists a basis for the state space such that $\hat{\mathbf{A}}$ and $\hat{\mathbf{A}}^T$ are contractions if the domains are, respectively, unobservable and uncontrollable subspaces of the state space.

6.3.3 Reduced-Order Dynamic Output Feedback

This section solves the covariance bounding control problem for the reduced-order dynamic output feedback case. The results presented in the previous sections can also be derived as special cases of the results in this section. The underlying Lyapunov inequality is

$$\mathbf{X} > (\mathbf{A} + \mathbf{BGM})\mathbf{X}(\mathbf{A} + \mathbf{BGM})^T + (\mathbf{D} + \mathbf{BGE})(\mathbf{D} + \mathbf{BGE})^T. \tag{6.40}$$

Theorem 6.3.3 *Let a symmetric matrix $\mathcal{U}$ be given and consider the system (4.32) with $\mathbf{B}_y = \mathbf{0}$. Then the following statements are equivalent:*

[3] We shall propose algorithms in Chapters 10 and 11.

(i) There exists a dynamic output feedback controller of order n_c which stabilizes the system and yields

$$\lim_{k\to\infty} \mathcal{E}\left[\mathbf{y}(k)\mathbf{y}^T(k)\right] < \mho.$$

(ii) There exists a positive definite matrix $\mathbf{X} \in \mathcal{R}^{(n_p+n_c)\times(n_p+n_c)}$ such that

$$\mathbf{B}^{\perp}(\mathbf{X} - \mathbf{A}\mathbf{X}\mathbf{A}^T - \mathbf{D}\mathbf{D}^T)\mathbf{B}^{\perp T} > \mathbf{0}, \quad \mathbf{C}\mathbf{X}\mathbf{C}^T + \mathbf{F}\mathbf{F}^T < \mho$$

$$\begin{bmatrix} \mathbf{M}^T \\ \mathbf{E}^T \end{bmatrix}^{\perp} \begin{bmatrix} \mathbf{X}^{-1} - \mathbf{A}^T\mathbf{X}^{-1}\mathbf{A} & -\mathbf{A}^T\mathbf{X}^{-1}\mathbf{D} \\ -\mathbf{D}^T\mathbf{X}^{-1}\mathbf{A} & \mathbf{I} - \mathbf{D}^T\mathbf{X}^{-1}\mathbf{D} \end{bmatrix} \begin{bmatrix} \mathbf{M}^T \\ \mathbf{E}^T \end{bmatrix}^{\perp T} > \mathbf{0}.$$

In this case, all such controllers are given by

$$\mathbf{G} = -(\mathbf{B}^T\Phi\mathbf{B})^{-1}\mathbf{B}^T\Phi\Lambda\mathbf{R}\Gamma^T(\Gamma\mathbf{R}\Gamma^T)^{-1} + (\mathbf{B}^T\Phi\mathbf{B})^{-1/2}\mathbf{L}\Psi^{1/2}$$

where $\mathbf{L}$ is an arbitrary matrix such that $\|\mathbf{L}\| < 1$ and

$$\Phi \triangleq [\mathbf{X} - \Lambda\mathbf{R}\Lambda^T + \Lambda\mathbf{R}\Gamma^T(\Gamma\mathbf{R}\Gamma^T)^{-1}\Gamma\mathbf{R}\Lambda^T]^{-1}$$

$$\Psi \triangleq \Omega - \Omega\Gamma\mathbf{R}\Lambda^T[\Phi - \Phi\mathbf{B}(\mathbf{B}^T\Phi\mathbf{B})^{-1}\mathbf{B}^T\Phi]\Lambda\mathbf{R}\Gamma^T\Omega$$

$$\Omega \triangleq (\Gamma\mathbf{R}\Gamma^T)^{-1}, \quad \Lambda \triangleq \begin{bmatrix} \mathbf{A} & \mathbf{D} \end{bmatrix}$$

$$\Gamma \triangleq \begin{bmatrix} \mathbf{M} & \mathbf{E} \end{bmatrix}, \quad \mathbf{R} \triangleq \begin{bmatrix} \mathbf{X} & \mathbf{0} \\ \mathbf{0} & \mathbf{I} \end{bmatrix}.$$

Proof. Noting that the Lyapunov inequality (6.40) can be equivalently written

$$\mathbf{X} > \left(\begin{bmatrix} \mathbf{A} & \mathbf{D} \end{bmatrix} + \mathbf{B}\mathbf{G}\begin{bmatrix} \mathbf{M} & \mathbf{E} \end{bmatrix} \right) \begin{bmatrix} \mathbf{X} & \mathbf{0} \\ \mathbf{0} & \mathbf{I} \end{bmatrix} \left(\begin{bmatrix} \mathbf{A}^T \\ \mathbf{D}^T \end{bmatrix} + \begin{bmatrix} \mathbf{M}^T \\ \mathbf{E}^T \end{bmatrix} \mathbf{G}^T\mathbf{B}^T \right),$$

the result immediately follows from Theorem 2.3.11. □

Corollary 6.3.3 *Let a matrix $\mathbf{X} > \mathbf{0}$ be given. Then the following statements are equivalent:*

(i) $\mathbf{X}$ is an assignable state covariance bound, i.e. $\mathbf{X}$ satisfies the matrix inequalities in statement (ii) of Theorem 6.3.3.

(ii) $\mathbf{X}$ satisfies

$$\mathbf{B}_p^{\perp}(\mathbf{X}_p - \mathbf{A}_p\mathbf{X}_p\mathbf{A}_p^T - \mathbf{D}_p\mathbf{D}_p^T)\mathbf{B}_p^{\perp T} > \mathbf{0}, \quad \mathbf{C}_p\mathbf{X}_p\mathbf{C}_p^T + \mathbf{D}_y\mathbf{D}_y^T < \mho$$

$$\begin{bmatrix} \mathbf{M}_p^T \\ \mathbf{D}_z^T \end{bmatrix}^{\perp} \begin{bmatrix} \mathbf{Y}_p - \mathbf{A}_p^T\mathbf{Y}_p\mathbf{A}_p & -\mathbf{A}_p^T\mathbf{Y}_p\mathbf{D}_p \\ -\mathbf{D}_p^T\mathbf{Y}_p\mathbf{A}_p & \mathbf{I} - \mathbf{D}_p^T\mathbf{Y}_p\mathbf{D}_p \end{bmatrix} \begin{bmatrix} \mathbf{M}_p^T \\ \mathbf{D}_z^T \end{bmatrix}^{\perp T} > \mathbf{0},$$

where $\mathbf{X}_p > \mathbf{0}$ and $\mathbf{Y}_p > \mathbf{0}$ are defined by

$$\mathbf{X} = \begin{bmatrix} \mathbf{X}_p & \mathbf{X}_{pc} \\ \mathbf{X}_{pc}^T & \mathbf{X}_c \end{bmatrix}, \quad \mathbf{Y}_p \triangleq (\mathbf{X}_p - \mathbf{X}_{pc}\mathbf{X}_c^{-1}\mathbf{X}_{pc}^T)^{-1}.$$

Proof. The result follows by specializing Theorem 6.3.3 where we utilize the structure of augmented matrices defined in (4.32). See the proof of Corollary 6.2.3. □

6.3.4 Full-Order Dynamic Output Feedback

We now remove the assumption $\mathbf{B}_y = \mathbf{0}$. In the general output feedback case, the best performance can be achieved by a full-order controller. That is, increasing the order of the controller does not help to improve performance if the controller order is larger than or equal to the plant order. Hence we may restrict our attention to full-order controllers to achieve a given performance bound. The following result does not require any assumptions on the plant and provides a state space formula for a full-order covariance upper bound controller.

Theorem 6.3.4 *Let a symmetric matrix $\mathcal{U}$ be given and consider the system (4.32). Then the following statements are equivalent:*

(i) There exists a dynamic output feedback controller which stabilizes the system and yields

$$\lim_{k\to\infty} \mathcal{E}\left[\mathbf{y}(k)\mathbf{y}^T(k)\right] < \mathcal{U}.$$

(ii) There exist matrices $\mathbf{X}_p$, $\mathbf{Y}_p$, $\mathbf{K}$ and $\mathbf{L}$ such that

$$\begin{bmatrix} \mathcal{U} & \mathbf{C}_p\mathbf{X}_p + \mathbf{B}_y\mathbf{K} & \mathbf{C}_p + \mathbf{B}_y\mathbf{L}\mathbf{M}_p & \mathbf{D}_y + \mathbf{B}_y\mathbf{L}\mathbf{D}_z \\ * & \mathbf{X}_p & \mathbf{I} & \mathbf{0} \\ * & * & \mathbf{Y}_p & \mathbf{0} \\ * & * & * & \mathbf{I} \end{bmatrix} > \mathbf{0} \qquad (6.41)$$

$$\begin{bmatrix} \mathbf{X}_p & \mathbf{A}_p\mathbf{X}_p + \mathbf{B}_p\mathbf{K} & \mathbf{A}_p + \mathbf{B}_p\mathbf{L}\mathbf{M}_p & \mathbf{D}_p + \mathbf{B}_p\mathbf{L}\mathbf{D}_z \\ * & \mathbf{X}_p & \mathbf{I} & \mathbf{0} \\ * & * & \mathbf{Y}_p & \mathbf{0} \\ * & * & * & \mathbf{I} \end{bmatrix} > \mathbf{0} \qquad (6.42)$$

$$\begin{bmatrix} \mathbf{M}_p^T \\ \mathbf{D}_z^T \end{bmatrix}^{\perp} \left(\begin{bmatrix} \mathbf{Y}_p & \mathbf{0} \\ \mathbf{0} & \mathbf{I} \end{bmatrix} - \begin{bmatrix} \mathbf{A}_p^T \\ \mathbf{D}_p^T \end{bmatrix} \mathbf{Y}_p \begin{bmatrix} \mathbf{A}_p & \mathbf{D}_p \end{bmatrix} \right) \begin{bmatrix} \mathbf{M}_p^T \\ \mathbf{D}_z^T \end{bmatrix}^{\perp T} > \mathbf{0} \qquad (6.43)$$

where $*$ denotes symmetric entries. In this case, one such controller is given by

$$\begin{aligned} \mathbf{A}_c &= (\mathbf{I} + \mathbf{\Omega}\mathbf{Z}^{-1})^{-1}(\hat{\mathbf{A}}_p + \mathbf{B}_p\mathbf{C}_c) - \mathbf{B}_c\mathbf{M}_p \\ \mathbf{B}_c &= (\mathbf{I} + \mathbf{\Omega}\mathbf{Z}^{-1})^{-1}(\hat{\mathbf{A}}_p\mathbf{Y}_p^{-1}\mathbf{M}_p^T + \hat{\mathbf{D}}_p\mathbf{D}_z^T)(\mathbf{M}_p\mathbf{Y}_p^{-1}\mathbf{M}_p^T + \mathbf{D}_z\mathbf{D}_z^T)^{-1} \\ \mathbf{C}_c &= (\mathbf{K} - \mathbf{D}_c\mathbf{M}_p\mathbf{X}_p)\mathbf{Z}^{-1} \\ \mathbf{D}_c &= \mathbf{L} \end{aligned}$$

where

$$\begin{aligned} \begin{bmatrix} \hat{\mathbf{A}}_p & \hat{\mathbf{D}}_p \end{bmatrix} &\triangleq \begin{bmatrix} \mathbf{A}_p & \mathbf{D}_p \end{bmatrix} + \mathbf{B}_p\mathbf{D}_c\begin{bmatrix} \mathbf{M}_p & \mathbf{D}_z \end{bmatrix} \\ \mathbf{Z} &\triangleq \mathbf{X}_p - \mathbf{Y}_p^{-1} \\ \mathbf{\Omega} &\triangleq \mathbf{Y}_p^{-1} - \hat{\mathbf{A}}_p\mathbf{Y}_p^{-1}\hat{\mathbf{A}}_p^T + \hat{\mathbf{D}}_p\hat{\mathbf{D}}_p^T \\ &\quad + (\hat{\mathbf{A}}_p\mathbf{Y}_p^{-1}\mathbf{M}_p^T + \hat{\mathbf{D}}_p\mathbf{D}_z^T)(\mathbf{M}_p\mathbf{Y}_p^{-1}\mathbf{M}_p^T + \mathbf{D}_z\mathbf{D}_z^T)^{-1} \\ &\quad \times(\hat{\mathbf{A}}_p\mathbf{Y}_p^{-1}\mathbf{M}_p^T + \hat{\mathbf{D}}_p\mathbf{D}_z^T)^T. \end{aligned}$$

Proof. Recall that the closed-loop system satisfies the specifications in statement (i) if and only if there exists $\mathbf{X} > \mathbf{0}$ such that

$$\mathcal{U} > \mathbf{C}_{c\ell}\mathbf{X}\mathbf{C}_{c\ell}^T + \mathbf{D}_{c\ell}\mathbf{D}_{c\ell}^T \qquad (6.44)$$

$$\mathbf{X} > \mathbf{A}_{c\ell}\mathbf{X}\mathbf{A}_{c\ell}^T + \mathbf{B}_{c\ell}\mathbf{B}_{c\ell}^T. \qquad (6.45)$$

Note that $\begin{bmatrix} \mathbf{A}_{c\ell} & \mathbf{B}_{c\ell} \end{bmatrix}$ can be written

$$\begin{bmatrix} \mathbf{A}_{c\ell} & \mathbf{B}_{c\ell} \end{bmatrix} = \underbrace{\begin{bmatrix} \hat{\mathbf{A}}_p & \mathbf{B}_p\mathbf{C}_c & \hat{\mathbf{D}}_p \\ \mathbf{0} & \mathbf{0} & \mathbf{0} \end{bmatrix}}_{\mathbf{A}_1} + \underbrace{\begin{bmatrix} \mathbf{0} \\ \mathbf{I} \end{bmatrix}}_{\mathbf{B}_1} \underbrace{\begin{bmatrix} \mathbf{A}_c & \mathbf{B}_c \end{bmatrix}}_{\mathbf{G}_1} \underbrace{\begin{bmatrix} \mathbf{0} & \mathbf{I} & \mathbf{0} \\ \mathbf{M}_p & \mathbf{0} & \mathbf{D}_z \end{bmatrix}}_{\mathbf{C}_1}.$$

Applying Theorem 2.3.11 to (6.45), there exists $\mathbf{G}_1$ such that

$$\mathbf{X} > (\mathbf{A}_1 + \mathbf{B}_1\mathbf{G}_1\mathbf{C}_1)\boldsymbol{\Psi}(\mathbf{A}_1 + \mathbf{B}_1\mathbf{G}_1\mathbf{C}_1)^T, \qquad \boldsymbol{\Psi} \triangleq \begin{bmatrix} \mathbf{X} & \mathbf{0} \\ \mathbf{0} & \mathbf{I} \end{bmatrix}$$

if and only if

$$\mathbf{B}_1^{\perp}(\mathbf{X} - \mathbf{A}_1\boldsymbol{\Psi}\mathbf{A}_1^T)\mathbf{B}_1^{\perp T} > \mathbf{0} \tag{6.46}$$

$$\mathbf{C}_1^{T\perp}(\boldsymbol{\Psi}^{-1} - \mathbf{A}_1^T\mathbf{X}^{-1}\mathbf{A}_1)\mathbf{C}_1^{T\perp T} > \mathbf{0} \tag{6.47}$$

hold. In this case, one such $\mathbf{G}_1$ is given by

$$\mathbf{G}_1 = -(\mathbf{B}_1^T\boldsymbol{\Phi}\mathbf{B}_1)^{-1}\mathbf{B}_1^T\boldsymbol{\Phi}\mathbf{A}_1\boldsymbol{\Psi}\mathbf{C}_1^T(\mathbf{C}_1\boldsymbol{\Psi}\mathbf{C}_1^T)^{-1} \tag{6.48}$$

where

$$\boldsymbol{\Phi} \triangleq (\mathbf{X} - \mathbf{A}_1\boldsymbol{\Psi}\mathbf{A}_1^T + \mathbf{A}_1\boldsymbol{\Psi}\mathbf{C}_1^T(\mathbf{C}_1\boldsymbol{\Psi}\mathbf{C}_1^T)^{-1}\mathbf{C}_1\boldsymbol{\Psi}\mathbf{A}_1^T)^{-1}.$$

Now, without loss of generality, we consider $\mathbf{X}$ of the following structure:

$$\mathbf{X} = \begin{bmatrix} \mathbf{X}_p & \mathbf{Z} \\ \mathbf{Z} & \mathbf{Z} \end{bmatrix}.$$

Defining

$$\mathbf{Y}_p \triangleq (\mathbf{X}_p - \mathbf{Z})^{-1},$$

it can be verified that (6.47) is equivalent to (6.43). Note that (6.46) is equivalent to

$$\mathbf{X}_p > \begin{bmatrix} \hat{\mathbf{A}}_p & \mathbf{B}_p\mathbf{C}_c \end{bmatrix} \begin{bmatrix} \mathbf{X}_p & \mathbf{Z} \\ \mathbf{Z} & \mathbf{Z} \end{bmatrix} \begin{bmatrix} \hat{\mathbf{A}}_p^T \\ \mathbf{C}_c^T\mathbf{B}_p^T \end{bmatrix} + \hat{\mathbf{D}}_p\hat{\mathbf{D}}_p^T \tag{6.49}$$

while (6.44) is equivalent to

$$\boldsymbol{\Omega} > \begin{bmatrix} \hat{\mathbf{C}}_p & \mathbf{B}_y\mathbf{C}_c \end{bmatrix} \begin{bmatrix} \mathbf{X}_p & \mathbf{Z} \\ \mathbf{Z} & \mathbf{Z} \end{bmatrix} \begin{bmatrix} \hat{\mathbf{C}}_p^T \\ \mathbf{C}_c^T\mathbf{B}_y^T \end{bmatrix} + \hat{\mathbf{D}}_y\hat{\mathbf{D}}_y^T \tag{6.50}$$

where

$$\begin{bmatrix} \hat{\mathbf{C}}_p & \hat{\mathbf{D}}_y \end{bmatrix} \triangleq \begin{bmatrix} \mathbf{C}_p & \mathbf{D}_y \end{bmatrix} + \mathbf{B}_y\mathbf{D}_c\begin{bmatrix} \mathbf{M}_p & \mathbf{D}_z \end{bmatrix}.$$

Clearly, (6.49) and (6.50) have exactly the same form. Hence, we only show the equivalence between (6.50) and (6.41); the equivalence between (6.49) and (6.42) follows in a similar manner.

Write (6.50) and $\mathbf{X} > \mathbf{0}$ as

$$\begin{bmatrix} \mathbf{I} & \mathbf{0} & \mathbf{0} \\ \mathbf{0} & \mathbf{T} & \mathbf{0} \\ \mathbf{0} & \mathbf{0} & \mathbf{I} \end{bmatrix} \begin{bmatrix} \boldsymbol{\mho} & \mathbf{C}_{c\ell}\mathbf{X} & \mathbf{D}_{c\ell} \\ \mathbf{X}\mathbf{C}_{c\ell}^T & \mathbf{X} & \mathbf{0} \\ \mathbf{D}_{c\ell} & \mathbf{0} & \mathbf{I} \end{bmatrix} \begin{bmatrix} \mathbf{I} & \mathbf{0} & \mathbf{0} \\ \mathbf{0} & \mathbf{T}^T & \mathbf{0} \\ \mathbf{0} & \mathbf{0} & \mathbf{I} \end{bmatrix} > \mathbf{0}$$

where $\mathbf{T}$ is defined in (6.20). Using the identity in (6.21), this inequality becomes

$$\begin{bmatrix} \mho & \hat{\mathbf{C}}_p\mathbf{X}_p + \mathbf{B}_y\mathbf{C}_c\mathbf{Z} & \hat{\mathbf{C}}_p\mathbf{Y}_p^{-1} & \hat{\mathbf{D}}_y \\ * & \mathbf{X}_p & \mathbf{Y}_p^{-1} & \mathbf{0} \\ * & * & \mathbf{Y}_p^{-1} & \mathbf{0} \\ * & * & * & \mathbf{I} \end{bmatrix} > \mathbf{0}$$

where $\mathbf{C}_p$ and $\mathbf{D}_y$ have been replaced by $\hat{\mathbf{C}}_p$ and $\hat{\mathbf{D}}_y$, respectively, due to the presence of nonzero $\mathbf{D}_c$. Defining

$$\mathbf{K} \triangleq \mathbf{C}_c\mathbf{Z} + \mathbf{D}_c\mathbf{M}_p\mathbf{X}_p, \qquad \mathbf{L} \triangleq \mathbf{D}_c \tag{6.51}$$

and using the congruent transformation involving $\mathbf{Y}_p$, we have (6.41).

Finally, the controller formulae for $\mathbf{A}_c$ and $\mathbf{B}_c$ can be derived by computing $\mathbf{G}_1$ in (6.48), and those for $\mathbf{C}_c$ and $\mathbf{D}_c$ from (6.51). □

We shall obtain a solution for the LQG problem based on Theorem 6.3.4. As in the continuous-time case, the LQG problem is to design a stabilizing controller that minimizes $tr(\mho)$ subject to the closed-loop constraints (6.44) and (6.45). In the following, we assume

$$\mathbf{D}_z\begin{bmatrix} \mathbf{D}_p^T & \mathbf{D}_z^T \end{bmatrix} = \begin{bmatrix} \mathbf{0} & \mathbf{V} \end{bmatrix}, \quad \mathbf{B}_y^T\begin{bmatrix} \mathbf{C}_p & \mathbf{B}_y \end{bmatrix} = \begin{bmatrix} \mathbf{0} & \mathbf{R} \end{bmatrix} \tag{6.52}$$

with $\mathbf{V} > \mathbf{0}$ and $\mathbf{R} > \mathbf{0}$ for simplicity. Note that $\mathbf{D}_y$ is not necessarily zero.

The following is the discrete-time LQG solution.

Theorem 6.3.5 *Consider the system given in (4.32). Suppose the assumptions in (6.52) hold. Then the controller that stabilizes (4.32) and solves*

$$\gamma_{\text{opt}} \triangleq \min\ tr\left(\lim_{k\to\infty} \mathcal{E}\left[\mathbf{y}(k)\mathbf{y}^T(k)\right]\right)$$

is given by

$$\begin{aligned}
\mathbf{A}_c &= \mathbf{A}_p + \mathbf{B}_p\mathbf{C}_c - \mathbf{B}_c\mathbf{M}_p + \mathbf{B}_p\mathbf{D}_c\mathbf{M}_p \\
\mathbf{B}_c &= \mathbf{A}_p\mathbf{Q}\mathbf{M}_p^T(\mathbf{M}_p\mathbf{Q}\mathbf{M}_p^T + \mathbf{V})^{-1} + \mathbf{B}_p\mathbf{D}_c \\
\mathbf{C}_c &= -(\mathbf{B}_p^T\mathbf{P}\mathbf{B}_p + \mathbf{R})^{-1}\mathbf{B}_p^T\mathbf{P}\mathbf{A}_p - \mathbf{D}_c\mathbf{M}_p \\
\mathbf{D}_c &= -(\mathbf{B}_p^T\mathbf{P}\mathbf{B}_p + \mathbf{R})^{-1}\begin{bmatrix} \mathbf{B}_p^T\mathbf{P} & \mathbf{B}_y^T \end{bmatrix}\begin{bmatrix} \mathbf{A}_p & \mathbf{D}_p \\ \mathbf{C}_p & \mathbf{D}_y \end{bmatrix}\begin{bmatrix} \mathbf{Q}\mathbf{M}_p^T \\ \mathbf{D}_z^T \end{bmatrix}(\mathbf{M}_p\mathbf{Q}\mathbf{M}_p^T + \mathbf{V})^{-1}
\end{aligned}$$

where $\mathbf{P}$ *and* $\mathbf{Q}$ *are the stabilizing solutions to*

$$\mathbf{P} = \mathbf{A}_p^T\mathbf{P}\mathbf{A}_p - \mathbf{A}_p^T\mathbf{P}\mathbf{B}_p(\mathbf{B}_p^T\mathbf{P}\mathbf{B}_p + \mathbf{R})^{-1}\mathbf{B}_p^T\mathbf{P}\mathbf{A}_p + \mathbf{C}_p^T\mathbf{C}_p \tag{6.53}$$

$$\mathbf{Q} = \mathbf{A}_p\mathbf{Q}\mathbf{A}_p^T - \mathbf{A}_p\mathbf{Q}\mathbf{M}_p^T(\mathbf{M}_p\mathbf{Q}\mathbf{M}_p^T + \mathbf{V})^{-1}\mathbf{M}_p\mathbf{Q}\mathbf{A}_p^T + \mathbf{D}_p\mathbf{D}_p^T. \tag{6.54}$$

Furthermore, the minimum value of the cost function is given by

$$\gamma_{\text{opt}} = \left\| \begin{bmatrix} \mathbf{P}^{1/2} & \mathbf{0} \\ \mathbf{0} & \mathbf{I} \end{bmatrix}\begin{bmatrix} \hat{\mathbf{A}}_p & \hat{\mathbf{D}}_p \\ \hat{\mathbf{C}}_p & \hat{\mathbf{D}}_y \end{bmatrix}\begin{bmatrix} \mathbf{Q}^{1/2} & \mathbf{0} \\ \mathbf{0} & \mathbf{I} \end{bmatrix} \right\|_F^2 - tr(\mathbf{P}\mathbf{Q})$$

where

$$\begin{bmatrix} \hat{\mathbf{A}}_p & \hat{\mathbf{D}}_p \\ \hat{\mathbf{C}}_p & \hat{\mathbf{D}}_y \end{bmatrix} \triangleq \begin{bmatrix} \mathbf{A}_p & \mathbf{D}_p \\ \mathbf{C}_p & \mathbf{D}_y \end{bmatrix} + \begin{bmatrix} \mathbf{B}_p \\ \mathbf{B}_y \end{bmatrix}\mathbf{D}_c\begin{bmatrix} \mathbf{M}_p & \mathbf{D}_z \end{bmatrix}.$$

Proof. Since $\mathbf{D}_z$ has full row rank due to assumption (6.52), a choice for the left annihilator in (6.43) is given by

$$\begin{bmatrix} \mathbf{M}_p^T \\ \mathbf{D}_z^T \end{bmatrix}^{\perp} = \begin{bmatrix} \mathbf{I} & -\mathbf{M}_p^T\mathbf{D}_z^{T+} \\ \mathbf{0} & \mathbf{D}_z^{T\perp} \end{bmatrix}.$$

With this choice, after some manipulation, (6.43) becomes

$$\mathbf{Q} > \mathbf{A}_p\mathbf{Q}\mathbf{A}_p^T - \mathbf{A}_p\mathbf{Q}\mathbf{M}_p^T(\mathbf{M}_p\mathbf{Q}\mathbf{M}_p^T + \mathbf{V})^{-1}\mathbf{M}_p\mathbf{Q}\mathbf{A}_p^T + \mathbf{D}_p\mathbf{D}_p^T$$

where $\mathbf{Q} \triangleq \mathbf{Y}_p^{-1}$. Since the other constraints (6.41) and (6.42) are more likely to be satisfied with larger $\mathbf{Y}_p$, we should search for the smallest $\mathbf{Q} \geq \mathbf{0}$ satisfying the above Riccati inequality. It turns out [107] that such $\mathbf{Q}$ is given by the stabilizing solution to the Riccati equation (6.54). Thus we choose $\mathbf{Y}_p^{-1} = \mathbf{Q}$ in (6.41) and (6.42) where $\mathbf{Q}$ is the stabilizing solution to (6.54).

By the Schur complement formula, it is readily verified that (6.41) and (6.42) are respectively equivalent to

$$\boldsymbol{\Omega} > (\hat{\mathbf{C}}_p + \mathbf{B}_y\mathbf{C}_c)\mathbf{Z}(\hat{\mathbf{C}}_p + \mathbf{B}_y\mathbf{C}_c)^T + (\hat{\mathbf{C}}_p\mathbf{Q}\hat{\mathbf{C}}_p^T + \hat{\mathbf{D}}_y\hat{\mathbf{D}}_y^T), \tag{6.55}$$

$$\mathbf{Z} > (\hat{\mathbf{A}}_p + \mathbf{B}_p\mathbf{C}_c)\mathbf{Z}(\hat{\mathbf{A}}_p + \mathbf{B}_p\mathbf{C}_c)^T + (-\mathbf{Q} + \hat{\mathbf{A}}_p\mathbf{Q}\hat{\mathbf{A}}_p^T + \hat{\mathbf{D}}_p\hat{\mathbf{D}}_p^T). \tag{6.56}$$

Using the dual characterization of the $\mathcal{H}_2$ norm, we see that there exist $\mathbf{Z} > \mathbf{0}$ and $\boldsymbol{\Omega} > \mathbf{0}$ satisfying (6.55), (6.56) and $tr(\boldsymbol{\Omega}) < \gamma$ if and only if there exists $\mathbf{P} > \mathbf{0}$ such that

$$\mathbf{P} > (\hat{\mathbf{A}}_p + \mathbf{B}_p\mathbf{C}_c)^T\mathbf{P}(\hat{\mathbf{A}}_p + \mathbf{B}_p\mathbf{C}_c) + (\hat{\mathbf{C}}_p + \mathbf{B}_y\mathbf{C}_c)^T(\hat{\mathbf{C}}_p + \mathbf{B}_y\mathbf{C}_c) \tag{6.57}$$

$$\gamma > tr\,[\,\mathbf{P}(-\mathbf{Q} + \hat{\mathbf{A}}_p\mathbf{Q}\hat{\mathbf{A}}_p^T + \hat{\mathbf{D}}_p\hat{\mathbf{D}}_p^T)\,] + tr(\hat{\mathbf{C}}_p\mathbf{Q}\hat{\mathbf{C}}_p^T + \hat{\mathbf{D}}_y\hat{\mathbf{D}}_y^T). \tag{6.58}$$

Noting that

$$\hat{\mathbf{A}}_p + \mathbf{B}_p\mathbf{C}_c = \mathbf{A}_p + \mathbf{B}_p\hat{\mathbf{C}}_c, \quad \hat{\mathbf{C}}_p + \mathbf{B}_y\mathbf{C}_c = \mathbf{C}_p + \mathbf{B}_y\hat{\mathbf{C}}_c, \quad \hat{\mathbf{C}}_c \triangleq \mathbf{C}_c + \mathbf{D}_c\mathbf{M}_p,$$

inequality (6.57) can be written

$$[\hat{\mathbf{C}}_c + (\mathbf{B}_p^T\mathbf{P}\mathbf{B}_p + \mathbf{R})^{-1}\mathbf{B}_p^T\mathbf{P}\mathbf{A}_p]^T(\mathbf{B}_p^T\mathbf{P}\mathbf{B}_p + \mathbf{R})[\hat{\mathbf{C}}_c + (\mathbf{B}_p^T\mathbf{P}\mathbf{B}_p + \mathbf{R})^{-1}\mathbf{B}_p^T\mathbf{P}\mathbf{A}_p]$$
$$-\mathbf{P} + \mathbf{A}_p^T\mathbf{P}\mathbf{A}_p - \mathbf{A}_p^T\mathbf{P}\mathbf{B}_p(\mathbf{B}_p^T\mathbf{P}\mathbf{B}_p + \mathbf{R})^{-1}\mathbf{B}_p^T\mathbf{P}\mathbf{A}_p + \mathbf{C}_p^T\mathbf{C}_p < \mathbf{0}.$$

Hence, $\mathbf{C}_c$ satisfying (6.57) exists if and only if

$$\mathbf{P} > \mathbf{A}_p^T\mathbf{P}\mathbf{A}_p - \mathbf{A}_p^T\mathbf{P}\mathbf{B}_p(\mathbf{B}_p^T\mathbf{P}\mathbf{B}_p + \mathbf{R})^{-1}\mathbf{B}_p^T\mathbf{P}\mathbf{A}_p + \mathbf{C}_p^T\mathbf{C}_p \tag{6.59}$$

holds, in which case, one such $\mathbf{C}_c$ is given by

$$\mathbf{C}_c = \hat{\mathbf{C}}_c - \mathbf{D}_c\mathbf{M}_p, \quad \hat{\mathbf{C}}_c = -(\mathbf{B}_p^T\mathbf{P}\mathbf{B}_p + \mathbf{R})^{-1}\mathbf{B}_p^T\mathbf{P}\mathbf{A}_p.$$

Now, since the matrix $(\cdot)$ on the right of $\mathbf{P}$ in (6.58) is positive (semi)definite due to (6.54), smaller $\mathbf{P}$ yields better performance (smaller γ). The smallest $\mathbf{P}$ such that (6.59) is given by the stabilizing solution to the Riccati equation (6.53) [107].

Finally, we find an optimal $\mathbf{D}_c$. Note that (6.58) can be written

$$\gamma > \left\| \begin{bmatrix} \mathbf{P}^{1/2} & \mathbf{0} \\ \mathbf{0} & \mathbf{I} \end{bmatrix} \begin{bmatrix} \hat{\mathbf{A}}_p & \hat{\mathbf{D}}_p \\ \hat{\mathbf{C}}_p & \hat{\mathbf{D}}_y \end{bmatrix} \begin{bmatrix} \mathbf{Q}^{1/2} & \mathbf{0} \\ \mathbf{0} & \mathbf{I} \end{bmatrix} \right\|_F^2 - tr(\mathbf{PQ})$$

where $\|\cdot\|_F$ is the Frobenius norm. Also note that the matrix inside $\|\cdot\|_F$ can be written

$$\begin{bmatrix} \mathbf{P}^{1/2}\mathbf{A}_p\mathbf{Q}^{1/2} & \mathbf{P}^{1/2}\mathbf{D}_p \\ \mathbf{C}_p\mathbf{Q}^{1/2} & \mathbf{D}_y \end{bmatrix} + \begin{bmatrix} \mathbf{P}^{1/2}\mathbf{B}_p \\ \mathbf{B}_y \end{bmatrix} \mathbf{D}_c \begin{bmatrix} \mathbf{M}_p\mathbf{Q}^{1/2} & \mathbf{D}_z \end{bmatrix}.$$

Hence, applying Theorem 2.3.2, matrix $\mathbf{D}_c$ that minimizes the above Frobenius norm is given by

$$\mathbf{D}_c = -\begin{bmatrix} \mathbf{P}^{1/2}\mathbf{B}_p \\ \mathbf{B}_y \end{bmatrix}^+ \begin{bmatrix} \mathbf{P}^{1/2}\mathbf{A}_p\mathbf{Q}^{1/2} & \mathbf{P}^{1/2}\mathbf{D}_p \\ \mathbf{C}_p\mathbf{Q}^{1/2} & \mathbf{D}_y \end{bmatrix} \begin{bmatrix} \mathbf{M}_p\mathbf{Q}^{1/2} & \mathbf{D}_z \end{bmatrix}^+.$$

After some manipulation this formula leads to the formula for $\mathbf{D}_c$ in Theorem 6.3.5. □

CHAPTER 6 SUMMARY

This chapter repeats the control design question of Chapter 5 except that the equality constraints of the covariance equations in Chapter 5 are replaced by inequality constraints in Chapter 6. This provides controllers that guarantee upper bounds on the actual covariance assigned by the controller. As shown later, the results obtained from this upper bound approach are computable via convex programming when state feedback is used, or when dynamic controllers of order equal to the plant are used. In all other cases the computation is nonconvex, and no algorithm is available which guarantees a solution when one exists. However, Chapters 10 and 11 will provide useful algorithms to obtain fixed-order controllers, even though the problems are not convex. All the results in this chapter are essentially from [55]. See [109, 162] for related results.

CHAPTER SEVEN

$\mathcal{H}_\infty$ Controllers

In classical control theory the peak of the frequency response of a closed-loop system can be determined from the intersection of the Nyquist diagram and the "M circles" (representing constant closed-loop magnitudes). The pioneering work of [159] allowed these ideas to be extended to MIMO systems, where analytical tests replaced the graphical work of Nyquist. The peak of the frequency response had other interpretations as well, generating fundamental results in robust control theory. This chapter characterizes all controllers which can yield specified upper bounds on the peak frequency response.

7.1 $\mathcal{H}_\infty$ CONTROL PROBLEM

Consider the linear time-invariant systems given in Sections 4.2 and 4.5. The closed-loop transfer matrix from disturbance $\mathbf{w}$ to regulated output $\mathbf{y}$ is given by $\mathbf{T}(\alpha) = \mathbf{C}_{c\ell}(\alpha\mathbf{I} - \mathbf{A}_{c\ell})^{-1}\mathbf{B}_{c\ell} + \mathbf{D}_{c\ell}$ where $\alpha = s$ for the continuous-time case and $\alpha = z$ for the discrete-time case. Recall that, if the transfer matrix $\mathbf{T}$ is stable, then its $\mathcal{H}_\infty$ norm is defined by

$$\|\mathbf{T}\|_{\mathcal{H}_\infty} \triangleq \sup_{\omega\in\mathcal{R}} \|\mathbf{T}(j\omega)\| \quad \text{(continuous-time case)}$$

$$\|\mathbf{T}\|_{\mathcal{H}_\infty} \triangleq \sup_{\theta\in[0,2\pi]} \|\mathbf{T}(e^{j\theta})\| \quad \text{(discrete-time case).}$$

Analysis results in Section 4.6 show that the $\mathcal{H}_\infty$ norm can be interpreted in the following two ways. One is a measure for the disturbance attenuation level. In particular, the energy-to-energy gain from $\mathbf{w}$ to $\mathbf{y}$ is exactly given by $\|\mathbf{T}\|_{\mathcal{H}_\infty}$. The other is a measure for robustness. Specifically, the closed-loop system remains stable for all perturbations $\mathbf{w} = \mathbf{\Delta y}$ such that $\|\mathbf{\Delta}\| \le \gamma^{-1}$ if $\|\mathbf{T}\|_{\mathcal{H}_\infty} < \gamma$. These properties associated with the $\mathcal{H}_\infty$ norm motivate us to consider the following $\mathcal{H}_\infty$ control problem:

> Let a performance bound $\gamma > 0$ be given. Determine if there exists a controller which stabilizes the system and yields the closed-loop transfer matrix such that $\|\mathbf{T}\|_\infty < \gamma$. Find all such controllers when one exists.

The following lemmas are useful for solving the above control problem. Both of these are just restatements of the analysis results in Section 4.6 and hence proofs are omitted.

Lemma 7.1.1 *[150, 160] Let a scalar $\gamma > 0$ be given and consider the linear time-invariant continuous-time system (4.14). Then the following statements are equivalent:*

(i) The controller $\mathbf{G}$ stabilizes the system and yields the closed-loop transfer matrix $\mathbf{T}(s)$ such that $\|\mathbf{T}\|_{\mathcal{H}_\infty} < \gamma$.

(ii) $\mathbf{R} \triangleq \gamma^2\mathbf{I} - \mathbf{D}_{c\ell}^T\mathbf{D}_{c\ell} > \mathbf{0}$ and there exists a matrix $\mathbf{Y} > \mathbf{0}$ such that

$$\mathbf{Y}\mathbf{A}_{c\ell} + \mathbf{A}_{c\ell}^T\mathbf{Y} + (\mathbf{Y}\mathbf{B}_{c\ell} + \mathbf{C}_{c\ell}^T\mathbf{D}_{c\ell})\mathbf{R}^{-1}(\mathbf{Y}\mathbf{B}_{c\ell} + \mathbf{C}_{c\ell}^T\mathbf{D}_{c\ell})^T + \mathbf{C}_{c\ell}^T\mathbf{C}_{c\ell} < \mathbf{0}.$$

(iii) There exists a matrix $\mathbf{Y} > \mathbf{0}$ such that

$$\begin{bmatrix} \mathbf{Y}\mathbf{A}_{c\ell} + \mathbf{A}_{c\ell}^T\mathbf{Y} & \mathbf{Y}\mathbf{B}_{c\ell} & \mathbf{C}_{c\ell}^T \\ \mathbf{B}_{c\ell}^T\mathbf{Y} & -\gamma^2\mathbf{I} & \mathbf{D}_{c\ell}^T \\ \mathbf{C}_{c\ell} & \mathbf{D}_{c\ell} & -\mathbf{I} \end{bmatrix} < \mathbf{0}.$$

Lemma 7.1.2 *[33, 96] Let a scalar $\gamma > 0$ be given and consider the linear time-invariant discrete-time system (4.32). Then the following statements are equivalent:*

(i) The controller $\mathbf{G}$ stabilizes the system and yields the closed-loop transfer matrix $\mathbf{T}(z)$ such that $\|\mathbf{T}\|_{\mathcal{H}_\infty} < \gamma$.

(ii) There exists a matrix $\mathbf{X} > \mathbf{0}$ such that

$$\mathbf{R} \triangleq \gamma^2\mathbf{I} - \mathbf{C}_{c\ell}\mathbf{X}\mathbf{C}_{c\ell}^T - \mathbf{D}_{c\ell}\mathbf{D}_{c\ell}^T > \mathbf{0}$$

$$\mathbf{X} > \mathbf{A}_{c\ell}\mathbf{X}\mathbf{A}_{c\ell}^T + (\mathbf{A}_{c\ell}\mathbf{X}\mathbf{C}_{c\ell}^T + \mathbf{B}_{c\ell}\mathbf{D}_{c\ell}^T)\mathbf{R}^{-1}(\mathbf{A}_{c\ell}\mathbf{X}\mathbf{C}_{c\ell}^T + \mathbf{B}_{c\ell}\mathbf{D}_{c\ell}^T)^T + \mathbf{B}_{c\ell}\mathbf{B}_{c\ell}^T.$$

(iii) There exists a matrix $\mathbf{X} > \mathbf{0}$ such that

$$\begin{bmatrix} \mathbf{X} & \mathbf{0} \\ \mathbf{0} & \gamma^2\mathbf{I} \end{bmatrix} > \begin{bmatrix} \mathbf{A}_{c\ell} & \mathbf{B}_{c\ell} \\ \mathbf{C}_{c\ell} & \mathbf{D}_{c\ell} \end{bmatrix}\begin{bmatrix} \mathbf{X} & \mathbf{0} \\ \mathbf{0} & \mathbf{I} \end{bmatrix}\begin{bmatrix} \mathbf{A}_{c\ell} & \mathbf{B}_{c\ell} \\ \mathbf{C}_{c\ell} & \mathbf{D}_{c\ell} \end{bmatrix}^T.$$

In view of these lemmas, the $\mathcal{H}_\infty$ control problem is equivalent to an algebraic problem of solving a certain matrix inequality for the controller parameter $\mathbf{G}$ (recall that closed-loop matrices $\mathbf{A}_{c\ell}$, $\mathbf{B}_{c\ell}$, etc. are functions of $\mathbf{G}$). In this case, solvability conditions are given in terms of some qualifications for the Lyapunov matrix $\mathbf{X}$ or $\mathbf{Y}$, which in turn define existence conditions for an $\mathcal{H}_\infty$ controller. All $\mathcal{H}_\infty$ controllers will be given explicitly as the general solution to the matrix inequality.

Throughout this chapter, we shall assume that there is no redundant actuator ($\mathbf{B}_p^T\mathbf{B}_p > \mathbf{0}$) and no redundant sensor ($\mathbf{M}_p\mathbf{M}_p^T > \mathbf{0}$), to facilitate the presentation. These assumptions reflect reasonable practical situation, and can easily be removed at the expense of a little more complicated controller formulae.

7.2 CONTINUOUS-TIME CASE

7.2.1 State Feedback

In this section, we consider the case where all the state variables can be measured without noises;

$$\mathbf{M}_p = \mathbf{I}, \quad \mathbf{D}_z = \mathbf{0}. \tag{7.1}$$

We shall introduce the following simplifying assumptions;

$$\mathbf{D}_y = \mathbf{0}, \quad \mathbf{B}_y^T\begin{bmatrix} \mathbf{C}_p & \mathbf{B}_y \end{bmatrix} = \begin{bmatrix} \mathbf{0} & \mathbf{I} \end{bmatrix}. \tag{7.2}$$

To provide another interpretation to this problem, define

$$J \triangleq \int_0^\infty \mathbf{y}^T(t)\mathbf{y}(t)dt = \int_0^\infty \left(\mathbf{x}_p^T(t)\mathbf{Q}\mathbf{x}_p(t) + \mathbf{u}^T(t)\mathbf{R}\mathbf{u}(t)\right) dt,$$

where $\mathbf{Q} \triangleq \mathbf{C}_p^T\mathbf{C}_p$ and $\mathbf{R} = \mathbf{I}$. Recall that the $\mathcal{H}_\infty$ norm constraint on the closed-loop transfer matrix $\|\mathbf{T}\|_{\mathcal{H}_\infty} < \gamma$ is equivalent to guaranteeing $J < \gamma^2$ for all disturbances $\mathbf{w}$ with its $\mathcal{L}_2$ norm bounded above by 1. Note also that the assumption (7.2) implies that the control input is fully penalized, i.e. $\mathbf{R} > \mathbf{0}$ and that there is no penalty on the disturbance $\mathbf{w}$, and there is no cross-weighting term between the state and the control input in the cost function. This cost function J is identical to that for the conventional Linear Quadratic Regulator (LQR) problem [1]. The difference is that the excitation $\mathbf{w}$ is impulsive (or equivalently, nonzero initial states with no external disturbances), which makes $J^{1/2}$ the $\mathcal{H}_2$ norm of $\mathbf{T}(s)$.

In fact, the assumption (7.2) is only as restrictive as the assumption

$$\mathbf{D}_y = \mathbf{0}, \quad \mathbf{U} \triangleq \mathbf{B}_y^T\mathbf{B}_y > \mathbf{0}, \tag{7.3}$$

since the $\mathcal{H}_\infty$ control problem for plants with the property (7.3) can be converted to that for plants with the property (7.2). This can be verified as follows. First note that

$$\begin{aligned} \|\mathbf{y}\|^2 &= (\mathbf{C}_p\mathbf{x}_p + \mathbf{B}_y\mathbf{u})^T(\mathbf{C}_p\mathbf{x}_p + \mathbf{B}_y\mathbf{u}) \\ &= \mathbf{x}_p^T\mathbf{C}_p^T(\mathbf{I} - \mathbf{B}_y\mathbf{B}_y^+)\mathbf{C}_p\mathbf{x}_p + (\mathbf{u} + \mathbf{B}_y^+\mathbf{C}_p\mathbf{x}_p)^T\mathbf{U}(\mathbf{u} + \mathbf{B}_y^+\mathbf{C}_p\mathbf{x}_p). \end{aligned}$$

Motivated by the above equality, let us define

$$\hat{\mathbf{C}}_p \triangleq (\mathbf{I} - \mathbf{B}_y\mathbf{B}_y^+)^{1/2}\mathbf{C}_p, \quad \hat{\mathbf{B}}_y \triangleq \mathbf{B}_y\mathbf{U}^{-1/2}$$

$$\hat{\mathbf{u}} \triangleq \mathbf{U}^{1/2}(\mathbf{u} + \mathbf{B}_y^+\mathbf{C}_p\mathbf{x}_p), \quad \hat{\mathbf{y}} \triangleq \hat{\mathbf{C}}_p\mathbf{x}_p + \hat{\mathbf{B}}_y\hat{\mathbf{u}}.$$

Then we see that $\|\mathbf{y}\| = \|\hat{\mathbf{y}}\|$ and hence the original $\mathcal{H}_\infty$ control problem can be converted to that for the following new system:

$$\begin{aligned} \dot{\mathbf{x}}_p &= \hat{\mathbf{A}}_p\mathbf{x}_p + \hat{\mathbf{B}}_p\hat{\mathbf{u}} + \mathbf{D}_p\mathbf{w} \\ \hat{\mathbf{y}} &= \hat{\mathbf{C}}_p\mathbf{x}_p + \hat{\mathbf{B}}_y\hat{\mathbf{u}} \end{aligned} \tag{7.4}$$

where

$$\hat{\mathbf{A}}_p \triangleq \mathbf{A}_p - \mathbf{B}_p\mathbf{B}_y^+\mathbf{C}_p, \quad \hat{\mathbf{B}}_p \triangleq \mathbf{B}_p\mathbf{U}^{-1/2}.$$

In this case, the control input for the original plant can be determined by

$$\mathbf{u} = \mathbf{U}^{-1/2}\hat{\mathbf{u}} - \mathbf{B}_y^+\mathbf{C}_p\mathbf{x}_p. \tag{7.5}$$

Note that the new system (7.4) satisfies the assumption (7.2). Finally, it should be clear that the above transformation may not be possible for the more general output feedback case since the controller implementation (7.5) requires the knowledge of the quantity $\mathbf{B}_y^+\mathbf{C}_p\mathbf{x}_p$.

The following result provides all state feedback $\mathcal{H}_\infty$ controllers with the above assumptions on the plant.

Theorem 7.2.1 *Let a scalar $\gamma > 0$ be given. Consider the system (4.14) and suppose the assumptions (7.1) and (7.2) hold. Then the following statements are equivalent:*

(i) There exists a (static) state feedback controller $u = Gx_p$ *which stabilizes the system and yields* $\|\mathbf{T}\|_\infty < \gamma$.
(ii) There exists a matrix $\mathbf{X} > \mathbf{0}$ *such that*

$$\begin{bmatrix} \mathbf{A}_p\mathbf{X} + \mathbf{X}\mathbf{A}_p^T + \mathbf{D}_p\mathbf{D}_p^T - \gamma^2\mathbf{B}_p\mathbf{B}_p^T & \mathbf{X}\mathbf{C}_p^T \\ \mathbf{C}_p\mathbf{X} & -\gamma^2\mathbf{I} \end{bmatrix} < \mathbf{0}.$$

(iii) There exists a matrix $\mathbf{Q} > \mathbf{0}$ *such that the Riccati equation*

$$\mathbf{Y}\mathbf{A}_p + \mathbf{A}_p^T\mathbf{Y} + \mathbf{Y}(\frac{1}{\gamma^2}\mathbf{D}_p\mathbf{D}_p^T - \mathbf{B}_p\mathbf{B}_p^T)\mathbf{Y} + \mathbf{C}_p^T\mathbf{C}_p + \mathbf{Q} = \mathbf{0}$$

has a solution $\mathbf{Y} > \mathbf{0}$.

In this case, all such state feedback controllers are given by

$$\mathbf{G} = -\mathbf{B}_p^T\mathbf{Y} + \mathbf{L}\mathbf{Q}^{1/2}$$

where $\mathbf{L}$ *is an arbitrary matrix such that* $\|\mathbf{L}\| < 1$.

Proof. From Lemma 7.1.1, a given state feedback gain $\mathbf{G}$ stabilizes the system and yields $\|\mathbf{T}\|_\infty < \gamma$ if and only if there exists a matrix $\mathbf{Y} > \mathbf{0}$ such that

$$\mathbf{Y}(\mathbf{A}_p + \mathbf{B}_p\mathbf{G}) + (\mathbf{A}_p + \mathbf{B}_p\mathbf{G})^T\mathbf{Y} + \frac{1}{\gamma^2}\mathbf{Y}\mathbf{D}_p\mathbf{D}_p^T\mathbf{Y} + (\mathbf{C}_p + \mathbf{B}_y\mathbf{G})^T(\mathbf{C}_p + \mathbf{B}_y\mathbf{G}) < \mathbf{0}.$$

After expanding, completing the square with respect to $\mathbf{G}$ yields

$$(\mathbf{G} + \mathbf{B}_p^T\mathbf{Y})^T(\mathbf{G} + \mathbf{B}_p^T\mathbf{Y}) < \mathbf{Q}$$

where

$$\mathbf{Q} \triangleq -(\mathbf{Y}\mathbf{A}_p + \mathbf{A}_p^T\mathbf{Y} + \mathbf{Y}(\frac{1}{\gamma^2}\mathbf{D}_p\mathbf{D}_p^T - \mathbf{B}_p\mathbf{B}_p^T)\mathbf{Y} + \mathbf{C}_p^T\mathbf{C}_p).$$

Then statement (iii) and the controller formula follow from the Schur complement formula and Corollary 2.3.7. The LMI in statement (ii) is related to $\gamma^2\mathbf{Y}^{-1}\mathbf{Q}\mathbf{Y}^{-1} > \mathbf{0}$ with $\mathbf{X} \triangleq \gamma^2\mathbf{Y}^{-1}$ by the Schur complement formula. □

7.2.2 Static Output Feedback

This section considers the static output feedback case. Let us introduce the following assumption on the plant:

$$\mathbf{D}_z\begin{bmatrix} \mathbf{D}_p^T & \mathbf{D}_z^T \end{bmatrix} = \begin{bmatrix} \mathbf{0} & \mathbf{I} \end{bmatrix}. \tag{7.6}$$

This assumption restricts how the disturbance enters the system; $\mathbf{D}_z\mathbf{D}_p^T = \mathbf{0}$ means that the process and measurement disturbances are independent (or in the stochastic setting, they are uncorrelated), and $\mathbf{D}_z\mathbf{D}_z^T = \mathbf{I}$ basically means that all the measurements are contaminated by the disturbance.

The following results present a solution to the $\mathcal{H}_\infty$ control problem with static output feedback controllers.

Theorem 7.2.2 *Let a scalar* $\gamma > 0$ *be given and consider the system (4.14). Suppose the assumptions (7.2) and (7.6) hold. Then the following statements are equivalent:*

(i) There exists a static output feedback controller $\mathbf{u} = \mathbf{G}\mathbf{z}$ *which stabilizes the system and yields* $\|\mathbf{T}\|_{\mathcal{H}_\infty} < \gamma$.

(ii) There exists a matrix pair $(\mathbf{X}, \mathbf{Y})$ *such that*

$$\mathbf{X} > \mathbf{0}, \quad \mathbf{Y} > \mathbf{0}, \quad \mathbf{X}\mathbf{Y} = \gamma^2\mathbf{I}$$

$$\mathbf{A}_p\mathbf{X} + \mathbf{X}\mathbf{A}_p^T + \mathbf{X}(\frac{1}{\gamma^2}\mathbf{C}_p^T\mathbf{C}_p - \mathbf{M}_p^T\mathbf{M}_p)\mathbf{X} + \mathbf{D}_p\mathbf{D}_p^T < \mathbf{0} \tag{7.7}$$

$$\mathbf{Y}\mathbf{A}_p + \mathbf{A}_p^T\mathbf{Y} + \mathbf{Y}(\frac{1}{\gamma^2}\mathbf{D}_p\mathbf{D}_p^T - \mathbf{B}_p\mathbf{B}_p^T)\mathbf{Y} + \mathbf{C}_p^T\mathbf{C}_p < \mathbf{0}. \tag{7.8}$$

(iii) There exists a matrix pair $(\mathbf{X}, \mathbf{Y})$ *such that*

$$\mathbf{X} > \mathbf{0}, \quad \mathbf{Y} > \mathbf{0}, \quad \mathbf{X}\mathbf{Y} = \gamma^2\mathbf{I}$$

$$\begin{bmatrix} \mathbf{A}_p\mathbf{X} + \mathbf{X}\mathbf{A}_p^T + \mathbf{D}_p\mathbf{D}_p^T - \gamma^2\mathbf{B}_p\mathbf{B}_p^T & \mathbf{X}\mathbf{C}_p^T \\ \mathbf{C}_p\mathbf{X} & -\gamma^2\mathbf{I} \end{bmatrix} < \mathbf{0} \tag{7.9}$$

$$\begin{bmatrix} \mathbf{Y}\mathbf{A}_p + \mathbf{A}_p^T\mathbf{Y} + \mathbf{C}_p\mathbf{C}_p^T - \gamma^2\mathbf{M}_p\mathbf{M}_p^T & \mathbf{Y}\mathbf{D}_p \\ \mathbf{D}_p^T\mathbf{Y} & -\gamma^2\mathbf{I} \end{bmatrix} < \mathbf{0}. \tag{7.10}$$

In this case, all such controllers are given by

$$\mathbf{G} = -\gamma^2\mathbf{B}_p^T\mathbf{Y}\mathbf{P}^{-1}\mathbf{M}_p^T + \mathbf{V}^{1/2}\mathbf{L}\mathbf{W}^{1/2} \tag{7.11}$$

where $\mathbf{L}$ *is an arbitrary matrix such that* $\|\mathbf{L}\| < 1$ *and*

$$\begin{aligned}
\mathbf{V} &\triangleq \mathbf{I} - \mathbf{B}_p^T\mathbf{Y}\mathbf{P}^{-1}\mathbf{Y}\mathbf{B}_p > \mathbf{0} \\
\mathbf{W} &\triangleq \gamma^2\mathbf{I} - \gamma^4\mathbf{M}_p\mathbf{P}^{-1}\mathbf{M}_p^T > \mathbf{0} \\
\mathbf{P} &\triangleq -[\mathbf{Y}\mathbf{A}_p + \mathbf{A}_p^T\mathbf{Y} + \mathbf{Y}(\frac{1}{\gamma^2}\mathbf{D}_p\mathbf{D}_p^T - \mathbf{B}_p\mathbf{B}_p^T)\mathbf{Y} + \mathbf{C}_p^T\mathbf{C}_p - \gamma^2\mathbf{M}_p^T\mathbf{M}_p] > \mathbf{0}.
\end{aligned}$$

Proof. Recall that, from Lemma 7.1.1, a given controller $\mathbf{G}$ stabilizes the system and yields $\|\mathbf{T}\|_\infty < \gamma$ if and only if $\mathbf{R} > \mathbf{0}$ and there exists $\mathbf{Y} > \mathbf{0}$ satisfying $\mathbf{Q} < \mathbf{0}$ where

$$\mathbf{R} \triangleq \gamma^2\mathbf{I} - \mathbf{D}_{c\ell}^T\mathbf{D}_{c\ell}$$

$$\mathbf{Q} \triangleq \mathbf{Y}\mathbf{A}_{c\ell} + \mathbf{A}_{c\ell}^T\mathbf{Y} + (\mathbf{Y}\mathbf{B}_{c\ell} + \mathbf{C}_{c\ell}^T\mathbf{D}_{c\ell})\mathbf{R}^{-1}(\mathbf{Y}\mathbf{B}_{c\ell} + \mathbf{C}_{c\ell}^T\mathbf{D}_{c\ell})^T + \mathbf{C}_{c\ell}^T\mathbf{C}_{c\ell}.$$

Using the matrix inversion lemma, after some manipulations, $\mathbf{Q}$ can be rewritten as

$$\mathbf{Q} = \begin{bmatrix} \mathbf{Y}\mathbf{B}_p & \gamma^2\mathbf{M}_p^T \end{bmatrix} \begin{bmatrix} \mathbf{I} & -\mathbf{G} \\ -\mathbf{G}^T & \gamma^2\mathbf{I} \end{bmatrix}^{-1} \begin{bmatrix} \mathbf{B}_p^T\mathbf{Y} \\ \gamma^2\mathbf{M}_p \end{bmatrix} - \mathbf{P}$$

where $\mathbf{P}$ is defined in Theorem 7.2.2. Using the Schur complement formula, it can be verified that $\mathbf{R} > \mathbf{0}$ if and only if

$$\begin{bmatrix} \mathbf{I} & -\mathbf{G} \\ -\mathbf{G}^T & \gamma^2\mathbf{I} \end{bmatrix} > \mathbf{0}.$$

Hence, $\mathbf{Q} < \mathbf{0}$ and $\mathbf{R} > \mathbf{0}$ are equivalent to $\mathbf{P} > \mathbf{0}$ and

$$\begin{bmatrix} \mathbf{I} & -\mathbf{G} \\ -\mathbf{G}^T & \gamma^2\mathbf{I} \end{bmatrix} > \begin{bmatrix} \mathbf{B}_p^T\mathbf{Y} \\ \gamma^2\mathbf{M}_p \end{bmatrix} \mathbf{P}^{-1} \begin{bmatrix} \mathbf{Y}\mathbf{B}_p & \gamma^2\mathbf{M}_p^T \end{bmatrix}.$$

Then, from Corollary 2.3.7, there exists a matrix $\mathbf{G}$ satisfying the above inequality if and only if $\mathbf{V} > \mathbf{0}$ and $\mathbf{W} > \mathbf{0}$, in which case, all such $\mathbf{G}$ are given by (7.11). Note that $\mathbf{P} > \mathbf{0}$ and $\mathbf{V} > \mathbf{0}$ are equivalent to

$$\mathbf{P} > \mathbf{Y}\mathbf{B}_p\mathbf{B}_p^T\mathbf{Y},$$

which leads to (7.7) with $\mathbf{X} \triangleq \gamma^2\mathbf{Y}^{-1}$. Similarly, $\mathbf{P} > \mathbf{0}$ and $\mathbf{W} > \mathbf{0}$ are equivalent to (7.8). Finally, multiplying (7.7) by $\gamma\mathbf{X}^{-1}$ from the left and right, we have

$$\mathbf{Y}\mathbf{A}_p + \mathbf{A}_p^T\mathbf{Y} + \frac{1}{\gamma^2}\mathbf{Y}\mathbf{D}_p\mathbf{D}_p^T\mathbf{Y} + \mathbf{C}_p^T\mathbf{C}_p - \gamma^2\mathbf{M}_p^T\mathbf{M}_p < \mathbf{0}$$

where $\mathbf{Y} \triangleq \gamma^2\mathbf{X}^{-1}$. Then (7.10) follows from the Schur complement formula. A similar manipulations shows (7.8) $\Leftrightarrow$ (7.9) for $\mathbf{X} \triangleq \gamma^2\mathbf{Y}^{-1}$. This completes the proof. □

Exercise 7.2.1 Provide a step by step derivation of Theorem 7.2.2 following the guideline given above.

7.2.3 Dynamic Output Feedback

In this section, we shall present a solution to the $\mathcal{H}_\infty$ control problem with (possibly) dynamic output feedback controllers. Recall that our derivations are based on the bounded real inequality given in Lemma 7.1.1. Since the closed-loop matrices for the (fixed-order) dynamic output feedback controller have exactly the same structure as those for the static output feedback controller (see Section 4.2), mathematical problems of solving the bounded real inequalities for both cases are essentially the same. However, for the dynamic output feedback case, the assumptions corresponding to (7.2) and (7.6), e.g., $\mathbf{H}^T\mathbf{H} > \mathbf{0}$ and $\mathbf{E}\mathbf{E}^T > \mathbf{0}$, cannot be made since matrix $\mathbf{H}$ ($\mathbf{E}$) is not full column (row) rank even if $\mathbf{B}_y$ ($\mathbf{D}_z$) is full column (row) rank, where $\mathbf{F}$, $\mathbf{H}$, $\mathbf{E}$ are defined in (4.17). Hence, the static result in the previous section cannot be specialized to the dynamic case. In the following, we shall impose no assumptions on matrices $\mathbf{E}$, $\mathbf{F}$ and $\mathbf{H}$.

Theorem 7.2.3 *Let a scalar $\gamma > 0$ be given and consider the system (4.14). Then the following statements are equivalent:*

(i) There exists a controller of order n_c which stabilizes the system and yields $\|\mathbf{T}\|_{\mathcal{H}_\infty} < \gamma$.

(ii) There exists a matrix pair $(\mathbf{X}, \mathbf{Y}) \in \mathcal{R}^{(n_p+n_c)\times(n_p+n_c)} \times \mathcal{R}^{(n_p+n_c)\times(n_p+n_c)}$ such that

$$\mathbf{X} > \mathbf{0}, \quad \mathbf{Y} > \mathbf{0}, \quad \mathbf{X}\mathbf{Y} = \gamma^2\mathbf{I}$$

$$\begin{bmatrix} \mathbf{B} \\ \mathbf{H} \end{bmatrix}^{\perp} \begin{bmatrix} \mathbf{A}\mathbf{X} + \mathbf{X}\mathbf{A}^T + \mathbf{D}\mathbf{D}^T & \mathbf{X}\mathbf{C}^T + \mathbf{D}\mathbf{F}^T \\ \mathbf{C}\mathbf{X} + \mathbf{F}\mathbf{D}^T & \mathbf{F}\mathbf{F}^T - \gamma^2\mathbf{I} \end{bmatrix} \begin{bmatrix} \mathbf{B} \\ \mathbf{H} \end{bmatrix}^{\perp T} < \mathbf{0} \tag{7.12}$$

$$\begin{bmatrix} \mathbf{M}^T \\ \mathbf{E}^T \end{bmatrix}^{\perp} \begin{bmatrix} \mathbf{Y}\mathbf{A} + \mathbf{A}^T\mathbf{Y} + \mathbf{C}^T\mathbf{C} & \mathbf{Y}\mathbf{D} + \mathbf{C}^T\mathbf{F} \\ \mathbf{D}^T\mathbf{Y} + \mathbf{F}^T\mathbf{C} & \mathbf{F}^T\mathbf{F} - \gamma^2\mathbf{I} \end{bmatrix} \begin{bmatrix} \mathbf{M}^T \\ \mathbf{E}^T \end{bmatrix}^{\perp T} < \mathbf{0}. \tag{7.13}$$

(iii) There exists a matrix pair $(\mathbf{X}_p, \mathbf{Y}_p) \in \mathcal{R}^{n_p\times n_p} \times \mathcal{R}^{n_p\times n_p}$ such that

$$\begin{bmatrix} \mathbf{X}_p & \gamma\mathbf{I} \\ \gamma\mathbf{I} & \mathbf{Y}_p \end{bmatrix} \geq \mathbf{0}, \quad \text{rank}\begin{bmatrix} \mathbf{X}_p & \gamma\mathbf{I} \\ \gamma\mathbf{I} & \mathbf{Y}_p \end{bmatrix} \leq n_p + n_c \tag{7.14}$$

$$\begin{bmatrix} \mathbf{B}_p \\ \mathbf{B}_y \end{bmatrix}^{\perp} \begin{bmatrix} \mathbf{A}_p\mathbf{X}_p + \mathbf{X}_p\mathbf{A}_p^T + \mathbf{D}_p\mathbf{D}_p^T & \mathbf{X}_p\mathbf{C}_p^T + \mathbf{D}_p\mathbf{D}_y^T \\ \mathbf{C}_p\mathbf{X}_p + \mathbf{D}_y\mathbf{D}_p^T & \mathbf{D}_y\mathbf{D}_y^T - \gamma^2\mathbf{I} \end{bmatrix} \begin{bmatrix} \mathbf{B}_p \\ \mathbf{B}_y \end{bmatrix}^{\perp T} < \mathbf{0} \quad (7.15)$$

$$\begin{bmatrix} \mathbf{M}_p^T \\ \mathbf{D}_z^T \end{bmatrix}^{\perp} \begin{bmatrix} \mathbf{Y}_p\mathbf{A}_p + \mathbf{A}_p^T\mathbf{Y}_p + \mathbf{C}_p^T\mathbf{C}_p & \mathbf{Y}_p\mathbf{D}_p + \mathbf{C}_p^T\mathbf{D}_y \\ \mathbf{D}_p^T\mathbf{Y}_p + \mathbf{D}_y^T\mathbf{C}_p & \mathbf{D}_y^T\mathbf{D}_y - \gamma^2\mathbf{I} \end{bmatrix} \begin{bmatrix} \mathbf{M}_p^T \\ \mathbf{D}_z^T \end{bmatrix}^{\perp T} < \mathbf{0}. \quad (7.16)$$

In this case, all such controllers are given by

$$\mathbf{G} = -\mathbf{R}^{-1}\mathbf{\Gamma}^T\mathbf{\Phi}\mathbf{\Lambda}^T(\mathbf{\Lambda}\mathbf{\Phi}\mathbf{\Lambda}^T)^{-1} + \mathbf{S}^{1/2}\mathbf{L}(\mathbf{\Lambda}\mathbf{\Phi}\mathbf{\Lambda}^T)^{-1/2} \quad (7.17)$$

where $\mathbf{L}$ *is an arbitrary matrix such that* $\|\mathbf{L}\| < 1$ *and* $\mathbf{R}$ *is an arbitrary positive definite matrix such that*

$$\mathbf{\Phi} \triangleq (\mathbf{\Gamma}\mathbf{R}^{-1}\mathbf{\Gamma}^T - \mathbf{\Theta})^{-1} > \mathbf{0}$$

and

$$\mathbf{S} \triangleq \mathbf{R}^{-1} - \mathbf{R}^{-1}\mathbf{\Gamma}^T[\mathbf{\Phi} - \mathbf{\Phi}\mathbf{\Lambda}^T(\mathbf{\Lambda}\mathbf{\Phi}\mathbf{\Lambda}^T)^{-1}\mathbf{\Lambda}\mathbf{\Phi}]\mathbf{\Gamma}\mathbf{R}^{-1}$$

$$\mathbf{\Theta} \triangleq \begin{bmatrix} \mathbf{Y}\mathbf{A} + \mathbf{A}^T\mathbf{Y} & \mathbf{Y}\mathbf{D} & \mathbf{C}^T \\ \mathbf{D}^T\mathbf{Y} & -\gamma^2\mathbf{I} & \mathbf{F}^T \\ \mathbf{C} & \mathbf{F} & -\mathbf{I} \end{bmatrix}, \quad \mathbf{\Gamma} \triangleq \begin{bmatrix} \mathbf{Y}\mathbf{B} \\ \mathbf{0} \\ \mathbf{H} \end{bmatrix}, \quad \mathbf{\Lambda} \triangleq \begin{bmatrix} \mathbf{M} & \mathbf{E} & \mathbf{0} \end{bmatrix}.$$

Proof. Using the definitions for the closed-loop matrices given in (4.16), it is easy to verify that the matrix inequality in statement (iii) of Lemma 7.1.1 can be written

$$\mathbf{\Gamma}\mathbf{G}\mathbf{\Lambda} + (\mathbf{\Gamma}\mathbf{G}\mathbf{\Lambda})^T + \mathbf{\Theta} < \mathbf{0}. \quad (7.18)$$

Then the result follows from Theorem 2.3.12 where we note that

$$\mathbf{\Gamma}^{\perp} = \begin{bmatrix} \begin{bmatrix} \mathbf{B} \\ \mathbf{H} \end{bmatrix}^{\perp} & \mathbf{0} \\ \mathbf{0} & \mathbf{I} \end{bmatrix} \left[\begin{array}{cc|c} \mathbf{Y}^{-1} & \mathbf{0} & \mathbf{0} \\ \mathbf{0} & \mathbf{0} & \mathbf{I} \\ \hline \mathbf{0} & \mathbf{I} & \mathbf{0} \end{array}\right]$$

$$\mathbf{\Lambda}^{T\perp} = \begin{bmatrix} \begin{bmatrix} \mathbf{M}^T \\ \mathbf{E}^T \end{bmatrix}^{\perp} & \mathbf{0} \\ \mathbf{0} & \mathbf{I} \end{bmatrix}$$

and use the Schur complement formula to reduce the dimensions of the LMIs describing the existence conditions. Then multiplying the reduced-order LMI for $\mathbf{\Gamma}^{\perp}\mathbf{\Theta}\mathbf{\Gamma}^{\perp T} < \mathbf{0}$ by γ^2, and defining $\mathbf{X} \triangleq \gamma^2\mathbf{Y}^{-1}$, we have the equivalence (i) $\Leftrightarrow$ (ii). To prove (ii) $\Leftrightarrow$ (iii), first suppose (ii) holds. Then defining partitioned blocks of $\mathbf{X}$ and $\mathbf{Y}$ by

$$\mathbf{X} = \begin{bmatrix} \mathbf{X}_p & \mathbf{X}_{pc} \\ \mathbf{X}_{pc}^T & \mathbf{X}_c \end{bmatrix}, \quad \mathbf{Y} = \begin{bmatrix} \mathbf{Y}_p & \mathbf{Y}_{pc} \\ \mathbf{Y}_{pc}^T & \mathbf{Y}_c \end{bmatrix}$$

and exploiting the structure of the augmented matrices defined in (4.17) and using

$$\begin{bmatrix} \mathbf{B} \\ \mathbf{H} \end{bmatrix}^{\perp} = \left[\begin{array}{cc} \mathbf{B}_p & \mathbf{0} \\ \mathbf{0} & \mathbf{I}_{n_c} \\ \hline \mathbf{B}_y & \mathbf{0} \end{array}\right]^{\perp} = \begin{bmatrix} \begin{bmatrix} \mathbf{B}_p \\ \mathbf{B}_y \end{bmatrix}^{\perp} & \mathbf{0} \end{bmatrix} \left[\begin{array}{cc|c} \mathbf{I}_{n_p} & \mathbf{0} & \mathbf{0} \\ \mathbf{0} & \mathbf{0} & \mathbf{I}_{n_z} \\ \hline \mathbf{0} & \mathbf{I}_{n_c} & \mathbf{0} \end{array}\right]$$

$$\begin{bmatrix} \mathbf{M}^T \\ \mathbf{E}^T \end{bmatrix}^{\perp} = \left[\begin{array}{cc} \mathbf{M}_p^T & \mathbf{0} \\ \mathbf{0} & \mathbf{I}_{n_c} \\ \hline \mathbf{D}_z^T & \mathbf{0} \end{array}\right]^{\perp} = \begin{bmatrix} \begin{bmatrix} \mathbf{M}_p^T \\ \mathbf{D}_z^T \end{bmatrix}^{\perp} & \mathbf{0} \end{bmatrix} \left[\begin{array}{cc|c} \mathbf{I}_{n_p} & \mathbf{0} & \mathbf{0} \\ \mathbf{0} & \mathbf{0} & \mathbf{I}_{n_w} \\ \hline \mathbf{0} & \mathbf{I}_{n_c} & \mathbf{0} \end{array}\right]$$

we see that (7.12) $\Rightarrow$ (7.15) and (7.13) $\Rightarrow$ (7.16). Note that $\mathbf{X} > \mathbf{0}$ and $\mathbf{Y} > \mathbf{0}$ imply $\mathbf{X}_p > \mathbf{0}$ and $\mathbf{Y}_p > \mathbf{0}$, and 11-block of $\mathbf{X} = \gamma^2\mathbf{Y}^{-1}$ is given by

$$\mathbf{X}_p = \gamma^2(\mathbf{Y}_p - \mathbf{Y}_{pc}\mathbf{Y}_c^{-1}\mathbf{Y}_{pc}^T)^{-1}.$$

Hence

$$\mathbf{Y}_p - \gamma^2\mathbf{X}_p^{-1} = \mathbf{Y}_{pc}\mathbf{Y}_c^{-1}\mathbf{Y}_{pc}^T \geq \mathbf{0} \tag{7.19}$$

or equivalently, using Lemma A.2.1

$$\begin{bmatrix} \mathbf{X}_p & \gamma\mathbf{I} \\ \gamma\mathbf{I} & \mathbf{Y}_p \end{bmatrix} \geq \mathbf{0}.$$

Moreover, we have

$$\begin{aligned}
\text{rank}\begin{bmatrix} \mathbf{X}_p & \gamma\mathbf{I} \\ \gamma\mathbf{I} & \mathbf{Y}_p \end{bmatrix} &= \text{rank}\begin{bmatrix} \mathbf{I} & \mathbf{0} \\ -\gamma\mathbf{X}_p^{-1} & \mathbf{I} \end{bmatrix}\begin{bmatrix} \mathbf{X}_p & \gamma\mathbf{I} \\ \gamma\mathbf{I} & \mathbf{Y}_p \end{bmatrix}\begin{bmatrix} \mathbf{I} & -\gamma\mathbf{X}_p^{-1} \\ \mathbf{0} & \mathbf{I} \end{bmatrix} \\
&= \text{rank}\begin{bmatrix} \mathbf{X}_p & \mathbf{0} \\ \mathbf{0} & \mathbf{Y}_p - \gamma^2\mathbf{X}_p^{-1} \end{bmatrix} \\
&= \text{rank}(\mathbf{Y}_p - \gamma^2\mathbf{X}_p^{-1}) + \text{rank}(\mathbf{X}_p).
\end{aligned}$$

Noting that

$$\text{rank}(\mathbf{X}_p) = n_p, \quad \text{rank}(\mathbf{Y}_p - \gamma^2\mathbf{X}_p^{-1}) = \text{rank}(\mathbf{Y}_{pc}\mathbf{Y}_c^{-1}\mathbf{Y}_{pc}^T) \leq n_c,$$

we have (ii) $\Rightarrow$ (iii). Conversely, if (iii) holds, then defining $\mathbf{Y}_{pc}$ and $\mathbf{Y}_c > \mathbf{0}$ by any matrices satisfying (7.19), it can be verified that matrix pair $(\mathbf{X}, \mathbf{Y})$ such that

$$\mathbf{Y} = \begin{bmatrix} \mathbf{Y}_p & \mathbf{Y}_{pc} \\ \mathbf{Y}_{pc}^T & \mathbf{Y}_c \end{bmatrix}, \quad \mathbf{X} = \gamma^2\mathbf{Y}^{-1} \tag{7.20}$$

satisfies the conditions in statement (ii). This completes the proof. □

To design an $\mathcal{H}_\infty$ controller, one must find a matrix pair $(\mathbf{X}, \mathbf{Y})$ satisfying the conditions in statement (ii) of Theorem 7.2.3. In this case, the controller order is fixed by the dimension of $\mathbf{X}$. As we will see in Chapters 10 and 11, it is extremely difficult to find such a matrix pair directly using the conditions in statement (ii). In view of the above proof, another way to construct $(\mathbf{X}, \mathbf{Y})$ is the following. First find a matrix pair $(\mathbf{X}_p, \mathbf{Y}_p)$ in statement (iii). Then compute a matrix factor $\mathbf{Y}_{pc}$ and $\mathbf{Y}_c > \mathbf{0}$ such that

$$\mathbf{Y}_{pc}\mathbf{Y}_c^{-1}\mathbf{Y}_{pc}^T = \mathbf{Y}_p - \gamma^2\mathbf{X}_p^{-1}, \tag{7.21}$$

and construct $\mathbf{Y}$ and $\mathbf{X}$ as in (7.20). Note that it is possible *not to fix* the controller order *a priori* by removing the rank condition in statement (ii). In this case, a matrix pair $(\mathbf{X}_p, \mathbf{Y}_p)$ can be found by convex programming,[1] and the resulting controller order can be chosen to be any integer such that

$$n_c \geq \text{rank}(\mathbf{Y}_p - \gamma^2\mathbf{X}_p^{-1}).$$

In particular, the controller order is minimal (equality holds in the above inequality) for a given pair $(\mathbf{X}_p, \mathbf{Y}_p)$ if the matrix factor in (7.21) is chosen so that $\mathbf{Y}_{pc}$ has full column rank. In this case, we have

$$n_c = \text{rank}(\mathbf{Y}_p - \gamma^2\mathbf{X}_p^{-1}) \leq n_p$$

where the inequality holds due to the dimension constraint. Thus we have the following.

[1] See Chapters 10 and 11.

Corollary 7.2.1 *Suppose there exists a stabilizing dynamic controller of some order (possibly larger than the plant order n_p) such that $\|\mathbf{T}\|_{\mathcal{H}_\infty} < \gamma$. Then there exists a stabilizing dynamic controller of order n_p such that $\|\mathbf{T}\|_{\mathcal{H}_\infty} < \gamma$.*

A similar statement holds for the state feedback case as follows. Suppose all the state variables are available for feedback, i.e. $\mathbf{M}_p = \mathbf{I}$ and $\mathbf{D}_z = \mathbf{0}$. Then the LMI (7.16) reduces to $\|\mathbf{D}_y\| < \gamma$. Hence, for any $\mathbf{X}_p > \mathbf{0}$ satisfying (7.15), we can always choose $\mathbf{Y}_p$ to be $\gamma^2\mathbf{X}_p^{-1}$ to satisfy the conditions in statement (iii). In this case, the controller order can be chosen

$$n_c = \text{rank}(\mathbf{Y}_p - \gamma^2\mathbf{X}_p^{-1}) = \mathbf{0}.$$

Thus we have the following.

Corollary 7.2.2 *Suppose all the state variables are measured without noise, i.e. $\mathbf{M}_p = \mathbf{I}$ and $\mathbf{D}_z = \mathbf{0}$. Suppose further that there exists a (possibly dynamic) controller which stabilizes the system and yields $\|\mathbf{T}\|_{\mathcal{H}_\infty} < \gamma$. Then there exists a stabilizing static state feedback controller $\mathbf{u} = \mathbf{G}\mathbf{x}_p$ such that $\|\mathbf{T}\|_{\mathcal{H}_\infty} < \gamma$.*

In other words, for the state feedback case, the (optimal) $\mathcal{H}_\infty$ performance cannot be improved by increasing the controller order; an optimal $\mathcal{H}_\infty$ state feedback controller can always be chosen to be a *static* feedback gain.

Once we find a pair $(\mathbf{X}_p, \mathbf{Y}_p)$, we have the following controller design freedoms; the factorization to $\mathbf{Y}_{pc}$ and $\mathbf{Y}_c$, the positive definite matrix $\mathbf{R} > \mathbf{0}$, and the matrix $\mathbf{L}$ such that $\|\mathbf{L}\| < 1$. The freedom in the choice of $\mathbf{Y}_{pc}$ and $\mathbf{Y}_c$ contributes only to the controller coordinate transformation. The positive definite matrix $\mathbf{R}$ can be restricted to have the structure of $\mathbf{R} = \mu\mathbf{I}$, where μ is a real number, without loss of generality (see the proof of Theorem 2.3.12). It can be shown by Theorem 2.3.10 that the admissible interval of the scalar μ is $0 < \mu < \mu_{\max}$ where $\mu_{\max}$ is a computable (without iteration) positive scalar or infinity. Hence, the essential controller design freedoms beyond the choice of $(\mathbf{X}_p, \mathbf{Y}_p)$ are $\|\mathbf{L}\| < 1$ and $\mu > 0$. We shall discuss intensively the use of the freedom due to $(\mathbf{X}_p, \mathbf{Y}_p)$ in a later section.

Finally, we shall show how our general result (Theorem 7.2.2) can be specialized with the standard assumptions (7.2) and (7.6). In this case, existence conditions are given by two Riccati inequalities and a spectral radius condition on the positive definite solutions to the Riccati inequalities. More importantly, we can eliminate the freedom $\mathbf{R} > \mathbf{0}$ in the controller parametrization, leaving $\mathbf{L}$ as the only free parameter.

Corollary 7.2.3 *Let $\gamma > 0$ be given and consider the system (4.14). Suppose the assumptions (7.2) and (7.6) hold. Then the following statements are equivalent:*

(i) There exists a controller of some order which stabilizes the system and yields $\|\mathbf{T}\|_\infty < \gamma$.

(ii) There exists a matrix pair $(\bar{\mathbf{X}}_p, \bar{\mathbf{Y}}_p)$ such that

$$\bar{\mathbf{X}}_p > \mathbf{0}, \quad \bar{\mathbf{Y}}_p > \mathbf{0}, \quad \rho(\bar{\mathbf{X}}_p\bar{\mathbf{Y}}_p) \le \gamma^2$$

$$\bar{\mathbf{Y}}_p\mathbf{A}_p + \mathbf{A}_p^T\bar{\mathbf{Y}}_p + \bar{\mathbf{Y}}_p(\frac{1}{\gamma^2}\mathbf{D}_p\mathbf{D}_p^T - \mathbf{B}_p\mathbf{B}_p^T)\bar{\mathbf{Y}}_p + \mathbf{C}_p^T\mathbf{C}_p < \mathbf{0}$$

$$\mathbf{A}_p\bar{\mathbf{X}}_p + \bar{\mathbf{X}}_p\mathbf{A}_p^T + \bar{\mathbf{X}}_p(\frac{1}{\gamma^2}\mathbf{C}_p^T\mathbf{C}_p - \mathbf{M}_p^T\mathbf{M}_p)\bar{\mathbf{X}}_p + \mathbf{D}_p\mathbf{D}_p^T < \mathbf{0}.$$

In this case, all such controllers are given by

$$\mathbf{G} = \mathbf{G}_1 + \mathbf{G}_2\mathbf{L}\mathbf{G}_3$$

where $\mathbf{L}$ *is an arbitrary matrix such that* $\|\mathbf{L}\| < 1$ *and*

$$\mathbf{G}_1 \triangleq (\mathbf{\Theta}_{12}\mathbf{\Theta}_{22}^{-1}\mathbf{\Lambda}_2^T - \mathbf{\Lambda}_1^T)\mathbf{G}_3^2$$

$$\mathbf{G}_2 \triangleq (\mathbf{\Theta}_{12}\mathbf{\Theta}_{22}^{-1}\mathbf{\Theta}_{12}^T - \mathbf{\Theta}_{11} - \mathbf{G}_1\mathbf{G}_3^{-2}\mathbf{G}_1^T)^{1/2}$$

$$\mathbf{G}_3 \triangleq (-\mathbf{\Lambda}_2\mathbf{\Theta}_{22}^{-1}\mathbf{\Lambda}_2^T)^{-1/2}$$

$$\begin{bmatrix} \mathbf{\Theta}_{11} & \mathbf{\Theta}_{12} \\ \mathbf{\Theta}_{12}^T & \mathbf{\Theta}_{22} \end{bmatrix} \triangleq \begin{bmatrix} \mathbf{\Gamma}^{+} \\ \mathbf{\Gamma}^{\perp} \end{bmatrix} \mathbf{\Theta} \begin{bmatrix} \mathbf{\Gamma}^{+T} & \mathbf{\Gamma}^{\perp T} \end{bmatrix}$$

$$\begin{bmatrix} \mathbf{\Lambda}_1 & \mathbf{\Lambda}_2 \end{bmatrix} \triangleq \mathbf{\Lambda} \begin{bmatrix} \mathbf{\Gamma}^{+T} & \mathbf{\Gamma}^{\perp T} \end{bmatrix}$$

and $\mathbf{\Theta}$, $\mathbf{\Gamma}$ *and* $\mathbf{\Lambda}$ *are defined in Theorem 7.2.3 in terms of the plant matrices and*

$$\mathbf{Y} \triangleq \begin{bmatrix} \mathbf{Y}_p & \mathbf{Y}_{pc} \\ \mathbf{Y}_{pc}^T & \mathbf{Y}_c \end{bmatrix} > \mathbf{0}, \quad \mathbf{Y}_p \triangleq \gamma^2\bar{\mathbf{X}}_p^{-1}, \quad \mathbf{Y}_{pc}\mathbf{Y}_c^{-1}\mathbf{Y}_{pc}^T \triangleq \gamma^2\bar{\mathbf{X}}_p^{-1} - \bar{\mathbf{Y}}_p.$$

Proof. With the assumptions (7.2) and (7.6), we can choose

$$\begin{bmatrix} \mathbf{B}_p \\ \mathbf{B}_y \end{bmatrix}^{\perp} = \begin{bmatrix} \mathbf{I}_{n_p} & -\mathbf{B}_p\mathbf{B}_y^{+} \\ \mathbf{0} & \mathbf{B}_y^{\perp} \end{bmatrix}, \quad \begin{bmatrix} \mathbf{M}_p^T \\ \mathbf{D}_z^T \end{bmatrix}^{\perp} = \begin{bmatrix} \mathbf{I}_{n_p} & -(\mathbf{D}_z^{+}\mathbf{M}_p)^T \\ \mathbf{0} & \mathbf{D}_z^{T\perp} \end{bmatrix},$$

and substitute into the LMIs (7.15) and (7.16) in statement (iii) of Theorem 7.2.3. Then using the Schur complement formula and defining $\bar{\mathbf{X}}_p \triangleq \gamma^2\mathbf{Y}_p^{-1}$ and $\bar{\mathbf{Y}}_p \triangleq \gamma^2\mathbf{X}_p^{-1}$, it is easy to obtain the above Riccati inequalities. The first LMI in statement (iii) of Theorem 7.2.3 is equivalent to $\mathbf{X}_p \geq \gamma^2\mathbf{Y}_p^{-1} > \mathbf{0}$, which is equivalent to $\mathbf{X}_p > \mathbf{0}$, $\mathbf{Y}_p > \mathbf{0}$ and $\rho(\mathbf{X}_p^{-1}\mathbf{Y}_p^{-1}) \leq \gamma^{-2}$. This proves the existence condition.

To prove the controller formula, recall that all $\mathcal{H}_\infty$ controllers are given as the solution to the matrix inequality (7.18). The result simply follows by applying Corollary 2.3.8 to (7.18), and we only need to show that $\mathbf{\Gamma}^T\mathbf{\Gamma} > \mathbf{0}$ and $\mathbf{\Lambda}\mathbf{\Gamma}^{\perp T}\mathbf{\Gamma}^{\perp}\mathbf{\Lambda}^T > 0$. To this end, note that

$$\mathbf{\Gamma}^T = \left[\begin{array}{cc|cc} \mathbf{B}_p^T & \mathbf{0} & \mathbf{0} & \mathbf{B}_y^T \\ \mathbf{0} & \mathbf{I}_{n_c} & \mathbf{0} & \mathbf{0} \end{array}\right] \begin{bmatrix} \mathbf{Y} & \mathbf{0} & \mathbf{0} \\ \mathbf{0} & \mathbf{I} & \mathbf{0} \\ \mathbf{0} & \mathbf{0} & \mathbf{I} \end{bmatrix}$$

has linearly independent rows ($\mathbf{\Gamma}^T\mathbf{\Gamma} > \mathbf{0}$) since $\mathbf{Y} > \mathbf{0}$ and $\mathbf{B}_y^T\mathbf{B}_y > \mathbf{0}$. Using a choice of $\mathbf{\Gamma}^{\perp}$:

$$\mathbf{\Gamma}^{\perp} = \left[\begin{array}{cc|cc} \mathbf{I}_{n_p} & \mathbf{0} & \mathbf{0} & -\mathbf{B}_p\mathbf{B}_y^{+} \\ \mathbf{0} & \mathbf{0} & \mathbf{0} & \mathbf{B}_y^{\perp} \\ \mathbf{0} & \mathbf{0} & \mathbf{I}_{n_w} & \mathbf{0} \end{array}\right] \begin{bmatrix} \mathbf{Y}^{-1} & \mathbf{0} & \mathbf{0} \\ \mathbf{0} & \mathbf{I} & \mathbf{0} \\ \mathbf{0} & \mathbf{0} & \mathbf{I} \end{bmatrix},$$

we have

$$\mathbf{\Lambda}\mathbf{\Gamma}^{\perp T} = \begin{bmatrix} \gamma^{-2}\mathbf{M}_p\mathbf{X}_p & \mathbf{0} & \mathbf{D}_z \\ \gamma^{-2}\mathbf{X}_{pc}^T & \mathbf{0} & \mathbf{0} \end{bmatrix}$$

where $\mathbf{X}_p$ and $\mathbf{X}_{pc}$ are the partitioned blocks of $\mathbf{X} \triangleq \gamma^2\mathbf{Y}^{-1}$. Clearly, $\mathbf{\Lambda}\mathbf{\Gamma}^{\perp T}$ has linearly independent rows ($\mathbf{\Lambda}\mathbf{\Gamma}^{\perp T}\mathbf{\Gamma}^{\perp}\mathbf{\Lambda}^T > 0$) if and only if $\mathbf{X}_{pc}^T\mathbf{X}_{pc} > 0$ since $\mathbf{D}_z\mathbf{D}_z^T > \mathbf{0}$ due to (7.6). But we can assume $\mathbf{X}_{pc}^T\mathbf{X}_{pc} > \mathbf{0}$ without loss of generality due to the following reason. Let an $\mathcal{H}_\infty$ controller $\mathbf{G}$ be given. Then there exists a matrix $\mathbf{X} > \mathbf{0}$ such that $\mathbf{Y} \triangleq \gamma^2\mathbf{X}^{-1}$ satisfies the matrix inequality (7.18). If $\mathbf{X}_{pc}^T\mathbf{X}_{pc} \not> \mathbf{0}$, let $\hat{\mathbf{X}}_{pc}$ be the matrix given by replacing the zero singular value(s) of $\mathbf{X}_{pc}$ with $\varepsilon > 0$ in the singular value decomposition of $\mathbf{X}_{pc}$. Correspondingly, define $\hat{\mathbf{X}}$ by replacing the 12-block $\mathbf{X}_{pc}$ and 21-block $\mathbf{X}_{pc}^T$ by $\hat{\mathbf{X}}_{pc}$ and $\hat{\mathbf{X}}_{pc}^T$, respectively. Then, for sufficiently small $\varepsilon > 0$, we have $\hat{\mathbf{X}}_{pc}^T\hat{\mathbf{X}}_{pc} > \mathbf{0}$, $\hat{\mathbf{X}} > \mathbf{0}$ and $\hat{\mathbf{X}}$ satisfies (7.18). Thus, all $\mathcal{H}_\infty$ controllers can still be captured even if we restrict our attention to the class of matrices $\mathbf{Y} > \mathbf{0}$ with the property $\mathbf{X}_{pc}^T\mathbf{X}_{pc} > \mathbf{0}$ for $\mathbf{X} \triangleq \gamma^2\mathbf{Y}^{-1}$. This completes the proof. □

Example 7.2.1 Consider the double integrator system given by

$$\begin{bmatrix} \mathbf{A}_p & \mathbf{D}_p & \mathbf{B}_p \\ \mathbf{C}_p & \mathbf{D}_y & \mathbf{B}_y \\ \mathbf{M}_p & \mathbf{D}_z & * \end{bmatrix} = \left[\begin{array}{cc|cc|c} 0 & 1 & 0 & 0 & 0 \\ 0 & 0 & 1 & 0 & 1 \\ \hline 1 & 0 & 0 & 0 & 0 \\ 0 & 0 & 0 & 0 & 1 \\ \hline 1 & 0 & 0 & 1 & * \end{array}\right].$$

Note that this system is exactly the same as the one considered in Example 6.2.2 except for the fact that the regulated output $\mathbf{y}$ now takes the control input into account.

We will first design a controller that achieves the $\mathcal{H}_\infty$ norm of the closed-loop transfer function bounded above by $\gamma = 3$. By minimizing $tr(\mathbf{X}_p + \mathbf{Y}_p)$ subject to (7.15), (7.16), and the first inequality in (7.14), we have

$$\mathbf{X}_p = \begin{bmatrix} 6.810 & -3.622 \\ -3.622 & 6.440 \end{bmatrix}, \quad \mathbf{Y}_p = \begin{bmatrix} 6.440 & -3.622 \\ -3.622 & 6.810 \end{bmatrix}.$$

See Example 6.2.2 for a motivation of the above minimization problem. Note that $\text{rank}(\mathbf{Y}_p - \gamma^2\mathbf{X}_p^{-1}) = 1$ and hence the $\mathcal{H}_\infty$ performance bound $\gamma = 3$ can be achieved by a first-order controller. Using the singular value decomposition of $\mathbf{Y}_p - \mathbf{X}_p^{-1}$, we have found matrices $\mathbf{Y}_{pc}$ and $\mathbf{Y}_c$ satisfying (7.19) and thus

$$\mathbf{Y} = \begin{bmatrix} \mathbf{Y}_p & \mathbf{Y}_{pc} \\ \mathbf{Y}_{pc}^T & \mathbf{Y}_c \end{bmatrix} = \begin{bmatrix} 6.440 & -3.622 & 0.697 \\ -3.622 & 6.810 & -0.717 \\ 0.697 & -0.717 & 0.107 \end{bmatrix}.$$

It can be verified that this $\mathbf{Y}$ and $\mathbf{X} := \gamma^2\mathbf{Y}^{-1}$ satisfy the conditions in statement (ii) of Theorem 7.2.3. Hence, a controller achieving the $\mathcal{H}_\infty$ norm bound γ is computed from (7.17) by choosing the free parameters $\mathbf{L} = \mathbf{0}$ and $\mathbf{R} = \varepsilon\mathbf{I}$ ($\varepsilon > 0$ sufficiently small) as follows:

$$\mathbf{G} = \begin{bmatrix} -2.615 & -0.238 \\ -22.814 & -2.522 \end{bmatrix}.$$

To see the effect of the $\mathcal{H}_\infty$ performance bound γ on the actual closed-loop performances, the above design procedure is repeated for $\gamma = 3,\ 4,\ 5$, yielding two additional controllers (We mention that there is no controller achieving $\gamma \le 2.5$). For each controller, the $\mathcal{H}_\infty$ norm of the closed-loop transfer function and the closed-loop poles are computed. The results are summarized in Table 7.1.

From this table, we see that there is a gap between γ and the achieved $\mathcal{H}_\infty$ norm for every case. As γ becomes smaller, the maximum (the smallest in magnitude) of the real-parts

Table 7.1 $\mathcal{H}_\infty$ controllers and their closed-loop properties

γ	Achieved $\mathcal{H}_\infty$ norm	Controller	Closed-loop poles
3	2.756	$-2.615 \times \frac{s+0.448}{s+2.522}$	$-0.699 \pm 0.745,\ -1.123$
4	3.511	$-2.884 \times \frac{s+0.775}{s+3.900}$	$-0.340 \pm 0.761,\ -3.220$
5	4.473	$-3.278 \times \frac{s+1.198}{s+5.999}$	$-0.232 \pm 0.810,\ -5.535$

of the closed-loop eigenvalues becomes smaller, and thus the transient response becomes faster.

Figure 7.1 shows the impulse responses of the closed-loop system for each of the three controllers, where the disturbance input is taken to be $w_1(t) = \delta(t)$ and $w_2(t) \equiv 0$. Note that the transient dies out faster if γ is smaller. In this case, the peak value of the control input remains small, as opposed to the result obtained in Example 6.2.2. This is not because of the difference between the performance measure (i.e. the output covariance or the $\mathcal{H}_2$ norm, and the $\mathcal{H}_\infty$ norm), but because of the fact that the control input $\mathbf{u}$ is contained in the performance output $\mathbf{z}$ in this example while it is not contained in $\mathbf{z}$ in Example 6.2.2.

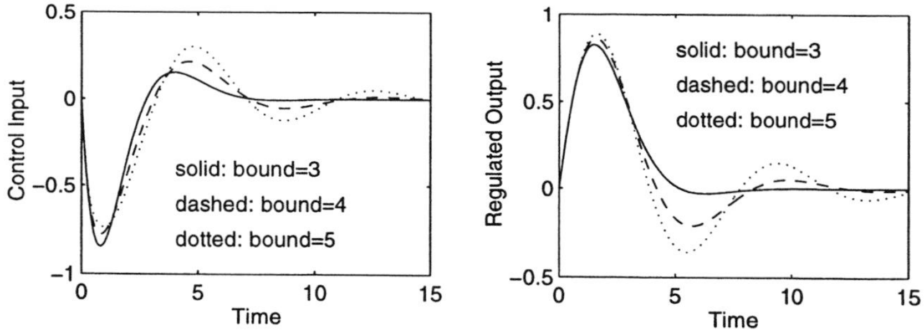

Figure 7.1 Closed-loop impulse responses.

7.3 DISCRETE-TIME CASE

7.3.1 State Feedback

In this section, we consider the $\mathcal{H}_\infty$ control problem for discrete-time systems (4.31) where all the state variables are assumed to be measurable without noises, i.e. the assumption (7.1) holds. As in the continuous-time case, we shall impose the standard assumption (7.2).

Theorem 7.3.1 *Let a scalar $\gamma > 0$ be given and consider the system (4.31). Suppose the assumption (7.2) holds. Then the following statements are equivalent:*

(i) There exists a (static) state feedback controller $\mathbf{u} = \mathbf{G}\mathbf{x}_p$ *which stabilizes the system and yields* $\|\mathbf{T}\|_\infty < \gamma$.
(ii) There exists a matrix $\mathbf{X} > \mathbf{0}$ *such that*

$$\mathbf{X} > \mathbf{D}_p\mathbf{D}_p^T$$

$$\begin{bmatrix} \mathbf{X} - \mathbf{A}_p\mathbf{X}\mathbf{A}_p^T - \mathbf{D}_p\mathbf{D}_p^T + \gamma^2\mathbf{B}_p\mathbf{B}_p^T & \mathbf{A}_p\mathbf{X}\mathbf{C}_p^T \\ \mathbf{C}_p\mathbf{X}\mathbf{A}_p^T & \gamma^2\mathbf{I} - \mathbf{C}_p\mathbf{X}\mathbf{C}_p^T \end{bmatrix} > \mathbf{0}.$$

(iii) There exists a matrix $\mathbf{Q} > \mathbf{0}$ *such that the Riccati equation*

$$\mathbf{Y} = \mathbf{A}_p^T\mathbf{Y}\mathbf{A}_p - \mathbf{A}_p^T\mathbf{Y}\mathbf{E}_p(\mathbf{E}_p^T\mathbf{Y}\mathbf{E}_p + \mathbf{J}_p)^{-1}\mathbf{E}_p^T\mathbf{Y}\mathbf{A}_p + \mathbf{C}_p^T\mathbf{C}_p + \mathbf{Q},$$

has a solution $\mathbf{Y} > \mathbf{0}$ *satisfying* $\|\mathbf{D}_p^T\mathbf{Y}\mathbf{D}_p\| < \gamma^2$, *where*

$$\mathbf{E}_p \triangleq \begin{bmatrix} \mathbf{B}_p & \mathbf{D}_p \end{bmatrix}, \quad \mathbf{J}_p \triangleq \begin{bmatrix} \mathbf{I}_{n_u} & \mathbf{0} \\ \mathbf{0} & -\gamma^2\mathbf{I}_{n_w} \end{bmatrix}.$$

In this case, all such controllers are given by

$$\mathbf{G} = -(\mathbf{B}_p^T\mathbf{P}\mathbf{B}_p + \mathbf{I})^{-1}\mathbf{B}_p^T\mathbf{P}\mathbf{A}_p + (\mathbf{B}_p^T\mathbf{P}\mathbf{B}_p + \mathbf{I})^{-1/2}\mathbf{L}\mathbf{Q}^{1/2}$$

where $\mathbf{L}$ *is an arbitrary matrix such that* $\|\mathbf{L}\| < 1$ *and*

$$\mathbf{P} \triangleq (\mathbf{Y}^{-1} - \frac{1}{\gamma^2}\mathbf{D}_p\mathbf{D}_p^T)^{-1} > \mathbf{0}.$$

Proof. From Lemma 7.1.2, a given state feedback controller G is stabilizing and yields $\|\mathbf{T}\|_\infty < \gamma$ if and only if there exists a matrix $\mathbf{X} > \mathbf{0}$ such that

$$\begin{bmatrix} \mathbf{X} & \mathbf{0} \\ \mathbf{0} & \gamma^2\mathbf{I} \end{bmatrix} > \begin{bmatrix} \mathbf{A}_p + \mathbf{B}_p\mathbf{G} & \mathbf{D}_p \\ \mathbf{C}_p + \mathbf{B}_y\mathbf{G} & \mathbf{0} \end{bmatrix} \begin{bmatrix} \mathbf{X} & \mathbf{0} \\ \mathbf{0} & \mathbf{I} \end{bmatrix} \begin{bmatrix} \mathbf{A}_p + \mathbf{B}_p\mathbf{G} & \mathbf{D}_p \\ \mathbf{C}_p + \mathbf{B}_y\mathbf{G} & \mathbf{0} \end{bmatrix}^T,$$

or equivalently,

$$\begin{bmatrix} \mathbf{X} - \mathbf{D}_p\mathbf{D}_p^T & \mathbf{0} \\ \mathbf{0} & \gamma^2\mathbf{I} \end{bmatrix} > \begin{bmatrix} \mathbf{A}_p + \mathbf{B}_p\mathbf{G} \\ \mathbf{C}_p + \mathbf{B}_y\mathbf{G} \end{bmatrix} \mathbf{X} \begin{bmatrix} \mathbf{A}_p^T + \mathbf{G}^T\mathbf{B}_p^T & \mathbf{C}_p^T + \mathbf{G}^T\mathbf{B}_y^T \end{bmatrix}.$$

Using the Schur complement formula, the above inequality and $\mathbf{X} > \mathbf{0}$ are equivalent to

$$\mathbf{X}^{-1} > \begin{bmatrix} \mathbf{A}_p^T + \mathbf{G}^T\mathbf{B}_p^T & \mathbf{C}_p^T + \mathbf{G}^T\mathbf{B}_y^T \end{bmatrix} \begin{bmatrix} \mathbf{X} - \mathbf{D}_p\mathbf{D}_p^T & \mathbf{0} \\ \mathbf{0} & \gamma^2\mathbf{I} \end{bmatrix}^{-1} \begin{bmatrix} \mathbf{A}_p + \mathbf{B}_p\mathbf{G} \\ \mathbf{C}_p + \mathbf{B}_y\mathbf{G} \end{bmatrix}$$

and $\hat{\mathbf{P}} \triangleq (\mathbf{X} - \mathbf{D}_p\mathbf{D}_p^T)^{-1} > \mathbf{0}$. After expanding, completing the square with respect to $\mathbf{G}$ yields

$$[\mathbf{G} + \hat{\mathbf{R}}^{-1}\mathbf{B}_p^T\hat{\mathbf{P}}\mathbf{A}_p]^T\hat{\mathbf{R}}[\mathbf{G} + \hat{\mathbf{R}}^{-1}\mathbf{B}_p^T\hat{\mathbf{P}}\mathbf{A}_p] < \hat{\mathbf{Q}}$$

where

$$\hat{\mathbf{R}} \triangleq \mathbf{B}_p^T\hat{\mathbf{P}}\mathbf{B}_p + \frac{1}{\gamma^2}\mathbf{I}$$

$$\hat{\mathbf{Q}} \triangleq \mathbf{X}^{-1} - \mathbf{A}_p^T\hat{\mathbf{P}}\mathbf{A}_p + \mathbf{A}_p^T\hat{\mathbf{P}}\mathbf{B}_p(\mathbf{B}_p^T\hat{\mathbf{P}}\mathbf{B}_p + \gamma^{-2}\mathbf{I})^{-1}\mathbf{B}_p^T\hat{\mathbf{P}}\mathbf{A}_p - \frac{1}{\gamma^2}\mathbf{C}_p^T\mathbf{C}_p.$$

Then, defining $\mathbf{P} \triangleq \gamma^2\hat{\mathbf{P}}$ and $\mathbf{Q} \triangleq \gamma^2\hat{\mathbf{Q}}$, Corollary 2.3.6 yields the existence condition $\mathbf{Q} > \mathbf{0}$ and the controller formula given above. Now, using the matrix inversion lemma, it can be verified that

$$\mathbf{Q} = \mathbf{Y} - \mathbf{A}_p^T\mathbf{Y}\mathbf{A}_p + \mathbf{A}_p^T\mathbf{Y}\mathbf{E}_p(\mathbf{E}_p^T\mathbf{Y}\mathbf{E}_p + \mathbf{J}_p)^{-1}\mathbf{E}_p^T\mathbf{Y}\mathbf{A}_p - \mathbf{C}_p^T\mathbf{C}_p$$

where $\mathbf{Y} \triangleq \gamma^2\mathbf{X}^{-1}$. Since $\hat{\mathbf{P}} > \mathbf{0}$, we have

$$\mathbf{P} = \gamma^2\hat{\mathbf{P}} = (\mathbf{Y}^{-1} - \frac{1}{\gamma^2}\mathbf{D}_p\mathbf{D}_p^T)^{-1} > \mathbf{0},$$

or equivalently, $\gamma^2\mathbf{I} - \mathbf{D}_p^T\mathbf{Y}\mathbf{D}_p > \mathbf{0}$. Thus we have (iii). To prove (ii), note that $\hat{\mathbf{Q}}$ can be rewritten

$$\hat{\mathbf{Q}} = \mathbf{X}^{-1} - \mathbf{A}_p^T(\mathbf{X} - \mathbf{D}_p\mathbf{D}_p^T + \gamma^2\mathbf{B}_p\mathbf{B}_p^T)^{-1}\mathbf{A}_p - \frac{1}{\gamma^2}\mathbf{C}_p^T\mathbf{C}_p > \mathbf{0}$$

and hence $\mathbf{X} > \mathbf{D}_p\mathbf{D}_p^T$ and $\mathbf{Q} > \mathbf{0}$ are equivalent to

$$\begin{bmatrix} \mathbf{X}^{-1} - \gamma^{-2}\mathbf{C}_p^T\mathbf{C}_p & \mathbf{A}_p^T \\ \mathbf{A}_p & \mathbf{X} - \mathbf{D}_p\mathbf{D}_p^T + \gamma^2\mathbf{B}_p\mathbf{B}_p^T \end{bmatrix} > \mathbf{0}, \quad \mathbf{X} > \mathbf{D}_p\mathbf{D}_p^T.$$

The first inequality holds if and only if

$$\mathbf{X} - \mathbf{D}_p\mathbf{D}_p^T + \gamma^2\mathbf{B}_p\mathbf{B}_p^T - \mathbf{A}_p(\mathbf{X}^{-1} - \frac{1}{\gamma^2}\mathbf{C}_p^T\mathbf{C}_p)^{-1}\mathbf{A}_p^T > \mathbf{0} \tag{7.22}$$

$$\mathbf{X}^{-1} - \frac{1}{\gamma^2}\mathbf{C}_p^T\mathbf{C}_p > \mathbf{0}. \tag{7.23}$$

By the matrix inversion lemma, (7.22) is equivalent to

$$\mathbf{X} - \mathbf{D}_p\mathbf{D}_p^T + \gamma^2\mathbf{B}_p\mathbf{B}_p^T - \mathbf{A}_p\mathbf{X}\mathbf{A}_p^T - \mathbf{A}_p\mathbf{X}\mathbf{C}_p^T(\gamma^2\mathbf{I} - \mathbf{C}_p\mathbf{X}\mathbf{C}_p^T)^{-1}\mathbf{C}_p\mathbf{X}\mathbf{A}_p^T > \mathbf{0}.$$

By the Schur complement formula, (7.23) is equivalent to

$$\gamma^2\mathbf{I} - \mathbf{C}_p\mathbf{X}\mathbf{C}_p^T > \mathbf{0}.$$

Thus, another use of the Schur complement formula yields (ii). □

7.3.2 Static Output Feedback

Consider the discrete-time system given by (4.31). This section provides all $\mathcal{H}_\infty$ controllers for the static output feedback case. The standard assumptions (7.2) and (7.6) are imposed throughout this section.

Theorem 7.3.2 *Let a scalar $\gamma > 0$ be given and consider the system (4.31). Suppose the assumptions (7.2) and (7.6) hold. Then the following statements are equivalent:*

(i) There exists a static output feedback controller $\mathbf{u} = \mathbf{G}\mathbf{z}$ which stabilizes the system and yields $\|\mathbf{T}\|_{\mathcal{H}_\infty} < \gamma$.

(ii) There exists a matrix pair $(\mathbf{X}, \mathbf{Y})$ *such that*

$$\mathbf{X} > \mathbf{0}, \quad \mathbf{Y} > \mathbf{0}, \quad \mathbf{XY} = \gamma^2\mathbf{I} \tag{7.24}$$

$$\begin{bmatrix} \mathbf{X} - \mathbf{A}_p\mathbf{X}\mathbf{A}_p^T - \mathbf{D}_p\mathbf{D}_p^T + \gamma^2\mathbf{B}_p\mathbf{B}_p^T & \mathbf{A}_p\mathbf{X}\mathbf{C}_p^T \\ \mathbf{C}_p\mathbf{X}\mathbf{A}_p^T & \gamma^2\mathbf{I} - \mathbf{C}_p\mathbf{X}\mathbf{C}_p^T \end{bmatrix} > \mathbf{0} \tag{7.25}$$

$$\begin{bmatrix} \mathbf{Y} - \mathbf{A}_p^T\mathbf{Y}\mathbf{A}_p - \mathbf{C}_p^T\mathbf{C}_p + \gamma^2\mathbf{M}_p^T\mathbf{M}_p & \mathbf{A}_p^T\mathbf{Y}\mathbf{D}_p \\ \mathbf{D}_p^T\mathbf{Y}\mathbf{A}_p & \gamma^2\mathbf{I} - \mathbf{D}_p^T\mathbf{Y}\mathbf{D}_p \end{bmatrix} > \mathbf{0}. \tag{7.26}$$

(iii) There exists a matrix pair $(\mathbf{X}, \mathbf{Y})$ *such that*

$$\mathbf{X} > \mathbf{0}, \quad \mathbf{Y} > \mathbf{0}, \quad \mathbf{XY} = \gamma^2\mathbf{I} \tag{7.27}$$

$$\mathbf{X} > \mathbf{A}_p\mathbf{X}\mathbf{A}_p^T - \mathbf{A}_p\mathbf{X}\mathbf{F}_p^T(\mathbf{F}_p\mathbf{X}\mathbf{F}_p^T + \mathbf{T}_p)^{-1}\mathbf{F}_p\mathbf{X}\mathbf{A}_p^T + \mathbf{D}_p\mathbf{D}_p^T \tag{7.28}$$

$$\mathbf{X}^{-1} + \mathbf{F}_p^T\mathbf{T}_p^{-1}\mathbf{F}_p > \mathbf{0} \tag{7.29}$$

$$\mathbf{Y} > \mathbf{A}_p^T\mathbf{Y}\mathbf{A}_p - \mathbf{A}_p^T\mathbf{Y}\mathbf{E}_p(\mathbf{E}_p^T\mathbf{Y}\mathbf{E}_p + \mathbf{J}_p)^{-1}\mathbf{E}_p^T\mathbf{Y}\mathbf{A}_p + \mathbf{C}_p^T\mathbf{C}_p \tag{7.30}$$

$$\mathbf{Y}^{-1} + \mathbf{E}_p\mathbf{J}_p^{-1}\mathbf{E}_p^T > \mathbf{0} \tag{7.31}$$

where

$$\mathbf{E}_p \triangleq \begin{bmatrix} \mathbf{B}_p & \mathbf{D}_p \end{bmatrix}, \quad \mathbf{J}_p \triangleq \begin{bmatrix} \mathbf{I}_{n_u} & \mathbf{0} \\ \mathbf{0} & -\gamma^2\mathbf{I}_{n_w} \end{bmatrix}$$

$$\mathbf{F}_p \triangleq \begin{bmatrix} \mathbf{M}_p \\ \mathbf{C}_p \end{bmatrix}, \quad \mathbf{T}_p \triangleq \begin{bmatrix} \mathbf{I}_{n_z} & \mathbf{0} \\ \mathbf{0} & -\gamma^2\mathbf{I}_{n_y} \end{bmatrix}.$$

In this case, all such controllers are given by

$$\mathbf{G} = \mathbf{\Gamma}\mathbf{\Phi}^{-1}\mathbf{\Lambda}^T + \mathbf{V}^{1/2}\mathbf{L}\mathbf{W}^{1/2}$$

where $\mathbf{L}$ *is an arbitrary matrix such that* $\|\mathbf{L}\| < 1$ *and*

$$\mathbf{\Phi} \triangleq \begin{bmatrix} \mathbf{X} - \mathbf{D}_p\mathbf{D}_p^T + \gamma^2\mathbf{B}_p\mathbf{B}_p^T & \mathbf{A}_p \\ \mathbf{A}_p^T & \mathbf{X}^{-1} - \gamma^{-2}\mathbf{C}_p^T\mathbf{C}_p + \mathbf{M}_p^T\mathbf{M}_p \end{bmatrix} > \mathbf{0}$$

$$\mathbf{\Gamma} \triangleq \begin{bmatrix} \gamma^2\mathbf{B}_p^T & \mathbf{0} \end{bmatrix}, \quad \mathbf{\Lambda} \triangleq \begin{bmatrix} \mathbf{0} & \mathbf{M}_p \end{bmatrix}$$

$$\mathbf{V} \triangleq \gamma^2\mathbf{I} - \mathbf{\Gamma}\mathbf{\Phi}^{-1}\mathbf{\Gamma}^T, \quad \mathbf{W} \triangleq \mathbf{I} - \mathbf{\Lambda}\mathbf{\Phi}^{-1}\mathbf{\Lambda}^T.$$

Proof. From the bounded real lemma, (i) holds if and only if there exist matrices $\mathbf{X} > \mathbf{0}$ and $\mathbf{G}$ satisfying

$$\begin{bmatrix} \mathbf{X} & \mathbf{0} \\ \mathbf{0} & \gamma^2\mathbf{I} \end{bmatrix} > \begin{bmatrix} \mathbf{A}_{c\ell} & \mathbf{B}_{c\ell} \\ \mathbf{C}_{c\ell} & \mathbf{D}_{c\ell} \end{bmatrix} \begin{bmatrix} \mathbf{X} & \mathbf{0} \\ \mathbf{0} & \mathbf{I} \end{bmatrix} \begin{bmatrix} \mathbf{A}_{c\ell} & \mathbf{B}_{c\ell} \\ \mathbf{C}_{c\ell} & \mathbf{D}_{c\ell} \end{bmatrix}^T, \tag{7.32}$$

or equivalently, using the Schur complement formula,

$$\mathbf{\Psi} \triangleq \begin{bmatrix} \mathbf{X} & \mathbf{0} & \mathbf{A}_{c\ell} & \mathbf{B}_{c\ell} \\ \mathbf{0} & \gamma^2\mathbf{I} & \mathbf{C}_{c\ell} & \mathbf{D}_{c\ell} \\ \mathbf{A}_{c\ell}^T & \mathbf{C}_{c\ell}^T & \mathbf{X}^{-1} & \mathbf{0} \\ \mathbf{B}_{c\ell}^T & \mathbf{D}_{c\ell}^T & \mathbf{0} & \mathbf{I} \end{bmatrix} > \mathbf{0}.$$

Let us introduce a square nonsingular matrix $\mathbf{T}$ defined by

$$\mathbf{T} \triangleq \begin{bmatrix} \mathbf{I}_{n_p} & \mathbf{0} & \mathbf{0} & \mathbf{0} \\ \mathbf{0} & \mathbf{0} & \mathbf{I}_{n_p} & \mathbf{0} \\ \mathbf{0} & \mathbf{B}_y^T & \mathbf{0} & \mathbf{0} \\ \mathbf{0} & \mathbf{0} & \mathbf{0} & \mathbf{D}_z \\ \mathbf{0} & \mathbf{B}_y^{\perp} & \mathbf{0} & \mathbf{0} \\ \mathbf{0} & \mathbf{0} & \mathbf{0} & \mathbf{D}_z^{T\perp} \end{bmatrix}$$

where the left annihilators are normalized such that $\mathbf{B}_y^{\perp}\mathbf{B}_y^{\perp T} = \mathbf{I}$ and $\mathbf{D}_z^{T\perp}\mathbf{D}_z^{T\perp T} = \mathbf{I}$. Then, by the congruent transformation with $\mathbf{T}$, we have $\mathbf{T\Psi T}^T > \mathbf{0}$ or

$$\left[\begin{array}{cccc|cc} \mathbf{X} & \mathbf{A}_p + \mathbf{B}_p\mathbf{GM}_p & \mathbf{0} & \mathbf{B}_p\mathbf{G} & \mathbf{0} & \mathbf{D}_p\mathbf{D}_z^{T\perp T} \\ (\mathbf{A}_p + \mathbf{B}_p\mathbf{GM}_p)^T & \mathbf{X}^{-1} & (\mathbf{GM}_p)^T & \mathbf{0} & (\mathbf{B}_y^{\perp}\mathbf{C}_p)^T & \mathbf{0} \\ \mathbf{0} & \mathbf{GM}_p & \gamma^2\mathbf{I} & \mathbf{G} & \mathbf{0} & \mathbf{0} \\ (\mathbf{B}_p\mathbf{G})^T & \mathbf{0} & \mathbf{G}^T & \mathbf{I} & \mathbf{0} & \mathbf{0} \\ \hline \mathbf{0} & \mathbf{B}_y^{\perp}\mathbf{C}_p & \mathbf{0} & \mathbf{0} & \gamma^2\mathbf{I} & \mathbf{0} \\ \mathbf{D}_z^{T\perp}\mathbf{D}_p^T & \mathbf{0} & \mathbf{0} & \mathbf{0} & \mathbf{0} & \mathbf{I} \end{array}\right] > \mathbf{0}.$$

Using the Schur complement formula for the partitioned blocks,

$$\left[\begin{array}{cc|cc} \mathbf{X} - \mathbf{D}_p\mathbf{D}_p^T & \mathbf{A}_p + \mathbf{B}_p\mathbf{GM}_p & \mathbf{0} & \mathbf{B}_p\mathbf{G} \\ (\mathbf{A}_p + \mathbf{B}_p\mathbf{GM}_p)^T & \mathbf{X}^{-1} - \gamma^{-2}\mathbf{C}_p^T\mathbf{C}_p & (\mathbf{GM}_p)^T & \mathbf{0} \\ \hline \mathbf{0} & \mathbf{GM}_p & \gamma^2\mathbf{I} & \mathbf{G} \\ (\mathbf{B}_p\mathbf{G})^T & \mathbf{0} & \mathbf{G}^T & \mathbf{I} \end{array}\right] > \mathbf{0}$$

where we note that

$$\mathbf{D}_p\mathbf{D}_z^{T\perp T}\mathbf{D}_z^{T\perp}\mathbf{D}_p^T = \mathbf{D}_p(\mathbf{I} - \mathbf{D}_z^T\mathbf{D}_z)\mathbf{D}_p^T = \mathbf{D}_p\mathbf{D}_p^T,$$

$$\mathbf{C}_p^T\mathbf{B}_y^{\perp T}\mathbf{B}_y^{\perp}\mathbf{C}_p = \mathbf{C}_p^T(\mathbf{I} - \mathbf{B}_y\mathbf{B}_y^T)\mathbf{C}_p = \mathbf{C}_p^T\mathbf{C}_p.$$

Using the Schur complement formula again for the partitioned blocks,

$$\begin{bmatrix} \mathbf{X} - \mathbf{D}_p\mathbf{D}_p^T & \mathbf{A}_p + \mathbf{B}_p\mathbf{GM}_p \\ (\mathbf{A}_p + \mathbf{B}_p\mathbf{GM}_p)^T & \mathbf{X}^{-1} - \gamma^{-2}\mathbf{C}_p^T\mathbf{C}_p \end{bmatrix}$$

$$> \begin{bmatrix} \mathbf{0} & \mathbf{B}_p\mathbf{G} \\ (\mathbf{GM}_p)^T & \mathbf{0} \end{bmatrix} \begin{bmatrix} \gamma^2 I & \mathbf{G} \\ \mathbf{G}^T & \mathbf{I} \end{bmatrix}^{-1} \begin{bmatrix} \mathbf{0} & \mathbf{B}_p\mathbf{G} \\ (\mathbf{GM}_p)^T & \mathbf{0} \end{bmatrix}^T \tag{7.33}$$

$$\begin{bmatrix} \gamma^2\mathbf{I} & \mathbf{G} \\ \mathbf{G}^T & \mathbf{I} \end{bmatrix} > \mathbf{0}. \tag{7.34}$$

Substituting the identity

$$\begin{bmatrix} \gamma^2\mathbf{I} & \mathbf{G} \\ \mathbf{G}^T & \mathbf{I} \end{bmatrix}^{-1} = \begin{bmatrix} \mathbf{R} & -\mathbf{RG} \\ -\mathbf{G}^T\mathbf{R} & \mathbf{I} + \mathbf{G}^T\mathbf{RG} \end{bmatrix}, \quad \mathbf{R} \triangleq (\gamma^2\mathbf{I} - \mathbf{GG}^T)^{-1}$$

into (7.33), and then collecting terms involving $\mathbf{G}$, we have

$$\begin{bmatrix} \mathbf{X} - \mathbf{D}_p\mathbf{D}_p^T + \gamma^2\mathbf{B}_p\mathbf{B}_p^T & \mathbf{A}_p \\ \mathbf{A}_p^T & \mathbf{X}^{-1} - \gamma^{-2}\mathbf{C}_p^T\mathbf{C}_p + \mathbf{M}_p^T\mathbf{M}_p \end{bmatrix}$$

$$> \begin{bmatrix} \gamma^2\mathbf{B}_p & \mathbf{0} \\ \mathbf{0} & \mathbf{M}_p^T \end{bmatrix} \begin{bmatrix} \gamma^2\mathbf{I} & \mathbf{G} \\ \mathbf{G}^T & \mathbf{I} \end{bmatrix}^{-1} \begin{bmatrix} \gamma^2\mathbf{B}_p & \mathbf{0} \\ \mathbf{0} & \mathbf{M}_p^T \end{bmatrix}^T. \tag{7.35}$$

Now we see that (7.32) $\Leftrightarrow$ (7.33)–(7.34) $\Leftrightarrow$ (7.34)–(7.35) $\Leftrightarrow$

$$\begin{bmatrix} \gamma^2\mathbf{I} & \mathbf{G} \\ \mathbf{G}^T & \mathbf{I} \end{bmatrix} > \begin{bmatrix} \mathbf{\Gamma} \\ \mathbf{\Lambda} \end{bmatrix} \mathbf{\Phi}^{-1} \begin{bmatrix} \mathbf{\Gamma}^T & \mathbf{\Lambda}^T \end{bmatrix}, \quad \mathbf{\Phi} > \mathbf{0}.$$

Then from Corollary 2.3.7, there exists $\mathbf{G}$ satisfying the above inequality if and only if $\mathbf{V} > \mathbf{0}$ and $\mathbf{W} > \mathbf{0}$, in which case, all such $\mathbf{G}$ are given by the formula in Theorem 7.3.2. Finally, note that $\mathbf{V} > \mathbf{0}$ and $\mathbf{\Phi} > \mathbf{0}$ are equivalent to $\mathbf{\Phi} > \gamma^{-2}\mathbf{\Gamma}^T\mathbf{\Gamma}$, or

$$\begin{bmatrix} \mathbf{X} - \mathbf{D}_p\mathbf{D}_p^T & \mathbf{A}_p \\ \mathbf{A}_p^T & \mathbf{X}^{-1} - \gamma^{-2}\mathbf{C}_p^T\mathbf{C}_p + \mathbf{M}_p^T\mathbf{M}_p \end{bmatrix} > \mathbf{0}. \tag{7.36}$$

Then by a similar procedure to the proof of Theorem 7.3.1, it is easy to verify that (7.36) $\Leftrightarrow$ (7.28)–(7.29) $\Leftrightarrow$ (7.26) with $\mathbf{Y} \triangleq \gamma^2\mathbf{X}^{-1}$. Similarly, $\mathbf{W} > \mathbf{0}$ and $\mathbf{\Phi} > \mathbf{0}$ are equivalent to

$$\begin{bmatrix} \mathbf{X} - \mathbf{D}_p\mathbf{D}_p^T + \gamma^2\mathbf{B}_p\mathbf{B}_p^T & \mathbf{A}_p \\ \mathbf{A}_p^T & \mathbf{X}^{-1} - \gamma^{-2}\mathbf{C}_p^T\mathbf{C}_p \end{bmatrix} > \mathbf{0} \tag{7.37}$$

and we have (7.37) $\Leftrightarrow$ (7.30)–(7.31) $\Leftrightarrow$ (7.25) with $\mathbf{Y} \triangleq \gamma^2\mathbf{X}^{-1}$. This completes the proof. □

7.3.3 Dynamic Output Feedback

This section considers the $\mathcal{H}_\infty$ control problem in a general setting, i.e. we shall not impose the standard assumptions (7.2) and (7.6). The the following theorem gives all dynamic (or static when $n_c = 0$) $\mathcal{H}_\infty$ controllers of order n_c.

Theorem 7.3.3 *Let a scalar $\gamma > 0$ be given and consider the system (4.31). The following statements are equivalent:*

(i) There exists a controller of order n_c which stabilizes the system and yields $\|\mathbf{T}\|_\infty < \gamma$.

(ii) There exists a matrix pair $(\mathbf{X}, \mathbf{Y}) \in \mathcal{R}^{(n_p+n_c)\times(n_p+n_c)} \times \mathcal{R}^{(n_p+n_c)\times(n_p+n_c)}$ such that

$$\mathbf{X} > \mathbf{0}, \quad \mathbf{Y} > \mathbf{0}, \quad \mathbf{XY} = \gamma^2\mathbf{I}$$

$$\begin{bmatrix} \mathbf{B} \\ \mathbf{H} \end{bmatrix}^\perp \left(\begin{bmatrix} \mathbf{X} & \mathbf{0} \\ \mathbf{0} & \gamma^2\mathbf{I} \end{bmatrix} - \begin{bmatrix} \mathbf{A} & \mathbf{D} \\ \mathbf{C} & \mathbf{F} \end{bmatrix} \begin{bmatrix} \mathbf{X} & \mathbf{0} \\ \mathbf{0} & \mathbf{I} \end{bmatrix} \begin{bmatrix} \mathbf{A} & \mathbf{D} \\ \mathbf{C} & \mathbf{F} \end{bmatrix}^T \right) \begin{bmatrix} \mathbf{B} \\ \mathbf{H} \end{bmatrix}^{\perp T} > \mathbf{0} \tag{7.38}$$

$$\begin{bmatrix} \mathbf{M}^T \\ \mathbf{E}^T \end{bmatrix}^\perp \left(\begin{bmatrix} \mathbf{Y} & \mathbf{0} \\ \mathbf{0} & \gamma^2\mathbf{I} \end{bmatrix} - \begin{bmatrix} \mathbf{A} & \mathbf{D} \\ \mathbf{C} & \mathbf{F} \end{bmatrix}^T \begin{bmatrix} \mathbf{Y} & \mathbf{0} \\ \mathbf{0} & \mathbf{I} \end{bmatrix} \begin{bmatrix} \mathbf{A} & \mathbf{D} \\ \mathbf{C} & \mathbf{F} \end{bmatrix} \right) \begin{bmatrix} \mathbf{M}^T \\ \mathbf{E}^T \end{bmatrix}^{\perp T} > \mathbf{0}. \tag{7.39}$$

(iii) There exists a matrix pair $(\mathbf{X}_p, \mathbf{Y}_p) \in \mathcal{R}^{n_p\times n_p} \times \mathcal{R}^{n_p\times n_p}$ such that

$$\begin{bmatrix} \mathbf{X}_p & \gamma\mathbf{I} \\ \gamma\mathbf{I} & \mathbf{Y}_p \end{bmatrix} \geq \mathbf{0}, \quad \text{rank} \begin{bmatrix} \mathbf{X}_p & \gamma\mathbf{I} \\ \gamma\mathbf{I} & \mathbf{Y}_p \end{bmatrix} \leq n_p + n_c$$

$$\begin{bmatrix} \mathbf{B}_p \\ \mathbf{B}_y \end{bmatrix}^\perp \left(\begin{bmatrix} \mathbf{X}_p & \mathbf{0} \\ \mathbf{0} & \gamma^2\mathbf{I} \end{bmatrix} - \begin{bmatrix} \mathbf{A}_p & \mathbf{D}_p \\ \mathbf{C}_p & \mathbf{D}_y \end{bmatrix} \begin{bmatrix} \mathbf{X}_p & \mathbf{0} \\ \mathbf{0} & \mathbf{I} \end{bmatrix} \begin{bmatrix} \mathbf{A}_p & \mathbf{D}_p \\ \mathbf{C}_p & \mathbf{D}_y \end{bmatrix}^T \right) \begin{bmatrix} \mathbf{B}_p \\ \mathbf{B}_y \end{bmatrix}^{\perp T} > \mathbf{0} \tag{7.40}$$

$$\begin{bmatrix} \mathbf{M}_p^T \\ \mathbf{D}_z^T \end{bmatrix}^{\perp} \left(\begin{bmatrix} \mathbf{Y}_p & \mathbf{0} \\ \mathbf{0} & \gamma^2\mathbf{I} \end{bmatrix} - \begin{bmatrix} \mathbf{A}_p & \mathbf{D}_p \\ \mathbf{C}_p & \mathbf{D}_y \end{bmatrix}^T \begin{bmatrix} \mathbf{Y}_p & \mathbf{0} \\ \mathbf{0} & \mathbf{I} \end{bmatrix} \begin{bmatrix} \mathbf{A}_p & \mathbf{D}_p \\ \mathbf{C}_p & \mathbf{D}_y \end{bmatrix} \right) \begin{bmatrix} \mathbf{M}_p^T \\ \mathbf{D}_z^T \end{bmatrix}^{\perp T} > \mathbf{0}. \tag{7.41}$$

In this case, all such controllers are given by

$$\mathbf{G} = -(\mathbf{\Gamma}^T\mathbf{\Phi}\mathbf{\Gamma})^{-1}\mathbf{\Gamma}^T\mathbf{\Phi}\mathbf{\Theta}\mathbf{R}\mathbf{\Lambda}^T(\mathbf{\Lambda}\mathbf{R}\mathbf{\Lambda}^T)^{-1} + (\mathbf{\Gamma}^T\mathbf{\Phi}\mathbf{\Gamma})^{-1/2}\mathbf{L}\mathbf{\Psi}^{1/2}$$

where $\mathbf{L}$ *is an arbitrary matrix such that* $\|\mathbf{L}\| < 1$ *and*

$$\mathbf{\Phi} \triangleq (\mathbf{Q} - \mathbf{\Theta}\mathbf{R}\mathbf{\Theta}^T + \mathbf{\Theta}\mathbf{R}\mathbf{\Lambda}^T(\mathbf{\Lambda}\mathbf{R}\mathbf{\Lambda}^T)^{-1}\mathbf{\Lambda}\mathbf{R}\mathbf{\Theta}^T)^{-1}$$

$$\mathbf{\Psi} \triangleq \mathbf{\Omega} - \mathbf{\Omega}\mathbf{\Lambda}\mathbf{R}\mathbf{\Theta}^T(\mathbf{\Phi} - \mathbf{\Phi}\mathbf{\Gamma}(\mathbf{\Gamma}^T\mathbf{\Phi}\mathbf{\Gamma})^{-1}\mathbf{\Gamma}^T\mathbf{\Phi})\mathbf{\Theta}\mathbf{R}\mathbf{\Lambda}^T\mathbf{\Omega}$$

$$\mathbf{\Omega} \triangleq (\mathbf{\Lambda}\mathbf{R}\mathbf{\Lambda}^T)^{-1}$$

$$\mathbf{Q} \triangleq \begin{bmatrix} \mathbf{X} & \mathbf{0} \\ \mathbf{0} & \gamma^2\mathbf{I} \end{bmatrix}, \quad \mathbf{R} \triangleq \begin{bmatrix} \mathbf{X} & \mathbf{0} \\ \mathbf{0} & \mathbf{I} \end{bmatrix}, \quad \mathbf{\Theta} \triangleq \begin{bmatrix} \mathbf{A} & \mathbf{D} \\ \mathbf{C} & \mathbf{F} \end{bmatrix}$$

$$\mathbf{\Gamma} \triangleq \begin{bmatrix} \mathbf{B} \\ \mathbf{H} \end{bmatrix}, \quad \mathbf{\Lambda} \triangleq [\,\mathbf{M} \quad \mathbf{E}\,].$$

Proof. From Lemma 7.1.2, a given controller satisfies the conditions in statement (i) if and only if there exists a matrix $\mathbf{X} > \mathbf{0}$ such that

$$(\mathbf{\Theta} + \mathbf{\Gamma}\mathbf{G}\mathbf{\Lambda})\mathbf{R}(\mathbf{\Theta} + \mathbf{\Gamma}\mathbf{G}\mathbf{\Lambda})^T < \mathbf{Q}.$$

Then statement (ii) and the controller formula directly follow from Theorem 2.3.11. Statement (iii) can be proved by a similar procedure to the proof of Theorem 7.2.3. □

Exercise 7.3.1 By specializing Theorem 7.3.3, obtain existence conditions

(a) for the state feedback case with assumptions (7.1) and (7.2), and
(b) for the static output feedback case with the assumptions (7.2) and (7.6).

CHAPTER 7 SUMMARY

Explicit formulae for all $\mathcal{H}_\infty$ controllers of any order are given in this chapter for both continuous- and discrete-time systems using state feedback or output feedback. The necessary and sufficient existence conditions involve linear matrix inequalities with a coupling constraint. The results presented here are essentially from [59]. Similar matrix inequality approaches may be found in [34, 82, 118, 119, 120]. Riccati equation approaches for the state feedback case can be found in [70, 104, 160]. Excellent, self-contained approaches include [25, 38, 140, 137, 138].

Chapter 4 shows the relationship between system gains and robust performance. The energy-to-energy gain also has use in nonlinear systems, although nonlinear problems are not treated here.

CHAPTER EIGHT

Model Reduction

Modeling of physical systems usually results in complex high-order models, and it is often desirable to replace these models with simpler reduced-order models without significant error. The model order reduction problem consists of approximating a high-order system $\mathbf{G}$ by a lower-order system $\hat{\mathbf{G}}$ according to some given criterion. In this chapter, necessary and sufficient conditions are derived for the solution of the $\mathcal{H}_\infty$ and the covariance bounded model reduction problems using a linear matrix inequality formulation. These approaches are consistent with the algebraic emphasis of this book. However, many other model reduction methods can be found in the literature, e.g. see [21, 125, 74, 145, 22, 91, 5, 73, 39, 27, 81].

8.1 $\mathcal{H}_\infty$ MODEL REDUCTION

The optimal $\mathcal{H}_\infty$ model reduction seeks to provide a reduced-order model that minimizes the $\mathcal{H}_\infty$ norm of the error between the full-order and the reduced-order model. The suboptimal $\mathcal{H}_\infty$ model reduction problem provides an upper bound on the $\mathcal{H}_\infty$ norm of the error system.

8.1.1 Continuous-Time Case

Consider a stable nth-order linear, time-invariant system $\mathbf{G}$ with a state space representation

$$\dot{\mathbf{x}} = \mathbf{A}\mathbf{x} + \mathbf{B}\mathbf{u} \tag{8.1}$$
$$\mathbf{y} = \mathbf{C}\mathbf{x} + \mathbf{D}\mathbf{u}. \tag{8.2}$$

The optimal $\mathcal{H}_\infty$ model reduction problem is to find a stable $\hat{n}$th-order system $\hat{\mathbf{G}}$ with state space representation

$$\dot{\hat{\mathbf{x}}} = \hat{\mathbf{A}}\hat{\mathbf{x}} + \hat{\mathbf{B}}\mathbf{u} \tag{8.3}$$
$$\hat{\mathbf{y}} = \hat{\mathbf{C}}\hat{\mathbf{x}} + \hat{\mathbf{D}}\mathbf{u} \tag{8.4}$$

where $\hat{n} < n$, such that the $\mathcal{H}_\infty$ norm error $\|\mathbf{G}(s) - \hat{\mathbf{G}}(s)\|_\infty$ is minimized, where $\hat{\mathbf{G}}(s) = \hat{\mathbf{C}}(s\mathbf{I}-\hat{\mathbf{A}})^{-1}\hat{\mathbf{B}}+\hat{\mathbf{D}}$. The γ-suboptimal $\mathcal{H}_\infty$ model reduction problem is to find $\hat{\mathbf{G}}$, if it exists, such that $\|\mathbf{G} - \hat{\mathbf{G}}\|_\infty < \gamma$ where γ is a given positive scalar. If $\hat{n} = 0$, the reduced-order system $\hat{\mathbf{G}}$ is a constant matrix $\hat{\mathbf{D}}$ and the model reduction problem is called the zero-order $\mathcal{H}_\infty$ approximation problem.

Appendix C describes a balanced realization and the balanced truncation method for model reduction [90]. The balanced truncation method consists of transforming the system $\mathbf{G}$ to a special realization such that the observability and controllability Gramians are equal and diagonal and truncating the states that correspond to the smallest diagonal elements (called Hankel singular values). This procedure provides a guaranteed twice-the-sum-of-the-tail $\mathcal{H}_\infty$ error bound [27]

$$\|\mathbf{G} - \hat{\mathbf{G}}\|_\infty \le \gamma_B = 2(\sigma_{\hat{n}+1} + \ldots + \sigma_n) \tag{8.5}$$

where the Hankel singular values $\sigma_1, \sigma_2, \ldots, \sigma_{\hat{n}}, \sigma_{\hat{n}+1}, \ldots, \sigma_n$ are ordered in descending order. The optimal Hankel norm model reduction can be chosen to satisfy the sum-of-the-tail error bound [37]

$$\|\mathbf{G} - \hat{\mathbf{G}}\|_\infty \le \gamma_H = \sigma_{\hat{n}+1} + \ldots + \sigma_n. \tag{8.6}$$

The Hankel model reduction is $\mathcal{H}_\infty$ optimal for $\hat{n} = n - 1$. However, it might be conservative in the general case.

The following results provide necessary and sufficient conditions for the solution of the γ-suboptimal $\mathcal{H}_\infty$ model reduction problem in terms of LMIs, and an explicit parametrization of all reduced-order models that correspond to a feasible solution.

Theorem 8.1.1 *The following statements are equivalent:*

(i) There exists an $\hat{n}$th-order system $\hat{\mathbf{G}}$ to solve the γ-suboptimal $\mathcal{H}_\infty$ model reduction problem.

(ii) There exist matrices $\mathbf{X} > \mathbf{0}$ and $\mathbf{Y} > \mathbf{0}$ such that the following conditions are satisfied

$$\mathbf{AX} + \mathbf{XA}^T + \mathbf{BB}^T < \mathbf{0} \tag{8.7}$$

$$\mathbf{YA} + \mathbf{A}^T\mathbf{Y} + \mathbf{C}^T\mathbf{C} < \mathbf{0} \tag{8.8}$$

$$\begin{bmatrix} \mathbf{X} & \gamma\mathbf{I} \\ \gamma\mathbf{I} & \mathbf{Y} \end{bmatrix} \ge \mathbf{0} \tag{8.9}$$

and

$$\text{rank}\begin{bmatrix} \mathbf{X} & \gamma\mathbf{I} \\ \gamma\mathbf{I} & \mathbf{Y} \end{bmatrix} \le n + \hat{n}\,. \tag{8.10}$$

If these conditions are satisfied, all γ-suboptimal $\hat{n}$th-order models that correspond to a feasible matrix pair $(\mathbf{X}, \mathbf{Y})$ are given by

$$\begin{bmatrix} \hat{\mathbf{D}} & \hat{\mathbf{C}} \\ \hat{\mathbf{B}} & \hat{\mathbf{A}} \end{bmatrix} = \hat{\mathbf{G}}_1 + \hat{\mathbf{G}}_2\mathbf{L}\hat{\mathbf{G}}_3 \tag{8.11}$$

where $\mathbf{L}$ is a $(p + \hat{n}) \times (m + \hat{n})$ matrix such that $\|\mathbf{L}\| < 1$, and

$$\begin{aligned} \hat{\mathbf{G}}_1 &= (\mathbf{M}_1 - \mathbf{Q}_{12}\mathbf{Q}_{22}^{-1}\mathbf{M}_2^T)(\mathbf{M}_2\mathbf{Q}_{22}^{-1}\mathbf{M}_2^T)^{-1} \\ \hat{\mathbf{G}}_2 &= (-\mathbf{Q}_{11} + \mathbf{Q}_{12}\mathbf{Q}_{22}^{-1}\mathbf{Q}_{12}^T - \mathbf{G}_1\mathbf{G}_3^2\mathbf{G}_1^T)^{1/2} \\ \hat{\mathbf{G}}_3 &= (-\mathbf{M}_2\mathbf{Q}_{22}^{-1}\mathbf{M}_2^T)^{1/2} \end{aligned} \tag{8.12}$$

where

$$\mathbf{M}_1 = \begin{bmatrix} \mathbf{0} & \mathbf{0} \\ \mathbf{0} & \mathbf{R}_x^2 \end{bmatrix}, \ \mathbf{M}_2 = \begin{bmatrix} \mathbf{0} & \mathbf{I} \\ \mathbf{R}_x\mathbf{L}_x^T & \mathbf{0} \end{bmatrix}$$

$$\mathbf{Q}_{11} = \begin{bmatrix} -\gamma^2\mathbf{I} & -\mathbf{C}\mathbf{L}_x\mathbf{R}_x \\ -\mathbf{R}_x\mathbf{L}_x^T\mathbf{C}^T & \mathbf{0} \end{bmatrix}, \ \mathbf{Q}_{12} = \begin{bmatrix} -\mathbf{C}\mathbf{X} & -\mathbf{D} \\ \mathbf{R}_x\mathbf{L}_x^T\mathbf{A}^T & \mathbf{0} \end{bmatrix} \qquad (8.13)$$

$$\mathbf{Q}_{22} = \begin{bmatrix} \mathbf{A}\mathbf{X} + \mathbf{X}\mathbf{A}^T & \mathbf{B} \\ \mathbf{B}^T & -\mathbf{I} \end{bmatrix}$$

and $\mathbf{R}_x$ *is an arbitrary positive definite matrix and* $\mathbf{L}_x$ *is an arbitrary matrix factor such that*

$$\mathbf{L}_x\mathbf{L}_x^T = \mathbf{X} - \gamma^2\mathbf{Y}^{-1}. \qquad (8.14)$$

Proof. The necessary and sufficient conditions (8.7)–(8.10) can be obtained using the state space representation of the error system $\mathbf{G} - \hat{\mathbf{G}}$

$$\dot{\bar{\mathbf{x}}} = (\bar{\mathbf{A}} + \bar{\mathbf{B}}\bar{\mathbf{G}}\bar{\mathbf{M}})\bar{\mathbf{x}} + (\bar{\mathbf{D}} + \bar{\mathbf{B}}\bar{\mathbf{G}}\bar{\mathbf{E}})\mathbf{u} \qquad (8.15)$$

$$\mathbf{z} = (\bar{\mathbf{C}} + \bar{\mathbf{H}}\bar{\mathbf{G}}\bar{\mathbf{M}})\bar{\mathbf{x}} + (\bar{\mathbf{F}} + \bar{\mathbf{H}}\bar{\mathbf{G}}\bar{\mathbf{E}})\mathbf{u} \qquad (8.16)$$

where

$$\bar{\mathbf{A}} = \begin{bmatrix} \mathbf{A} & \mathbf{0} \\ \mathbf{0} & \mathbf{0} \end{bmatrix}, \ \bar{\mathbf{B}} = \bar{\mathbf{M}} = \begin{bmatrix} \mathbf{0} & \mathbf{0} \\ \mathbf{0} & \mathbf{I} \end{bmatrix}, \ \bar{\mathbf{E}} = \begin{bmatrix} \mathbf{I} \\ \mathbf{0} \end{bmatrix}, \bar{\mathbf{D}} = \begin{bmatrix} \mathbf{B} \\ \mathbf{0} \end{bmatrix} \qquad (8.17)$$

$$\bar{\mathbf{C}} = \begin{bmatrix} \mathbf{C} & \mathbf{0} \end{bmatrix}, \ \bar{\mathbf{H}} = \begin{bmatrix} -\mathbf{I} & \mathbf{0} \end{bmatrix}, \ \bar{\mathbf{F}} = \mathbf{D} \qquad (8.18)$$

$$\bar{\mathbf{G}} = \begin{bmatrix} \hat{\mathbf{D}} & \hat{\mathbf{C}} \\ \hat{\mathbf{B}} & \hat{\mathbf{A}} \end{bmatrix}, \quad \mathbf{z} = \mathbf{y} - \hat{\mathbf{y}} \qquad (8.19)$$

and applying the conditions of Theorem 7.2.3 in Chapter 7, to guarantee that the transfer function of (8.15) and (8.16) from $\mathbf{u}$ to $\mathbf{z}$ has $\mathcal{H}_\infty$ norm less than γ. The parametrization (8.11)–(8.14) of all reduced-order models that correspond to a feasible solution can be obtained using the corresponding parametrization in Theorem 7.2.3. □

Hence, the γ-suboptimal $\mathcal{H}_\infty$ model reduction problem is characterized as a feasibility problem of finding a pair of positive definite matrices $(\mathbf{X}, \mathbf{Y})$ in the intersection of the constraint sets (8.7), (8.8), (8.9) and (8.10).

Particularly simple necessary and sufficient conditions and a parametrization of all solutions can be obtained for the zero-order $\mathcal{H}_\infty$ approximation problem.

Theorem 8.1.2 *The following statements are equivalent:*

(i) There exists a constant matrix $\hat{\mathbf{D}}$ *to solve the zero-order* γ*-suboptimal* $\mathcal{H}_\infty$ *approximation problem.*

(ii) There exists a matrix $\mathbf{X} > \mathbf{0}$ *such that*

$$\mathbf{A}\mathbf{X} + \mathbf{X}\mathbf{A}^T + \mathbf{B}\mathbf{B}^T < \mathbf{0} \qquad (8.20)$$

$$\begin{bmatrix} \mathbf{A}\mathbf{X} + \mathbf{X}\mathbf{A}^T & \mathbf{X}\mathbf{C}^T \\ \mathbf{C}\mathbf{X} & -\gamma^2\mathbf{I} \end{bmatrix} < \mathbf{0}. \qquad (8.21)$$

All zero-order γ-suboptimal $\mathcal{H}_\infty$ solutions $\hat{\mathbf{D}}$ that correspond to a feasible matrix $\mathbf{X}$ are given by

$$\hat{\mathbf{D}} = \hat{\mathbf{D}}_1 + \hat{\mathbf{D}}_2\mathbf{L}\hat{\mathbf{D}}_3 \tag{8.22}$$

where

$$\begin{aligned} \hat{\mathbf{D}}_1 &= \mathbf{D} - \mathbf{C}\mathbf{X}(\mathbf{A}\mathbf{X} + \mathbf{X}\mathbf{A}^T)^{-1}\mathbf{B} \\ \hat{\mathbf{D}}_2 &= \left[\gamma^2\mathbf{I} + \mathbf{C}\mathbf{X}(\mathbf{A}\mathbf{X} + \mathbf{X}\mathbf{A}^T)^{-1}\mathbf{X}\mathbf{C}^T\right]^{1/2} \\ \hat{\mathbf{D}}_3 &= \left[\mathbf{I} + \mathbf{B}^T(\mathbf{A}\mathbf{X} + \mathbf{X}\mathbf{A}^T)^{-1}\mathbf{B}\right]^{1/2} \end{aligned} \tag{8.23}$$

and $\mathbf{L}$ *is any* $p \times m$ *matrix such that* $\|\mathbf{L}\| < 1$.

Proof. For $\hat{n} = 0$, condition (8.10) of Theorem 8.1.1 is equivalent to rank $(\gamma^2\mathbf{I} - \mathbf{X}\mathbf{Y}) = 0$, i.e. $\mathbf{X}\mathbf{Y} = \gamma^2\mathbf{I}$, where $\mathbf{X} > \mathbf{0}$, $\mathbf{Y} > \mathbf{0}$. Hence, it can easily be shown that condition (8.9) is trivially satisfied. Multiplying (8.8) on the left-hand and right-hand side by $\mathbf{Y}^{-1} = \frac{1}{\gamma^2}\mathbf{X}$ provides

$$\mathbf{A}\mathbf{X} + \mathbf{X}\mathbf{A}^T + \frac{1}{\gamma^2}\mathbf{X}\mathbf{C}^T\mathbf{C}\mathbf{X} < 0$$

which according to the Schur complement formula is equivalent to condition (8.21).

To obtain the parametrization of all zero-order γ-suboptimal $\mathcal{H}_\infty$ approximations note that for $\hat{n} = 0$ the expressions (8.12)–(8.14) become

$$\begin{aligned} \hat{\mathbf{G}}_1 &= -\mathbf{Q}_{12}\mathbf{Q}_{22}^{-1}\mathbf{M}^T(\mathbf{M}\mathbf{Q}_{22}^{-1}\mathbf{M}^T)^{-1} \\ \hat{\mathbf{G}}_2 &= (\gamma^2\mathbf{I} + \mathbf{Q}_{12}\mathbf{Q}_{22}^{-1}\mathbf{Q}_{12}^T - \hat{\mathbf{G}}_1\hat{\mathbf{G}}_3^{-2}\hat{\mathbf{G}}_1^T)^{1/2} \\ \hat{\mathbf{G}}_3 &= (-\mathbf{M}\mathbf{Q}_{22}\mathbf{M}^T)^{-1/2} \end{aligned} \tag{8.24}$$

where

$$\mathbf{M} = \begin{bmatrix} \mathbf{0} & \mathbf{I} \end{bmatrix}$$

$$\mathbf{Q}_{12} = \begin{bmatrix} \mathbf{C}\mathbf{X} & \mathbf{D} \end{bmatrix}, \quad \mathbf{Q}_{22} = \begin{bmatrix} \mathbf{A}\mathbf{X} + \mathbf{X}\mathbf{A}^T & \mathbf{B} \\ \mathbf{B}^T & -\gamma^2\mathbf{I} \end{bmatrix}.$$

By defining

$$\mathbf{Q}_{22}^{-1} = \mathbf{\Psi} = \begin{bmatrix} \mathbf{\Psi}_{11} & \mathbf{\Psi}_{12} \\ \mathbf{\Psi}_{12}^T & \mathbf{\Psi}_{22} \end{bmatrix}$$

we obtain after some matrix algebra

$$\begin{aligned} \hat{\mathbf{G}}_1 &= \mathbf{C}\mathbf{X}\mathbf{\Psi}_{12}\mathbf{\Psi}_{22}^{-1} + \mathbf{D} \\ \hat{\mathbf{G}}_2 &= \left[\gamma^2\mathbf{I} + \mathbf{C}\mathbf{X}(\mathbf{\Psi}_{11} - \mathbf{\Psi}_{12}\mathbf{\Psi}_{22}^{-1}\mathbf{\Psi}_{12}^T)\mathbf{X}\mathbf{C}^T\right] \\ \hat{\mathbf{G}}_3 &= (-\mathbf{\Psi}_{22})^{-1/2}. \end{aligned} \tag{8.25}$$

Using the block matrix inverse expression for $\mathbf{Q}_{22}^{-1}$ we find after several matrix algebraic manipulations

$$\begin{aligned} \mathbf{\Psi}_{12}\mathbf{\Psi}_{22}^{-1} &= (\mathbf{A}\mathbf{X} + \mathbf{X}\mathbf{A}^T)^{-1}\mathbf{B} \\ \mathbf{\Psi}_{11} - \mathbf{\Psi}_{12}\mathbf{\Psi}_{22}^{-1}\mathbf{\Psi}_{12}^T &= (\mathbf{A}\mathbf{X} + \mathbf{X}\mathbf{A}^T)^{-1} \\ \mathbf{\Psi}_{22} &= -\mathbf{I} - \mathbf{B}^T(\mathbf{A}\mathbf{X} + \mathbf{X}\mathbf{A}^T)^{-1}\mathbf{B}. \end{aligned}$$

Substituting these expressions in (8.25) we obtain (8.23). □

8.1.2 Discrete-Time Case

Next, we consider the discrete-time $\mathcal{H}_\infty$ model reduction problem: Given a stable, nth-order discrete-time system $\mathbf{G}$ described by

$$\mathbf{x}_{k+1} = \mathbf{A}\mathbf{x}_k + \mathbf{B}\mathbf{u}_k \tag{8.26}$$
$$\mathbf{y}_k = \mathbf{C}\mathbf{x}_k + \mathbf{D}\mathbf{u}_k \tag{8.27}$$

where $\mathbf{G}(z) = \mathbf{C}(z\mathbf{I} - \mathbf{A})^{-1}\mathbf{B} + \mathbf{D}$, find a stable $\hat{n}$th-order discrete-time system $\hat{\mathbf{G}}$

$$\hat{\mathbf{x}}_{k+1} = \hat{\mathbf{A}}\hat{\mathbf{x}}_k + \hat{\mathbf{B}}\mathbf{u}_k \tag{8.28}$$
$$\hat{\mathbf{y}}_k = \hat{\mathbf{C}}\hat{\mathbf{x}}_k + \hat{\mathbf{D}}\mathbf{u}_k \tag{8.29}$$

such that the $\mathcal{H}_\infty$ norm $\|\mathbf{G} - \hat{\mathbf{G}}\|_\infty$ is minimized, where $\hat{\mathbf{G}}(z) = \hat{\mathbf{C}}(z\mathbf{I} - \hat{\mathbf{A}})^{-1}\hat{\mathbf{B}} + \hat{\mathbf{D}}$. The discrete-time γ-suboptimal $\mathcal{H}_\infty$ model reduction and the discrete-time zero-order $\mathcal{H}_\infty$ approximation problems are defined accordingly.

The following results provide necessary and sufficient conditions for the existence of a solution to the discrete-time γ-suboptimal $\mathcal{H}_\infty$ model reduction problem and a state space parametrization of all reduced-order models.

Theorem 8.1.3 *The following statements are equivalent:*

(i) There exists an $\hat{n}$th-order system $\hat{\mathbf{G}}$ to solve the discrete-time γ-suboptimal $\mathcal{H}_\infty$ model reduction problem.

(ii) There exist matrices $\mathbf{X} > \mathbf{0}$ and $\mathbf{Y} > \mathbf{0}$ such that the following conditions are satisfied:

$$\mathbf{X} - \mathbf{A}\mathbf{X}\mathbf{A}^T - \mathbf{B}\mathbf{B}^T > \mathbf{0} \tag{8.30}$$
$$\mathbf{Y} - \mathbf{A}^T\mathbf{Y}\mathbf{A} - \mathbf{C}^T\mathbf{C} > \mathbf{0} \tag{8.31}$$
$$\begin{bmatrix} \mathbf{X} & \gamma\mathbf{I} \\ \gamma\mathbf{I} & \mathbf{Y} \end{bmatrix} \geq \mathbf{0} \tag{8.32}$$

and

$$\operatorname{rank}\begin{bmatrix} \mathbf{X} & \gamma\mathbf{I} \\ \gamma\mathbf{I} & \mathbf{Y} \end{bmatrix} \leq n + \hat{n}\,. \tag{8.33}$$

All γ-suboptimal $\hat{n}$th-order models that correspond to a feasible matrix pair $(\mathbf{X}, \mathbf{Y})$ are given by

$$\begin{bmatrix} \hat{\mathbf{D}} & \hat{\mathbf{C}} \\ \hat{\mathbf{B}} & \hat{\mathbf{A}} \end{bmatrix} = \hat{\mathbf{G}}_1 + \hat{\mathbf{G}}_2\mathbf{L}\hat{\mathbf{G}}_3 \tag{8.34}$$

where $\mathbf{L}$ is any $(p + \hat{n}) \times (m + \hat{n})$ matrix such that $\|\mathbf{L}\| < 1$, and

$$\begin{aligned} \hat{\mathbf{G}}_1 &= -(\mathbf{\Gamma}^T\mathbf{\Phi}\mathbf{\Gamma})^{-1}\mathbf{\Gamma}^T\mathbf{\Phi}\mathbf{\Theta}\mathbf{R}\mathbf{\Lambda}^T(\mathbf{\Lambda}\mathbf{R}\mathbf{\Lambda}^T)^{-1} \\ \hat{\mathbf{G}}_2 &= (\mathbf{\Gamma}^T\mathbf{\Phi}\mathbf{\Gamma})^{-1/2} \\ \hat{\mathbf{G}}_3 &= \{\mathbf{\Omega} - \mathbf{\Omega}\mathbf{\Lambda}\mathbf{R}\mathbf{\Theta}^T\left[\mathbf{\Phi} - \mathbf{\Phi}\mathbf{\Gamma}(\mathbf{\Gamma}^T\mathbf{\Phi}\mathbf{\Gamma})^{-1}\mathbf{\Gamma}^T\mathbf{\Phi}\right]\mathbf{\Theta}\mathbf{R}\mathbf{\Lambda}^T\mathbf{\Omega}\}^{1/2} \end{aligned} \tag{8.35}$$

where

$$\mathbf{\Phi} = (\mathbf{Q} - \mathbf{\Theta}\mathbf{R}\mathbf{\Theta}^T + \mathbf{\Theta}\mathbf{R}\mathbf{\Lambda}^T(\mathbf{\Lambda}\mathbf{R}\mathbf{\Lambda}^T)^{-1}\mathbf{\Lambda}\mathbf{R}\mathbf{\Theta}^T)^{-1}$$

$$\mathbf{Q} = \begin{bmatrix} \bar{\mathbf{X}} & \mathbf{0} \\ \mathbf{0} & \gamma^2\mathbf{I} \end{bmatrix}, \quad \mathbf{R} = \begin{bmatrix} \bar{\mathbf{X}} & \mathbf{0} \\ \mathbf{0} & \mathbf{I} \end{bmatrix}, \quad \mathbf{\Omega} = \begin{bmatrix} \mathbf{I} & \mathbf{0} \\ \mathbf{0} & \mathbf{X}_c^{-1} \end{bmatrix}$$

$$\bar{\mathbf{X}} = \begin{bmatrix} \mathbf{X} & \mathbf{X}_{pc} \\ \mathbf{X}_{pc}^T & \mathbf{X}_c \end{bmatrix}, \ \Lambda = \begin{bmatrix} \mathbf{0} & \mathbf{0} & \mathbf{I} \\ \mathbf{0} & \mathbf{I} & \mathbf{0} \end{bmatrix} \quad (8.36)$$

$$\Theta = \begin{bmatrix} \mathbf{A} & \mathbf{0} & \mathbf{B} \\ \mathbf{0} & \mathbf{0} & \mathbf{0} \\ \mathbf{C} & \mathbf{0} & \mathbf{D} \end{bmatrix}, \ \Gamma = \begin{bmatrix} \mathbf{0} & \mathbf{0} \\ \mathbf{0} & \mathbf{I} \\ -\mathbf{I} & \mathbf{0} \end{bmatrix}$$

and $\mathbf{X}_{pc}$, $\mathbf{X}_c$ *are arbitrary matrices such that* $\bar{\mathbf{X}} > \mathbf{0}$.

Proof. The necessary and sufficient conditions for the solvability of the discrete-time γ-suboptimal $\mathcal{H}_\infty$ model reduction problem and the parametrization of all reduced-order models can be obtained from the characterization of all solutions of the discrete-time $\mathcal{H}_\infty$ problem via LMIs. That is, the discrete-time equivalent of the error system (8.15), yields the same structures (8.17)–(8.19). Then Theorem 7.3.3 gives the appropriate $\mathcal{H}_\infty$ control solution for $\bar{\mathbf{G}}$, where substitutions of the structure (8.17)–(8.19) readily verify the theorem. □

The conditions (8.30)–(8.33) and the parametrization of the reduced-order models (8.34)–(8.36) can be significantly simplified for the case of the discrete-time zero-order γ-suboptimal $\mathcal{H}_\infty$ approximation problem.

Theorem 8.1.4 *The following statements are equivalent:*

(i) There exists a constant matrix $\hat{\mathbf{D}}$ *to solve the discrete-time zero-order* γ*-suboptimal* $\mathcal{H}_\infty$ *approximation problem.*

(ii) There exists a matrix $\mathbf{X} > \mathbf{0}$ *such that the following conditions are satisfied*

$$\mathbf{X} - \mathbf{A}\mathbf{X}\mathbf{A}^T - \mathbf{B}\mathbf{B}^T > \mathbf{0} \quad (8.37)$$

$$\begin{bmatrix} \mathbf{X} - \mathbf{A}\mathbf{X}\mathbf{A}^T & -\mathbf{A}\mathbf{X}\mathbf{C}^T \\ -\mathbf{C}\mathbf{X}\mathbf{A}^T & \gamma^2\mathbf{I} - \mathbf{C}\mathbf{X}\mathbf{C}^T \end{bmatrix} > \mathbf{0}. \quad (8.38)$$

All constant matrices $\hat{\mathbf{D}}$ *that correspond to a feasible solution* $\mathbf{X}$ *are given by*

$$\hat{\mathbf{D}} = \hat{\mathbf{D}}_1 + \hat{\mathbf{D}}_2\mathbf{L}\hat{\mathbf{D}}_3 \quad (8.39)$$

where

$$\begin{aligned} \hat{\mathbf{D}}_1 &= \mathbf{D} + \mathbf{C}\mathbf{X}\mathbf{A}^T(\mathbf{X} - \mathbf{A}\mathbf{X}\mathbf{A}^T)^{-1}\mathbf{B} \\ \hat{\mathbf{D}}_2 &= \left[\gamma^2\mathbf{I} - \mathbf{C}\mathbf{X}\mathbf{C}^T - \mathbf{C}\mathbf{X}\mathbf{A}^T(\mathbf{X} - \mathbf{A}\mathbf{X}\mathbf{A}^T)^{-1}\mathbf{A}\mathbf{X}\mathbf{C}^T\right]^{1/2} \\ \hat{\mathbf{D}}_3 &= \left[\mathbf{I} - \mathbf{B}^T(\mathbf{X} - \mathbf{A}\mathbf{X}\mathbf{A}^T)^{-1}\mathbf{B}\right]^{1/2} \end{aligned} \quad (8.40)$$

and $\mathbf{L}$ *is any* $p \times m$ *matrix such that* $\|\mathbf{L}\| < 1$.

Proof. As in the continuous-time case, for $\hat{n} = 0$, condition (8.33) of Theorem 8.1.3 provides $\mathbf{X}\mathbf{Y} = \gamma^2\mathbf{I}$, where $\mathbf{X} > \mathbf{0}$, $\mathbf{Y} > \mathbf{0}$. Hence, (8.31) is equivalent to

$$\mathbf{X}^{-1} - \mathbf{A}^T\mathbf{X}^{-1}\mathbf{A} - \frac{1}{\gamma^2}\mathbf{C}^T\mathbf{C} > 0$$

which provides, using the Schur complement formula,

$$\begin{bmatrix} \mathbf{X}^{-1} & \mathbf{A}^T \\ \mathbf{A} & \mathbf{X} \end{bmatrix} > \begin{bmatrix} \frac{1}{\gamma}\mathbf{C}^T \\ \mathbf{0} \end{bmatrix} \begin{bmatrix} \frac{1}{\gamma}\mathbf{C} & \mathbf{0} \end{bmatrix}.$$

Using the Schur complement formula again we obtain

$$\begin{bmatrix} \mathbf{X}^{-1} & \mathbf{A}^T & \frac{1}{\gamma}\mathbf{C}^T \\ \mathbf{A} & \mathbf{X} & \mathbf{0} \\ \frac{1}{\gamma}\mathbf{C} & \mathbf{0} & \mathbf{I} \end{bmatrix} > \mathbf{0}\,.$$

Using the dual formula we get

$$\begin{bmatrix} \mathbf{X} & \mathbf{0} \\ \mathbf{0} & \mathbf{I} \end{bmatrix} > \begin{bmatrix} \mathbf{A} \\ \frac{1}{\gamma}\mathbf{C} \end{bmatrix} \mathbf{X} \begin{bmatrix} \mathbf{A}^T & \frac{1}{\gamma}\mathbf{C}^T \end{bmatrix}$$

which provides (8.38).

The parametrization of all constant matrices $\hat{\mathbf{D}}$ that solve the discrete-time zero-order $\mathcal{H}_\infty$ approximation problem is obtained following similar steps as in the continuous-time case. □

Example 8.1.1 Consider the zero-order $\mathcal{H}_\infty$ approximation problem for a system $\mathbf{G}(s)$ with state space matrices

$$\mathbf{A} = \begin{bmatrix} -2 & 3 & -1 & 1 \\ 0 & -1 & 1 & 0 \\ 0 & 0 & -3 & 12 \\ 0 & 0 & 0 & -4 \end{bmatrix}, \quad \mathbf{B} = \begin{bmatrix} -2.5 & 0 & -1.2 \\ 1.3 & -1 & 1 \\ 1.6 & 2 & 0 \\ -3.4 & 0.1 & 2 \end{bmatrix}$$

$$\mathbf{C} = \begin{bmatrix} -2.5 & 1.3 & 1.6 & -3.4 \\ 0 & -1 & 2 & 0.1 \\ -1.2 & 1 & 0 & 2 \end{bmatrix}, \quad \mathbf{D} = 0.$$

It can be shown that the following matrix $\mathbf{X}$ satisfies conditions (8.20) and (8.21) of Theorem 8.1.2 for $\gamma = 6.2824$:

$$X = \begin{bmatrix} 28.3554 & -4.3175 & 16.2043 & 0.9239 \\ -4.3175 & 9.7310 & -3.5798 & 0.5693 \\ 16.2043 & -3.5798 & 31.7152 & -0.7418 \\ 0.9239 & 0.5693 & -0.7418 & 3.3769 \end{bmatrix}.$$

Hence, according to Theorem 8.1.2, there exists a constant matrix $\hat{\mathbf{D}}$ to satisfy the zero-order $\mathcal{H}_\infty$ approximation

$$\|\mathbf{G}(s) - \hat{\mathbf{D}}\|_\infty \le 6.2824.$$

Numerical algorithms to obtain the feasible matrix $\mathbf{X}$ and the $\mathcal{H}_\infty$ approximation norm bound γ will be presented in Chapters 10 and 11. Using the techniques in these chapters, it can be shown that the above value of γ is the optimal $\mathcal{H}_\infty$ norm bound for this problem.

The zero-order H_∞ approximates that correspond to the feasible solution $\mathbf{X}$ are parametrized as follows:

$$\hat{\mathbf{D}} = \hat{\mathbf{D}}_1 + \hat{\mathbf{D}}_2 \mathbf{L} \hat{\mathbf{D}}_3$$

where $\hat{\mathbf{D}}_1$, $\hat{\mathbf{D}}_2$ and $\hat{\mathbf{D}}_3$ are computed from 8.23 as follows:

$$\hat{\mathbf{D}}_1 = \begin{bmatrix} 1.6731 & 0.7830 & 0.7233 \\ -1.9232 & 1.3841 & 2.7552 \\ -0.0738 & -0.4314 & 2.2064 \end{bmatrix}$$

$$\hat{\mathbf{D}}_2 = \begin{bmatrix} 1.8600 & -1.1183 & -1.8145 \\ -1.1183 & 1.3841 & -0.2787 \\ -1.8145 & -0.2787 & 4.7054 \end{bmatrix}$$

$$\hat{\mathbf{D}}_3 = \begin{bmatrix} 0.3627 & 0.2575 & 0.3165 \\ 0.2575 & 0.7781 & -0.0302 \\ 0.3165 & -0.0302 & 0.4987 \end{bmatrix}$$

and $\mathbf{L}$ is any matrix with $\|\mathbf{L}\| < 1$.

8.2 MODEL REDUCTION WITH COVARIANCE ERROR BOUNDS

The covariance error bound model reduction problem seeks to provide a reduced-order model to bound the covariance of the output difference of the full and the reduced-order models.

8.2.1 Continuous-Time Case

Consider a stable nth-order strictly proper system with state space representation

$$\dot{\mathbf{x}} = \mathbf{A}\mathbf{x} + \mathbf{B}\mathbf{w} \tag{8.41}$$

$$\mathbf{y} = \mathbf{C}\mathbf{x} \tag{8.42}$$

where $\mathbf{w}$ is a zero-mean white noise excitation with unit intensity. The suboptimal covariance upper bound model reduction problem is to find a reduced-order strictly proper system

$$\dot{\hat{\mathbf{x}}} = \hat{\mathbf{A}}\hat{\mathbf{x}} + \hat{\mathbf{B}}\mathbf{w} \tag{8.43}$$

$$\hat{\mathbf{y}} = \hat{\mathbf{C}}\hat{\mathbf{x}} \tag{8.44}$$

of order $\hat{n} < n$ such that the covariance matrix $\tilde{\mathbf{Y}} = \lim_{t\to\infty} \mathcal{E}\tilde{\mathbf{y}}\tilde{\mathbf{y}}^T$ of the output error $\tilde{\mathbf{y}} = \mathbf{y} - \hat{\mathbf{y}}$ satisfies a bound $\tilde{\mathbf{Y}} < \bar{\mathbf{Y}}$, where $\bar{\mathbf{Y}}$ is a specified symmetric positive definite matrix. For simplicity we will consider the case $\bar{\mathbf{Y}} = \epsilon\mathbf{I}$, where ϵ is a given positive scalar. Following the system performance analysis discussion in Chapter 4, a deterministic interpretation of this model reduction problem is to find a reduced-order model (8.43)–(8.44) such that the peak value of the output error $\|\tilde{\mathbf{y}}\|_{\mathcal{L}_\infty}$ is less than $\sqrt{\epsilon}$, when $\mathbf{w}$ is any bounded energy input with $\|\mathbf{w}\|_{\mathcal{L}_2} \leq 1$.

The solution to the upper bound covariance model reduction problem is given by the following result:

Theorem 8.2.1 *Let ϵ be a given positive scalar. Consider the stable linear time-invariant system (8.41)–(8.42) where $\mathbf{w}$ is a stochastic white noise process uncorrelated with $\mathbf{x}(0)$, and with covariance $\mathbf{I}$. Then the following statements are equivalent:*

(i) There exists a reduced-order model (8.43)–(8.44) of order $\hat{n}$ such that the output covariance error is bounded above by ϵ, that is $\tilde{\mathbf{Y}} < \epsilon\mathbf{I}$.

(ii) There exists a matrix pair $(\mathbf{X}, \mathbf{Y})$ such that

$$\begin{bmatrix} \mathbf{X} & \mathbf{I} \\ \mathbf{I} & \mathbf{Y} \end{bmatrix} \geq \mathbf{0}, \quad \text{rank}\begin{bmatrix} \mathbf{X} & \mathbf{I} \\ \mathbf{I} & \mathbf{Y} \end{bmatrix} \leq n + \hat{n} \tag{8.45}$$

$$\begin{aligned} \mathbf{AX} + \mathbf{XA}^T + \mathbf{BB}^T &< \mathbf{0} &(8.46)\\ \epsilon\mathbf{Y} - \mathbf{C}^T\mathbf{C} &> \mathbf{0} &(8.47)\\ \mathbf{YA} + \mathbf{A}^T\mathbf{Y} &< \mathbf{0}. &(8.48)\end{aligned}$$

In this case all such reduced-order models are given by

$$\begin{aligned} [\ \hat{\mathbf{B}}\ \ \hat{\mathbf{A}}\ ] &= -\mathbf{R}^{-1}\mathbf{\Gamma}\Phi\Lambda^T(\Lambda\Phi\Lambda^T)^{-1} + \mathbf{S}^{1/2}\mathbf{L}_1(\Lambda\Phi\Lambda^T)^{-1/2}\\ \hat{\mathbf{C}} &= \mathbf{C}\mathbf{X}_{21}\mathbf{X}_2^{-1} + (\epsilon\mathbf{I} - \mathbf{C}\mathbf{Y}^{-1}\mathbf{C}^T)^{1/2}\mathbf{L}_2\mathbf{X}_2^{-1/2}\end{aligned}$$

where

$$\mathbf{\Gamma} = \begin{bmatrix} \mathbf{0}\\ \mathbf{I}\\ \mathbf{0}\end{bmatrix},\ \mathbf{\Theta} = \begin{bmatrix} \mathbf{AX}+\mathbf{XA}^T & \mathbf{AX}_{21} & \mathbf{B}\\ \mathbf{X}_{21}^T\mathbf{A}^T & \mathbf{0} & \mathbf{0}\\ \mathbf{B}^T & \mathbf{0} & -\mathbf{I}\end{bmatrix},\ \Lambda^T = \begin{bmatrix} \mathbf{0} & \mathbf{X}_{21}\\ \mathbf{0} & \mathbf{X}_2\\ \mathbf{I} & \mathbf{0}\end{bmatrix} \qquad (8.49)$$

$$\begin{aligned} \mathbf{S} &= \mathbf{R}^{-1} - \mathbf{R}^{-1}\mathbf{\Gamma}^T[\mathbf{\Phi} - \mathbf{\Phi}\Lambda^T(\Lambda\mathbf{\Phi}\Lambda^T)^{-1}\Lambda\mathbf{\Phi}]\mathbf{\Gamma}\mathbf{R}^{-1}\\ \mathbf{\Phi} &= (\mathbf{\Gamma}\mathbf{R}^{-1}\mathbf{\Gamma}^T - \mathbf{\Theta})^{-1} > \mathbf{0}.\end{aligned}$$

$\mathbf{X}_{21}$ *and* $\mathbf{X}_2$ *are any matrix factors satisfying*

$$\mathbf{X} - \mathbf{Y}^{-1} = \mathbf{X}_{21}^T\mathbf{X}_2^{-1}\mathbf{X}_{21} > \mathbf{0}.$$

$\mathbf{R}$ *is an arbitrary positive definite matrix and* $\mathbf{L}_1$ *and* $\mathbf{L}_2$ *satisfy* $\|\mathbf{L}_1\| < 1$ *and* $\|\mathbf{L}_2\| < 1$.

Proof. Consider the augmented system

$$\begin{aligned} \dot{\tilde{\mathbf{x}}} &= \tilde{\mathbf{A}}\tilde{x} + \tilde{\mathbf{B}}\tilde{w} &(8.50)\\ \tilde{\mathbf{y}} &= \tilde{\mathbf{C}}\tilde{x} &(8.51)\end{aligned}$$

where

$$\begin{aligned} \tilde{\mathbf{A}} &= \mathbf{A}_1 + \mathbf{B}_1\hat{\mathbf{G}}_1\mathbf{M}_1\\ \tilde{\mathbf{B}} &= \mathbf{B}_2 + \mathbf{B}_1\hat{\mathbf{G}}_1\mathbf{M}_2\\ \tilde{\mathbf{C}} &= \mathbf{C}_1 - \hat{\mathbf{C}}B_1^T\end{aligned}$$

and

$$\begin{aligned} \mathbf{A}_1 &= \begin{bmatrix} \mathbf{A} & \mathbf{0}\\ \mathbf{0} & \mathbf{0}\end{bmatrix},\ \mathbf{B}_1 = \begin{bmatrix} \mathbf{0}\\ \mathbf{I}\end{bmatrix},\ \mathbf{B}_2 = \begin{bmatrix} \mathbf{B}\\ \mathbf{I}\end{bmatrix}\\ \mathbf{M}_1 &= \begin{bmatrix} \mathbf{0} & \mathbf{0}\\ \mathbf{0} & \mathbf{I}\end{bmatrix},\ \mathbf{M}_2 = \begin{bmatrix} \mathbf{I}\\ \mathbf{0}\end{bmatrix}\\ \mathbf{C}_1 &= [\ \mathbf{C}\ \ \mathbf{0}\ ],\ \hat{\mathbf{G}}_1 = [\ \hat{\mathbf{B}}\ \ \hat{\mathbf{A}}\ ].\end{aligned}$$

From Lemma 8.1.1 it follows that the upper bound model reduction problem has a solution if and only if there exists a matrix $\tilde{\mathbf{X}} > \mathbf{0}$ such that

$$\begin{aligned} \tilde{\mathbf{A}}\tilde{\mathbf{X}} + \tilde{\mathbf{X}}\tilde{\mathbf{A}}^T + \tilde{\mathbf{B}}\tilde{\mathbf{B}}^T &< \mathbf{0}\\ \tilde{\mathbf{C}}\tilde{\mathbf{X}}\tilde{\mathbf{C}}^T &< \epsilon\mathbf{I}\end{aligned}$$

that is

$$(\mathbf{A}_1 + \mathbf{B}_1\hat{\mathbf{G}}_1\mathbf{M}_1)\tilde{\mathbf{X}} + \tilde{\mathbf{X}}(\mathbf{A}_1 + \mathbf{B}_1\hat{\mathbf{G}}_1\mathbf{M}_1)^T + \tilde{\mathbf{B}}\tilde{B}^T < \mathbf{0} \tag{8.52}$$

$$(\mathbf{C}_1 - \hat{\mathbf{C}}B_1^T)\tilde{\mathbf{X}}(\mathbf{C}_1 - \hat{\mathbf{C}}B_1^T)^T < \epsilon\mathbf{I}. \tag{8.53}$$

The matrix inequality (8.52) is equivalent to the linear matrix inequality

$$\mathbf{\Gamma}\hat{G}_1\mathbf{\Lambda} + (\mathbf{\Gamma}\hat{G}_1\Lambda)^T + \mathbf{\Theta} < \mathbf{0}$$

where $\mathbf{\Gamma}$, $\mathbf{\Lambda}$ and $\mathbf{\Theta}$ are given by

$$\mathbf{\Gamma} = \begin{bmatrix} \mathbf{B}_1 \\ \mathbf{0} \end{bmatrix}, \ \mathbf{\Theta} = \begin{bmatrix} \mathbf{A}_1\tilde{\mathbf{X}} + \tilde{\mathbf{X}}A_1^T & \mathbf{B}_2 \\ \mathbf{B}_2^T & -\mathbf{I} \end{bmatrix}, \ \mathbf{\Lambda}^T = \begin{bmatrix} \tilde{\mathbf{X}}\mathbf{M}_1^T \\ \mathbf{M}_2^T \end{bmatrix}.$$

Using Theorem 2.3.12, it is easy to show that there exists a solution $\hat{\mathbf{G}}_1$ if and only if

$$\mathbf{B}_1^{\perp}(\mathbf{A}_1\tilde{\mathbf{X}} + \tilde{\mathbf{X}}\mathbf{A}_1^T + \mathbf{B}_2\mathbf{B}_2^T)\mathbf{B}_1^{\perp T} < 0$$

$$\begin{bmatrix} \mathbf{M}_1^T \\ \mathbf{M}_2^T \end{bmatrix}^{\perp} \begin{bmatrix} \tilde{\mathbf{X}}^{-1}\mathbf{A}_1 + \mathbf{A}_1^T\tilde{\mathbf{X}}^{-1} & \tilde{\mathbf{X}}^{-1}\mathbf{B}_2 \\ \mathbf{B}_2^T\tilde{\mathbf{X}}^{-1} & -\mathbf{I} \end{bmatrix} \begin{bmatrix} \mathbf{M}_1^T \\ \mathbf{M}_2^T \end{bmatrix}^{\perp T} < 0.$$

Partitioning $\tilde{\mathbf{X}}$ and $\tilde{\mathbf{X}}^{-1} = \tilde{\mathbf{Y}}$ as

$$\tilde{\mathbf{X}} = \begin{bmatrix} \mathbf{X} & \mathbf{X}_{21}^T \\ \mathbf{X}_{21} & \mathbf{X}_2 \end{bmatrix}, \ \tilde{\mathbf{Y}} = \begin{bmatrix} \mathbf{Y} & \mathbf{Y}_{21}^T \\ \mathbf{Y}_{21} & \mathbf{Y}_2 \end{bmatrix}$$

and utilizing the definitions of $\mathbf{A}_1$, $\mathbf{B}_1$,$\mathbf{B}_2$, $\mathbf{M}_1$ and $\mathbf{M}_2$ we get (8.46) and (8.48).

The solvability condition of (8.53) is obtained from Theorem 2.3.11 as follows

$$\mathbf{B}_1^{\perp}(\tilde{\mathbf{X}}^{-1} - \epsilon^{-1}\mathbf{C}_1^T\mathbf{C}_1)\mathbf{B}_1^{\perp T} > 0$$

which provides (8.46). Moreover, from $\tilde{\mathbf{X}}^{-1} = \tilde{\mathbf{Y}}$ we get that

$$\mathbf{X} - \mathbf{Y}^{-1} = \mathbf{X}_{21}^T\mathbf{X}_2^{-1}\mathbf{X}_{21} > \mathbf{0}$$

and also

$$\text{rank}\begin{bmatrix} \mathbf{X} & \mathbf{I} \\ \mathbf{I} & \mathbf{Y} \end{bmatrix} = \text{rank}(\mathbf{X} - \mathbf{Y}^{-1}) + \text{rank}(\mathbf{Y}) \leq n + \hat{n}$$

which provide (8.45).

The parametrization of all reduced-order models is obtained from the parametrization of all solutions in Theorems 2.3.11 and 2.3.12. □

Example 8.2.1 Consider the following continuous-time system

$$\begin{aligned} \dot{\mathbf{x}} &= \begin{bmatrix} -0.005 & -0.99 \\ -0.99 & -5000 \end{bmatrix}\mathbf{x} + \begin{bmatrix} 1 \\ 100 \end{bmatrix} w \\ y &= \begin{bmatrix} 1 & 100 \end{bmatrix}\mathbf{x}. \end{aligned}$$

We seek to obtain a first-order model that approximates the given system in a covariance error bound sense.

The following pair of matrices $(\mathbf{X}, \mathbf{Y})$ satisfy the conditions (8.45)–(8.48) of Theorem 8.2.1 for $\epsilon = 96.0787$

$$\mathbf{X} = \begin{bmatrix} 0.1161 & 0.0312 \\ 0.0312 & 108.9409 \end{bmatrix}, \quad \mathbf{Y} = \begin{bmatrix} 8.6139 & 0.0003 \\ 0.0003 & 1.3048 \end{bmatrix}.$$

Hence, there exists a first-order system to satisfy the output covariance error bound $\tilde{\mathbf{Y}} < 96.0787\mathbf{I}$. Using the formulae of Theorem 8.2.1 we obtain the following reduced-order system:

$$\begin{aligned} \dot{\hat{\mathbf{x}}} &= -4999.8\hat{\mathbf{x}} - 100.0\mathbf{w} \\ \hat{\mathbf{y}} &= -100.0\hat{\mathbf{x}}. \end{aligned}$$

The values of $\mathbf{X}$, $\mathbf{Y}$ and ϵ can be obtained utilizing the algorithms that will be presented in Chapters 10 and 11. In fact it can be shown that the above value of the covariance error bound ϵ is the minimum one.

For comparison, it is noted that a balanced model reduction method produces the following reduced-order model

$$\begin{aligned} \dot{\hat{\mathbf{x}}} &= -0.005\hat{\mathbf{x}} + \mathbf{w} \\ \hat{\mathbf{y}} &= \hat{\mathbf{x}} \end{aligned}$$

and the covariance error bound that corresponds to this model is $\bar{\mathbf{Y}} = 10^4\mathbf{I}$. Hence, balanced model reduction could result in reduced-order models with poor covariance error bounds.

8.2.2 Discrete-Time Case

We consider here the discrete-time covariance bounded model reduction problem defined as follows. Let the model, its reduction, and the corresponding output error $\tilde{\mathbf{y}} = \mathbf{y} - \hat{\mathbf{y}}$, be described by (8.26)–(8.27). Assume $\mathbf{w}$ is a zero-mean white noise with covariance $\mathbf{I}$. Define the output covariance by $\tilde{\mathbf{Y}} = \lim_{t\to\infty} \mathcal{E}(\mathbf{y} - \hat{\mathbf{y}})(\mathbf{y} - \hat{\mathbf{y}})^T$. Then the covariance bounded model reduction problem is to find a realization of specified order $\hat{n}$ such that $\tilde{\mathbf{Y}} \leq \bar{\mathbf{Y}}$ where $\bar{\mathbf{Y}}$ is a specified symmetric positive definite matrix. The solution to this problem is described as follows, where we set $\bar{\mathbf{Y}} = \epsilon\mathbf{I}$ for simplicity.

Consider a stable nth-order LTI discrete-time system

$$\begin{bmatrix} \mathbf{y}_k \\ \mathbf{x}_{k+1} \end{bmatrix} = \begin{bmatrix} \mathbf{D} & \mathbf{C} \\ \mathbf{B} & \mathbf{A} \end{bmatrix} \begin{bmatrix} \mathbf{w}_k \\ \mathbf{x}_k \end{bmatrix}. \tag{8.54}$$

We want to find a reduced-order system of order $\hat{n} \leq n$

$$\begin{bmatrix} \hat{\mathbf{y}}_k \\ \hat{\mathbf{x}}_{k+1} \end{bmatrix} = \hat{\mathbf{G}} \begin{bmatrix} \mathbf{w}_k \\ \hat{\mathbf{x}}_k \end{bmatrix} \tag{8.55}$$

such that the output covariance of $\mathbf{y} - \hat{\mathbf{y}}$ is bounded by $\epsilon\mathbf{I}$.

By augmenting the given model (8.54) to the reduced model (8.55), defining $\tilde{\mathbf{X}}$ as a state covariance upper bound of the augmented system, namely,

$$\tilde{\mathbf{X}} \geq \mathcal{E}_\infty \begin{bmatrix} \mathbf{x} \\ \hat{\mathbf{x}} \end{bmatrix} \begin{bmatrix} \mathbf{x}^T & \hat{\mathbf{x}}^T \end{bmatrix},$$

then the upper bound $\tilde{\mathbf{X}}$ and the covariance $\tilde{\mathbf{Y}}$ satisfy

$$\tilde{\mathbf{A}}\tilde{\mathbf{X}}\tilde{\mathbf{A}}^T + \tilde{\mathbf{B}}\tilde{\mathbf{B}}^T < \tilde{\mathbf{X}} \tag{8.56}$$

$$\tilde{\mathbf{Y}} \leq \tilde{\mathbf{C}}\tilde{\mathbf{X}}\tilde{\mathbf{C}}^T + \tilde{\mathbf{D}}\tilde{\mathbf{D}}^T < \epsilon\mathbf{I} \tag{8.57}$$

where

$$\hat{\mathbf{G}} = \left[\begin{array}{c} \hat{\mathbf{G}}_1 \\ \hline \hat{\mathbf{G}}_2 \end{array}\right] = \left[\begin{array}{cc} \hat{\mathbf{D}} & \hat{\mathbf{C}} \\ \hline \hat{\mathbf{B}} & \hat{\mathbf{A}} \end{array}\right]$$

and

$$\tilde{\mathbf{A}} = \mathbf{A}_1 + \mathbf{B}_1\hat{\mathbf{G}}_2\mathbf{M}_1, \quad \tilde{\mathbf{B}} = \mathbf{B}_2 + \mathbf{B}_1\hat{\mathbf{G}}_2\mathbf{M}_2$$
$$\tilde{\mathbf{C}} = \mathbf{C}_1 - \hat{\mathbf{G}}_1\mathbf{M}_1, \quad \tilde{\mathbf{D}} = \mathbf{D} - \hat{\mathbf{G}}_1\mathbf{M}_2$$
$$\mathbf{A}_1 = \begin{bmatrix} \mathbf{A} & \mathbf{0} \\ \mathbf{0} & \mathbf{0} \end{bmatrix}, \quad \mathbf{B}_1 = \begin{bmatrix} \mathbf{0} \\ \mathbf{I} \end{bmatrix}, \quad \mathbf{M}_1 = \begin{bmatrix} \mathbf{0} & \mathbf{0} \\ \mathbf{0} & \mathbf{I} \end{bmatrix}$$
$$\mathbf{B}_2 = \begin{bmatrix} \mathbf{B} \\ \mathbf{0} \end{bmatrix}, \quad \mathbf{M}_2 = \begin{bmatrix} \mathbf{I} \\ \mathbf{0} \end{bmatrix}, \quad \mathbf{C}_1 = \begin{bmatrix} \mathbf{C} & \mathbf{0} \end{bmatrix}.$$

The solution for the covariance bounded model reduction is given as follows.

Theorem 8.2.2 *The following statements are equivalent.*

(i) There exists an $\hat{n}$th-order system (8.55) that solves the covariance bounded model reduction problem, $\tilde{\mathbf{Y}} < \epsilon\mathbf{I}$.

(ii) There exists a pair of $n \times n$ matrices $(\mathbf{X}, \mathbf{Y})$ such that

$$\begin{bmatrix} \mathbf{X} & \mathbf{I} \\ \mathbf{I} & \mathbf{Y} \end{bmatrix} \geq \mathbf{0}, \qquad \text{rank}\begin{bmatrix} \mathbf{X} & \mathbf{I} \\ \mathbf{I} & \mathbf{Y} \end{bmatrix} \leq n + \hat{n}$$
$$\mathbf{A}\mathbf{X}\mathbf{A}^T - \mathbf{X} + \mathbf{B}\mathbf{B}^T < \mathbf{0}$$
$$\mathbf{C}^T\mathbf{C} - \epsilon\mathbf{Y} < \mathbf{0}$$
$$\mathbf{A}^T\mathbf{Y}\mathbf{A} - \mathbf{Y} < \mathbf{0}.$$

If the above conditions hold for some $\mathbf{X}$ and $\mathbf{Y}$, then all the solutions can be expressed by:

$$\hat{\mathbf{G}}_1 = \hat{\mathbf{A}} + \hat{\mathbf{Q}}^{\frac{1}{2}}\hat{\mathbf{L}}\mathbf{R}^{-\frac{1}{2}}, \quad \hat{\mathbf{G}} = \begin{bmatrix} \hat{\mathbf{G}}_1^T & \hat{\mathbf{G}}_2^T \end{bmatrix}^T, \quad \|\hat{\mathbf{L}}\| < 1$$
$$\hat{\mathbf{G}}_2 = -(\mathbf{B}_1^T\mathbf{Q}^{-1}\mathbf{B}_1)^{-1}\mathbf{B}_1^T\mathbf{Q}^{-1}\bar{\mathbf{A}} + (\mathbf{B}_1^T\mathbf{Q}^{-1}\mathbf{B}_1)^{-\frac{1}{2}}\mathbf{L}\Psi^{\frac{1}{2}}, \quad \|\mathbf{L}\| < 1$$

where $\tilde{\mathbf{X}}_{21}$ and $\tilde{\mathbf{X}}_2$ are any matrix factors satisfying

$$\mathbf{X} - \mathbf{Y}^{-1} = \tilde{\mathbf{X}}_{21}^T\tilde{\mathbf{X}}_2^{-1}\tilde{\mathbf{X}}_{21}$$

and $\hat{\mathbf{X}}$ is defined by,

$$\tilde{\mathbf{X}} = \begin{bmatrix} \mathbf{X} & \tilde{\mathbf{X}}_{21}^T \\ \tilde{\mathbf{X}}_{21} & \tilde{\mathbf{X}}_2 \end{bmatrix},$$

where

$$\hat{\mathbf{Q}} = \epsilon\mathbf{I} - \mathbf{C}\mathbf{Y}^{-1}\mathbf{C}^T$$

$$\mathbf{Q} = \begin{bmatrix} \mathbf{X} - \mathbf{A}\mathbf{Y}^{-1}\mathbf{A}^T & \tilde{\mathbf{X}}_{21}^T \\ \tilde{\mathbf{X}}_{21} & \tilde{\mathbf{X}}_2 \end{bmatrix}$$
$$\bar{\mathbf{A}} = \begin{bmatrix} \mathbf{B} & \mathbf{A}\tilde{\mathbf{X}}_{21}^T\tilde{\mathbf{X}}_2^{-1} \\ \mathbf{0} & \mathbf{0} \end{bmatrix}$$
$$\hat{\mathbf{A}} = \begin{bmatrix} \mathbf{D} & \mathbf{C}\tilde{\mathbf{X}}_{21}^T\tilde{\mathbf{X}}_2^{-1} \end{bmatrix}$$
$$\mathbf{R} = \begin{bmatrix} \mathbf{I} & \mathbf{0} \\ \mathbf{0} & \tilde{\mathbf{X}}_2 \end{bmatrix}$$
$$\mathbf{\Psi} = \mathbf{R}^{-1} - \bar{\mathbf{A}}^T\mathbf{Q}^{-1}\bar{\mathbf{A}} + \bar{\mathbf{A}}^T\mathbf{Q}^{-1}\mathbf{B}_1(\mathbf{B}_1\mathbf{Q}^{-1}\mathbf{B}_1)^{-1}\mathbf{B}_1^T\mathbf{Q}^{-1}\bar{\mathbf{A}}.$$

Proof. (8.56) and (8.57) can be written as

$$(\mathbf{A}_1 + \mathbf{B}_1\hat{\mathbf{G}}_2\bar{\mathbf{M}})\tilde{\mathbf{X}}(\mathbf{A}_1 + \mathbf{B}_1\hat{\mathbf{G}}_2\bar{\mathbf{M}})^T + (\bar{\mathbf{D}} + \mathbf{B}_1\hat{\mathbf{G}}_2\bar{\mathbf{E}})(\bar{\mathbf{D}} + \mathbf{B}_1\hat{\mathbf{G}}_2\bar{\mathbf{E}})^T < \tilde{\mathbf{X}} \tag{8.58}$$
$$(\bar{\mathbf{C}} - \hat{\mathbf{G}}_1\bar{\mathbf{M}})\tilde{\mathbf{X}}(\bar{\mathbf{C}} - \hat{\mathbf{G}}_1\bar{\mathbf{M}})^T + (\mathbf{D} - \hat{\mathbf{G}}_1\bar{\mathbf{E}})(\mathbf{D} - \hat{\mathbf{G}}_1)^T < \epsilon\mathbf{I} \tag{8.59}$$

i.e. the above two matrix inequalities are decoupled with respect to the unknowns $\hat{\mathbf{G}}_1$ and $\hat{\mathbf{G}}_2$. After completing the square in $\hat{\mathbf{G}}_2$, Theorem 2.3.11 can be used to show that there exists a $\hat{\mathbf{G}}_2$ solving (8.58) if and only if there exists a $\tilde{\mathbf{X}} = \tilde{\mathbf{X}}^T > \mathbf{0}$ with $\tilde{\mathbf{X}} \in \mathcal{R}^{(n+\hat{n})\times(n+\hat{n})}$ such that

$$\mathbf{B}_1^\perp(\tilde{\mathbf{X}} - \mathbf{A}_1\tilde{\mathbf{X}}\mathbf{A}_1^T - \mathbf{B}_2\mathbf{B}_2^T)\mathbf{B}_1^{\perp T} > \mathbf{0}$$
$$\begin{bmatrix} \mathbf{M}_1^T \\ \mathbf{M}_2^T \end{bmatrix}^\perp \begin{bmatrix} \tilde{\mathbf{X}}^{-1} - \mathbf{A}_1^T\tilde{\mathbf{X}}^{-1}\mathbf{A}_1 & -\mathbf{A}_1\tilde{\mathbf{X}}^{-1}\mathbf{B}_2 \\ -\mathbf{B}_2\tilde{\mathbf{X}}^{-1}\mathbf{A}_1 & \mathbf{I} - \mathbf{B}_2^T\tilde{\mathbf{X}}^{-1}\mathbf{B}_2 \end{bmatrix} \begin{bmatrix} \mathbf{M}_1^T \\ \mathbf{M}_2^T \end{bmatrix}^{\perp T} > \mathbf{0}.$$

Partitioning $\tilde{\mathbf{X}}$ and $\tilde{\mathbf{Y}} = \tilde{\mathbf{X}}^{-1} > \mathbf{0}$ as

$$\tilde{\mathbf{X}} = \begin{bmatrix} \mathbf{X} & \tilde{\mathbf{X}}_{21}^T \\ \tilde{\mathbf{X}}_{21} & \tilde{\mathbf{X}}_2 \end{bmatrix}, \quad \tilde{\mathbf{Y}} = \begin{bmatrix} \mathbf{Y} & \tilde{\mathbf{Y}}_{21}^T \\ \tilde{\mathbf{Y}}_{21} & \tilde{\mathbf{Y}}_2 \end{bmatrix}$$

and substituting the known matrices $\mathbf{A}_1, \mathbf{B}_1, \mathbf{B}_2, \mathbf{M}_1$ and $\mathbf{M}_2$ together with the partition of $\tilde{\mathbf{X}}$ and $\tilde{\mathbf{Y}}$ into the above inequalities leads to

$$\mathbf{X} - \mathbf{A}\mathbf{X}\mathbf{A}^T - \mathbf{B}\mathbf{B}^T > \mathbf{0}$$
$$\mathbf{Y} - \mathbf{A}^T\mathbf{Y}\mathbf{A} > \mathbf{0}.$$

The same technique for solving (8.58) for $\hat{\mathbf{G}}_2$ can also solve (8.59) for $\hat{\mathbf{G}}_1$ by substituting the parameters $\mathbf{A}_1 \to \bar{\mathbf{C}}, \mathbf{B}_1 \to -\mathbf{I}, \bar{\mathbf{D}} \to \mathbf{D}$, etc. (Note that $\mathbf{B}_1^\perp$ is void in this case.) Hence there exists a $\hat{\mathbf{G}}_1$ solving (8.59) if and only if there exists $\tilde{\mathbf{X}}$ satisfying

$$\begin{bmatrix} \mathbf{M}_1^T \\ \mathbf{M}_2^T \end{bmatrix}^\perp \begin{bmatrix} \tilde{\mathbf{X}}^{-1} - \mathbf{C}_1^T\epsilon^{-1}\mathbf{C}_1 & -\mathbf{C}_1^T\epsilon^{-1}\mathbf{D} \\ -\mathbf{D}^T\epsilon^{-1}\mathbf{C}_1 & \mathbf{I} - \mathbf{D}^T\epsilon^{-1}\mathbf{D} \end{bmatrix} \begin{bmatrix} \mathbf{M}_1^T \\ \mathbf{M}_2^T \end{bmatrix}^{\perp T} > \mathbf{0}$$

which is equivalent to

$$\mathbf{Y} - \mathbf{C}^T\mathbf{C}/\epsilon > \mathbf{0}.$$

Notice that, by Schur complement, $\tilde{\mathbf{Y}} = \tilde{\mathbf{X}}^{-1}$ implies

$$\mathbf{Y} = (\mathbf{X} - \tilde{\mathbf{X}}_{21}^T\tilde{\mathbf{X}}_2^{-1}\tilde{\mathbf{X}}_{21})^{-1}$$

or

$$\mathbf{X} - \mathbf{Y}^{-1} = \tilde{\mathbf{X}}_{21}^T\tilde{\mathbf{X}}_2^{-1}\tilde{\mathbf{X}}_{21} \geq \mathbf{0}.$$

By another Schur complement, the above leads to

$$\begin{bmatrix} \mathbf{X} & \mathbf{I} \\ \mathbf{I} & \mathbf{Y} \end{bmatrix} \geq \mathbf{0}.$$

Since $\tilde{\mathbf{X}}_2 \in \mathcal{R}^{\hat{n}\times\hat{n}}$ we have

$$\text{rank}(\mathbf{X} - \mathbf{Y}^{-1}) \leq \hat{n}.$$

Also consider

$$\begin{aligned}
\text{rank}\begin{bmatrix} \mathbf{X} & \mathbf{I} \\ \mathbf{I} & \mathbf{Y} \end{bmatrix} &= \text{rank}\begin{bmatrix} \mathbf{I} & -\mathbf{Y}^{-1} \\ \mathbf{0} & \mathbf{I} \end{bmatrix}\begin{bmatrix} \mathbf{X} & \mathbf{I} \\ \mathbf{I} & \mathbf{Y} \end{bmatrix}\begin{bmatrix} \mathbf{I} & \mathbf{0} \\ -\mathbf{Y}^{-1} & \mathbf{I} \end{bmatrix} \\
&= \text{rank}\begin{bmatrix} \mathbf{X} - \mathbf{Y}^{-1} & \mathbf{0} \\ \mathbf{0} & \mathbf{Y} \end{bmatrix} \\
&= \text{rank}\,(\mathbf{X} - \mathbf{Y}^{-1}) + \text{rank}(\mathbf{Y}) \leq \hat{n} + n.
\end{aligned}$$

This proves the existence condition in (ii). The rest of the theorem comes from direct application of the other part of Theorem 2.3.11. □

CHAPTER 8 SUMMARY

This chapter provides necessary and sufficient conditions for the solution of two kinds of model reduction problems. All reduced-order models are parametrized that have modeling error less than a specified "value". The two mesurements for "value" are the frequency response errors and covariance errors. One cannot know what is a good model without knowing its purpose. Hence, the modeling and control problems are not independent. Modeling for control design is still an active research topic. This chapter only gives results that lend themselves to the matrix inequality methods of this book. See the special journal issue [128] devoted to control-related modeling approaches.

CHAPTER NINE

Unified Perspective

This chapter shows a unified perspective based on Linear Matrix Inequalities (LMIs) for designing controllers with various specifications including stability, performance and robustness. In particular, many control design problems can be reduced to just one linear algebra problem of solving an LMI for the controller parameter $\mathbf{G}$. In fact, the covariance control problem and the $\mathcal{H}_\infty$ control problem treated in the previous chapters are examples of such control problems. The purpose of this chapter is to show explicitly the common mathematical structure of these and other control problems hidden in the proofs of the previous synthesis results. We shall list 17 control problems that all reduce to a *single* linear algebra problem to solve

$$\mathbf{\Gamma G \Lambda} + (\mathbf{\Gamma G \Lambda})^T + \mathbf{\Theta} < \mathbf{0}$$

for the matrix $\mathbf{G}$.

Consider the linear time-invariant plant and the controller

$$\begin{bmatrix} \partial \mathbf{x}_p \\ \mathbf{y}_1 \\ \mathbf{y}_2 \\ \mathbf{z} \end{bmatrix} = \begin{bmatrix} \mathbf{A}_p & \mathbf{D}_{p1} & \mathbf{D}_{p2} & \mathbf{B}_p \\ \mathbf{C}_{p1} & \mathbf{D}_{y11} & \mathbf{D}_{y12} & \mathbf{B}_{y1} \\ \mathbf{C}_{p2} & \mathbf{D}_{y21} & \mathbf{D}_{y22} & \mathbf{B}_{y2} \\ \mathbf{M}_p & \mathbf{D}_{z1} & \mathbf{D}_{z2} & \mathbf{0} \end{bmatrix} \begin{bmatrix} \mathbf{x}_p \\ \mathbf{w}_1 \\ \mathbf{w}_2 \\ \mathbf{u} \end{bmatrix}, \quad \begin{bmatrix} \partial \mathbf{x}_c \\ \mathbf{u} \end{bmatrix} = \begin{bmatrix} \mathbf{A}_c & \mathbf{B}_c \\ \mathbf{C}_c & \mathbf{D}_c \end{bmatrix} \begin{bmatrix} \mathbf{x}_c \\ \mathbf{z} \end{bmatrix} \tag{9.1}$$

where ∂ is the differentiation/delay operator for the continuous-time/discrete-time case, $\mathbf{x}_p$ and $\mathbf{x}_c$ are the plant and the controller states, respectively, $\mathbf{z}$ and $\mathbf{u}$ are the measured output and the control input, and $\mathbf{y}_1$, $\mathbf{y}_2$, $\mathbf{w}_1$ and $\mathbf{w}_2$ are the exogenous signals that will be used to describe control design specifications. The closed-loop system can be described by

$$\begin{bmatrix} \partial \mathbf{x} \\ \mathbf{y}_1 \\ \mathbf{y}_2 \end{bmatrix} = \begin{bmatrix} \mathbf{A}_{c\ell} & \mathbf{B}_{c\ell 1} & \mathbf{B}_{c\ell 2} \\ \mathbf{C}_{c\ell 1} & \mathbf{D}_{c\ell 11} & \mathbf{D}_{c\ell 12} \\ \mathbf{C}_{c\ell 2} & \mathbf{D}_{c\ell 21} & \mathbf{D}_{c\ell 22} \end{bmatrix} \begin{bmatrix} \mathbf{x} \\ \mathbf{w}_1 \\ \mathbf{w}_2 \end{bmatrix}, \tag{9.2}$$

where $\mathbf{x} \triangleq [\ \mathbf{x}_p^T\ \mathbf{x}_c^T\ ]^T$ is the closed-loop state and

$$\begin{bmatrix} \mathbf{A}_{c\ell} & \mathbf{B}_{c\ell 1} & \mathbf{B}_{c\ell 2} \\ \mathbf{C}_{c\ell 1} & \mathbf{D}_{c\ell 11} & \mathbf{D}_{c\ell 12} \\ \mathbf{C}_{c\ell 2} & \mathbf{D}_{c\ell 21} & \mathbf{D}_{c\ell 22} \end{bmatrix} \triangleq \begin{bmatrix} \mathbf{A} & \mathbf{D}_1 & \mathbf{D}_2 \\ \mathbf{C}_1 & \mathbf{F}_{11} & \mathbf{F}_{12} \\ \mathbf{C}_2 & \mathbf{F}_{21} & \mathbf{F}_{22} \end{bmatrix} + \begin{bmatrix} \mathbf{B} \\ \mathbf{H}_1 \\ \mathbf{H}_2 \end{bmatrix} \mathbf{G} [\ \mathbf{M} \quad \mathbf{E}_1 \quad \mathbf{E}_2\ ] \tag{9.3}$$

$$\begin{bmatrix} \mathbf{A} & \mathbf{D}_1 & \mathbf{D}_2 & \mathbf{B} \\ \mathbf{C}_1 & \mathbf{F}_{11} & \mathbf{F}_{12} & \mathbf{H}_1 \\ \mathbf{C}_2 & \mathbf{F}_{21} & \mathbf{F}_{22} & \mathbf{H}_2 \\ \mathbf{M} & \mathbf{E}_1 & \mathbf{E}_2 & \mathbf{G}^T \end{bmatrix} \triangleq \left[\begin{array}{cc|c|c|cc} \mathbf{A}_p & \mathbf{0} & \mathbf{D}_{p1} & \mathbf{D}_{p2} & \mathbf{B}_p & \mathbf{0} \\ \mathbf{0} & \mathbf{0} & \mathbf{0} & \mathbf{0} & \mathbf{0} & \mathbf{I}_{n_c} \\ \hline \mathbf{C}_{p1} & \mathbf{0} & \mathbf{D}_{y11} & \mathbf{D}_{y12} & \mathbf{B}_{y1} & \mathbf{0} \\ \hline \mathbf{C}_{p2} & \mathbf{0} & \mathbf{D}_{y21} & \mathbf{D}_{y22} & \mathbf{B}_{y2} & \mathbf{0} \\ \hline \mathbf{M}_p & \mathbf{0} & \mathbf{D}_{z1} & \mathbf{D}_{z2} & \mathbf{D}_c^T & \mathbf{B}_c^T \\ \mathbf{0} & \mathbf{I}_{n_c} & \mathbf{0} & \mathbf{0} & \mathbf{C}_c^T & \mathbf{A}_c^T \end{array}\right]. \tag{9.4}$$

These definitions will be used in the subsequent sections where we show how to solve different control problems in a unified manner.

9.1 CONTINUOUS-TIME CASE

In this section, we shall present a unified perspective for the continuous-time case. The main result of this section shows that certain control problems with stability, performance and robustness specifications can be reduced to a mathematical problem of solving an LMI

$$\mathbf{\Gamma G \Lambda} + (\mathbf{\Gamma G \Lambda})^T + \mathbf{\Theta} < \mathbf{0} \tag{9.5}$$

for the controller parameter $\mathbf{G}$. Specifically, we show appropriate matrices $\mathbf{\Gamma}$, $\mathbf{\Lambda}$, and $\mathbf{\Theta}$ for each control problems in terms of the plant data and a Lyapunov matrix (and possibly a scaling matrix). The linear algebra problem (9.5) can be solved by using Theorem 2.3.12, which provides necessary and sufficient conditions for the existence of $\mathbf{G}$ satisfying (9.5);

$$\mathbf{\Gamma}^{\perp}\mathbf{\Theta}\mathbf{\Gamma}^{\perp T} < \mathbf{0}, \quad \mathbf{\Lambda}^{T\perp}\mathbf{\Theta}\mathbf{\Lambda}^{T\perp T} < \mathbf{0}. \tag{9.6}$$

As has been shown for the covariance upper bound control problem and the $\mathcal{H}_\infty$ control problem in Chapters 6 and 7, these existence conditions reduce to LMIs in terms of Lyapunov matrix $\mathbf{X}$ and its inverse $\mathbf{X}^{-1}$, and possibly a scaling matrix $\mathbf{S}$, which can further be specialized to Riccati inequalities with certain assumptions on the plant. Finally, when (9.6) holds, one such $\mathbf{G}$ is given by

$$\mathbf{G} = -\rho\mathbf{\Gamma}^T\mathbf{\Phi}\mathbf{\Lambda}^T(\mathbf{\Lambda\Phi\Lambda}^T)^{-1} \tag{9.7}$$

where $\rho > 0$ is an arbitrary scalar such that

$$\mathbf{\Phi} \triangleq (\rho\mathbf{\Gamma\Gamma}^T - \mathbf{\Theta})^{-1} > \mathbf{0}.$$

All such $\mathbf{G}$ are also available in Theorem 2.3.12.

9.1.1 Stabilizing Control

The simplest control problem is the one with only the stability specification. Consider a linear time-invariant plant:

$$\begin{bmatrix} \dot{\mathbf{x}}_p \\ \mathbf{z} \end{bmatrix} = \begin{bmatrix} \mathbf{A}_p & \mathbf{B}_p \\ \mathbf{M}_p & \mathbf{0} \end{bmatrix} \begin{bmatrix} \mathbf{x}_p \\ \mathbf{u} \end{bmatrix} \tag{9.8}$$

where $\mathbf{x}_p$ is the state, $\mathbf{u}$ is the control input and $\mathbf{z}$ is the measured output. The stabilization problem is the following.

> Determine whether or not there exists a controller in (9.1) which asymptotically stabilizes the system (9.8). Parametrize all such controllers when one exists.

This problem can be reduced to the above linear algebra problem (9.5) as follows.

Theorem 9.1.1 *Let a controller* **G** *be given. Then the following statements are equivalent:*

(i) The controller **G** *solves the stabilization problem.*
(ii) There exists a matrix $\mathbf{X} > \mathbf{0}$ *such that (9.5) holds where*

$$\left[\ \boldsymbol{\Gamma} \quad \boldsymbol{\Lambda}^T \quad \boldsymbol{\Theta}\ \right] \triangleq \left[\ \mathbf{B} \ \middle|\ \mathbf{X}\mathbf{M}^T \ \middle|\ \mathbf{A}\mathbf{X} + \mathbf{X}\mathbf{A}^T\ \right].$$

Proof. The result directly follows from Lyapunov's stability theory which states that (i) is equivalent to the existence of a Lyapunov matrix $\mathbf{X} > \mathbf{0}$ satisfying

$$(\mathbf{A} + \mathbf{B}\mathbf{G}\mathbf{M})\mathbf{X} + \mathbf{X}(\mathbf{A} + \mathbf{B}\mathbf{G}\mathbf{M})^T < \mathbf{0}$$

where matrices **A**, **B** and **M** are the augmented matrices defined in (9.4). □

9.1.2 Covariance Upper Bound Control

Next, we shall consider the covariance upper bound control problem discussed in Chapter 6. We explicitly show that the problem can be reduced to the LMI of the form (9.5) although this reduction process has already been done implicitly in the proofs given in Chapter 6. To this end, we shall restate the covariance upper bound control problem for the reader's convenience. Consider the linear time-invariant system

$$\begin{bmatrix} \dot{\mathbf{x}}_p \\ \mathbf{y} \\ \mathbf{z} \end{bmatrix} = \begin{bmatrix} \mathbf{A}_p & \mathbf{D}_p & \mathbf{B}_p \\ \mathbf{C}_p & \mathbf{0} & \mathbf{0} \\ \mathbf{M}_p & \mathbf{D}_z & \mathbf{0} \end{bmatrix} \begin{bmatrix} \mathbf{x}_p \\ \mathbf{w} \\ \mathbf{u} \end{bmatrix} \tag{9.9}$$

where $\mathbf{x}_p$ is the state, **w** is the white noise with intensity **I**, **u** is the control input, **y** is the output of interest and **z** is the measured output. The covariance upper bound control problem is the following.

> Let an output covariance bound $\mho > \mathbf{0}$ be given. Determine whether or not there exists a controller in (9.1) which asymptotically stabilizes the system (9.9) and yields an output covariance satisfying
>
> $$\lim_{t\to\infty} \mathcal{E}\,[\mathbf{y}(t)\mathbf{y}^T(t)] < \mho.$$
>
> Parametrize all such controllers when one exists.

This problem can be reduced to the following.

Theorem 9.1.2 *Let a controller* **G** *and an output covariance bound* $\mho > \mathbf{0}$ *be given. Then the following statements are equivalent:*

(i) The controller **G** *solves the covariance upper bound control problem.*
(ii) There exists a matrix $\mathbf{X} > \mathbf{0}$ *such that* $\mathbf{C}\mathbf{X}\mathbf{C}^T < \mho$ *and (9.5) hold where*

$$\left[\ \boldsymbol{\Gamma} \quad \boldsymbol{\Lambda}^T \quad \boldsymbol{\Theta}\ \right] \triangleq \left[\begin{array}{c|c|cc} \mathbf{B} & \mathbf{X}\mathbf{M}^T & \mathbf{A}\mathbf{X} + \mathbf{X}\mathbf{A}^T & \mathbf{D} \\ \mathbf{0} & \mathbf{E}^T & \mathbf{D}^T & -\mathbf{I} \end{array}\right].$$

Proof. Recall from Chapter 6 that statement (i) holds if and only if there exists a state covariance upper bound $\mathbf{X} > \mathbf{0}$ such that $\mathbf{CXC}^T < \mathcal{U}$ and

$$(\mathbf{A}+\mathbf{BGM})\mathbf{X}+\mathbf{X}(\mathbf{A}+\mathbf{BGM})^T+(\mathbf{D}+\mathbf{BGE})(\mathbf{D}+\mathbf{BGE})^T < \mathbf{0}$$

or equivalently, by the Schur complement formula,

$$\begin{bmatrix} (\mathbf{A}+\mathbf{BGM})\mathbf{X}+\mathbf{X}(\mathbf{A}+\mathbf{BGM})^T & \mathbf{D}+\mathbf{BGE} \\ (\mathbf{D}+\mathbf{BGE})^T & -\mathbf{I} \end{bmatrix} < \mathbf{0}.$$

It is trivial to verify the equivalence of the above statements and (ii). □

9.1.3 Linear Quadratic Regulator

Another control problem which falls into the framework of (9.5) is the Linear Quadratic Regulator (LQR) problem. Consider the linear time-invariant system

$$\begin{bmatrix} \dot{\mathbf{x}}_p \\ \mathbf{y} \\ \mathbf{z} \end{bmatrix} = \begin{bmatrix} \mathbf{A}_p & \mathbf{D}_p & \mathbf{B}_p \\ \mathbf{C}_p & \mathbf{0} & \mathbf{B}_y \\ \mathbf{M}_p & \mathbf{0} & \mathbf{0} \end{bmatrix} \begin{bmatrix} \mathbf{x}_p \\ \mathbf{w} \\ \mathbf{u} \end{bmatrix} \tag{9.10}$$

where $\mathbf{x}_p$ is the state, $\mathbf{w}$ is the impulsive disturbance $\mathbf{w}(t) = \mathbf{w}_0\delta(t)$ where $\delta(\cdot)$ is the Dirac delta function, $\mathbf{u}$ is the control input, $\mathbf{y}$ is the output of interest and $\mathbf{z}$ is the measured output. The LQR problem is to guarantee an upper bound on the square integral of the output signal as follows.

> Let a performance bound $\gamma > 0$ be given. Determine whether or not there exists a controller in (9.1) which asymptotically stabilizes the system (9.10) and yields the zero initial state response $\mathbf{y}$ such that $\|\mathbf{y}\|_{\mathcal{L}_2} < \gamma$ for all directions of the impulsive disturbance $\|\mathbf{w}_0\| \le 1$. Parametrize all such controllers when one exists.

It turns out that this problem can be reduced to a mathematical problem which is the dual of that for the covariance upper bound control problem.

Theorem 9.1.3 *Let a controller* $\mathbf{G}$ *and a performance bound* $\gamma > 0$ *be given. Then the following statements are equivalent:*

(i) The controller $\mathbf{G}$ *solves the LQR problem.*
(ii) There exists a matrix $\mathbf{Y} > \mathbf{0}$ *such that* $\|\mathbf{D}^T\mathbf{Y}\mathbf{D}\| < \gamma^2$ *and (9.5) hold where*

$$\begin{bmatrix} \mathbf{\Gamma} & \mathbf{\Lambda}^T & \mathbf{\Theta} \end{bmatrix} \triangleq \left[\begin{array}{c|c|cc} \mathbf{YB} & \mathbf{M}^T & \mathbf{YA}+\mathbf{A}^T\mathbf{Y} & \mathbf{M}^T \\ \mathbf{H} & \mathbf{0} & \mathbf{M} & -\mathbf{I} \end{array}\right].$$

Proof. Using the analysis result given in section 4.6, statement (i) is equivalent to the existence of a Lyapunov matrix $\mathbf{Y} > \mathbf{0}$ such that $\|\mathbf{D}^T\mathbf{Y}\mathbf{D}\| < \gamma^2$ and

$$\mathbf{Y}(\mathbf{A}+\mathbf{BGM})+(\mathbf{A}+\mathbf{BGM})^T\mathbf{Y}+(\mathbf{C}+\mathbf{HGM})^T(\mathbf{C}+\mathbf{HGM}) < \mathbf{0}.$$

Then the result follows from the Schur complement formula. □

9.1.4 $\mathcal{L}_\infty$ Control

Consider the linear time-invariant system

$$\begin{bmatrix} \dot{\mathbf{x}}_p \\ \mathbf{y} \\ \mathbf{z} \end{bmatrix} = \begin{bmatrix} \mathbf{A}_p & \mathbf{D}_p & \mathbf{B}_p \\ \mathbf{C}_p & \mathbf{0} & \mathbf{0} \\ \mathbf{M}_p & \mathbf{D}_z & \mathbf{0} \end{bmatrix} \begin{bmatrix} \mathbf{x}_p \\ \mathbf{w} \\ \mathbf{u} \end{bmatrix} \tag{9.11}$$

where $\mathbf{x}_p$ is the state, $\mathbf{w}$ is the disturbance with finite energy, $\mathbf{y}$ is the output of interest, and $\mathbf{z}$ and $\mathbf{u}$ are the measured output and the control input, respectively. The $\mathcal{L}_\infty$ control problem is to find a controller that guarantees a bound on the peak value of the output $\mathbf{y}$ in response to any unit energy disturbance. This problem can be given as follows.

> Let a performance bound $\gamma > 0$ be given. Determine whether or not there exists a controller in (9.1) which asymptotically stabilizes the system (9.11) and yields the output $\mathbf{y}$ satisfying $\|\mathbf{y}\|_{\mathcal{L}_\infty} < \gamma$ for all disturbances $\mathbf{w}$ such that $\|\mathbf{w}\|_{\mathcal{L}_2} \leq 1$. Parametrize all such controllers when one exists.

This problem is mathematically equivalent to the covariance upper bound control problem with covariance bound $\mho \triangleq \gamma^2\mathbf{I}$. This fact can readily be verified from the analysis results in Chapter 4, and hence, we give a characterization of the $\mathcal{L}_\infty$ controllers without proof as follows.

Theorem 9.1.4 *Let a controller* $\mathbf{G}$ *and a performance bound* $\gamma > 0$ *be given. Then the following statements are equivalent:*

(i) The controller $\mathbf{G}$ *solves the* $\mathcal{L}_\infty$ *control problem.*
(ii) There exists a matrix $\mathbf{X} > \mathbf{0}$ *such that* $\mathbf{CXC}^T < \gamma^2\mathbf{I}$ *and (9.5) hold where*

$$\begin{bmatrix} \boldsymbol{\Gamma} & \boldsymbol{\Lambda}^T & \boldsymbol{\Theta} \end{bmatrix} \triangleq \left[\begin{array}{c|c|cc} \mathbf{B} & \mathbf{XM}^T & \mathbf{AX}+\mathbf{XA}^T & \mathbf{D} \\ \mathbf{0} & \mathbf{E}^T & \mathbf{D}^T & -\mathbf{I} \end{array}\right].$$

9.1.5 $\mathcal{H}_\infty$ Control

Consider the linear time-invariant system

$$\begin{bmatrix} \dot{\mathbf{x}}_p \\ \mathbf{y} \\ \mathbf{z} \end{bmatrix} = \begin{bmatrix} \mathbf{A}_p & \mathbf{D}_p & \mathbf{B}_p \\ \mathbf{C}_p & \mathbf{D}_y & \mathbf{B}_y \\ \mathbf{M}_p & \mathbf{D}_z & \mathbf{0} \end{bmatrix} \begin{bmatrix} \mathbf{x}_p \\ \mathbf{w} \\ \mathbf{u} \end{bmatrix} \tag{9.12}$$

where $\mathbf{x}_p$ is the state, $\mathbf{y}$ and $\mathbf{w}$ are the regulated output and the disturbance, and $\mathbf{z}$ and $\mathbf{u}$ are the measured output and the control input, respectively. Let the closed-loop transfer matrix from $\mathbf{w}$ to $\mathbf{y}$ with the controller in (9.1) be denoted by $\mathbf{T}(s)$;

$$\mathbf{T}(s) \triangleq \mathbf{C}_{c\ell}(s\mathbf{I} - \mathbf{A}_{c\ell})^{-1}\mathbf{B}_{c\ell} + \mathbf{D}_{c\ell}.$$

The $\mathcal{H}_\infty$ control problem can be stated as follows.

> Let a performance bound $\gamma > 0$ be given. Determine whether or not there exists a controller in (9.1) which asymptotically stabilizes the system (9.12) and yields the closed-loop transfer matrix such that $\|\mathbf{T}\|_{\mathcal{H}_\infty} < \gamma$. Parametrize all such controllers when one exists.

This problem may be interpreted in the following two ways: First, as we discussed in Chapter 4, the energy-to-energy gain Γ_{ee} defined in Section 4.6.1 is equal to the $\mathcal{H}_\infty$ norm of the corresponding transfer matrix, i.e. $\Gamma_{ee} = \|\mathbf{T}\|_{\mathcal{H}_\infty}$. Hence, a controller that solves the $\mathcal{H}_\infty$ control problem guarantees that the energy ($\mathcal{L}_2$ norm) of the output $\mathbf{y}$ is less than γ for all disturbances $\mathbf{w}$ with $\|\mathbf{w}\|_{\mathcal{L}_2} \leq 1$. Thus, γ is the worst-case performance bound. Another interpretation is to view the $\mathcal{H}_\infty$ control problem as a robust stabilization problem. As shown in Section 4.7.1, by the small gain theorem, the condition $\|\mathbf{T}\|_{\mathcal{H}_\infty} < \gamma$ guarantees robust stability with respect to norm-bounded uncertainty of size less than or equal to γ^{-1}. Hence, in this case, γ is a robustness bound.

For the $\mathcal{H}_\infty$ control problem, we have the following result.

Theorem 9.1.5 *Let a controller* $\mathbf{G}$ *and a performance (or robustness) bound* $\gamma > 0$ *be given. Then the following statements are equivalent:*

(i) The controller $\mathbf{G}$ *solves the* $\mathcal{H}_\infty$ *control problem.*

(ii) There exists a matrix $\mathbf{X} > \mathbf{0}$ *such that (9.5) holds where*

$$\left[\begin{array}{ccc} \boldsymbol{\Gamma} & \boldsymbol{\Lambda}^T & \boldsymbol{\Theta} \end{array}\right] \triangleq \left[\begin{array}{c|c|ccc} \mathbf{B} & \mathbf{X}\mathbf{M}^T & \mathbf{A}\mathbf{X}+\mathbf{X}\mathbf{A}^T & \mathbf{X}\mathbf{C}^T & \mathbf{D} \\ \mathbf{H} & \mathbf{0} & \mathbf{C}\mathbf{X} & -\gamma\mathbf{I} & \mathbf{F} \\ \mathbf{0} & \mathbf{E}^T & \mathbf{D}^T & \mathbf{F}^T & -\gamma\mathbf{I} \end{array}\right].$$

Proof. The result simply follows from substituting the definitions of the closed-loop matrices in (9.3) into the LMI in statement (iv) of Theorem 4.6.3, where we note that $\Gamma_{ee} = \|\mathbf{T}\|_{\mathcal{H}_\infty}$. □

9.1.6 Positive Real Control

Consider the linear time-invariant system

$$\left[\begin{array}{c} \dot{\mathbf{x}}_p \\ \mathbf{y} \\ \mathbf{z} \end{array}\right] = \left[\begin{array}{ccc} \mathbf{A}_p & \mathbf{D}_p & \mathbf{B}_p \\ \mathbf{C}_p & \mathbf{D}_y & \mathbf{B}_y \\ \mathbf{M}_p & \mathbf{D}_z & \mathbf{0} \end{array}\right] \left[\begin{array}{c} \mathbf{x}_p \\ \mathbf{w} \\ \mathbf{u} \end{array}\right] \tag{9.13}$$

where $\mathbf{x}_p$ is the state, $\mathbf{z}$ and $\mathbf{u}$ are the measured output and the control input, and $\mathbf{y}$ and $\mathbf{w}$ are the exogenous signals to describe the design specification. Let the closed-loop transfer matrix from $\mathbf{w}$ to $\mathbf{y}$ with the controller in (9.1) be given by

$$\mathbf{T}(s) \triangleq \mathbf{C}_{c\ell}(s\mathbf{I} - \mathbf{A}_{c\ell})^{-1}\mathbf{B}_{c\ell} + \mathbf{D}_{c\ell}.$$

The transfer matrix $\mathbf{T}(s)$ is said to be *strongly positive real* if it is asymptotically stable and

$$\mathbf{T}(j\omega) + \mathbf{T}^T(-j\omega) > 0$$

for all frequencies ω, including infinity. Using this notion, the positive real control problem can be stated as follows.

> Determine whether or not there exists a controller in (9.1) which asymptotically stabilizes the system (9.13) and yields the strongly positive real closed-loop transfer matrix $\mathbf{T}(s)$. Parametrize all such controllers when one exists.

This problem can be considered as a robust stabilization problem for systems with positive real (or passive) uncertainties [47, 141]. The positive real control problem can also be reduced to a problem of the type given by (9.5) as follows.

Theorem 9.1.6 *Let a controller* $\mathbf{G}$ *be given. Then the following statements are equivalent:*

(i) The controller $\mathbf{G}$ *solves the positive real control problem.*

(ii) There exists a matrix $\mathbf{X} > \mathbf{0}$ *such that (9.5) holds where*

$$\left[\begin{array}{ccc} \boldsymbol{\Gamma} & \boldsymbol{\Lambda}^T & \boldsymbol{\Theta} \end{array}\right] \triangleq \left[\begin{array}{c|c|cc} \mathbf{B} & \mathbf{X}\mathbf{M}^T & \mathbf{A}\mathbf{X}+\mathbf{X}\mathbf{A}^T & \mathbf{X}\mathbf{C}^T-\mathbf{D} \\ \mathbf{H} & -\mathbf{E}^T & \mathbf{C}\mathbf{X}-\mathbf{D}^T & -\mathbf{F}-\mathbf{F}^T \end{array}\right].$$

Proof. It can be shown that the transfer matrix $\mathbf{T}(s)$ is internally stable and is strongly positive real if and only if there exists a matrix $\mathbf{X} > \mathbf{0}$ such that

$$\begin{bmatrix} \mathbf{A}_{c\ell} & \mathbf{B}_{c\ell} \\ \mathbf{C}_{c\ell} & \mathbf{D}_{c\ell} \end{bmatrix}\begin{bmatrix} \mathbf{X} & \mathbf{0} \\ \mathbf{0} & -\mathbf{I} \end{bmatrix}+\begin{bmatrix} \mathbf{X} & \mathbf{0} \\ \mathbf{0} & -\mathbf{I} \end{bmatrix}\begin{bmatrix} \mathbf{A}_{c\ell} & \mathbf{B}_{c\ell} \\ \mathbf{C}_{c\ell} & \mathbf{D}_{c\ell} \end{bmatrix}^T < \mathbf{0}.$$

Then the result follows by substituting the closed-loop matrices in (9.3) into the above matrix inequality. □

9.1.7 Robust $\mathcal{H}_2$ Control

Consider the uncertain system described by

$$\begin{bmatrix} \dot{\mathbf{x}}_p \\ \mathbf{y}_1 \\ \mathbf{y}_2 \\ \mathbf{z} \end{bmatrix} = \begin{bmatrix} \mathbf{A}_p & \mathbf{D}_{p1} & \mathbf{D}_{p2} & \mathbf{B}_p \\ \mathbf{C}_{p1} & \mathbf{D}_{y11} & \mathbf{0} & \mathbf{B}_{y1} \\ \mathbf{C}_{p2} & \mathbf{D}_{y21} & \mathbf{0} & \mathbf{B}_{y2} \\ \mathbf{M}_p & \mathbf{D}_{z1} & \mathbf{0} & \mathbf{0} \end{bmatrix}\begin{bmatrix} \mathbf{x}_p \\ \mathbf{w}_1 \\ \mathbf{w}_2 \\ \mathbf{u} \end{bmatrix}, \quad \mathbf{w}_1 = \boldsymbol{\Delta}\mathbf{y}_1 \qquad (9.14)$$

where $\mathbf{x}_p$ is the state, $\mathbf{y}_1$ and $\mathbf{w}_1$ are the exogenous signals to describe the uncertainty $\boldsymbol{\Delta}$, $\mathbf{y}_2$ and $\mathbf{w}_2$ are the output of interest and the impulsive disturbance, and $\mathbf{z}$ and $\mathbf{u}$ are the measured output and the control input. The nominal system ($\boldsymbol{\Delta} \equiv \mathbf{0}$) is linear time-invariant, while the uncertainty $\boldsymbol{\Delta}$ is assumed to belong to the following set of norm-bounded time-varying, structured uncertainties:

$$\mathcal{BU}_C \triangleq \{\, \boldsymbol{\Delta} : \mathcal{R} \to \mathcal{R}^{m\times m}, \ \|\boldsymbol{\Delta}(t)\| \le 1, \ \boldsymbol{\Delta}(t) \in \mathcal{U} \,\} \qquad (9.15)$$

where

$$\mathcal{U} \triangleq \{\text{block diag}(\delta_1\mathbf{I}_{k_1}, \cdots, \delta_s\mathbf{I}_{k_s}, \ \boldsymbol{\Delta}_1, \cdots, \boldsymbol{\Delta}_f) : \ \delta_i \in \mathcal{R}, \ \ \boldsymbol{\Delta}_i \in \mathcal{R}^{k_{s+i}\times k_{s+i}} \,\}. \qquad (9.16)$$

In the above, we have implicitly assumed for simplicity that the uncertainty $\boldsymbol{\Delta}$ is square. Define a subset of positive definite matrices that commute with $\boldsymbol{\Delta} \in \mathcal{U}$:

$$\begin{aligned} \mathcal{S} \ \triangleq \ & \{\text{block diag } (\mathbf{S}_1 \cdots \mathbf{S}_s \ s_1\mathbf{I}_{k_{s+1}} \cdots s_f\mathbf{I}_{k_{s+f}}) : \\ & \mathbf{S}_i \in \mathcal{R}^{k_i\times k_i}, \ \ s_i \in \mathcal{R}, \ \ \mathbf{S}_i > \mathbf{0}, \ \ s_i > 0 \,\}. \end{aligned} \qquad (9.17)$$

We consider the following robust performance problem to guarantee a bound on the energy of the output $\mathbf{y}_2$ in response to the worst-case impulsive disturbance $\mathbf{w}_2$, for all uncertainties $\boldsymbol{\Delta} \in \mathcal{BU}_C$.

> Let a robust performance bound $\gamma > 0$ be given. Find a controller in (9.1), for the uncertain system (9.14), such that the closed-loop system is robustly stable and the output $\mathbf{y}_2$ satisfies $\|\mathbf{y}_2\|_{\mathcal{L}_2} < \gamma$ for all impulsive disturbances $\mathbf{w}_2(t) = \mathbf{w}_0\delta(t)$ with $\|\mathbf{w}_0\| \le 1$, and for all possible uncertainties $\boldsymbol{\Delta} \in \mathcal{BU}_C$.

This problem can be approached using the robust $\mathcal{H}_2$ performance bound given in Theorem 4.7.2 as follows.

Theorem 9.1.7 *Let a controller* $\mathbf{G}$ *and a robust performance bound* $\gamma > 0$ *be given. Suppose there exist a Lyapunov matrix* $\mathbf{Y} > \mathbf{0}$ *and a scaling matrix* $\mathbf{S} \in \mathcal{S}$ *such that* $\mathbf{D}_2^T\mathbf{Y}\mathbf{D}_2 < \gamma^2\mathbf{I}$ *and (9.5) hold where*

$$[\ \Gamma \quad \Lambda^T \quad \Theta\ ] \triangleq \left[\begin{array}{c|c|cccc} \mathbf{YB} & \mathbf{M}^T & \mathbf{YA}+\mathbf{A}^T\mathbf{Y} & \mathbf{YD}_1 & \mathbf{C}_1^T & \mathbf{C}_2^T \\ \mathbf{0} & \mathbf{E}_1^T & \mathbf{D}_1^T\mathbf{Y} & -\mathbf{S} & \mathbf{F}_{11}^T & \mathbf{F}_{21}^T \\ \mathbf{H}_1 & \mathbf{0} & \mathbf{C}_1 & \mathbf{F}_{11} & -\mathbf{S}^{-1} & \mathbf{0} \\ \mathbf{H}_2 & \mathbf{0} & \mathbf{C}_2 & \mathbf{F}_{21} & \mathbf{0} & -\mathbf{I} \end{array}\right].$$

Then the controller $\mathbf{G}$ *solves the robust* $\mathcal{H}_2$ *control problem.*

Proof. The result follows from Theorem 4.7.2 and the definitions for the closed-loop matrices given in (9.3). □

9.1.8 Robust $\mathcal{L}_\infty$ Control

Consider the uncertain system given by

$$\begin{bmatrix} \dot{\mathbf{x}}_p \\ \mathbf{y}_1 \\ \mathbf{y}_2 \\ \mathbf{z} \end{bmatrix} = \begin{bmatrix} \mathbf{A}_p & \mathbf{D}_{p1} & \mathbf{D}_{p2} & \mathbf{B}_p \\ \mathbf{C}_{p1} & \mathbf{D}_{y11} & \mathbf{D}_{y12} & \mathbf{B}_{y1} \\ \mathbf{C}_{p2} & \mathbf{0} & \mathbf{0} & \mathbf{0} \\ \mathbf{M}_p & \mathbf{D}_{z1} & \mathbf{D}_{z2} & \mathbf{0} \end{bmatrix} \begin{bmatrix} \mathbf{x}_p \\ \mathbf{w}_1 \\ \mathbf{w}_2 \\ \mathbf{u} \end{bmatrix}, \qquad \mathbf{w}_1 = \mathbf{\Delta}\mathbf{y}_1 \tag{9.18}$$

where $\mathbf{x}_p$ is the state, $\mathbf{y}_1$ and $\mathbf{w}_1$ are the exogenous signals used to describe the uncertainty $\mathbf{\Delta}$, $\mathbf{y}_2$ and $\mathbf{w}_2$ are the output of interest and the finite energy disturbance, and $\mathbf{z}$ and $\mathbf{u}$ are the measured output and the control input. The uncertainty $\mathbf{\Delta}$ is assumed to belong to the set $\mathcal{BU}_C$ defined in (9.15). The robust $\mathcal{L}_\infty$ control problem is to guarantee a bound on the peak value of the output $\mathbf{y}_2$ for any unit energy disturbance $\mathbf{w}_2$ in the presence of uncertainty $\mathbf{\Delta} \in \mathcal{BU}_C$.

> Let a robust performance bound $\gamma > 0$ be given. Find a controller in (9.1), for the uncertain system (9.18), such that the closed-loop system is robustly stable and the output $\mathbf{y}_2$ satisfies $\|\mathbf{y}_2\|_{\mathcal{L}_\infty} < \gamma$ for any disturbance $\mathbf{w}_2$ such that $\|\mathbf{w}_2\|_{\mathcal{L}_2} \leq 1$, and for all possible uncertainties $\mathbf{\Delta} \in \mathcal{BU}_C$.

We may try to solve this problem using the robust $\mathcal{L}_\infty$ performance bound given in Theorem 4.7.3 as follows.

Theorem 9.1.8 *Let a controller* $\mathbf{G}$ *and a robust performance bound* $\gamma > 0$ *be given. Suppose there exist a Lyapunov matrix* $\mathbf{X} > \mathbf{0}$ *and a scaling matrix* $\mathbf{S} \in \mathcal{S}$ *such that* $\mathbf{C}_2\mathbf{X}\mathbf{C}_2^T < \gamma^2\mathbf{I}$ *and (9.5) hold where*

$$[\ \Gamma \quad \Lambda^T \quad \Theta\ ] \triangleq \left[\begin{array}{c|c|cccc} \mathbf{XM}^T & \mathbf{B} & \mathbf{AX}+\mathbf{XA}^T & \mathbf{XC}_1^T & \mathbf{D}_1 & \mathbf{D}_2 \\ \mathbf{0} & \mathbf{H}_1 & \mathbf{C}_1\mathbf{X} & -\mathbf{S} & \mathbf{F}_{11} & \mathbf{F}_{12} \\ \mathbf{E}_1^T & \mathbf{0} & \mathbf{D}_1^T & \mathbf{F}_{11}^T & -\mathbf{S}^{-1} & \mathbf{0} \\ \mathbf{E}_2^T & \mathbf{0} & \mathbf{D}_2^T & \mathbf{F}_{12}^T & \mathbf{0} & -\mathbf{I} \end{array}\right].$$

Then the controller $\mathbf{G}$ *solves the robust* $\mathcal{L}_\infty$ *control problem.*

Proof. The result follows from Theorem 4.7.3 and the definitions for the closed-loop matrices in (9.3). □

9.1.9 Robust $\mathcal{H}_\infty$ Control

Consider the uncertain system

$$\begin{bmatrix} \dot{\mathbf{x}}_p \\ \mathbf{y}_1 \\ \mathbf{y}_2 \\ \mathbf{z} \end{bmatrix} = \begin{bmatrix} \mathbf{A}_p & \mathbf{D}_{p1} & \mathbf{D}_{p2} & \mathbf{B}_p \\ \mathbf{C}_{p1} & \mathbf{D}_{y11} & \mathbf{D}_{y12} & \mathbf{B}_{y1} \\ \mathbf{C}_{p2} & \mathbf{D}_{y21} & \mathbf{D}_{y22} & \mathbf{B}_{y2} \\ \mathbf{M}_p & \mathbf{D}_{z1} & \mathbf{D}_{z2} & \mathbf{0} \end{bmatrix} \begin{bmatrix} \mathbf{x}_p \\ \mathbf{w}_1 \\ \mathbf{w}_2 \\ \mathbf{u} \end{bmatrix}, \quad \mathbf{w}_1 = \mathbf{\Delta}\mathbf{y}_1 \tag{9.19}$$

where $\mathbf{x}_p$ is the state, $\mathbf{y}_1$ and $\mathbf{w}_1$ are the exogenous signals used to describe the uncertainty $\mathbf{\Delta} \in \mathcal{BU}_C$, $\mathbf{y}_2$ and $\mathbf{w}_2$ are the output of interest and the finite energy disturbance, and $\mathbf{z}$ and $\mathbf{u}$ are the measured output and the control input. The robust $\mathcal{H}_\infty$ control problem is to guarantee a bound on the energy of the output $\mathbf{y}_2$ in response to any unit energy disturbance $\mathbf{w}_2$ in the presence of uncertainty $\mathbf{\Delta} \in \mathcal{BU}_C$.

> Let a robust performance bound $\gamma > 0$ be given. Find a controller in (9.1), for the uncertain system (9.19), such that the closed-loop system is robustly stable and the output $\mathbf{y}_2$ satisfies $\|\mathbf{y}_2\|_{\mathcal{L}_2} < \gamma$ for any disturbance $\mathbf{w}_2$ such that $\|\mathbf{w}_2\|_{\mathcal{L}_2} \leq 1$, and for all possible uncertainties $\mathbf{\Delta} \in \mathcal{BU}_C$.

This problem can be reduced to the following.

Theorem 9.1.9 *Let a controller* $\mathbf{G}$ *and a robust performance bound* $\gamma > 0$ *be given. Suppose there exist a Lyapunov matrix* $\mathbf{Y} > \mathbf{0}$ *and a scaling matrix* $\mathbf{S} \in \mathcal{S}$ *such that (9.5) holds where*

$$\begin{bmatrix} \mathbf{\Gamma} & \mathbf{\Lambda}^T & \mathbf{\Theta} \end{bmatrix} \triangleq \left[\begin{array}{c|c|ccccc} \mathbf{YB} & \mathbf{M}^T & \mathbf{YA}+\mathbf{A}^T\mathbf{Y} & \mathbf{YD}_1 & \mathbf{YD}_2 & \mathbf{C}_1^T & \mathbf{C}_2^T \\ \mathbf{0} & \mathbf{E}_1^T & \mathbf{D}_1^T\mathbf{Y} & -\mathbf{S} & \mathbf{0} & \mathbf{F}_{11}^T & \mathbf{F}_{21}^T \\ \mathbf{0} & \mathbf{E}_2^T & \mathbf{D}_2^T\mathbf{Y} & \mathbf{0} & -\gamma\mathbf{I} & \mathbf{F}_{12}^T & \mathbf{F}_{22}^T \\ \mathbf{H}_1 & \mathbf{0} & \mathbf{C}_1 & \mathbf{F}_{11} & \mathbf{F}_{12} & -\mathbf{S}^{-1} & \mathbf{0} \\ \mathbf{H}_2 & \mathbf{0} & \mathbf{C}_2 & \mathbf{F}_{21} & \mathbf{F}_{22} & \mathbf{0} & -\gamma\mathbf{I} \end{array}\right].$$

Then the controller $\mathbf{G}$ *solves the robust* $\mathcal{H}_\infty$ *control problem.*

Proof. Using the definitions for the closed-loop matrices (9.3), the result can be verified from Theorem 4.7.4. □

9.2 DISCRETE-TIME CASE

This section provides a unified perspective for the discrete-time case. We shall show that, for the discrete-time case, the control design problems considered in the previous section can be reduced to the problem of solving a Quadratic Matrix Inequality (QMI)

$$(\mathbf{\Theta} + \mathbf{\Gamma}\mathbf{G}\mathbf{\Lambda})\mathbf{R}(\mathbf{\Theta} + \mathbf{\Gamma}\mathbf{G}\mathbf{\Lambda})^T < \mathbf{Q} \tag{9.20}$$

for the controller parameter $\mathbf{G}$ where matrices $\mathbf{\Theta}$, $\mathbf{\Gamma}$, $\mathbf{\Lambda}$, $\mathbf{R}$ and $\mathbf{Q}$ are appropriately defined for each control problem. Note that the above linear algebra problem is a special case of

(9.5) which arose in the continuous-time case. To see this, using the Schur complement formula, write (9.20) as

$$\begin{bmatrix} \mathbf{\Gamma} \\ \mathbf{0} \end{bmatrix} \mathbf{G} \begin{bmatrix} \mathbf{0} & -\mathbf{\Lambda} \end{bmatrix} + \begin{bmatrix} \mathbf{0} \\ -\mathbf{\Lambda}^T \end{bmatrix} \mathbf{G}^T \begin{bmatrix} \mathbf{\Gamma}^T & \mathbf{0} \end{bmatrix} - \begin{bmatrix} \mathbf{Q} & \mathbf{\Theta} \\ \mathbf{\Theta}^T & \mathbf{R}^{-1} \end{bmatrix} < \mathbf{0}, \quad (9.21)$$

where we used $\mathbf{R} > \mathbf{0}$ (which holds for all the control problems considered below). Thus, *both* continuous-time and discrete-time control problems reduce to the LMI (9.5). In this regard, the LMI (9.5) defines a fundamental algebraic problem in the LMI formulation of control problems.

To solve the QMI (9.20) for $\mathbf{G}$, we could apply Theorem 2.3.12 to the LMI (9.21). However, the QMI (9.20) can be directly solved by Theorem 2.3.11. In this case, necessary and sufficient conditions for the existence of $\mathbf{G}$ satisfying (9.20) are given by

$$\mathbf{\Gamma}^{\perp}(\mathbf{Q} - \mathbf{\Theta}\mathbf{R}\mathbf{\Theta}^T)\mathbf{\Gamma}^{\perp T} > \mathbf{0}$$

$$\mathbf{\Lambda}^{T\perp}(\mathbf{R}^{-1} - \mathbf{\Theta}^T\mathbf{Q}^{-1}\mathbf{\Theta})\mathbf{\Lambda}^{T\perp T} > \mathbf{0}$$

and all such $\mathbf{G}$ are explicitly parametrized by a free parameter $\mathbf{L}$ such that $\|\mathbf{L}\| < 1$. As in the continuous-time case, these existence conditions lead to LMIs in terms of the Lyapunov matrix $\mathbf{X}$ and its inverse $\mathbf{X}^{-1}$, which can further be reduced to a convex LMI problem for certain cases.

9.2.1 Stabilization Problem

Let us first consider the simplest control problem of stabilizing a linear time-invariant plant

$$\begin{bmatrix} \mathbf{x}_p(k+1) \\ \mathbf{z}(k) \end{bmatrix} = \begin{bmatrix} \mathbf{A}_p & \mathbf{B}_p \\ \mathbf{M}_p & \mathbf{0} \end{bmatrix} \begin{bmatrix} \mathbf{x}_p(k) \\ \mathbf{u}(k) \end{bmatrix} \quad (9.22)$$

where $\mathbf{x}_p$ is the state, $\mathbf{u}$ is the control input and $\mathbf{z}$ is the measured output. The stabilization problem is the following.

> Determine whether or not there exists a controller in (9.1) which asymptotically stabilizes the system (9.22). Parametrize all such controllers when one exists.

This problem can be reduced to the following linear algebra problem.

Theorem 9.2.1 *Let a controller* $\mathbf{G}$ *be given. Then the following statements are equivalent:*

(i) The controller $\mathbf{G}$ *solves the stabilization problem.*

(ii) There exists a matrix $\mathbf{X} > \mathbf{0}$ *such that (9.20) holds where*

$$\begin{bmatrix} \mathbf{\Theta} & \mathbf{\Gamma} & \mathbf{Q} \end{bmatrix} \triangleq \begin{bmatrix} \mathbf{A} \mid \mathbf{B} \mid \mathbf{X} \end{bmatrix}$$

$$\begin{bmatrix} \mathbf{\Lambda}^T & \mathbf{R} \end{bmatrix} \triangleq \begin{bmatrix} \mathbf{M}^T \mid \mathbf{X} \end{bmatrix}.$$

Proof. The result is just a restatement of Lyapunov's stability theory for discrete-time systems, which says that statement (i) holds if and only if there exists a Lyapunov matrix $\mathbf{X} > \mathbf{0}$ such that

$$\mathbf{X} > (\mathbf{A} + \mathbf{BGM})\mathbf{X}(\mathbf{A} + \mathbf{BGM})^T.$$

This completes the proof. □

9.2.2 Covariance Upper Bound Control

Next, we shall consider the covariance upper bound control problem discussed in Chapter 6. The state space model of the plant is given by

$$\begin{bmatrix} \mathbf{x}_p(k+1) \\ \mathbf{y}(k) \\ \mathbf{z}(k) \end{bmatrix} = \begin{bmatrix} \mathbf{A}_p & \mathbf{D}_p & \mathbf{B}_p \\ \mathbf{C}_p & \mathbf{D}_y & \mathbf{0} \\ \mathbf{M}_p & \mathbf{D}_z & \mathbf{0} \end{bmatrix} \begin{bmatrix} \mathbf{x}_p(k) \\ \mathbf{w}(k) \\ \mathbf{u}(k) \end{bmatrix} \tag{9.23}$$

where $\mathbf{x}_p$ is the state, $\mathbf{w}$ is the white noise with covariance $\mathbf{I}$, $\mathbf{u}$ is the control input, $\mathbf{y}$ is the output of interest and $\mathbf{z}$ is the measured output. The covariance upper bound control problem is the following.

Let an output covariance bound $\mathcal{U} > \mathbf{0}$ be given. Determine whether or not there exists a controller in (9.1) which asymptotically stabilizes the system (9.23) and yields an output covariance satisfying

$$\lim_{k\to\infty} \mathcal{E}\,[\mathbf{y}(k)\mathbf{y}^T(k)] < \mathcal{U}.$$

Parametrize all such controllers when one exists.

For this problem, we have the following result.

Theorem 9.2.2 *Let a controller* $\mathbf{G}$ *and an output covariance bound* $\mathcal{U} > \mathbf{0}$ *be given. Then the following statements are equivalent:*

(i) The controller $\mathbf{G}$ *solves the covariance upper bound control problem.*

(ii) There exists a matrix $\mathbf{X} > \mathbf{0}$ *such that* $\mathbf{C}\mathbf{X}\mathbf{C}^T + \mathbf{F}\mathbf{F}^T < \mathcal{U}$ *and (9.20) hold where*

$$[\,\Theta \quad \Gamma \quad \mathbf{Q}\,] \triangleq [\,\mathbf{A}\ \ \mathbf{D} \mid \mathbf{B} \mid \mathbf{X}\,],$$

$$[\,\Lambda^T \quad \mathbf{R}\,] \triangleq \left[\begin{array}{c|cc} \mathbf{M}^T & \mathbf{X} & \mathbf{0} \\ \mathbf{E}^T & \mathbf{0} & \mathbf{I} \end{array}\right].$$

Proof. The result simply follows from Lemma 6.1.2. □

9.2.3 Linear Quadratic Regulator

We consider the Linear Quadratic Regulator (LQR) problem for the linear time-invariant system

$$\begin{bmatrix} \mathbf{x}_p(k+1) \\ \mathbf{y}(k) \\ \mathbf{z}(k) \end{bmatrix} = \begin{bmatrix} \mathbf{A}_p & \mathbf{D}_p & \mathbf{B}_p \\ \mathbf{C}_p & \mathbf{D}_y & \mathbf{B}_y \\ \mathbf{M}_p & \mathbf{0} & \mathbf{0} \end{bmatrix} \begin{bmatrix} \mathbf{x}_p(k) \\ \mathbf{w}(k) \\ \mathbf{u}(k) \end{bmatrix} \tag{9.24}$$

where $\mathbf{x}_p$ is the state, $\mathbf{w}$ is the pulse disturbance $\mathbf{w}(k) = \mathbf{w}_0\delta(k)$ where $\delta(k)$ is the Kronecker delta function, $\mathbf{u}$ is the control input, $\mathbf{y}$ is the output of interest and $\mathbf{z}$ is the measured output. The LQR problem is defined by an upper bound on the square summation of the output signal as follows.

Let a performance bound $\gamma > 0$ be given. Determine whether or not there exists a controller in (9.1) which asymptotically stabilizes the system (9.24) and yields the zero initial state response $\mathbf{y}$ satisfying $\|\mathbf{y}\|_{\ell_2} < \gamma$ for all directions of the pulse disturbance $\|\mathbf{w}_0\| \le 1$. Parametrize all such controllers when one exists.

This problem can be reduced to the following.

Theorem 9.2.3 *Let a controller* $\mathbf{G}$ *and a performance bound* $\gamma > 0$ *be given. Then the following statements are equivalent:*

(i) The controller $\mathbf{G}$ *solves the LQR problem.*
(ii) There exists a matrix $\mathbf{Y} > \mathbf{0}$ *such that* $\|\mathbf{D}^T\mathbf{Y}\mathbf{D}+\mathbf{F}^T\mathbf{F}\| < \gamma^2$ *and (9.20) hold where*

$$\begin{bmatrix} \Theta & \Gamma & \mathbf{Q} \end{bmatrix} \triangleq \begin{bmatrix} \mathbf{A}^T & \mathbf{C}^T & | & \mathbf{M}^T & | & \mathbf{Y} \end{bmatrix}$$

$$\begin{bmatrix} \Lambda^T & \mathbf{R} \end{bmatrix} \triangleq \left[\begin{array}{c|cc} \mathbf{B} & \mathbf{Y} & \mathbf{0} \\ \mathbf{H} & \mathbf{0} & \mathbf{I} \end{array}\right].$$

Proof. From the analysis result in Chapter 4, statement (i) holds if and only if there exists a matrix $\mathbf{Y} > \mathbf{0}$ such that $\|\mathbf{D}^T\mathbf{Y}\mathbf{D}+\mathbf{F}^T\mathbf{F}\| < \gamma^2$ and

$$\mathbf{Y} > (\mathbf{A}+\mathbf{BGM})^T\mathbf{Y}(\mathbf{A}+\mathbf{BGM}) + (\mathbf{C}+\mathbf{HGM})^T(\mathbf{C}+\mathbf{HGM}). \tag{9.25}$$

Then it is straightforward to verify the result. □

9.2.4 ℓ_∞ Control

Consider the linear time-invariant system

$$\begin{bmatrix} \mathbf{x}_p(k+1) \\ \mathbf{y}(k) \\ \mathbf{z}(k) \end{bmatrix} = \begin{bmatrix} \mathbf{A}_p & \mathbf{D}_p & \mathbf{B}_p \\ \mathbf{C}_p & \mathbf{D}_y & \mathbf{0} \\ \mathbf{M}_p & \mathbf{D}_z & \mathbf{0} \end{bmatrix} \begin{bmatrix} \mathbf{x}_p(k) \\ \mathbf{w}(k) \\ \mathbf{u}(k) \end{bmatrix} \tag{9.26}$$

where $\mathbf{x}_p$ is the state, $\mathbf{y}$ is the output of interest, $\mathbf{w}$ is the finite energy disturbance, $\mathbf{z}$ and $\mathbf{u}$ are the measured output and the control input, respectively. The ℓ_∞ control problem can be stated as follows.

> Let a performance bound $\gamma > 0$ be given. Determine whether or not there exists a controller in (9.1) which asymptotically stabilizes the system (9.26) and yields the output $\mathbf{y}$ such that $\|\mathbf{y}\|_{\ell_\infty} < \gamma$ for any disturbances $\mathbf{w}$ with $\|\mathbf{w}\|_{\ell_2} \le 1$. Parametrize all such controllers when one exists.

This problem reduces to the following.

Theorem 9.2.4 *Let a controller* $\mathbf{G}$ *and a performance bound* $\gamma > 0$ *be given. Then the following statements are equivalent:*

(i) The controller $\mathbf{G}$ *solves the* ℓ_∞ *control problem.*
(ii) There exists a matrix $\mathbf{X} > \mathbf{0}$ *such that* $\|\mathbf{C}\mathbf{X}\mathbf{C}^T+\mathbf{F}\mathbf{F}^T\| < \gamma^2$ *and (9.20) hold where*

$$\begin{bmatrix} \Theta & \Gamma & \mathbf{Q} \end{bmatrix} \triangleq \begin{bmatrix} \mathbf{A} & \mathbf{D} & | & \mathbf{B} & | & \mathbf{X} \end{bmatrix}$$

$$\begin{bmatrix} \Lambda^T & \mathbf{R} \end{bmatrix} \triangleq \left[\begin{array}{c|cc} \mathbf{M}^T & \mathbf{X} & \mathbf{0} \\ \mathbf{E}^T & \mathbf{0} & \mathbf{I} \end{array}\right].$$

Proof. The result follows from Theorem 4.6.5. □

9.2.5 $\mathcal{H}_\infty$ Control

Consider the linear time-invariant system

$$\begin{bmatrix} \mathbf{x}_p(k+1) \\ \mathbf{y}(k) \\ \mathbf{z}(k) \end{bmatrix} = \begin{bmatrix} \mathbf{A}_p & \mathbf{D}_p & \mathbf{B}_p \\ \mathbf{C}_p & \mathbf{D}_y & \mathbf{B}_y \\ \mathbf{M}_p & \mathbf{D}_z & \mathbf{0} \end{bmatrix} \begin{bmatrix} \mathbf{x}_p(k) \\ \mathbf{w}(k) \\ \mathbf{u}(k) \end{bmatrix} \tag{9.27}$$

where $\mathbf{x}_p$ is the state, $\mathbf{y}$ and $\mathbf{w}$ are the performance signals that we use to describe the design specification, and $\mathbf{z}$ and $\mathbf{u}$ are the measured output and the control input. We denote the closed-loop transfer matrix from $\mathbf{w}$ to $\mathbf{y}$, with the controller in (9.1), by

$$\mathbf{T}(z) \triangleq \mathbf{C}_{c\ell}(z\mathbf{I} - \mathbf{A}_{c\ell})^{-1}\mathbf{B}_{c\ell} + \mathbf{D}_{c\ell}.$$

The $\mathcal{H}_\infty$ control problem is the following.

> Let a performance bound $\gamma > 0$ be given. Determine whether or not there exists a controller in (9.1) which asymptotically stabilizes the system (9.27) and yields the closed-loop transfer matrix such that $\|\mathbf{T}\|_{\mathcal{H}_\infty} < \gamma$. Parametrize all such controllers when one exists.

As in the continuous-time case, this problem has two distinct physical significances; namely, robustness with respect to norm-bounded perturbations, and the disturbance attenuation level measured by the energy-to-energy gain, (see Chapter 4). The $\mathcal{H}_\infty$ control problem for the discrete-time case reduces to the following.

Theorem 9.2.5 *Let a controller* $\mathbf{G}$ *and a robustness/performance bound* $\gamma > 0$ *be given. Then the following statements are equivalent:*

(i) The controller $\mathbf{G}$ *solves the* $\mathcal{H}_\infty$ *control problem.*

(ii) There exists a matrix $\mathbf{X} > \mathbf{0}$ *such that (9.20) holds where*

$$\begin{bmatrix} \mathbf{\Theta} & \mathbf{\Gamma} & \mathbf{Q} \end{bmatrix} \triangleq \left[\begin{array}{cc|c|cc} \mathbf{A} & \mathbf{D} & \mathbf{B} & \mathbf{X} & \mathbf{0} \\ \mathbf{C} & \mathbf{F} & \mathbf{H} & \mathbf{0} & \gamma^2\mathbf{I} \end{array}\right]$$

$$\begin{bmatrix} \mathbf{\Lambda}^T & \mathbf{R} \end{bmatrix} \triangleq \left[\begin{array}{c|cc} \mathbf{M}^T & \mathbf{X} & \mathbf{0} \\ \mathbf{E}^T & \mathbf{0} & \mathbf{I} \end{array}\right].$$

Proof. The result can be verified using Theorem 4.6.6 and noting that $\Upsilon_{ee} = \|\mathbf{T}\|_{\mathcal{H}_\infty}$. □

9.2.6 Robust $\mathcal{H}_2$ Control

Consider the uncertain system described by

$$\begin{bmatrix} \mathbf{x}_p(k+1) \\ \mathbf{y}_1(k) \\ \mathbf{y}_2(k) \\ \mathbf{z}(k) \end{bmatrix} = \begin{bmatrix} \mathbf{A}_p & \mathbf{D}_{p1} & \mathbf{D}_{p2} & \mathbf{B}_p \\ \mathbf{C}_{p1} & \mathbf{D}_{y11} & \mathbf{0} & \mathbf{B}_{y1} \\ \mathbf{C}_{p2} & \mathbf{D}_{y21} & \mathbf{D}_{y22} & \mathbf{B}_{y2} \\ \mathbf{M}_p & \mathbf{D}_{z1} & \mathbf{0} & \mathbf{0} \end{bmatrix} \begin{bmatrix} \mathbf{x}_p(k) \\ \mathbf{w}_1(k) \\ \mathbf{w}_2(k) \\ \mathbf{u}(k) \end{bmatrix}, \quad \mathbf{w}_1 = \mathbf{\Delta}\mathbf{y}_1 \tag{9.28}$$

where $\mathbf{x}_p$ is the state, $\mathbf{y}_1$ and $\mathbf{w}_1$ are the exogenous signals to describe the uncertainty $\mathbf{\Delta}$, $\mathbf{y}_2$ and $\mathbf{w}_2$ are the output of interest and the impulsive disturbance, and $\mathbf{z}$ and $\mathbf{u}$ are the measured output and the control input. We assume that the uncertainty $\mathbf{\Delta}$ belongs to the following set of norm-bounded, time-varying, structured uncertainties:

$$\mathcal{BU}_D \triangleq \{ \mathbf{\Delta} : \mathcal{I} \rightarrow \mathcal{R}^{m\times m},\ \|\mathbf{\Delta}(k)\| \leq 1,\ \mathbf{\Delta}(k) \in \mathcal{U} \} \tag{9.29}$$

where $\mathcal{I}$ is the set of integers and $\mathcal{U}$ is the subset of structured positive definite matrices defined by (9.16). In the following, we will use the set of scaling matrices $\mathbf{S}$ in (9.17), corresponding to the uncertainty structure (9.16). The robust $\mathcal{H}_2$ control problem for the discrete-time case is analogous to the continuous-time counterpart, and can be stated as follows.

> Let a robust performance bound $\gamma > 0$ be given. Find a controller in (9.1) for the uncertain system (9.28), such that the closed-loop system is robustly stable and the output $\mathbf{y}_2$ satisfies $\|\mathbf{y}_2\|_{\ell_2} < \gamma$ for all pulse disturbances $\mathbf{w}_2(k) = \mathbf{w}_0\delta(k)$ with $\|\mathbf{w}_0\| \leq 1$, and for all uncertainties $\mathbf{\Delta} \in \mathcal{BU}_D$.

To address this problem in the framework of Theorem 4.7.6, we need a technical assumption $\mathbf{w}_1(0) = \mathbf{0}$. With this assumption, we have the following.

Theorem 9.2.6 *Let a controller* $\mathbf{G}$ *and a robust performance bound* $\gamma > 0$ *be given. Suppose there exist a Lyapunov matrix* $\mathbf{Y} > 0$ *and a scaling matrix* $\mathbf{S} \in \mathcal{S}$ *such that* $\mathbf{D}_2^T\mathbf{Y}\mathbf{D}_2 + \mathbf{F}_{22}^T\mathbf{F}_{22} < \gamma^2\mathbf{I}$ *and (9.20) hold where*

$$\left[\ \mathbf{\Theta}\quad \mathbf{\Gamma}\quad \mathbf{Q}\ \right] \triangleq \left[\begin{array}{ccc|c|cc} \mathbf{A}^T & \mathbf{C}_1^T & \mathbf{C}_2^T & \mathbf{M}^T & \mathbf{Y} & \mathbf{0} \\ \mathbf{D}_1^T & \mathbf{F}_{11}^T & \mathbf{F}_{21}^T & \mathbf{E}_1^T & \mathbf{0} & \mathbf{S} \end{array}\right]$$

$$\left[\ \mathbf{\Lambda}^T\quad \mathbf{R}\ \right] \triangleq \left[\begin{array}{c|ccc} \mathbf{B} & \mathbf{Y} & \mathbf{0} & \mathbf{0} \\ \mathbf{H}_1 & \mathbf{0} & \mathbf{S} & \mathbf{0} \\ \mathbf{H}_2 & \mathbf{0} & \mathbf{0} & \mathbf{I} \end{array}\right].$$

Then the controller $\mathbf{G}$ *solves the robust* $\mathcal{H}_2$ *control problem.*

Proof. The result follows from Theorem 4.7.6 by substituting the closed-loop matrices defined in (9.3). □

9.2.7 Robust ℓ_∞ Control

Consider the uncertain system described by

$$\begin{bmatrix} \mathbf{x}_p(k+1) \\ \mathbf{y}_1(k) \\ \mathbf{y}_2(k) \\ \mathbf{z}(k) \end{bmatrix} = \begin{bmatrix} \mathbf{A}_p & \mathbf{D}_{p1} & \mathbf{D}_{p2} & \mathbf{B}_p \\ \mathbf{C}_{p1} & \mathbf{D}_{y11} & \mathbf{D}_{y12} & \mathbf{B}_{y1} \\ \mathbf{C}_{p2} & \mathbf{0} & \mathbf{D}_{y22} & \mathbf{0} \\ \mathbf{M}_p & \mathbf{D}_{z1} & \mathbf{D}_{z2} & \mathbf{0} \end{bmatrix} \begin{bmatrix} \mathbf{x}_p(k) \\ \mathbf{w}_1(k) \\ \mathbf{w}_2(k) \\ \mathbf{u}(k) \end{bmatrix}, \quad \mathbf{w}_1 = \mathbf{\Delta}\mathbf{y}_1 \tag{9.30}$$

where $\mathbf{x}_p$ is the state, $\mathbf{y}_1$ and $\mathbf{w}_1$ are the exogenous signals to describe the uncertainty $\mathbf{\Delta} \in \mathcal{BU}_D$, $\mathbf{y}_2$ and $\mathbf{w}_2$ are the output of interest and the finite energy disturbance, and $\mathbf{z}$ and $\mathbf{u}$ are the measured output and the control input. We consider the following robust ℓ_∞ control problem.

> Let a robust performance bound $\gamma > 0$ be given. Find a controller in (9.1), for the uncertain system (9.30), such that the closed-loop system is robustly stable and the output $\mathbf{y}_2$ satisfies $\|\mathbf{y}_2\|_{\ell_\infty} < \gamma$ for all disturbances such that $\|\mathbf{w}_2\|_{\ell_2} \leq 1$, and for all uncertainties $\mathbf{\Delta} \in \mathcal{BU}_D$.

This problem may be (conservatively) approached using the following theorem.

Theorem 9.2.7 *Let a controller* $\mathbf{G}$ *and a robust performance bound* $\gamma > 0$ *be given. Suppose there exist a Lyapunov matrix* $\mathbf{X} > 0$ *and a scaling matrix* $\mathbf{S} \in \mathcal{S}$ *such that* $\mathbf{C}_2\mathbf{X}\mathbf{C}_2^T + \mathbf{F}_{22}\mathbf{F}_{22}^T < \gamma^2\mathbf{I}$ *and (9.20) hold where*

$$\begin{bmatrix} \Theta & \Gamma & \mathbf{Q} \end{bmatrix} \triangleq \left[\begin{array}{ccc|c|cc} \mathbf{A} & \mathbf{D}_1 & \mathbf{D}_2 & \mathbf{B} & \mathbf{X} & \mathbf{0} \\ \mathbf{C}_1 & \mathbf{F}_{11} & \mathbf{F}_{12} & \mathbf{H}_1 & \mathbf{0} & \mathbf{S} \end{array}\right]$$

$$\begin{bmatrix} \Lambda^T & \mathbf{R} \end{bmatrix} \triangleq \left[\begin{array}{c|ccc} \mathbf{M}^T & \mathbf{X} & \mathbf{0} & \mathbf{0} \\ \mathbf{E}_1^T & \mathbf{0} & \mathbf{S} & \mathbf{0} \\ \mathbf{E}_2^T & \mathbf{0} & \mathbf{0} & \mathbf{I} \end{array}\right].$$

Then the controller $\mathbf{G}$ *solves the robust* ℓ_∞ *control problem.*

Proof. The result follows by replacing matrices in Theorem 4.7.7 by the closed-loop matrices defined in (9.3). □

9.2.8 Robust $\mathcal{H}_\infty$ Control

Consider the uncertain system described by

$$\begin{bmatrix} \mathbf{x}_p(k+1) \\ \mathbf{y}_1(k) \\ \mathbf{y}_2(k) \\ \mathbf{z}(k) \end{bmatrix} = \begin{bmatrix} \mathbf{A}_p & \mathbf{D}_{p1} & \mathbf{D}_{p2} & \mathbf{B}_p \\ \mathbf{C}_{p1} & \mathbf{D}_{y11} & \mathbf{D}_{y12} & \mathbf{B}_{y1} \\ \mathbf{C}_{p2} & \mathbf{D}_{y21} & \mathbf{D}_{y22} & \mathbf{B}_{y2} \\ \mathbf{M}_p & \mathbf{D}_{z1} & \mathbf{D}_{z2} & \mathbf{0} \end{bmatrix} \begin{bmatrix} \mathbf{x}_p(k) \\ \mathbf{w}_1(k) \\ \mathbf{w}_2(k) \\ \mathbf{u}(k) \end{bmatrix}, \quad \mathbf{w}_1 = \Delta\mathbf{y}_1 \quad (9.31)$$

where $\mathbf{x}_p$ is the state, $\mathbf{y}_1$ and $\mathbf{w}_1$ are the exogenous signals to describe the uncertainty $\Delta \in \mathcal{BU}_D$, $\mathbf{y}_2$ and $\mathbf{w}_2$ are the signals to describe the performance specification, and $\mathbf{z}$ and $\mathbf{u}$ are the measured output and the control input. The robust $\mathcal{H}_\infty$ control problem can be stated as follows.

> Let a robust performance bound $\gamma > 0$ be given. Find a controller in (9.1) for the uncertain system (9.31), such that the closed-loop system is robustly stable and the output $\mathbf{y}_2$ satisfies $\|\mathbf{y}_2\|_{\ell_2} < \gamma$ for all disturbances such that $\|\mathbf{w}_2\|_{\ell_2} \le 1$, and for all uncertainties $\Delta \in \mathcal{BU}_D$.

then the following formulation of the robust $\mathcal{H}_\infty$ control problem is called the state space upper bound μ-synthesis [26, 96].

Theorem 9.2.8 *Let a controller* $\mathbf{G}$ *and a robust performance bound* $\gamma > 0$ *be given. Suppose there exist a Lyapunov matrix* $\mathbf{X} > 0$ *and a scaling matrix* $\mathbf{S} \in \mathcal{S}$ *satisfying inequality (9.20) with*

$$\begin{bmatrix} \Theta & \Gamma & \mathbf{Q} \end{bmatrix} \triangleq \left[\begin{array}{ccc|c|ccc} \mathbf{A} & \mathbf{D}_1 & \mathbf{D}_2 & \mathbf{B} & \mathbf{X} & \mathbf{0} & \mathbf{0} \\ \mathbf{C}_1 & \mathbf{F}_{11} & \mathbf{F}_{12} & \mathbf{H}_1 & \mathbf{0} & \mathbf{S} & \mathbf{0} \\ \mathbf{C}_2 & \mathbf{F}_{21} & \mathbf{F}_{22} & \mathbf{H}_2 & \mathbf{0} & \mathbf{0} & \gamma^2\mathbf{I} \end{array}\right]$$

$$\begin{bmatrix} \Lambda^T & \mathbf{R} \end{bmatrix} \triangleq \left[\begin{array}{c|ccc} \mathbf{M}^T & \mathbf{X} & \mathbf{0} & \mathbf{0} \\ \mathbf{E}_1^T & \mathbf{0} & \mathbf{S} & \mathbf{0} \\ \mathbf{E}_2^T & \mathbf{0} & \mathbf{0} & \mathbf{I} \end{array}\right].$$

Then the controller $\mathbf{G}$ *solves the robust* $\mathcal{H}_\infty$ *control problem.*

Proof. The result follows from Theorem 4.7.8, using the closed-loop matrices defined in (9.3). □

CHAPTER 9 SUMMARY

We have shown that many linear control design problems, with stability, performance, and robustness specifications, can be reduced to a single problem of solving a matrix inequality

$$\mathbf{\Gamma G \Lambda} + (\mathbf{\Gamma G \Lambda})^T + \mathbf{\Theta} < \mathbf{0}$$

for the controller parameter $\mathbf{G}$, where the other matrices are appropriately defined for each control problem in terms of the plant data, a Lyapunov matrix $\mathbf{X}$ (or $\mathbf{Y}$), and possibly with a scaling matrix $\mathbf{S}$.

When designing a controller based on this approach, one must find a Lyapunov matrix $\mathbf{X}$ (or $\mathbf{Y}$) satisfying the existence conditions

$$\mathbf{\Gamma}^{\perp}\mathbf{\Theta}\mathbf{\Gamma}^{\perp T} < \mathbf{0}, \quad \mathbf{\Lambda}^{T\perp}\mathbf{\Theta}\mathbf{\Lambda}^{T\perp T} < \mathbf{0}.$$

Once such an $\mathbf{X}$ is found, then a controller can be computed by the explicit formula given in either Theorem 2.3.12 (for the continuous-time case) or Theorem 2.3.11 (for the discrete-time case). In the subsequent chapters, we shall give some algorithms to find a Lyapunov matrix $\mathbf{X}$. As will be shown, this Lyapunov matrix search can also be unified for each control problem.

The unified approach in this chapter is essentially from [55], and has appeared in a conference paper [61]. An LMI solution and detailed discussion for each control problem considered here can be found in the literature including [62, 64, 59]. The impact of the unified approach on control education is discussed in [127]. Unified perspective for control design based on algebraic problem formulation can be traced back to [130]. Note, however, that the formulation in [130] is based on matrix equations rather than matrix inequalities.

The unified approach presented here is based on a single algebraic problem of the type (9.5). A complete solution (existence conditions and an explicit formula) for this problem was first given in [58] (see [55, 59] for a proof). The existence condition also appeared in [8, 34]. The problem (9.20), which is a special case of (9.5) has been solved [20, 102] and the solution has been applied to the scaled $\mathcal{H}_\infty$ control problem [98].

CHAPTER TEN

Projection Methods

The main objectives of Chapters 10 and 11 are to provide effective computational tools for the numerical solution of the control design problems discussed in the previous chapters. To this end, the concept of convexity, defined later in this chapter, will play a fundamental role in our discussion. Convexity will allow existing computational techniques for our design purposes, but also it will enable development of some new tools specifically designed to find solutions for these design problems. Convexity is the most important property to seek in the control design problems formulated in the previous chapters, since effective techniques exist to solve such problems.

In this chapter, we will follow a geometric approach in our computational design. The design problems will often be formulated as feasibility problems, where the desired solution lies in the intersection of a family of constraint sets, or as minimum distance problems where the solution is the point in the intersection that minimizes the distance from a given point, or as problems involving the minimum distance between disjoint sets. The constraint sets will often have a simple geometric structure, such as planes, cones or polygons and the reader is encouraged to visualize the geometry of the problems in simple examples. The simple geometry of some of the design problems will motivate the use of *Alternating Convex Projection* techniques to obtain a numerical solution. The reader should be familiar with Appendix A before reading this chapter.

10.1 ALTERNATING CONVEX PROJECTION TECHNIQUES

10.1.1 Convexity

Consider a set $\mathcal{K}$ in a vector space. The set $\mathcal{K}$ is called *convex* if for any two vectors $\mathbf{x}$ and $\mathbf{y}$ in $\mathcal{K}$ the vector $(1-\lambda)\mathbf{x}+\lambda\mathbf{y}$ is also in $\mathcal{K}$ for $0 \leq \lambda \leq 1$. This definition says that given two points in a convex set, the line segment between them is also in the set. For example, a subspace is a convex set. Also, the set of matrices $\mathbf{X}$ satisfying $\mathbf{X} \geq \mathbf{P}$, for a given $\mathbf{P}$, forms a convex set.

Example 10.1.1 Show that the set of positive semidefinite matrices forms a convex set.
Solution: Let $\mathbf{X}$ and $\mathbf{Y}$ be $n \times n$ positive semidefinite matrices. Define the matrix $\mathbf{C} = (1-\lambda)\mathbf{X}+\lambda\mathbf{Y}$ where $0 \leq \lambda \leq 1$. We have to show that $\mathbf{C}$ is positive semidefinite. To

this end, consider any vector $\mathbf{z} \in \mathcal{C}^n$ and observe that $\mathbf{z}^*\mathbf{C}\mathbf{z} = (1-\lambda)\mathbf{z}^*\mathbf{X}\mathbf{z} + \lambda\mathbf{z}^*\mathbf{Y}\mathbf{z}$. However, $\mathbf{z}^*\mathbf{X}\mathbf{z} \geq 0$ and $\mathbf{z}^*\mathbf{Y}\mathbf{z} \geq 0$ for any $\mathbf{z}$. Hence, $\mathbf{z}^*\mathbf{C}\mathbf{z} \geq 0$ for any $\mathbf{z}$, i.e. $\mathbf{C}$ is positive semidefinite.

10.1.2 Feasibility, Optimization and Infeasible Optimization Problems

We begin our discussion with a description of the type of computational problems we will encounter. Generally, our unknowns will be symmetric matrices constrained to satisfy given matrix equality or inequality constraints. Each one of these matrix constraints defines a set in the space of symmetric matrices. For example, consider the matrix equality constraint

$$\mathbf{A}\mathbf{X} + \mathbf{X}\mathbf{A}^T + \mathbf{Q} = \mathbf{0} \tag{10.1}$$

where

$$\mathbf{A} = \begin{bmatrix} 0 & 1 \\ 0 & 0 \end{bmatrix}, \quad \mathbf{Q} = \begin{bmatrix} 1 & 0 \\ 0 & 0 \end{bmatrix}.$$

It is simple to verify that (10.1) defines the following set in the space of 2×2 symmetric matrices

$$\{\mathbf{X} : X_{11} = arbitrary, \quad X_{12} = X_{21} = -\frac{1}{2}, \quad X_{22} = 0\}. \tag{10.2}$$

In the (X_{11}, X_{22}, X_{12}) space, this set corresponds to a line parallel to the X_{11} axis which crosses the X_{12} axis at $-1/2$. Note that when $\mathbf{Q} = \mathbf{I}$, the set defined by (10.1) is the empty set, since in this case no matrix $\mathbf{X}$ exists to satisfy (10.1). In our computational problems we will often seek a matrix in the intersection of these types of matrix constraint sets.

To begin, consider a family $\mathcal{C}_1, \mathcal{C}_2, \ldots, \mathcal{C}_m$ of m sets in the space of symmetric matrices. We assume that these sets are convex and closed (i.e. they contain their limit points [114]). For our purposes, any set defined by a matrix equality constraint, as in (10.1), or a matrix semidefinite constraint, as the constraint $\mathbf{A}\mathbf{X} + \mathbf{X}\mathbf{A}^T \leq \mathbf{0}$, is closed. Note that the concept of closedness of a set should not be confused with the concept of boundedness. A set can be closed but unbounded, as for example the set (10.2).

We now define the type of problems we will solve:

Feasibility problem: Suppose that the sets $\mathcal{C}_1, \mathcal{C}_2, \ldots, \mathcal{C}_m$ have a nonempty intersection. Then, the feasibility problem is to find a symmetric matrix $\mathbf{X}$ in this intersection, i.e. to find $\mathbf{X}$ such that

$$\mathbf{X} \in \mathcal{C}_1 \bigcap \mathcal{C}_2 \bigcap \cdots \bigcap \mathcal{C}_m. \tag{10.3}$$

Note that there might be no solution or an infinite number of solutions in a feasibility problem (see Figure 10.1).

Optimization problem: Again suppose that the sets $\mathcal{C}_1, \mathcal{C}_2, \ldots, \mathcal{C}_m$ have nonempty intersection and consider a given $n \times n$ symmetric matrix $\mathbf{X}_o$. The optimization problem we will solve is to find the matrix in the intersection of the sets, which is closest to the matrix $\mathbf{X}_o$. In mathematical terms, we seek to solve the minimization problem

$$minimize \;\; \|\mathbf{X} - \mathbf{X}_o\| \quad subject\ to \;\; \mathbf{X} \in \mathcal{C}_1 \bigcap \mathcal{C}_2 \bigcap \cdots \bigcap \mathcal{C}_m. \tag{10.4}$$

According to the projection theorem for convex sets [83], this minimization problem has a unique solution, given by the projection of the matrix $\mathbf{X}_o$ on the intersection of the sets (see Figure 10.2).

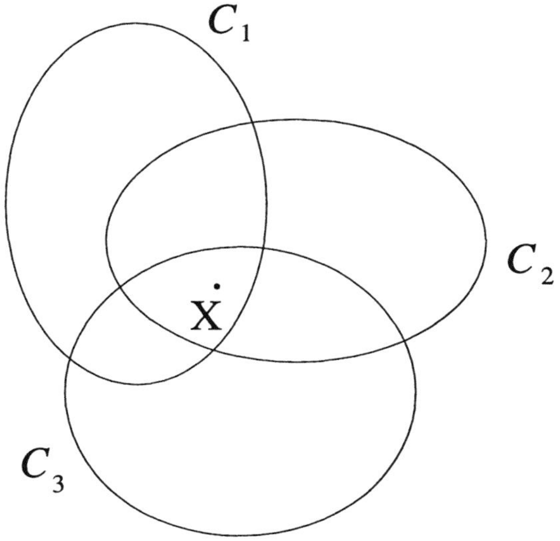

Figure 10.1 Feasibility problem.

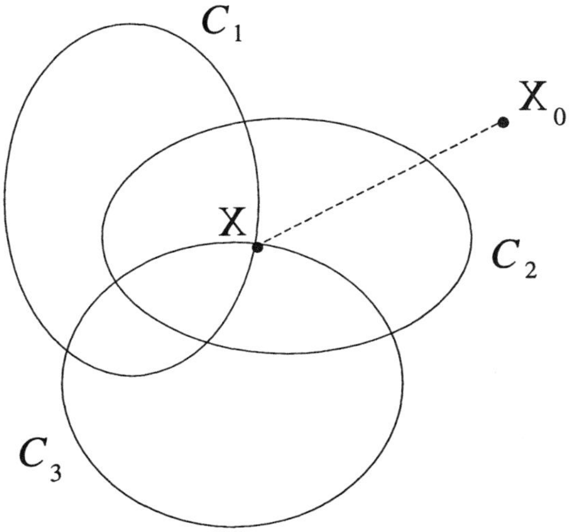

Figure 10.2 Optimization problem.

Infeasible optimization problem: Now consider the case of two nonempty constraint sets $\mathcal{C}_1$ and $\mathcal{C}_2$ and suppose that they are disjoint, i.e. their intersection is empty. The infeasible optimization problem is to find a symmetric matrix $\mathbf{X}$ in the set $\mathcal{C}_1$ which is closest to the

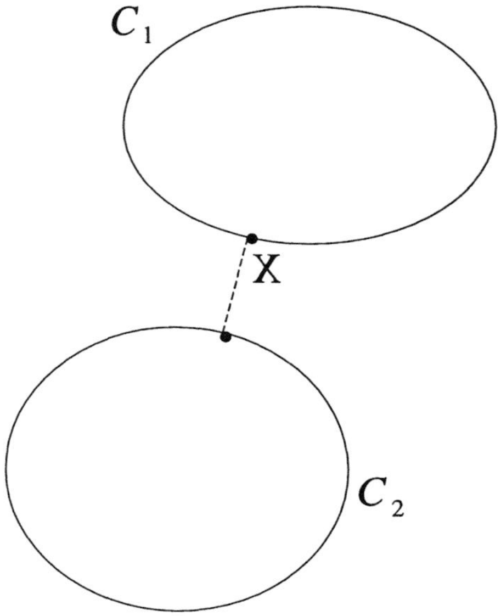

Figure 10.3 Infeasible optimization problem.

set C_2 (see Figure 10.3). In mathematical terms we seek to solve the minimization problem

$$\textit{minimize} \quad dist(\mathbf{X}, C_2) \quad \textit{subject to} \quad \mathbf{X} \in C_1 \tag{10.5}$$

where $dist(\mathbf{X}, C_2)$ is the distance of the matrix $\mathbf{X}$ from the set C_2, defined by

$$dist(\mathbf{X}, C_2) \triangleq \inf\{\|\mathbf{X} - \mathbf{Y}\| : \mathbf{Y} \in C_2\}.$$

The solution to the infeasible optimization problem might not be unique. For example, when C_1 and C_2 are two parallel planes, any point of C_1 provides a solution to the minimization problem (10.5).

In the following, we will describe numerical techniques to solve the above types of problems using Alternating Convex Projection (ACP) techniques.

10.1.3 The Standard ACP Method

Consider a family $C_1, C_2, \ldots, C_m$ of closed, convex sets in the space of symmetric matrices. We suppose that the sets have a nonempty intersection and we seek to solve the feasibility problem defined in section 12.1.2. Let P_{C_i} denote the orthogonal projection operator onto the set C_i where $i = 1, \ldots, m$. That is, for any $n \times n$ symmetric matrix $\mathbf{X}$, the matrix $P_{C_i}(\mathbf{X})$ denotes the orthogonal projection of $\mathbf{X}$ onto C_i, i.e. the matrix in C_i which has minimum distance from the matrix $\mathbf{X}$. The orthogonal projection theorem ([83] and Appendix A) guarantees that this projection is uniquely defined. We assume that the sets $C_1, C_2, \ldots, C_m$ are of simple geometric structure (for example planes, cones, spheres, polygons, etc.) so that an analytical expression for each projection operator P_{C_i} can be derived. The question we would like to answer is the following: Is it possible to provide a solution to the feasibility problem by making use of the orthogonal projections onto each constraint set? The answer is yes, and is provided by the following result ([12], [46]) which we call the *Standard Alternating Projection Theorem.*

Theorem 10.1.1 *Let $\mathbf{X}_0$ be any $n \times n$ symmetric matrix, and $C_1, C_2, \ldots, C_m$ be a family of closed, convex sets in the space of symmetric matrices. Then, if there exists an intersection, the sequence of alternating projections*

$$\begin{aligned}
\mathbf{X}_1 &= P_{C_1}\mathbf{X}_0 \\
\mathbf{X}_2 &= P_{C_2}\mathbf{X}_1 \\
&\vdots \\
\mathbf{X}_m &= P_{C_m}\mathbf{X}_{m-1} \\
&\vdots \\
\mathbf{X}_{2m} &= P_{C_{2m}}\mathbf{X}_{2m-1} \\
\mathbf{X}_{2m+1} &= P_{C_{2m+1}}\mathbf{X}_{2m} \\
&\vdots \\
\mathbf{X}_{3m} &= P_{C_{3m}}\mathbf{X}_{3m-1} \\
&\vdots
\end{aligned} \tag{10.6}$$

converges to a point in the intersection of the sets, i.e. $\mathbf{X}_i \rightarrow \mathbf{X}$ where $\mathbf{X} \in C_1 \bigcap C_2 \bigcap \ldots \bigcap C_m$. If no intersection exists, the sequence converges to a limit cycle (a periodic iteration between the disjoint sets).

Hence, starting from any symmetric matrix, the sequence of alternating projections onto the constraint sets converges to a solution of the feasibility problem, if one exists. A schematic representation of the Standard Alternating Projection Method is shown in Figure 10.4. It can be easily verified that the limit $\mathbf{X}$ of the alternating projection sequence depends on the starting point $\mathbf{X}_0$, as well as the order of the projections. Hence, by rearranging the sequence of projections we can obtain a different feasible point. See [73] for some examples.

10.1.4 The Optimal ACP Method

Our next objective is to provide a technique to solve the optimization problem (10.4), using an alternating projection approach. We first observe that the Standard ACP algorithm (10.6) does not necessarily converge to the solution of the optimization problem. In fact, a simple example where C_1 is a disc in the plane and C_2 is a line which intersects the disc, can easily show that the sequence of alternating convex projections (10.6) is not adequate in this case. However, a simple modification of the Standard ACP method provides an algorithm which solves the optimization problem. To describe this result, consider the closed, convex sets $C_1, C_2, \ldots, C_m$ and a given $n \times n$ matrix $\mathbf{X}_0$. The following result ([9], [48]), which we call the *Optimal Alternating Convex Projection* Theorem, provides an algorithm which converges to the solution of the optimization problem (10.4).

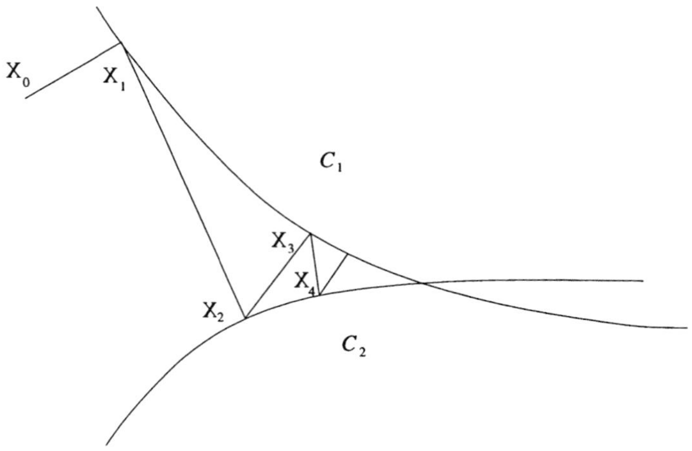

Figure 10.4 Standard Alternating Convex Projection Algorithm.

Theorem 10.1.2 *Consider the sequence of matrices $\{\mathbf{X}_i\}$, $i = 1, 2, \ldots, \infty$ computed as follows:*

$$\begin{array}{ll}
\mathbf{X}_1 = P_{C_1}\mathbf{X}_0 & \mathbf{Z}_1 = \mathbf{X}_1 - \mathbf{X}_0 \\
\mathbf{X}_2 = P_{C_2}\mathbf{X}_1 & \mathbf{Z}_2 = \mathbf{X}_2 - \mathbf{X}_1 \\
\vdots & \vdots \\
\mathbf{X}_m = P_{C_m}\mathbf{X}_{m-1} & \mathbf{Z}_m = \mathbf{X}_m - \mathbf{X}_{m-1} \\
\mathbf{X}_{m+1} = P_{C_1}(\mathbf{X}_m - \mathbf{Z}_1) & \mathbf{Z}_{m+1} = \mathbf{Z}_1 + \mathbf{X}_{m+1} - \mathbf{X}_m \\
\mathbf{X}_{m+2} = P_{C_2}(\mathbf{X}_{m+1} - \mathbf{Z}_2) & \mathbf{Z}_{m+2} = \mathbf{Z}_2 + \mathbf{X}_{m+2} - \mathbf{X}_{m+1} \\
\vdots & \vdots \\
\mathbf{X}_{2m} = P_{C_m}(\mathbf{X}_{2m-1} - \mathbf{Z}_m) & \mathbf{Z}_{2m} = \mathbf{Z}_m + \mathbf{X}_{2m} - \mathbf{X}_{2m-1} \\
\mathbf{X}_{2m+1} = P_{C_1}(\mathbf{X}_{2m} - \mathbf{Z}_{m+1}) & \mathbf{Z}_{2m+1} = \mathbf{Z}_{m+1} + \mathbf{X}_{2m+1} - \mathbf{X}_{2m} \\
\vdots & \vdots
\end{array} \tag{10.7}$$

Then the sequence $\{\mathbf{X}_i\}$ converges to a matrix $\mathbf{X}$ that solves the optimization problem (10.4).

Note that the algorithm (10.7) is a modified alternating projection algorithm, where in each step, an increment $\mathbf{Z}_i$ is removed before projection to the corresponding convex set. This forces the algorithm to converge to the solution $\mathbf{X}$ of the optimization problem.

An important feature of the two projection algorithms (10.6) and (10.7) is that, when the analytical expressions for the projections onto the constraint sets are available, then the algorithms can be implemented very easily and the amount of calculations in one iteration is very small. However, in some cases the algorithms may suffer from slow convergence. For example, consider the case of two planes intersecting with a small angle. In this case the standard ACP algorithm (10.6) might oscillate for many iterations between the two sets before it converges to a point in the intersection. An effective remedy for the case of the feasibility problem might be to use the *Directional Alternating Convex Projection* Algorithm, described below [46].

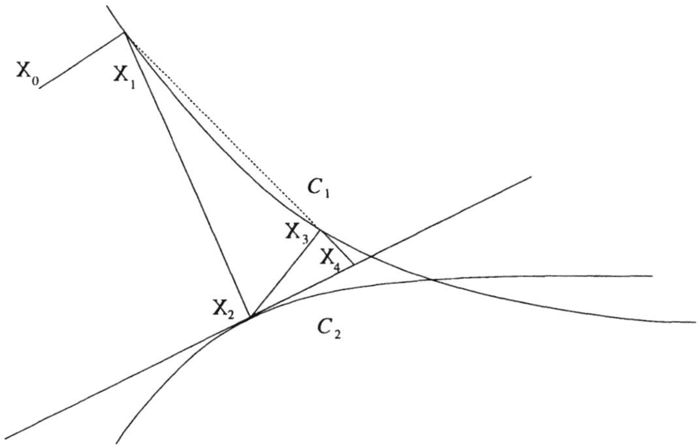

Figure 10.5 Directional Alternating Projection Algorithm.

10.1.5 The Directional ACP Method

The directional ACP method uses information about the geometry of the constraint sets to provide an algorithm with accelerated convergence to solve the feasibility problem (10.3). The basic idea behind this approach is to utilize in each iteration the tangent plane of one of the constraint sets, so that the sequence of points we obtain approaches the intersection of the sets more rapidly (see Figure 10.5). For simplicity, we will consider the case of two closed and convex constraint sets $\mathcal{C}_1$ and $\mathcal{C}_2$. The *Directional Alternating Convex Projection* Theorem is described next, where $\langle \mathbf{X}, \mathbf{Y} \rangle$ denotes the inner product of two matrices **A** and **B** (see Appendix A.5).

Theorem 10.1.3 *Let $\mathbf{X}_0$ be any $n \times n$ symmetric matrix. Then the sequence of matrices $\{\mathbf{X}_i\}$, $i = 1, 2, \ldots, \infty$ given by*

$$\begin{aligned}
&\mathbf{X}_1 = P_{\mathcal{C}_1}\mathbf{X}_0, \quad \mathbf{X}_2 = P_{\mathcal{C}_2}\mathbf{X}_1, \quad \mathbf{X}_3 = P_{\mathcal{C}_1}\mathbf{X}_1 \\
&\mathbf{X}_4 = \mathbf{X}_1 + \lambda_1(\mathbf{X}_3 - \mathbf{X}_1), \quad \lambda_1 = \frac{\|\mathbf{X}_1 - \mathbf{X}_2\|^2}{\langle \mathbf{X}_1 - \mathbf{X}_3, \mathbf{X}_1 - \mathbf{X}_2 \rangle} \\
&\mathbf{X}_5 = P_{\mathcal{C}_1}\mathbf{X}_4, \quad \mathbf{X}_6 = P_{\mathcal{C}_2}\mathbf{X}_5, \quad \mathbf{X}_7 = P_{\mathcal{C}_1}\mathbf{X}_6 \\
&\mathbf{X}_8 = \mathbf{X}_5 + \lambda_2(\mathbf{X}_7 - \mathbf{X}_5), \quad \lambda_1 = \frac{\|\mathbf{X}_5 - \mathbf{X}_6\|_F^2}{\langle \mathbf{X}_5 - \mathbf{X}_7, \mathbf{X}_5 - \mathbf{X}_6 \rangle} \\
&\vdots
\end{aligned} \tag{10.8}$$

converges to a point in the intersection of the sets $\mathcal{C}_1$ and $\mathcal{C}_2$.

Hence, starting from any symmetric matrix, the sequence of directional alternating projections (10.8) provides an accelerated numerical algorithm to solve the Feasibility Problem (10.3). In fact, it can be easily verified that when the two sets $\mathcal{C}_1$ and $\mathcal{C}_2$ are hyperplanes in the space of symmetric matrices then the alternating projection algorithm converges to a feasible point in one cycle, independently of the angle between the two hyperplanes.

Exercise 10.1.1 Based on the geometry of Figure 10.5 derive the expression for $\mathbf{X}_4$ and λ_1 in (10.8), in terms of $\mathbf{X}_1$, $\mathbf{X}_2$ and $\mathbf{X}_3$.

Example 10.1.2 We consider the state space model of a fighter aircraft provided in MATLAB Robust Control Toolbox (1988).

$$\mathbf{A}_p = \begin{bmatrix} -0.0226 & -36.6170 & -18.8970 & -32.0900 & 3.2509 & -0.7626 \\ 0.0001 & -1.8997 & 0.9831 & -0.0007 & -0.1708 & -0.0050 \\ 0.0123 & 11.7200 & -2.6316 & 0.0009 & -31.6040 & 22.3960 \\ 0 & 0 & 1.0000 & 0 & 0 & 0 \\ 0 & 0 & 0 & 0 & -30.0000 & 0 \\ 0 & 0 & 0 & 0 & 0 & -30.0000 \end{bmatrix}$$

$$\mathbf{B}_p = \mathbf{D}_p = \begin{bmatrix} 0 & 0 \\ 0 & 0 \\ 0 & 0 \\ 0 & 0 \\ 0 & 0 \\ 30 & 0 \\ 0 & 30 \end{bmatrix}, \quad \mathbf{C}_p = \begin{bmatrix} 0 & 1 & 0 & 0 & 0 & 0 \\ 0 & 0 & 0 & 1 & 0 & 0 \end{bmatrix}.$$

We assume that all states are available for feedback, i.e. $\mathbf{z} = \mathbf{x}_p$. We look for a static state feedback controller to satisfy the following output performance constraints

$$\mathcal{E}y_1^2 \leq 0.01, \quad \mathcal{E}y_2^2 \leq 0.01$$

i.e. the output covariance constraint set $\mathcal{V}$ has the form

$$\mathcal{V} = \{\mathbf{X} \in \mathcal{S} : X_{22} \leq 0.01, \quad X_{44} \leq 0.01\}.$$

We seek to solve the covariance feasibility problem (10.27) i.e. to find an assignable covariance matrix $\mathbf{X}_p$ such that $\mathbf{X}_p \in \mathcal{V}$, and the covariance optimization problem (10.28)–(10.29) where we look for an assignable covariance matrix $\mathbf{X}_p$ such that $\mathbf{X}_p \in \mathcal{V}$ and $\|\mathbf{X}_p - 2\mathbf{I}\|$ is minimized. For comparison, we will apply all three alternating projection methods described in Section 12.1. The algorithms are initialized from the same starting point $(\mathbf{X}_p)_0 = 2\mathbf{I}$ and we will require an error bound for the solution (sum of the distances from the constraint sets) less than 10^{-5}. The standard alternating projection technique (10.6) provides the following feasible covariance matrix:

$$\mathbf{X}_p = \begin{bmatrix} 2.0037 & 0.0138 & -0.0277 & -0.0638 & -0.5342 & 0.3760 \\ 0.0138 & 0.0028 & 0.0058 & -0.0038 & 0.0042 & -0.0531 \\ -0.0277 & 0.0058 & 1.3283 & 0.0000 & -0.0596 & 0.0689 \\ -0.0638 & -0.0038 & 0.0000 & 0.0100 & 0.0752 & 0.0488 \\ -0.5342 & 0.0042 & -0.0596 & 0.0752 & 2.0009 & -0.0014 \\ 0.3760 & -0.0531 & 0.0689 & 0.0488 & -0.0014 & 2.0044 \end{bmatrix}.$$

This method required 15855 iterations (a number of $6.3607\ 10^7$ floating point operations (flops)) to converge to the feasible solution. The distance of the starting point $(\mathbf{X}_p)_0 = 2\mathbf{I}$

to the feasible covariance $\mathbf{X}_p$ is $\|\mathbf{X}_p - (\mathbf{X}_p)_0\| = 3.0499$. The algorithm (10.7) which provides the feasible covariance of minimum distance from $(\mathbf{X}_p)_0 = 2\mathbf{I}$ converged to the following solution:

$$\mathbf{X}_p = \begin{bmatrix} 2.0005 & 0.0207 & -0.0144 & -0.0795 & -0.5332 & 0.3762 \\ 0.0207 & 0.0028 & 0.0054 & -0.0037 & 0.0022 & -0.0561 \\ -0.0144 & 0.0054 & 1.3301 & 0.0000 & -0.0599 & 0.0690 \\ -0.0795 & -0.0037 & 0.0000 & 0.0100 & 0.0713 & 0.0432 \\ -0.5332 & 0.0022 & -0.0599 & 0.0713 & 2.0014 & -0.0020 \\ 0.3762 & -0.0561 & 0.0690 & 0.0432 & -0.0020 & 2.0029 \end{bmatrix}.$$

This method required 16259 iterations (9.3526 10^7 flops) and the distance from the starting point is $\|\mathbf{X}_p - (\mathbf{X}_p)_0\| = 3.0496$. The directional alternating projection algorithm (10.8) provided the following answer:

$$\mathbf{X}_p = \begin{bmatrix} 2.0056 & 0.0101 & -0.0328 & -0.0567 & -0.5350 & 0.3757 \\ 0.0101 & 0.0028 & 0.0060 & -0.0039 & 0.0046 & -0.0525 \\ -0.0328 & 0.0060 & 1.3274 & 0.0000 & -0.0595 & 0.0688 \\ -0.0567 & -0.0039 & -0.0000 & 0.0100 & 0.0767 & 0.0511 \\ -0.5350 & 0.0046 & -0.0595 & 0.0767 & 2.0009 & -0.0010 \\ 0.3757 & -0.0525 & 0.0688 & 0.0511 & -0.0010 & 2.0055 \end{bmatrix}.$$

This method required 6 iterations (1.7070 10^5 flops) and the distance from the starting point is $\|\mathbf{X}_p - (\mathbf{X}_p)_0\| = 3.0502$. Note that in all cases the output variance constraint which corresponds to the first output $\mathbf{y}_1$ (2×2 element of $\mathbf{X}_p$) is not binding, although the one which corresponds to $\mathbf{y}_2$ (4×4 element of $\mathbf{X}_p$) is binding (i.e. is reaching the allowed bound 0.01).

A controller which assigns to the closed-loop system the covariance matrix $\mathbf{X}_p$ which resulted from the directional projection algorithm (10.8) (setting the free parameters to minimize the required control effort as in [42]) is the following:

$$\mathbf{G} = \begin{bmatrix} -0.0103 & 26.5768 & 1.4216 & 14.0812 & 0.3636 & 0.2426 \\ 0.0402 & -10.6659 & -1.0383 & -4.9792 & 0.2426 & -0.8011 \end{bmatrix}.$$

We observe that for our example, the directional alternating projection algorithm converges to a feasible solution much faster than the other two algorithms.

10.2 GEOMETRIC FORMULATION OF COVARIANCE CONTROL

In this section we will provide a geometric formulation of the covariance control design problem, as a feasibility, optimization or infeasible optimization problem in the space of $n_p \times n_p$ symmetric matrices. Hence, later we will apply the ACP techniques of the previous section for a numerical solution. The results of this section can also be found in [43] and [44].

10.2.1 State Feedback

Consider the static state feedback covariance control design problem described in section 6.2.1. Let $\mathcal{S}$ be the set (vector space) of $n_p \times n_p$ real symmetric matrices and define the following subsets of $\mathcal{S}$

$$\mathcal{A} = \{\mathbf{X} \in \mathcal{S} : (\mathbf{I} - \mathbf{B}\mathbf{B}^+)(\mathbf{A}\mathbf{X} + \mathbf{X}\mathbf{A}^T + \mathbf{W})(\mathbf{I} - \mathbf{B}\mathbf{B}^+) = \mathbf{0}\} \tag{10.9}$$

$$\mathcal{P} = \{\mathbf{X} \in \mathcal{S} : \mathbf{X} > \mathbf{0}\}. \tag{10.10}$$

Then, according to Theorem 6.2.1, a matrix $\mathbf{X}$ is assignable to the closed-loop system as a state covariance if and only if $\mathbf{X}$ is in the intersection of the two sets $\mathcal{A}$ and $\mathcal{P}$. Hence, we have the following geometric interpretation of the assignability theory.

Corollary 10.2.1 *A matrix $\mathbf{X}$ is an assignable plant state covariance if and only if $\mathbf{X} \in \mathcal{A} \bigcap \mathcal{P}$.*

Note that the sets $\mathcal{A}$ and $\mathcal{P}$ have special structure. Specifically, since the assignability equation in (10.9) is a linear equation, the set $\mathcal{A}$ is a plane (an *affine subspace*) in the space of symmetric matrices $\mathcal{S}$. On the other hand, the set $\mathcal{P}$ of positive definite matrices is a convex cone in $\mathcal{S}$. The set $\mathcal{A}$ is obviously closed and convex, however $\mathcal{P}$ is not closed. Since it will be easier for us to deal with closed sets we define the following closed ϵ-approximation of the set $\mathcal{P}$

$$\mathcal{P}_\epsilon = \{\mathbf{X} \in \mathcal{S} : \mathbf{X} \geq \epsilon\mathbf{I}\} \tag{10.11}$$

where ϵ is an arbitrarily small positive number. We call any matrix $\mathbf{X} \in \mathcal{A} \bigcap \mathcal{P}_\epsilon$ an ϵ-assignable plant covariance. Since we will choose ϵ to be very small, we will not distinguish between ϵ-assignability and assignability.

Our geometric formulation implies that the static state feedback covariance assignability problem can be seen as a feasibility problem: Find a matrix $\mathbf{X}$ such that

$$\mathbf{X} \in \mathcal{A} \bigcap \mathcal{P}_\epsilon. \tag{10.12}$$

This formulation of the covariance control problem will allow us to use the alternating convex projection techniques for a numerical solution.

10.2.2 Dynamic Output Feedback with Measurement Noise

Now we consider a more complex problem, that of full-order dynamic output feedback covariance control. The necessary and sufficient conditions for a matrix $\mathbf{X}_p$ to be assignable as a plant state covariance by a full-order dynamic controller are provided in Corollary 5.2.2. Define the following set in the space $\mathcal{S}$ of $n_p \times n_p$ symmetric matrices.

$$\mathcal{P} = \{\mathbf{X}_p \in \mathcal{S} : \mathbf{X}_p > \mathbf{P}\}. \tag{10.13}$$

where $\mathbf{P}$ is defined by

$$\mathbf{A}_p\mathbf{P} + \mathbf{P}\mathbf{A}_p^T - \mathbf{P}\mathbf{M}_p^T\mathbf{V}^{-1}\mathbf{M}_p\mathbf{P} + \mathbf{W} = \mathbf{0}. \tag{10.14}$$

Note that the set $\mathcal{P}$ in (10.13) is similar to the one in (10.10), by setting $\mathbf{P} = \mathbf{0}$. Hence, we give both sets the same symbol, and the reader must distinguish the different cases of static state feedback and full-order, dynamic feedback. We conclude that the following result is true.

Corollary 10.2.2 *A matrix* $\mathbf{X}_p$ *is an assignable plant state covariance by full-order dynamic output feedback if and only if* $\mathbf{X}_p \in \mathcal{A} \bigcap \mathcal{P}$.

As earlier, we define the following closed ϵ-approximation of $\mathcal{P}$

$$\mathcal{P}_\epsilon = \{\mathbf{X}_p \in \mathcal{S} : \mathbf{X}_p \leq \mathbf{P} + \epsilon \mathbf{I}\} \tag{10.15}$$

where ϵ is an arbitrarily small positive number.

Therefore, the full-order dynamic covariance control problem can be posed as a feasibility problem: Find a matrix $\mathbf{X}_p$ such that

$$\mathbf{X}_p \in \mathcal{A} \bigcap \mathcal{P}_\epsilon. \tag{10.16}$$

Hence, both the static state feedback and the full-order dynamic output feedback covariance control problems can be formulated as similar feasibility problems of finding a symmetric matrix in the intersection of a plane with a convex cone.

10.2.3 Output Performance Constraints

To make the covariance control problems of the previous sections more practical, one can add some output performance constraints to the design problem. To this end, consider the following output variance constraints (or $\mathcal{L}_2$ to $\mathcal{L}_\infty$ constraints) on the system outputs

$$(\mathbf{C}\mathbf{X}\mathbf{C}^T)_{ii} \leq \sigma_i, \quad i = 1, 2, \ldots, n_y \tag{10.17}$$

where $\mathbf{C} = \begin{bmatrix}\mathbf{C}_p, \mathbf{0}\end{bmatrix}$ is the system output matrix and σ_i, $i = 1, 2, \ldots, n_y$, are the desired bounds on the output variances. We suppose that

$$\mathbf{C}_p^T = \begin{bmatrix}\mathbf{c}_1, \mathbf{c}_2, \ldots, \mathbf{c}_{n_y}\end{bmatrix}$$

where $\mathbf{c}_i$ are the columns of the $\mathbf{C}_p^T$. Then, by defining the following subsets of $\mathcal{S}$:

$$\mathcal{V}_i = \{\mathbf{X}_p \in \mathcal{S} : \mathbf{c}_i^T \mathbf{X}_p \mathbf{c}_i \leq \sigma_i\} \tag{10.18}$$

each one of the output variance constraints (10.17) is equivalent to the feasibility condition

$$\mathbf{X}_p \in \mathcal{V}_i. \tag{10.19}$$

In a similar way we consider the output quadratic cost constraint

$$tr(\mathbf{X}_p \mathbf{C}_p^T \mathbf{Q} \mathbf{C}_p) \leq \gamma \tag{10.20}$$

where $\mathbf{Q}$ is a positive definite weighting matrix and γ is a given positive bound on the output cost. We define the following constraint set in $\mathcal{S}$

$$\mathcal{C} = \{\mathbf{X}_p \in \mathcal{S} : tr(\mathbf{X}_p \mathbf{C}_p^T \mathbf{Q} \mathbf{C}_p) \leq \gamma\}. \tag{10.21}$$

Then, the output cost constraint (10.20) is equivalent to the feasibility condition

$$\mathbf{X}_p \in \mathcal{C}. \tag{10.22}$$

We can also treat block output covariance constraints as follows. Suppose that the output matrix $\mathbf{C}_p$ is decomposed into block matrices as follows

$$\mathbf{C}_p^T = [\mathbf{C}_1, \mathbf{C}_2, \ldots, \mathbf{C}_k]$$

where the decomposition is compatible with the block output covariance constraints of interest. That is, we assume that we want to satisfy matrix covariance constraints of the form

$$\mathbf{C}_i^T \mathbf{X}_p \mathbf{C}_i \leq \mathbf{Y}_i, \quad i = 1, 2, \ldots, k \tag{10.23}$$

where $\mathbf{Y}_i$ are given bounding matrices. By defining the constraint sets

$$\mathcal{O}_i = \{\mathbf{X}_p \in \mathcal{S} : \mathbf{C}_i^T \mathbf{X}_p \mathbf{C}_i \leq \mathbf{Y}_i\} \tag{10.24}$$

the ith block covariance constraint is equivalent to the feasibility condition

$$\mathbf{X}_p \in \mathcal{O}_i. \tag{10.25}$$

From the above geometric formulation we conclude that the covariance control design problem, subject to output variance, output cost and output matrix covariance constraints can be formulated as a feasibility problem of the general form: Find a matrix $\mathbf{X}_p$ such that

$$\mathbf{X}_p \in \mathcal{A} \bigcap \mathcal{P}_\epsilon \bigcap \mathcal{V}_1 \bigcap \mathcal{V}_2 \bigcap \cdots \bigcap \mathcal{V}_{n_y} \bigcap \mathcal{O}_1 \bigcap \mathcal{O}_2 \ldots \bigcap \mathcal{O}_k \bigcap \mathcal{C}. \tag{10.26}$$

We examine next the fundamental covariance control problems we wish to solve, using our geometric approach.

10.2.4 Covariance Control Problems

Following the abstract mathematical formulation of Section 12.1.2. we can define the following covariance control design problems using the geometric approach of the previous section.

Covariance Feasibility Problem

As we have already seen, the covariance control design problem, subject to output covariance and output cost constraints, can be formulated as a feasibility problem in $\mathcal{S}$, as follows: Find a matrix $\mathbf{X}_p$ such that

$$\mathbf{X}_p \in \mathcal{A} \bigcap \mathcal{P}_\epsilon \bigcap \mathcal{V}_1 \bigcap \mathcal{V}_2 \bigcap \cdots \bigcap \mathcal{V}_{n_y} \bigcap \mathcal{O}_1 \bigcap \mathcal{O}_2 \ldots \bigcap \mathcal{O}_k \bigcap \mathcal{C} \tag{10.27}$$

or determine that none exists. Note that stabilizability of the system implies that the intersection of the sets $\mathcal{A}$ and $\mathcal{P}_\epsilon$ is nonempty. However, when the output performance constraints, determined by the sets $\mathcal{V}_i$, $\mathcal{O}_i$ and $\mathcal{C}$, are too stringent, then it is possible that no matrix $\mathbf{X}_p$ exists to satisfy the feasibility condition (10.27).

Covariance Optimization Problem

Alternatively, we can formulate the covariance design problem as an optimization problem, as follows: Suppose that $\mathbf{X}_*$ denotes a desirable but nonassignable plant covariance matrix. Hence, $\mathbf{X}_*$ contains the desired closed-loop system properties we would like our system to possess (we know from Chapter 4 that many closed-loop properties can be characterized in terms of the plant covariance matrix, for example robustness, pole location, output performance, etc.) We seek an assignable plant covariance $\mathbf{X}_p$ which also satisfies output variance and output cost constraints, and minimizes the distance from the desired covariance $\mathbf{X}_*$. In mathematical terms, the problem can be posed as: Given $\mathbf{X}_*$, find $\mathbf{X}_p$ to solve the minimization problem

$$\text{minimize } \|\mathbf{X}_p - \mathbf{X}_*\| \tag{10.28}$$

subject to

$$\mathbf{X}_p \in \mathcal{A} \bigcap \mathcal{P}_\epsilon \bigcap \mathcal{V}_1 \bigcap \mathcal{V}_2 \bigcap \cdots \bigcap \mathcal{V}_{n_y} \bigcap \mathcal{O}_1 \bigcap \mathcal{O}_2 \bigcap \ldots \bigcap \mathcal{O}_k \bigcap \mathcal{C}. \tag{10.29}$$

Note that when the intersection of the constraint sets is nonempty, this problem has a unique solution.

Infeasible Optimization Problem

The natural question which arises from the above discussion is what happens when the intersection of the constraint sets is empty? Can we obtain a satisfactory answer by relaxing some of the constraints? Note that this is a very important question in practical design problems since usually the design engineer does not know *a priori* if all the design constraints can be met. In the covariance design problem the assignability constraints $\mathcal{A}$ and $\mathcal{P}_\epsilon$ are *hard* constraints (i.e. they must be satisfied in the design). Hence, in the case of infeasible constraints we seek an assignable covariance which approximates as closely as possible the desired, but unachievable, performance constraints. In mathematical terms, we look to solve the following optimization problem: Find a matrix $\mathbf{X}_p$ to solve the minimization problem

$$\text{minimize } dist(\mathbf{X}_p, \mathcal{V} \bigcap \mathcal{O} \bigcap \mathcal{C}) \quad \text{subject to } \mathbf{X}_p \in \mathcal{A} \bigcap \mathcal{P}_\epsilon \tag{10.30}$$

where $\mathcal{V} = \mathcal{V}_1 \bigcap \mathcal{V}_2 \bigcap \cdots \bigcap \mathcal{V}_{n_y}$ and $\mathcal{O} = \mathcal{O}_1 \bigcap \mathcal{O}_2 \bigcap \cdots \bigcap \mathcal{O}_k$. The function $dist(\mathbf{X}_p, \mathcal{V} \bigcap \mathcal{O} \bigcap \mathcal{C})$ is the distance of the matrix X_p from the intersection of the sets $\mathcal{V}$, $\mathcal{O}$ and $\mathcal{C}$.

When we obtain a matrix X_p that solves the covariance design problem then all controllers which assign this covariance to the closed-loop system can be obtained using the covariance control parametrizations of Chapter 5 or Chapter 6.

10.3 PROJECTIONS FOR COVARIANCE CONTROL

In this section we develop the analytic expressions for the projection operators onto the constraints sets of the covariance design problem, formulated in the previous section. Using these expressions we will be able to compute numerical solutions for the covariance design problem, utilizing the alternating convex projection techniques of Section 12.1. We start with the assignability constraint set $\mathcal{A}$.

10.3.1 Projection onto the Assignability Set

Recall that the assignability set $\mathcal{A}$ is a plane (or an affine space) in the space of symmetric matrices $\mathcal{S}$. Hence, we can derive an expression for the orthogonal projection onto this set using the projection theorem. Given any matrix $\mathbf{X}_0$ in $\mathcal{S}$, the orthogonal projection $P_{\mathcal{A}}$ can be calculated using the following result [43].

Theorem 10.3.1 *The projection $P_{\mathcal{A}}\mathbf{X}_0$ of the matrix $\mathbf{X}_0$ onto the assignability constraint set $\mathcal{A}$ is given by*

$$P_{\mathcal{A}}\mathbf{X}_0 = vec^{-1}\{\mathbf{X}_0 - \mathbf{\Delta}(\mathbf{K}\mathbf{\Delta})^{+}vec(\mathbf{E}_p\mathbf{Q}_p\mathbf{E}_p)\} \tag{10.31}$$

where $\mathbf{E}_p = \mathbf{I} - \mathbf{B}_p\mathbf{B}_p^{+}$, $\mathbf{K} = \mathbf{E}_p \bigotimes (\mathbf{E}_p\mathbf{A}_p) + (\mathbf{E}_p\mathbf{A}_p) \bigotimes \mathbf{E}_p$, $\mathbf{Q}_p = \mathbf{A}_p\mathbf{X}_p + \mathbf{X}_p\mathbf{A}_p^T + \mathbf{W}$ *and $\mathbf{\Delta}$ is the $n_p^2 \times n_p(n_p+1)/2$ matrix whose columns form an orthonormal basis for $\mathcal{S}$.*

Recall that the Kronecker product $\mathbf{M} \bigotimes \mathbf{N}$ of two matrices $\mathbf{M} \in \mathcal{R}^{m\times m}$ and $\mathbf{N} \in \mathcal{R}^{k\times k}$ is the $mk \times mk$ matrix $\mathbf{L} = \left[M_{ij}\mathbf{N}\right]$. Note that the generalized inverse $(\mathbf{K}\mathbf{\Delta})^{+}$ is independent of $\mathbf{X}_0$, hence this pseudoinverse must be calculated only once in the alternating projection algorithm.

10.3.2 Projection onto the Positivity Set

The orthogonal projection of a matrix $\mathbf{X}_0$ onto the positivity set $\mathcal{P}_\epsilon$, defined by (10.11) or (10.15) can be easily computed using the following result [49].

Theorem 10.3.2 *Let $\mathbf{X}_0$ be a given matrix in $\mathcal{S}$, and let $\mathbf{X}_0 - \mathbf{P} = \mathbf{U}\mathbf{L}\mathbf{U}^T$ be the eigenvalue–eigenvector decomposition of $\mathbf{X}_0 - \mathbf{P}$ where $\mathbf{L}$ is a diagonal matrix of eigenvalues and $\mathbf{U}$ is the orthogonal matrix of eigenvectors. The projection $P_{\mathcal{P}_\epsilon}\mathbf{X}_0$ of the matrix $\mathbf{X}_0$ onto the constraint set $\mathcal{P}_\epsilon$ is given by*

$$P_{\mathcal{P}_\epsilon}\mathbf{X}_0 = \mathbf{U}\mathbf{L}_{+}\mathbf{U}^T + \mathbf{P} \tag{10.32}$$

where $\mathbf{L}_{+}$ is the diagonal matrix obtained by replacing the negative eigenvalues in $\mathbf{L}$ by zero.

Hence, the numerical computation of this projection requires an eigenvalue–eigenvector decomposition of a symmetric matrix.

10.3.3 Projection onto the Variance Constraint Set

The projection of a matrix $\mathbf{X}_0$ onto the variance constraint set $\mathcal{V}_i$ is provided by the following result.

Theorem 10.3.3 *Let $\mathbf{X}_0$ be a given matrix in $\mathcal{S}$. The orthogonal projection $P_{\mathcal{V}_i}\mathbf{X}_0$ is given by*

$$P_{\mathcal{V}_i}\mathbf{X}_0 = \frac{\sigma_i^* - \mathbf{c}_i^T\mathbf{X}_0\mathbf{c}_i}{\|\mathbf{c}_i\|^4}\mathbf{c}_i\mathbf{c}_i^T + \mathbf{X}_0 \tag{10.33}$$

where $\sigma_i^* = min(\sigma_i, \mathbf{c}_i^T\mathbf{X}_0\mathbf{c}_i)$.

The proof of the above result is left as an exercise for the reader.

10.3.4 Projection onto the Block Covariance Constraint Set

The following result provides the expression for the projection onto the set of the ith block output covariance constraint $\mathcal{O}_i$.

Theorem 10.3.4 *Let $\mathbf{X}_0$ be a given matrix in $\mathcal{S}$. Consider the singular value decomposition*

$$\mathbf{C}_i^T = \mathbf{U}_i\,[\boldsymbol{\Sigma}_i \;\; \mathbf{0}]\,\mathbf{V}_i^T \tag{10.34}$$

and define

$$\bar{\mathbf{X}}_{pi} \triangleq \mathbf{V}_i^T\mathbf{X}_p\mathbf{V}_i = \begin{bmatrix} \bar{\mathbf{X}}_{pi11} & \bar{\mathbf{X}}_{pi12} \\ \bar{\mathbf{X}}_{pi12}^T & \bar{\mathbf{X}}_{pi22} \end{bmatrix}, \quad \bar{\mathbf{X}}_{pi11} \in \mathcal{R}^{n_{yi}\times n_{yi}}. \tag{10.35}$$

Consider the eigenvalue–eigenvector decomposition

$$\bar{\mathbf{X}}_{pi11} - \boldsymbol{\Sigma}_i^{-1}\mathbf{U}_i^T\mathbf{Y}_i\mathbf{U}_i\boldsymbol{\Sigma}_i^{-1} = \mathbf{W}_i\boldsymbol{\Lambda}_i\mathbf{W}_i^T \tag{10.36}$$

where $\boldsymbol{\Lambda}_i$ is a diagonal matrix which contains the eigenvalues of the matrix $\bar{\mathbf{X}}_{pi11} - \boldsymbol{\Sigma}_i^{-1}\mathbf{U}_i^T\mathbf{Y}_i\mathbf{U}_i\boldsymbol{\Sigma}_i^{-1}$ and $\mathbf{W}_i$ is the orthogonal matrix of eigenvectors. The orthogonal projection $P_{\mathcal{O}_i}\mathbf{X}_0$ is given by

$$P_{\mathcal{O}_i}\mathbf{X}_0 = \mathbf{V}_i \begin{bmatrix} \bar{\mathbf{X}}_{pi11}^* & \bar{\mathbf{X}}_{pi12} \\ \bar{\mathbf{X}}_{pi12}^T & \bar{\mathbf{X}}_{pi22} \end{bmatrix} \mathbf{V}_i^T \tag{10.37}$$

where

$$\bar{\mathbf{X}}_{pi11}^* \triangleq \mathbf{W}_i\boldsymbol{\Lambda}_i^-\mathbf{W}_i^T + \boldsymbol{\Sigma}_i^{-1}\mathbf{U}_i^T\mathbf{Y}_i\mathbf{U}_i\boldsymbol{\Sigma}_i^{-1} \tag{10.38}$$

and $\boldsymbol{\Lambda}_i^-$ is the diagonal matrix obtained by replacing the positive eigenvalues in $\boldsymbol{\Lambda}_i$ by zero.

Proof. Let

$$\hat{\mathbf{X}}_{pi} = \begin{bmatrix} \hat{\mathbf{X}}_{pi11} & \hat{\mathbf{X}}_{pi12} \\ \hat{\mathbf{X}}_{pi12}^T & \hat{\mathbf{X}}_{pi22} \end{bmatrix}, \quad \hat{\mathbf{X}}_{pi11} \in \mathcal{R}^{n_{yi}\times n_{yi}}.$$

Consider the inner product

$$tr\left[(\mathbf{X}_{pi}^* - \mathbf{X}_p)(\mathbf{X}_{pi}^* - \hat{\mathbf{X}}_{pi})\right].$$

Since $\mathbf{V}_i$ is an orthogonal matrix, this inner product is equal to

$$tr\left[(\mathbf{V}_i^T\mathbf{X}_{pi}^*\mathbf{V}_i - \mathbf{V}_i^T\mathbf{X}_p\mathbf{V}_i)(\mathbf{V}_i^T\mathbf{X}_{pi}^*\mathbf{V}_i - \mathbf{V}_i^T\hat{\mathbf{X}}_{pi}\mathbf{V}_i)\right]. \tag{10.39}$$

Define

$$\mathbf{V}_i^T\hat{\mathbf{X}}_{pi}\mathbf{V}_i = \begin{bmatrix} \tilde{\mathbf{X}}_{pi11} & \tilde{\mathbf{X}}_{pi12} \\ \tilde{\mathbf{X}}_{pi12}^T & \tilde{\mathbf{X}}_{pi22} \end{bmatrix}$$

and note that (10.37) implies that

$$\mathbf{V}_i^T\mathbf{X}_{pi}^*\mathbf{V}_i = \begin{bmatrix} \bar{\mathbf{X}}_{pi11}^* & \bar{\mathbf{X}}_{pi12} \\ \bar{\mathbf{X}}_{pi12}^T & \bar{\mathbf{X}}_{pi22} \end{bmatrix}.$$

Hence, the inner product (10.39) is equal to

$$tr\left(\left[\begin{array}{cc}\bar{\mathbf{X}}^*_{pi11} & \bar{\mathbf{X}}_{pi12}\\ \bar{\mathbf{X}}^T_{pi12} & \bar{\mathbf{X}}_{pi22}\end{array}\right]-\left[\begin{array}{cc}\bar{\mathbf{X}}_{pi11} & \bar{\mathbf{X}}_{pi12}\\ \bar{\mathbf{X}}^T_{pi12} & \bar{\mathbf{X}}_{pi22}\end{array}\right]\right)\left(\left[\begin{array}{cc}\bar{\mathbf{X}}^*_{pi11} & \bar{\mathbf{X}}_{pi12}\\ \bar{\mathbf{X}}^T_{pi12} & \bar{\mathbf{X}}_{pi22}\end{array}\right]-\left[\begin{array}{cc}\tilde{\mathbf{X}}_{pi11} & \tilde{\mathbf{X}}_{pi12}\\ \tilde{\mathbf{X}}^T_{pi12} & \tilde{\mathbf{X}}_{pi22}\end{array}\right]\right)$$

$$= tr\left[(\bar{\mathbf{X}}^*_{pi11} - \bar{\mathbf{X}}_{pi11})(\bar{\mathbf{X}}^*_{pi11} - \tilde{\mathbf{X}}_{pi11})\right]. \tag{10.40}$$

Now observe that since $\hat{\mathbf{X}}_{pi} \in \mathcal{O}_i$ we have

$$\mathbf{C}_i^T\hat{\mathbf{X}}_{pi}\mathbf{C}_i \le \mathbf{Y}_i$$

and by substituting in this inequality the singular value decomposition (10.34), and pre- and post-multiplying by $\mathbf{U}_i^T$ and $\boldsymbol{\Sigma}_i^{-1}$ we obtain

$$[\mathbf{I}\ \ \mathbf{0}]\,\mathbf{V}_i^T\hat{\mathbf{X}}_{pi}\mathbf{V}_i\left[\begin{array}{c}\mathbf{I}\\ \mathbf{0}\end{array}\right] \le \boldsymbol{\Sigma}_i^{-1}\mathbf{U}_i^T\mathbf{Y}_i\mathbf{U}_i\boldsymbol{\Sigma}_i^{-1}$$

or

$$\hat{\mathbf{X}}_{pi11} \le \boldsymbol{\Sigma}_i^{-1}\mathbf{U}_i^T\mathbf{Y}_i\mathbf{U}_i\boldsymbol{\Sigma}_i^{-1}.$$

Hence, $\hat{\mathbf{X}}_{pi11}$ is an element of the set

$$\{\hat{\mathbf{X}}_{pi11} : \hat{\mathbf{X}}_{pi11} \le \boldsymbol{\Sigma}_i^{-1}\mathbf{U}_i^T\mathbf{Y}_i\mathbf{U}_i\boldsymbol{\Sigma}_i^{-1}\}.$$

According to Theorem 12.3.3, the orthogonal projection of a matrix $\bar{\mathbf{X}}_{pi11}$ onto this set is provided by the expression (10.32). Hence, the minimum distance condition implies that the inner product (10.40) is nonpositive. Hence the inner product (10.39) is nonpositive, and this completes the proof. □

This projection requires an eigenvalue–eigenvector decomposition and matrix manipulation to compute the projection matrix.

10.3.5 Projection onto the Output Cost Constraint Set

Next, the projection onto the output cost constraint set $\mathcal{C}$ is provided. This result is a special case of Theorem 12.3.3.

Theorem 10.3.5 *Let $\mathbf{X}_0$ be a given matrix in $\mathcal{S}$ and $\gamma > 0$. The projection $P_{\mathcal{C}}\mathbf{X}_0$ is given by*

$$P_{\mathcal{C}}\mathbf{X}_0 = \frac{\gamma^* - tr(\mathbf{X}_0\mathbf{R})}{\|\mathbf{R}\|^2}\mathbf{R} + \mathbf{X}_0 \tag{10.41}$$

where $\mathbf{R} = \mathbf{C}^T\mathbf{Q}\mathbf{C}$ and $\gamma^* = \min\left[\gamma, tr(\mathbf{X}_p\mathbf{R})\right]$.

The above expressions for the projections can be used to solve the covariance feasibility problem (10.27), the covariance optimization problem (10.28)–(10.29) or the infeasible covariance optimization problem (10.30) using the alternating convex projection methods described in Section 12.1. An application of alternating projection techniques for the controller redesign problem of the Hubble space telescope can be found in [161]. An application to structural control appears in [73].

Exercise 10.3.1 Consider the state space model of a double integrator:

$$\begin{aligned}\dot{x}_1 &= x_2\\ \dot{x}_2 &= u + w\\ y &= x_1\end{aligned}$$

where $w(t)$ is white noise with unit intensity. We seek to find a static state feedback control law to stabilize the system and satisfy the output variance constraint

$$\mathcal{E}(y^2) \leq 0.1.$$

Show that the assignable covariance matrix $\mathbf{X}$ should be in the intersection of the following constraint sets

$$\begin{aligned}\mathcal{A} &= \{\mathbf{X} \in \mathcal{S} : X_{12} = 0\}\\ \mathcal{P} &= \{\mathbf{X} \in \mathcal{S} : X_{11} \geq 0,\ X_{22} \geq 0,\ X_{12} \leq \sqrt{X_{11}X_{22}}\}\\ \mathcal{V} &= \{\mathbf{X} \in \mathcal{S} : X_{11} \leq 0.1\}.\end{aligned}$$

Draw a sketch of these sets in the (X_{11}, X_{22}, X_{12}) space. Show that the standard alternating projection method will converge to a feasible solution in at most 3 steps. Show that all covariance controllers are parametrized as

$$\mathbf{G} = \begin{bmatrix} -\dfrac{X_{22}}{X_{11}} & -\dfrac{1}{2X_{12}} \end{bmatrix}.$$

10.4 GEOMETRIC FORMULATION OF LMI CONTROL DESIGN

In this section we reformulate the LMI control design problems described in Sections 8.2.3 and 9.2.3 as matrix feasibility problems of a simpler geometry. This allows us to explicitly derive the expressions for the orthogonal projection operators onto the constrained sets and use alternating projection techniques for solution. For details see [45].

To start our discussion, we consider the following set of real symmetric matrices:

$$\mathcal{L} = \{\mathbf{X} \in \mathcal{S} : \mathbf{EXF} + \mathbf{F}^T\mathbf{XE}^T + \mathbf{Q} < \mathbf{0}\} \tag{10.42}$$

where the matrices $\mathbf{E}$, $\mathbf{F}$ and $\mathbf{Q} \in \mathcal{S}$ are given real matrices of compatible dimensions. It can be easily verified that $\mathcal{L}$ is a convex subset in the space of symmetric matrices. Note that each one of the LMI constraints (7.15) and (7.16) can be written as in (10.42) by appropriate choices of the matrices $\mathbf{E}$, $\mathbf{F}$ and $\mathbf{Q}$. For example, the LMI (7.15) obtains the form in (10.42) by defining

$$\mathbf{E} \triangleq \begin{bmatrix} \mathbf{B}_p \\ \mathbf{B}_y \end{bmatrix}^{\perp} \begin{bmatrix} \mathbf{A}_p \\ \mathbf{C}_p \end{bmatrix}, \quad \mathbf{F} \triangleq [\mathbf{I}\ \ \mathbf{0}] \begin{bmatrix} \mathbf{B}_p \\ \mathbf{B}_y \end{bmatrix}^{\perp T} \tag{10.43}$$

$$\mathbf{Q} \triangleq \begin{bmatrix} \mathbf{B}_p \\ \mathbf{B}_y \end{bmatrix}^{T} \begin{bmatrix} \mathbf{D}_p\mathbf{D}_p^T & \mathbf{D}_p\mathbf{D}_y^T \\ \mathbf{D}_y\mathbf{D}_p^T & \mathbf{D}_y\mathbf{D}_y^T - \gamma^2\mathbf{I} \end{bmatrix} \begin{bmatrix} \mathbf{B}_p \\ \mathbf{B}_y \end{bmatrix}^{\perp T}.$$

For computational purposes, we prefer to have closed sets (i.e. sets which include their limit points), so we consider the following closed "ϵ-approximation" of the set $\mathcal{L}$:

$$\mathcal{L}_\epsilon \triangleq \{\mathbf{X} \in \mathcal{S} : \mathbf{EXF} + \mathbf{F}^T\mathbf{XE}^T \leq -\mathbf{Q} - \epsilon\mathbf{I}\} \tag{10.44}$$

where ϵ is a small positive constant. Hence, the set $\mathcal{L}_\epsilon$ is closed, convex, it is contained in $\mathcal{L}$ and it approaches $\mathcal{L}$ as $\epsilon \to 0$. We will use $\mathcal{L}_\epsilon$ to represent (in an ϵ-approximation sense) any of the constraint sets defined by LMIs, by an appropriate choice of the matrices **E**, **F**, **Q** and **X**.

The next result, provides a decomposition of the set $\mathcal{L}_\epsilon$ into two closed, convex sets of simpler geometry by increasing the dimension of the parameter space.

Theorem 10.4.1 *Define the following sets:*

$$\mathcal{J}_\epsilon \triangleq \{\mathbf{W} \in \mathcal{S}_{2n} : [\mathbf{E}\ \ \mathbf{F}^T]\mathbf{W}\begin{bmatrix} \mathbf{E}^T \\ \mathbf{F} \end{bmatrix} \le -\mathbf{Q}_\epsilon\} \tag{10.45}$$

$$\mathcal{T} \triangleq \{\mathbf{W} \in \mathcal{S}_{2n} : \mathbf{W} = \begin{bmatrix} \mathbf{W}_{11} & \mathbf{W}_{12} \\ \mathbf{W}_{12}^T & \mathbf{W}_{22} \end{bmatrix},\ \ \mathbf{W}_{11} = \mathbf{W}_{22} = \mathbf{0},\ \ \mathbf{W}_{12} \in \mathcal{S}_n\} \tag{10.46}$$

where $\mathbf{Q}_\epsilon \triangleq \mathbf{Q} + \epsilon\mathbf{I}$*. Then the following statements are equivalent.*

(i) $\mathbf{X} \in \mathcal{L}_\epsilon$

(ii) $\mathbf{X} = \mathbf{W}_{12}$ *where* $\mathbf{W} \in \mathcal{J}_\epsilon \bigcap \mathcal{T}$.

Proof. Let **X** be in $\mathcal{L}_\epsilon$. Then **X** satisfies

$$\mathbf{EXF} + \mathbf{F}^T\mathbf{XE}^T \le -\mathbf{Q} - \epsilon\mathbf{I} \tag{10.47}$$

which is equivalent to

$$[\mathbf{E}\ \ \mathbf{F}^T]\mathbf{W}\begin{bmatrix} \mathbf{E}^T \\ \mathbf{F} \end{bmatrix} \le -\mathbf{Q}_\epsilon. \tag{10.48}$$

This provides condition (ii). Conversely, if **X** satisfies (ii) then simple calculations reveal that (10.47) holds, hence **X** is in $\mathcal{L}_\epsilon$. □

Therefore the two closed convex sets $\mathcal{J}_\epsilon$ and $\mathcal{T}$ can be used to provide an equivalent description of the LMI constraint set $\mathcal{L}_\epsilon$. The advantage of describing the set $\mathcal{L}_\epsilon$ in terms of the intersection $\mathcal{J}_\epsilon \bigcap \mathcal{T}$ follows from the fact that we can obtain explicit expressions for the orthogonal projection operators onto the sets $\mathcal{J}_\epsilon$ and $\mathcal{T}$. These expressions are computed next.

The following proposition provides the orthogonal projection onto the constraint set $\mathcal{J}_\epsilon$.

Theorem 10.4.2 *Let* $\mathbf{W} \in \mathcal{S}_{2n}$*. Consider the singular value decomposition of* $[\mathbf{E}\ \ \mathbf{F}^T]$

$$[\mathbf{E}\ \ \mathbf{F}^T] = \mathbf{U}[\boldsymbol{\Sigma}\ \ \mathbf{0}]\mathbf{V}^T \tag{10.49}$$

where **U** *and* **V** *are orthogonal matrices, and define*

$$\bar{\mathbf{W}} \triangleq \mathbf{V}^T\mathbf{W}\mathbf{V} = \begin{bmatrix} \bar{\mathbf{W}}_{11} & \bar{\mathbf{W}}_{12} \\ \bar{\mathbf{W}}_{12}^T & \bar{\mathbf{W}}_{22} \end{bmatrix},\quad \bar{\mathbf{W}}_{11} \in \mathcal{S}_n. \tag{10.50}$$

Consider the eigenvalue–eigenvector decomposition

$$\bar{\mathbf{W}}_{11} + \boldsymbol{\Sigma}^{-1}\mathbf{U}^T\mathbf{Q}_\epsilon\mathbf{U}\boldsymbol{\Sigma}^{-1} = \mathbf{L}\boldsymbol{\Lambda}\mathbf{L}^T \tag{10.51}$$

where Λ *is a diagonal matrix that contains the eigenvalues of the matrix* $\bar{\mathbf{W}}_{11} + \Sigma^{-1}\mathbf{U}^T\mathbf{Q}_\epsilon\mathbf{U}\Sigma^{-1}$ *and* $\mathbf{L}$ *is the corresponding orthogonal matrix of the normalized eigenvectors. The projection* $\bar{\mathbf{W}}^* = P_{\mathcal{J}_\epsilon}\mathbf{W}$ *of the matrix* $\mathbf{W}$ *onto the set* $\mathcal{J}_\epsilon$ *is given by*

$$\mathbf{W}^* = \mathbf{V}\begin{bmatrix} \bar{\mathbf{W}}_{11}^* & \bar{\mathbf{W}}_{12} \\ \bar{\mathbf{W}}_{12}^T & \bar{\mathbf{W}}_{22} \end{bmatrix}\mathbf{V}^T \tag{10.52}$$

where

$$\bar{\mathbf{W}}_{11}^* = \mathbf{L}\Lambda_-\mathbf{L}^T\Sigma^{-1}\mathbf{U}^T\mathbf{Q}_\epsilon\mathbf{U}\Sigma^{-1} \tag{10.53}$$

where Λ_- *is the diagonal matrix obtained by replacing the positive eigenvalues of* Λ *by zero.*

The proof of this theorem is similar to the proof of Theorem 12.3.4. The following result provides the projection onto the set $\mathcal{T}$.

Theorem 10.4.3 *Let* $\mathbf{W} \in \mathcal{S}_{2n}$. *The orthogonal projection* $\mathbf{W}^* = P_{\mathcal{T}}\mathbf{W}$ *of the matrix* $\mathbf{W}$ *onto the set* $\mathcal{T}$ *is provided by*

$$\mathbf{W}^* = \begin{bmatrix} \mathbf{0} & \mathbf{X}^* \\ \mathbf{X}^* & \mathbf{0} \end{bmatrix} \tag{10.54}$$

where $\mathbf{X}^* = (\mathbf{W}_{12} + \mathbf{W}_{12}^T)/2$.

Proof. It is clear that $\mathcal{T}$ is a subspace of $\mathcal{S}_{2n}$ and $\mathbf{W}^*$ is in $\mathcal{T}$. Now let

$$\mathbf{W} = \begin{bmatrix} \mathbf{W}_{11} & \mathbf{W}_{12} \\ \mathbf{W}_{12}^T & \mathbf{W}_{22} \end{bmatrix} \tag{10.55}$$

be in $\mathcal{S}_{2n}$ and let

$$\hat{\mathbf{W}} = \begin{bmatrix} \mathbf{0} & \hat{\mathbf{X}} \\ \hat{\mathbf{X}} & \mathbf{0} \end{bmatrix} \tag{10.56}$$

be any element of $\mathcal{T}$. Then simple calculations reveal that

$$\langle \mathbf{W} - \mathbf{W}^*, \hat{\mathbf{W}} - \mathbf{W}^* \rangle =$$

$$tr\left\{\begin{bmatrix} \mathbf{W}_{11} & \frac{\mathbf{W}_{12}-\mathbf{W}_{12}^T}{2} \\ \frac{\mathbf{W}_{12}^T-\mathbf{W}_{12}}{2} & \mathbf{W}_{22} \end{bmatrix}\begin{bmatrix} \mathbf{0} & \hat{\mathbf{X}} - \frac{\mathbf{W}_{12}+\mathbf{W}_{12}^T}{2} \\ \hat{\mathbf{X}} - \frac{\mathbf{W}_{12}+\mathbf{W}_{12}^T}{2} & \mathbf{0} \end{bmatrix}\right\} = 0. \tag{10.57}$$

Hence, $\mathbf{W}^*$ is the orthogonal projection of $\mathbf{W}$ onto $\mathcal{T}$. □

In addition to the LMI constraints sets, we seek explicit expressions for the orthogonal projection onto the positivity and the rank constraint sets corresponding to the conditions (7.14). To this end, define the following sets in $\mathcal{S}_{2n}$

$$\mathcal{D} \triangleq \left\{\mathbf{Z} \in \mathcal{S}_{2n} : \mathbf{Z} = \begin{bmatrix} \mathbf{X} & \mathbf{0} \\ \mathbf{0} & \mathbf{Y} \end{bmatrix},\ \mathbf{X}, \mathbf{Y} \in \mathcal{S}_n\right\} \tag{10.58}$$

$$\mathcal{P} \triangleq \{\mathbf{Z} \in \mathcal{S}_{2n} : \mathbf{Z} \geq -\mathbf{J}\} \tag{10.59}$$

$$\mathcal{R} \triangleq \{\mathbf{Z} \in \mathcal{S}_{2n} : \text{rank}(\mathbf{Z} + \mathbf{J}) \leq k\} \tag{10.60}$$

where k is a given integer such that $n \leq k \leq 2n$, and

$$\mathbf{J} = \begin{bmatrix} \mathbf{0} & \mathbf{I}_n \\ \mathbf{I}_n & \mathbf{0} \end{bmatrix} \in \mathcal{S}_{2n}. \tag{10.61}$$

Notice that the sets $\mathcal{D}$ and $\mathcal{P}$ are closed convex sets. The expressions for the orthogonal projections onto these sets are provided next.

Theorem 10.4.4 *Let*

$$\mathbf{Z} = \begin{bmatrix} \mathbf{Z}_{11} & \mathbf{Z}_{12} \\ \mathbf{Z}_{12}^T & \mathbf{Z}_{22} \end{bmatrix} \in \mathcal{S}_{2n}. \tag{10.62}$$

The orthogonal projection, $\mathbf{Z}^* = P_{\mathcal{D}}\mathbf{Z}$, of $\mathbf{Z}$ onto the set $\mathcal{D}$ is given by

$$\mathbf{Z}^* = \begin{bmatrix} \mathbf{Z}_{11} & \mathbf{0} \\ \mathbf{0} & \mathbf{Z}_{22} \end{bmatrix} \in \mathcal{S}_{2n}. \tag{10.63}$$

The proof of the above result is simple and it is left to the reader. The orthogonal projection onto the set $\mathcal{P}$ is provided by the following result which follows from [49].

Theorem 10.4.5 *Let $\mathbf{Z} \in \mathcal{S}_n$ and let $\mathbf{Z} + \mathbf{J} = \mathbf{L}\boldsymbol{\Lambda}\mathbf{L}^T$ be the eigenvalue–eigenvector decomposition of $\mathbf{Z} + \mathbf{J}$ where $\boldsymbol{\Lambda}$ is the diagonal matrix of the eigenvalues and $\mathbf{L}$ is the orthogonal matrix of the normalized eigenvectors. The orthogonal projection, $\mathbf{Z}^* = P_{\mathcal{P}}\mathbf{Z}$, of $\mathbf{Z}$ onto the set $\mathcal{P}$ is given by*

$$\mathbf{Z}^* = \mathbf{L}\boldsymbol{\Lambda}_-\mathbf{L}^T - \mathbf{J} \tag{10.64}$$

where $\boldsymbol{\Lambda}_-$ is the diagonal matrix obtained by replacing the negative eigenvalues in $\boldsymbol{\Lambda}$ by zero.

Hence, this projection requires an eigenvalue–eigenvector decomposition of the $2n \times 2n$ symmetric matrix $\mathbf{Z} + \mathbf{J}$.

We note that the rank constraint set $\mathcal{R}$, defined by (10.60), is a closed set, but it is not convex. Therefore, given a matrix $\mathbf{Z}$ in $\mathcal{S}_{2n}$, there might be several matrices in $\mathcal{R}$ which minimize the distance from $\mathbf{Z}$. We will call any such matrix a projection of $\mathbf{Z}$ on $\mathcal{R}$. The following result (see [50]) provides a projection onto the set $\mathcal{R}$.

Theorem 10.4.6 *Let $\mathbf{Z} \in \mathcal{S}_n$ and let $\mathbf{Z} + \mathbf{J} = \mathbf{U}\boldsymbol{\Sigma}\mathbf{V}^T$ be a singular value decomposition of $\mathbf{Z} + \mathbf{J}$. The orthogonal projection, $\mathbf{Z}^* = P_{\mathcal{R}}\mathbf{Z}$, of $\mathbf{Z}$ onto the set $\mathcal{R}$ is given by*

$$\mathbf{Z}^* = \mathbf{U}\boldsymbol{\Sigma}_k\mathbf{V}^T - \mathbf{J} \tag{10.65}$$

where $\boldsymbol{\Sigma}_k$ is the diagonal matrix obtained by replacing the smallest $2n - k$ singular values in $\mathbf{Z} + \mathbf{J}$ by zero.

To summarize this section, the constraint sets of the linear matrix inequality control design problems of Chapters 6 to 9 are reformulated as matrix constraint sets of simpler geometric structure. This allows analytical expressions for the projection operators onto these sets. Hence, alternating projection algorithms can be used to provide numerical solutions.

10.5 FIXED-ORDER CONTROL DESIGN

For fixed-order control design, the linear matrix inequality problems involve a nonconvex set $\mathcal{R}$. We seek a solution to the feasibility problem (10.3) for the case where some of the sets $\mathcal{C}_i$ are not convex. To extend our projection techniques to this case, we define the orthogonal projection to a nonconvex set $\mathcal{C}_i$ to be given by any solution of the minimum distance problem (10.4). We notice that a given point might have several projections (possibly infinite) to a nonconvex set $\mathcal{C}_i$. Unfortunately, when some of the sets $\mathcal{C}_i$ are nonconvex, the alternating projection algorithms of Theorems 12.1.1 and 12.1.3 are not guaranteed to converge to a feasible solution, even when such a solution exists. However, it can be shown that convergence is still guaranteed in a local sense, i.e. when the starting point of the algorithm is in a neighborhood of a feasible solution.

The following procedure is proposed to address the low-order control design problem:

STEP 1 Solve the feasibility problem (10.3) that corresponds to a full-order controller $n_c = n_p$. This is a convex problem and convergence of the alternating projection algorithms to a feasible solution $(\mathbf{X}, \mathbf{Y})$ is guaranteed.

STEP 2 Consider the problem where the controller order is reduced by one, i.e. set $n_c = n_c - 1$. Using as the initial condition the solution of the previous step, solve the low-order controller feasibility problem using the proposed algorithms.

STEP 3 Return to STEP 2, until the controller order n_c is the desired one, or the alternating projection algorithm does not converge. In this last case, a different initial condition might be tried for solving STEP 1, and the process can be repeated.

We conclude this section by summarizing the fact that alternating projection techniques provide simple and easy to implement algorithms for solution of feasibility problems. These algorithms require expressions for the projection operators onto the constraint sets, therefore these sets require a simple geometry so that such expressions can be analytically derived. For convex feasibility problems these algorithms converge globally, however for nonconvex problems the convergence is guaranteed only locally.

Example 10.5.1 In this example we use the alternating convex projection techniques for control design for a 2-mass-spring system. We assume that the two bodies have equal mass $m_1 = m_2 = 1$ and they are connected by a spring with stiffness $k = 1$. We assume that only the position of body 2 is measured and a control force acts on body 1, i.e. the problem is noncollocated.

A state space representation of the system, for the noise-free case ($\mathbf{w} = \mathbf{0}$), is the following:

$$\begin{aligned} \dot{\mathbf{x}}_p &= \mathbf{A}_p\mathbf{x}_p + \mathbf{B}_p\mathbf{u} \\ \mathbf{z} &= \mathbf{M}_p\mathbf{x}_p \end{aligned} \tag{10.66}$$

where

$$\mathbf{A}_p = \begin{bmatrix} 0 & 0 & 1 & 0 \\ 0 & 0 & 0 & 1 \\ -1 & 1 & 0 & 0 \\ 1 & -1 & 0 & 0 \end{bmatrix}, \quad \mathbf{B}_p = \begin{bmatrix} 0 \\ 0 \\ 1 \\ 0 \end{bmatrix}, \quad \mathbf{M}_p = \begin{bmatrix} 0 & 1 & 0 & 0 \end{bmatrix} \tag{10.67}$$

and the state variables x_1 and x_3 are the position and velocity respectively, of body 1; and x_2 and x_4 are the position and velocity of body 2.

We seek a stabilizing controller for this system, which places the closed-loop poles to the left of a vertical line $Re(z) = -\alpha$ in the complex plane. To this end, according to the results of Chapter 6, we seek a feasible solution to the following set of matrix inequalities:

$$\mathbf{B}_p^{\perp}((\mathbf{A}_p + \alpha\mathbf{I})\mathbf{X}_p + \mathbf{X}_p(\mathbf{A}_p^T + \alpha\mathbf{I}))\mathbf{B}_p^{\perp T} < \mathbf{0} \quad (10.68)$$

$$\mathbf{M}_p^{T\perp}(\mathbf{Y}_p(\mathbf{A}_p + \alpha\mathbf{I}) + (\mathbf{A}_p^T + \alpha\mathbf{I})\mathbf{Y}_p)\mathbf{M}_p^{T\perp T} < \mathbf{0} \quad (10.69)$$

$$\begin{bmatrix} \mathbf{X}_p & \mathbf{I} \\ \mathbf{I} & \mathbf{Y}_p \end{bmatrix} \leq \mathbf{0} \quad \text{and} \quad \text{rank}\begin{bmatrix} \mathbf{X}_p & \mathbf{I} \\ \mathbf{I} & \mathbf{Y}_p \end{bmatrix} \leq n_p + n_c \quad (10.70)$$

where the matrix $\mathbf{A}_p + \alpha\mathbf{I}$ has been used to guarantee that the closed-loop poles are to the left of $Re(z) = -\alpha$. We set $\alpha = 0.2$, and we begin with the search for a full-order controller ($n_c = 4$). The feasibility problem (10.68)–(10.70) is a convex problem and our alternating projection methods are guaranteed to converge to a feasible solution. Using the directional alternating projection algorithm of Section 12.1.5 along with the orthogonal projection expressions developed in Section 12.4, we obtain the following feasible solution:

$$\mathbf{X}_p = \begin{bmatrix} 2.4864 & 1.3467 & -0.4973 & -1.1690 \\ 1.3467 & 3.4575 & 0.6304 & -0.6915 \\ -0.4973 & 0.6304 & 2.9937 & -0.6721 \\ -1.1690 & -0.6915 & -0.6721 & 2.3874 \end{bmatrix}$$

$$\mathbf{Y}_p = \begin{bmatrix} 2.3874 & -0.6721 & -0.6915 & -1.1690 \\ -0.6721 & 2.9937 & 0.6304 & -0.4973 \\ -0.6915 & 0.6304 & 3.4575 & 1.3467 \\ -1.1690 & -0.4973 & 1.3467 & 2.4864 \end{bmatrix}.$$

It can be easily verified that these matrices satisfy the conditions (10.68)–(10.70). The directional alternating projection algorithm needed 606 iterations to converge to this solution (this corresponds to a CPU time of 152.66 sec on a Sun Sparc II Workstation). Using the results of Chapter 6, we can obtain the following stabilizing controller which corresponds to this feasible set $(\mathbf{X}_p, \mathbf{Y}_p)$:

$$\dot{\mathbf{x}}_c = \begin{bmatrix} -0.5231 & 4.4631 & 7.5167 & 0.0000 \\ 0.7067 & -0.4077 & 1.0140 & 0.0000 \\ -0.3783 & -5.5355 & -5.7237 & 0.0000 \\ 0.0000 & 0.0000 & 0.0000 & -33.5333 \end{bmatrix}\mathbf{x}_c + \begin{bmatrix} -8.4938 \\ 0.8348 \\ 9.0487 \\ 0.0000 \end{bmatrix}\mathbf{z}$$

$$\mathbf{u} = \begin{bmatrix} -0.2128 & 3.4045 & 4.4224 & 0.0000 \end{bmatrix}\mathbf{x}_c - 6.2692\,\mathbf{y}.$$

This controller provides the following closed-loop poles:

$$-0.2028 \pm 0.4625i, \qquad -0.2257 \pm 1.1387i$$

$$-0.2982 \pm 1.6737i, \qquad -5.2012, \qquad -33.5333$$

which obviously satisfy the required settling time constraint.

In the following, we seek a second-order controller ($n_c = 2$) to satisfy the same requirements. In this case we include the nonconvex rank condition (10.70). The directional alternating projection algorithm converged to the following feasible solution:

$$\mathbf{X}_p = \begin{bmatrix} 2.7374 & 1.9351 & -0.5475 & -1.4226 \\ 1.9351 & 4.3499 & 0.6468 & -0.8700 \\ -0.5475 & 0.6486 & 1.5012 & -0.2332 \\ -1.4226 & -0.8700 & -0.2332 & 2.7629 \end{bmatrix}$$

$$\mathbf{Y}_p = \begin{bmatrix} 2.7629 & -0.2332 & -0.8700 & -1.4226 \\ -0.2332 & 1.5012 & 0.6486 & -0.5475 \\ -0.8700 & 0.6486 & 4.3499 & 1.9351 \\ -1.4226 & -0.5475 & 1.9351 & 2.7374 \end{bmatrix}.$$

It can be easily checked that these matrices satisfy the conditions (10.68)–(10.70) for $n_c = 2$. In this case the directional alternating projection method required 901 iterations (the corresponding CPU time is 202.76 sec) for convergence. A second-order controller which corresponds to the feasible solution $(\mathbf{X}_p, \mathbf{Y}_p)$ is

$$\begin{aligned} \dot{\mathbf{x}}_c &= \begin{bmatrix} -0.5667 & 0.6145 \\ -0.6954 & -0.6373 \end{bmatrix} \mathbf{x}_c + \begin{bmatrix} 1.2966 \\ 1.3876 \end{bmatrix} \mathbf{z} \\ \mathbf{u} &= \begin{bmatrix} -0.2271 & -0.5598 \end{bmatrix} \mathbf{x}_c + 1.1671\mathbf{y}. \end{aligned}$$

This controller provides the following closed-loop poles:

$$-0.2006 \pm 0.3778i, \quad -0.2007 \pm 1.9358i, \quad -0.2007 \pm 1.3343i$$

which satisfy the desired pole regional constraint. Note that a second-order controller is the lowest order stabilizing controller for the plant (10.66)–(10.67).

Next we consider the H_∞ control design problem for the case where there is a plant disturbance w_2 acting on body 2 and a sensor measurement noise v; i.e. the state space representation of the system is

$$\begin{aligned} \dot{\mathbf{x}}_p &= \mathbf{A}_p\mathbf{x}_p + \mathbf{B}_p\mathbf{u} + \mathbf{D}_p\mathbf{w}_p \\ \mathbf{z} &= \mathbf{M}_p\mathbf{x}_p + \mathbf{v} \\ \mathbf{y} &= \mathbf{C}_p\mathbf{x}_p \end{aligned} \qquad (10.71)$$

where $\mathbf{A}_p$, $\mathbf{B}_p$, $\mathbf{M}_p$ are as before and

$$\mathbf{D}_p = \begin{bmatrix} 0 \\ 0 \\ 0 \\ 1 \end{bmatrix}, \quad \mathbf{C}_p = \mathbf{M}_p, \quad \mathbf{w}_p = w_2, \quad \mathbf{v} = v.$$

We seek a second-order stabilizing controller to guarantee that the H_∞ norm of the closed-loop transfer function from the disturbance $\mathbf{w} = [w_2 \;\; v]^T$ to the regulated output $\mathbf{y} = x_2$ is less than $\gamma = 8$. To this end, we use the low-order alternating projection algorithm to solve the feasibility problem (7.14)–(7.16). The directional alternating projection algorithm provides a feasible pair $(\mathbf{X}_p, \mathbf{Y}_p)$ in 2173 iterations (which required 745.96 sec of CPU time). A second-order controller which corresponds to this feasible pair is

$$\begin{aligned} \dot{\mathbf{x}}_c &= \begin{bmatrix} -0.6113 & 0.4507 \\ -1.3614 & -3328 \end{bmatrix} \mathbf{x}_c + \begin{bmatrix} 0.2624 \\ 0.3165 \end{bmatrix} \mathbf{z} \\ \mathbf{u} &= \begin{bmatrix} -0.4880 & -0.3328 \end{bmatrix} \mathbf{x}_c + 0.3100\,\mathbf{y}. \end{aligned}$$

This controller provides a closed-loop H_∞ norm equal to $4.96 < 8$. Hence, the required disturbance attenuation properties are satisfied.

CHAPTER 10 SUMMARY

Alternating projection algorithms are proposed to solve: i) covariance control design problems, and ii) linear matrix inequality control design problems. These problems are formulated as matrix feasibility problems of finding matrix parameters in the intersection of a family of matrix constraint sets. Analytical expressions for the orthogonal projections onto these constraint sets have been developed. Alternating projection methods utilize these projections in an iterative fashion to obtain matrix parameters that satisfy the design constraints. The full-order covariance control and linear matrix inequality control problems are described via convex matrix equality and inequality constraints and convergence of the alternating projection algorithms is guaranteed when a solution exists.

Also, fixed-order control design problems have been considered. However, the fixed-order control design is characterized by adding a matrix rank constraint in the matrix feasibility design problem. Projections onto this rank constraint set can be easily obtained and alternating projection methods can be used for a solution, but global convergence of the alternating projection algorithms is not guaranteed in this case because of the nonconvexity of this rank constraint.

Algorithms are given here to find the intersection of convex sets. The method is called alternating projection since it alternately projects from a given point to the surface of a convex set and repeats this process, projecting onto the next set, etc. The computational advantage of this process is that for the problems of interest in this book, it is possible to derive analytical solutions to the orthogonal projection, avoiding the need that most convex problem solvers have of numerically solving a least squares problem to compute the minimum distance to the desired set. Geometrically speaking, the orthogonal projection onto one of the member sets is not the best direction (the intersection sought may be in another direction). While gradient calculations might give better research directions, gradient calculations provide no clue on the step size. Alternating projection methods know exactly how far to go, hence a convergence proof is possible.

Analytical expressions for the projections onto "rank constrained" nonconvex sets are also given. This is useful in solving fixed-order control problems. In fact, this is the most important use of the alternating projection method.

CHAPTER ELEVEN

Successive Centering Methods

11.1 CONTROL DESIGN WITH UNSPECIFIED CONTROLLER ORDER

This section addresses computational aspects of our approach to the linear control design based on the unified LMI formulation discussed in Chapter 9. Recall that our unified approach to control system design yields analytical solutions which are of the same mathematical nature. Specifically, many of the control problems considered in the earlier chapters can be formulated as a convex feasibility/optimization problem *if the controller order is not fixed.* The objective of this section is to give a computational algorithm to solve this convex problem.

11.1.1 Problem Formulation

The results given in Chapters 6 to 9 show that many suboptimal control problems[1] can be reduced to the following type of convex problem:

The Suboptimal Control Problem:

Find a matrix pair $(\mathbf{X}_p, \mathbf{Y}_p) \in \mathcal{C}$ where $\mathcal{C}$ is a convex set defined by LMIs as follows:

$$\mathcal{C} \triangleq \{\ (\mathbf{X}, \mathbf{Y}) :\ \mathbf{\Phi}(\mathbf{X}) > \mathbf{0},\ \ \mathbf{\Omega}(\mathbf{Y}) > \mathbf{0},\ \ \mathbf{\Psi}(\mathbf{X}, \mathbf{Y}) \geq \mathbf{0}\ \}$$

where $\mathbf{\Phi}(\cdot)$ and $\mathbf{\Omega}(\cdot)$ are affine mappings on the set of real symmetric matrices, and

$$\mathbf{\Psi}(\mathbf{X}, \mathbf{Y}) \triangleq \begin{bmatrix} \mathbf{X} & \mathbf{I} \\ \mathbf{I} & \mathbf{Y} \end{bmatrix}. \tag{11.1}$$

For instance, it can be easily verified using Theorems 9.2.3 and 2.3.11 that there exists a controller of some order which solves the discrete-time LQR problem described in section 9.2.3 if and only if $\mathcal{C} \neq \phi$ where

$$\mathbf{\Phi}(\mathbf{X}) \triangleq \begin{bmatrix} \mathbf{B}_p \\ \mathbf{B}_y \end{bmatrix}^{\perp} \begin{bmatrix} \mathbf{X} - \mathbf{A}_p\mathbf{X}\mathbf{A}_p^T & -\mathbf{A}_p\mathbf{X}\mathbf{C}_p^T \\ -\mathbf{C}_p\mathbf{X}\mathbf{A}_p^T & \mathbf{I} - \mathbf{C}_p\mathbf{X}\mathbf{C}_p^T \end{bmatrix} \begin{bmatrix} \mathbf{B}_p \\ \mathbf{B}_y \end{bmatrix}^{\perp T}, \tag{11.2}$$

[1] By "suboptimal", we mean that we do not optimize a performance measure but merely guarantee an upper bound on it.

$$\boldsymbol{\Omega}(\mathbf{Y}) \triangleq \begin{bmatrix} \mathbf{M}_p^{T\perp}(\mathbf{Y} - \mathbf{A}_p^T\mathbf{Y}\mathbf{A}_p - \mathbf{C}_p^T\mathbf{C}_p)\mathbf{M}_p^{T\perp T} & \mathbf{0} \\ \mathbf{0} & \gamma\mathbf{I} - \mathbf{D}_p^T\mathbf{Y}\mathbf{D}_p - \mathbf{D}_y^T\mathbf{D}_y \end{bmatrix}.$$

In fact, this existence condition can be directly derived by dualizing Theorem 6.3.3 since the covariance control problem and the LQR problem are dual to each other as shown in Chapter 9.

Once we find a matrix pair $(\mathbf{X}_p, \mathbf{Y}_p) \in \mathcal{C}$, a Lyapunov matrix $\mathbf{Y} \in \mathcal{R}^{(n_p+n_c)\times(n_p+n_c)}$ in Theorem 9.2.3 can be constructed by taking any matrix factor of $\mathbf{Y}_p - \mathbf{X}_p^{-1}$ to obtain $\mathbf{Y}_{pc}$ and $\mathbf{Y}_c$ as follows:

$$\mathbf{Y} = \begin{bmatrix} \mathbf{Y}_p & \mathbf{Y}_{pc} \\ \mathbf{Y}_{pc}^T & \mathbf{Y}_c \end{bmatrix},$$

where

$$\mathbf{Y}_{pc}\mathbf{Y}_c^{-1}\mathbf{Y}_{pc}^T \triangleq \mathbf{Y}_p - \mathbf{X}_p^{-1}, \quad \mathbf{Y}_c > \mathbf{0}.$$

Then all feasible controllers can be given by the explicit formula in Theorem 2.3.11.

It should be clear from the unified LMI formulations that all the control problems in Chapter 9 can be solved by the same design procedure. Namely, we first compute that pair $(\mathbf{X}_p, \mathbf{Y}_p) \in \mathcal{C}$, find a Lyapunov matrix $\mathbf{X}$ or $\mathbf{Y}$, then obtain a controller by the explicit formulae. Note that, since all feasible controllers $\mathbf{G}$ are given *explicitly*, we do not require an iterative computation for the last step of the control design process. The only iterations we need are to find the matrix pair $(\mathbf{X}_p, \mathbf{Y}_p) \in \mathcal{C}$.

In view of the above example for the LQR problem, the set $\mathcal{C}$ depends on the performance bound γ in general. So let us express the dependence explicitly by $\mathcal{C}(\gamma)$. As shown above, *suboptimal* control problems naturally lead to the convex *feasibility* problem of finding a matrix pair $(\mathbf{X}_p, \mathbf{Y}_p) \in \mathcal{C}(\gamma)$ *for a given value of* $\gamma > 0$. On the other hand, *optimal* control problems can be formulated in the following way.

The Optimal Control Problem:

Solve[2]

$$\gamma^* \triangleq \min\{\gamma : \mathcal{C}(\gamma) \neq \phi\}$$

where

$$\mathcal{C}(\gamma) \triangleq \{(\mathbf{X}, \mathbf{Y}) : \boldsymbol{\Phi}(\mathbf{X}, \gamma) > \mathbf{0}, \quad \boldsymbol{\Omega}(\mathbf{Y}, \gamma) > \mathbf{0}, \quad \boldsymbol{\Psi}(\mathbf{X}, \mathbf{Y}) \geq \mathbf{0} \}$$

where $\boldsymbol{\Phi}(\cdot,\cdot)$ and $\boldsymbol{\Omega}(\cdot,\cdot)$ are affine with respect to the first arguments, and $\boldsymbol{\Psi}(\cdot,\cdot)$ is defined in (11.1).

For the LQR problem, only $\boldsymbol{\Omega}(\cdot,\cdot)$ depends on γ and $\boldsymbol{\Phi}(\cdot,\cdot)$ does not. However, in general, both of these may depend on γ as for the $\mathcal{H}_\infty$ control problem. Note also that, according to Theorems 7.2.3 and 7.3.3, the LMI corresponding to $\boldsymbol{\Psi}(\mathbf{X}, \mathbf{Y}) \geq \mathbf{0}$ also depends on γ as follows;

$$\begin{bmatrix} \mathbf{X} & \gamma\mathbf{I} \\ \gamma\mathbf{I} & \mathbf{Y} \end{bmatrix} \geq \mathbf{0}.$$

However, a simple change of variables $\hat{\mathbf{X}} \triangleq \gamma^{-1}\mathbf{X}$, $\hat{\mathbf{Y}} \triangleq \gamma^{-1}\mathbf{Y}$ leads to the constraint $\boldsymbol{\Psi}(\hat{\mathbf{X}}, \hat{\mathbf{Y}}) \geq \mathbf{0}$, and thus the optimal control problem stated above can also treat this class of problems. For the covariance upper bound control problem, we may fix the structure of the output covariance bound as in $\mathbf{CXC}^T < \gamma\boldsymbol{\mho}$, then minimize γ.

[2] Strictly speaking, the "min" should be replaced by "inf" since $\mathcal{C}(\gamma)$ is an open set and the optimal solution may lie on the boundary of $\mathcal{C}(\gamma)$. However, we prefer the use of "min" since we are interested in finding a matrix pair $(\mathbf{X}_p, \mathbf{Y}_p) \in \mathcal{C}(\gamma)$ where γ is arbitrarily close to the infimum γ^*. Note that γ^* may not be attained.

It should be noted that the suboptimal control problem for a fixed performance level γ can also be formulated as an optimization problem of the form given above. For instance, we can minimize α over $(\mathbf{X}, \mathbf{Y})$ subject to

$$\mathbf{\Phi}(X) + \alpha\mathbf{I} > \mathbf{0}, \quad \Omega(\mathbf{Y}) + \alpha\mathbf{I} > \mathbf{0}, \quad \mathbf{\Psi}(\mathbf{X}, \mathbf{Y}) \geq \mathbf{0}.$$

In this case, the suboptimal control problem is feasible if and only if the minimum value of α is such that $\alpha^* \leq 0$.

The optimal control problem can equivalently be written

$$(\mathbf{X}_p^*, \mathbf{Y}_p^*, \gamma^*) \triangleq \arg\min\{\gamma : \mathbf{\Phi}(\mathbf{X}, \gamma) > \mathbf{0},\ \mathbf{\Omega}(\mathbf{Y}, \gamma) > \mathbf{0},\ \mathbf{\Psi}(X, Y) \geq \mathbf{0}\}$$

where $*$ denotes the optimal solution (not the complex conjugate transpose of a matrix). In this case, γ^* is the optimal value of the performance measure, and the pair $(\mathbf{X}_p^*, \mathbf{Y}_p^*)$ will be used to construct an optimal Lyapunov matrix in the next control design step. In many cases, the feasible domain of the above problem, i.e. the set

$$\mathcal{D} \triangleq \{(\mathbf{X}, \mathbf{Y}, \gamma) : (\mathbf{X}, \mathbf{Y}) \in \mathcal{C}(\gamma)\}$$

is convex and hence a globally optimal solution can be found.

If we want to fix the controller order to be n_c, then the following additional nonconvex constraint on $(\mathbf{X}_p, \mathbf{Y}_p) \in \mathcal{C}(\gamma)$ is necessary (and sufficient):

$$n_c = \text{rank}(\mathbf{X}_p - \mathbf{Y}_p^{-1})$$

or equivalently (see the proof of Theorem 7.2.3),

$$n_p + n_c = \text{rank}\begin{bmatrix} \mathbf{X}_p & \mathbf{I} \\ \mathbf{I} & \mathbf{Y}_p \end{bmatrix}.$$

This additional rank constraint destroys the convexity of the problem and makes it much harder to solve. Note that, if we do not fix the controller order *a priori*, then the computational problem is convex and the resulting controller can always be chosen to be of order equal to or less than the plant order since the rank of matrix $\mathbf{X}_p - \mathbf{Y}_p^{-1}$ can never exceed n_p due to its dimension constraint. Note that for the static output feedback case ($n_c = 0$), the above rank condition and $\Psi(\mathbf{X}_p, \mathbf{Y}_p) \geq 0$ reduce to $\mathbf{X}_p = \mathbf{Y}_p^{-1} > \mathbf{0}$. For the general fixed-order control problem ($n_c > 0$), we have an alternative formulation leading to the conditions which have exactly the same structure as those for the static output feedback case (e.g. Theorems 6.3.3 and 7.2.3) where the Lyapunov matrices $\mathbf{X}$ and $\mathbf{Y}$ are constrained by $\mathbf{X} = \mathbf{Y}^{-1} > \mathbf{0}$. We shall address this nonconvex problem in later sections. The following two sections present an algorithm to solve the optimal control problem stated above.

11.1.2 Analytic Center

This section summarizes preliminary materials for the algorithm given in the next section. Specifically, we shall define the analytic center of an LMI and provide necessary steps to compute it. The concept of the analytic center is very simple, and in fact, is essential to all the algorithms given in this chapter.

Consider an LMI

$$\mathbf{F}(x) \triangleq \mathbf{F}_0 + \sum_{i=1}^{m} x_i \mathbf{F}_i > \mathbf{0}$$

where $\mathbf{F}_i = \mathbf{F}_i^T \in \mathcal{R}^{n\times n}$ for $i = 0, 1, \cdots, m$, and the corresponding feasible set $\mathcal{F}$:

$$\mathcal{F} \triangleq \{ \mathbf{x} : \ \mathbf{F}(\mathbf{x}) > \mathbf{0} \},$$

where $\mathbf{x} \triangleq (x_1 \cdots x_m)^T$. A barrier function $\phi : \mathcal{F} \rightarrow \mathcal{R}$ for the set $\mathcal{F}$ can be given by

$$\phi(\mathbf{x}) \triangleq \log \ \det \mathbf{F}(\mathbf{x})^{-1}.$$

Note that $\phi(\mathbf{x})$ approaches infinity when $\mathbf{x}$ approaches the boundary of $\mathcal{F}$ (where $\mathbf{F}(\mathbf{x})$ becomes singular). The function is strictly convex on $\mathcal{F}$. Hence, if $\mathcal{F}$ is nonempty and bounded, then it has a unique minimizer $\mathbf{x}^*$

$$\mathbf{x}^* \triangleq \arg\min\{\phi(\mathbf{x}) : \ \mathbf{x} \in \mathcal{F}\},$$

or equivalently,

$$\mathbf{x}^* = \arg\max\{ \det \mathbf{F}(\mathbf{x}) : \ \mathbf{x} \in \mathcal{F} \}.$$

We call $\mathbf{x}^*$ the *analytic center* of the LMI $\mathbf{F}(\mathbf{x}) > \mathbf{0}$. Intuitively, the analytic center is the "most feasible point" of the set $\mathcal{F}$ in the sense that $\mathbf{x}^*$ is the point where the "distance from the boundary of $\mathcal{F}$" ($\det \mathbf{F}(\mathbf{x})$) is maximum.

Given an initial feasible point $\mathbf{x}_0 \in \mathcal{F}$, the analytic center can be computed by Newton's method

$$\mathbf{x}_{k+1} \triangleq \mathbf{x}_k - \zeta_k \mathbf{H}(\mathbf{x}_k)^{-1}\mathbf{g}(\mathbf{x}_k)$$

where $\mathbf{g}(\mathbf{x}_k)$ and $\mathbf{H}(\mathbf{x}_k)$ are the gradient and Hessian of ϕ at $\mathbf{x}_k \in \mathcal{F}$, and can be computed by

$$g_i(\mathbf{x}) = -tr \ \mathbf{F}(\mathbf{x})^{-1}\mathbf{F}_i$$

$$\mathbf{H}_{ij}(\mathbf{x}) = tr \ \mathbf{F}(\mathbf{x})^{-1}\mathbf{F}_i\mathbf{F}(\mathbf{x})^{-1}\mathbf{F}_j$$

and ζ_k is the damping factor of the kth iteration:

$$\zeta_k \triangleq \begin{cases} 1 & \text{if } \delta(\mathbf{x}_k) \leq 1/4 \\ 1/(1+\delta(\mathbf{x}_k)) & \text{if } \delta(\mathbf{x}_k) > 1/4 \end{cases}$$

where

$$\delta(\mathbf{x}_k) \triangleq \sqrt{\mathbf{g}(\mathbf{x}_k)^T\mathbf{H}(\mathbf{x}_k)^{-1}\mathbf{g}(\mathbf{x}_k)}.$$

11.1.3 The Method of Centers

This section provides a comprehensive summary of the "method of centers" for solving a class of (quasi)convex minimization problem involving a Linear Matrix Inequality (LMI).

The optimization problem, for which an algorithm based on the notion of the analytic center will be given, is the following:

$$\gamma^* \triangleq \min\{ \gamma : \ \mathcal{C}(\gamma) \neq \phi \}$$

where

$$\mathcal{C}(\gamma) \triangleq \{ \mathbf{x} : \ \mathbf{F}(\mathbf{x}, \gamma) > \mathbf{0} \},$$

and $\mathbf{F}(\cdot,\cdot)$ is a matrix-valued function which maps $\mathcal{R}^n \times \mathcal{R}$ into $\mathcal{P}_s^{m\times m}$, where $\mathcal{P}_s$ denotes the set of real symmetric matrices. Note that the optimal control problem stated in Section 11.1.1 falls into this type of problem where

$$\mathbf{F}(\mathbf{x},\gamma) \triangleq \begin{bmatrix} \mathbf{\Phi}(\mathbf{X},\gamma) & \mathbf{0} & \mathbf{0} \\ \mathbf{0} & \mathbf{\Omega}(\mathbf{Y},\gamma) & \mathbf{0} \\ \mathbf{0} & \mathbf{0} & \mathbf{\Psi}(\mathbf{X},\mathbf{Y}) \end{bmatrix}$$

where the vector $\mathbf{x} \in \mathcal{R}^n$ $(n \triangleq n_p(n_p+1))$ consists of $n_p(n_p+1)/2$ elements of symmetric matrices $\mathbf{X}$ and $\mathbf{Y}$. Notice that this formulation enforces the constraint $\mathbf{\Psi}(\mathbf{X},\mathbf{Y}) > \mathbf{0}$ which is stronger than the constraint $\mathbf{\Psi}(\mathbf{X},\mathbf{Y}) \geq \mathbf{0}$ of the original problem. However, this difference is immaterial since we can always come as close to the boundary of $\mathbf{\Psi}(\mathbf{X},\mathbf{Y}) \geq \mathbf{0}$ as desired while satisfying $\mathbf{\Psi}(\mathbf{X},\mathbf{Y}) > \mathbf{0}$. We shall consider the class of problems for which $\mathbf{F}(\cdot,\cdot)$ has the following properties:

(a) $\mathcal{C}(\gamma) \neq \phi$ for sufficiently large $\gamma > 0$ and

$$\mathcal{C}(\gamma_0) \neq \phi \;\Rightarrow\; \mathcal{C}(\gamma) \neq \phi, \;\; \forall\, \gamma \geq \gamma_0.$$

(b) $\mathcal{C}(\gamma) = \phi$ for $\gamma < 0$.

(c) $\mathcal{C}(\gamma)$ is bounded for each γ such that $\mathcal{C}(\gamma) \neq \phi$, i.e. there exists a scalar $\sigma(\gamma)$ such that

$$\|\mathbf{x}\| \leq \sigma(\gamma), \quad \forall\, \mathbf{x} \in \mathcal{C}(\gamma).$$

(d) $\mathbf{F}(\cdot,\cdot)$ is affine with respect to the first argument, and hence it can be written in the form

$$\mathbf{F}(x,\gamma) = \mathbf{F}_0(\gamma) + \sum_{i=1}^{n} x_i \mathbf{F}_i(\gamma)$$

where $\mathbf{F}_i(\gamma) = \mathbf{F}_i(\gamma)^T$ for all $\gamma > 0$ and $i = 0, 1, \ldots, n$.

Items (a) and (b) basically guarantee that the optimization problem is well-posed, i.e. the feasible domain is nonempty and the value of the objective function is bounded below in the feasible domain. Items (c) and (d) imply that $\mathcal{C}(\gamma)$ is a bounded convex set whenever it is nonempty. Thus the existence of the analytic center of $\mathbf{F}(\mathbf{x},\gamma) > \mathbf{0}$ for each fixed (large enough) γ is guaranteed. In the context of control design, the first statement of item (a) corresponds to the requirement that the plant be stabilizable (by a possibly dynamic output feedback controller of some order)[3] since, for sufficiently large γ, the performance requirement is effectively removed, leaving the closed-loop stability as the only control design specification. The meaning of the second statement of item (a) becomes clear once we recall that, for a given performance bound γ, there exists a controller which guarantees the performance level γ if and only if $\mathcal{C}(\gamma)$ is nonempty. Clearly, if there exists a controller with performance γ_0, then such a controller guarantees the worse performance $\gamma \geq \gamma_0$. Item (b) is a reasonable assumption since we usually define a performance measure such that zero is the best possible performance. Thus, there should not exist a controller with a negative performance level.

We are now ready to state the algorithm based on the notion of the analytic center for solving the above LMI optimization problem.

---

[3]The system $\dot{\mathbf{x}}_p = \mathbf{A}_p\mathbf{x}_p + \mathbf{B}_p\mathbf{u}$, $\mathbf{z} = \mathbf{M}_p\mathbf{x}_p$ can be stabilized by a dynamic controller if and only if the triple $(\mathbf{A}_p, \mathbf{B}_p, \mathbf{M}_p)$ is stabilizable and detectable.

The Algorithm:

1. Initialize γ_0 and $\mathbf{x}_0$ such that $\mathbf{F}(\mathbf{x}_0, \gamma_0) > \mathbf{0}$ and let $k = 0$. Choose a reduction factor $0 < \theta < 1$ and an error tolerance $\varepsilon > 0$.
2. Update γ_k by
$$\gamma_{k+1} \triangleq (1-\theta) f(\mathbf{x}_k) + \theta \gamma_k$$
where $f(\mathbf{x})$ is the smallest value of γ such that $\mathbf{F}(\mathbf{x}, \gamma) \geq \mathbf{0}$
$$f(\mathbf{x}) \triangleq \inf \{ \gamma : \mathbf{F}(\mathbf{x}, \gamma) > \mathbf{0} \}.$$
3. Let x_{k+1} be the analytic center of $\mathbf{F}(\mathbf{x}, \gamma_{k+1}) > \mathbf{0}$
$$\mathbf{x}_{k+1} \triangleq \arg\max \{\det \mathbf{F}(\mathbf{x}, \gamma_{k+1}) : \mathbf{F}(\mathbf{x}, \gamma_{k+1}) > \mathbf{0}\}.$$
4. If $\gamma_k - f(\mathbf{x}_k) \leq \varepsilon$, then stop. Otherwise, let $k \leftarrow k+1$ and go to 2.

The idea of the above algorithm is to update $\mathbf{x}$ so that a better (strictly smaller) value of γ can be found at each iteration, which leads to a strictly decreasing sequence of γ_k. One way to update $\mathbf{x}$ to achieve this purpose is to let $\mathbf{x}_k$ be the analytic center of $\mathbf{F}(\mathbf{x}, \gamma_k) > \mathbf{0}$. Since this inequality is strict (i.e. the pair $(\mathbf{x}_k, \gamma_k)$ is an interior point of the closed set defined by $\mathbf{F}(\mathbf{x}, \gamma) \geq \mathbf{0}$), small perturbation in γ_k does not push the point to the boundary of the set. Therefore, we can always find a better γ. Intuitively, this improvement for the value of γ_{k+1} over γ_k will be large if the point $(\mathbf{x}_k, \gamma_k)$ is "deep inside" the set. Thus, it makes sense to choose $\mathbf{x}_k$ to be the analytic center.

This algorithm generates a decreasing sequence of γ_k
$$\gamma_0 > \gamma_1 > \cdots .$$
If $\mathbf{F}$ is given by
$$\mathbf{F}(\mathbf{x}, \gamma) = \begin{bmatrix} \gamma \mathbf{A}(\mathbf{x}) - \mathbf{B}(\mathbf{x}) & \mathbf{0} & \mathbf{0} \\ \mathbf{0} & \mathbf{A}(\mathbf{x}) & \mathbf{0} \\ \mathbf{0} & \mathbf{0} & \mathbf{C}(\mathbf{x}) \end{bmatrix}$$
then the problem is a quasiconvex optimization problem and the sequence γ_k converges to the globally optimal solution γ^* [7].

An initial feasible point $(\mathbf{x}_0, \gamma_0)$ can be found by solving a Lyapunov or Riccati equation. For instance, consider the LQR problem for the discrete-time case. Recall that, in our formulation, γ is the performance bound and $\mathbf{x}$ is the vector formed by stacking the elements of $(\mathbf{X}_p, \mathbf{Y}_p) \in \mathcal{C}(\gamma)$, which is directly related to the Lyapunov matrix $\mathbf{Y}$ solving the Lyapunov inequality (9.25). Noting that the performance measure is finite if and only if the closed-loop system is stable, we can solve
$$\mathbf{Y} = \mathbf{A}_{c\ell}^T \mathbf{Y} \mathbf{A}_{c\ell} + \mathbf{C}_{c\ell}^T \mathbf{C}_{c\ell} + \mathbf{Q}$$
for some stabilizing (possibly full-order) controller $\mathbf{G}$ and an arbitrary $\mathbf{Q} > \mathbf{0}$, and compute the value of the performance measure $\|\mathbf{B}^T\mathbf{Y}\mathbf{B} + \mathbf{F}^T\mathbf{F}\|$. Since this $\mathbf{Y}$ satisfies (9.25), the matrix pair $(\mathbf{X}_p, \mathbf{Y}_p)$ defined by
$$\mathbf{Y} = \begin{bmatrix} \mathbf{Y}_p & \mathbf{Y}_{pc} \\ \mathbf{Y}_{pc}^T & \mathbf{Y}_c \end{bmatrix}, \quad \mathbf{X}_p = (\mathbf{Y}_p - \mathbf{Y}_{pc}\mathbf{Y}_c^{-1}\mathbf{Y}_{pc}^T)^{-1}$$
must be such that $(\mathbf{X}_p, \mathbf{Y}_p) \in \mathcal{C}(\gamma)$ for any γ satisfying $\|\mathbf{B}^T\mathbf{Y}\mathbf{B} + \mathbf{F}^T\mathbf{F}\| < \gamma$. Thus, using the triple $(\mathbf{X}_p, \mathbf{Y}_p, \gamma)$, an initial point $(\mathbf{x}_0, \gamma_0)$ can be determined. In the above, the

stabilizing controller can be designed using any method such as LQG, pole assignment, etc. Finally, for the covariance upper bound control problem, the dual of the above discussion can be applied. For the $\mathcal{H}_\infty$ control problem, we need to solve the $\mathcal{H}_\infty$ Riccati equations in Lemmas 7.1.1 and 7.1.2 for sufficiently large $\gamma > 0$ instead of the Lyapunov equation.

The choice of the parameter θ does not alter the convergence property (i.e. any θ such that $0 < \theta < 1$ will result in convergent sequences γ_k and $\mathbf{x}_k$). However, the value of θ does affect the speed of convergence. In general, a smaller θ yields a smaller number of iteration steps for convergence, but in this case, the computational load for the analytic center determination at each iteration will be more demanding.

If we use Newton's method for computing the analytic center at each iteration, we need an initial feasible point. Clearly, $\mathbf{x}_k$ is a feasible point for the $(k+1)$th iteration since $\mathbf{F}(\mathbf{x}_k, \gamma_{k+1}) > \mathbf{0}$ due to the property (a) of $\mathbf{F}(\cdot, \cdot)$ given above. Thus initialization for the analytic center computation does not introduce additional computational burdens.

We have described an algorithm for solving a class of LMI optimization problems. The algorithm is based on the notion of the analytic center, and utilizes the special structure (LMI constraints) of the problem. In this sense, the algorithm is more specialized than other general convex programming methods such as the cutting plane algorithm and the ellipsoid algorithm, and hence one can expect a better performance for solving the particular LMI problem. Indeed, numerical experiences suggest that the algorithm converges much faster than the above mentioned methods.

11.2 CONTROL DESIGN WITH FIXED CONTROLLER ORDER

This section addresses computational aspects of the *fixed-order* control design. As mentioned in the previous section, the fixed-order control problem is much harder than the problem of designing controllers of unspecified order. Specifically, the fixed-order control problem involves a nonconvex coupling constraint and hence it is extremely difficult to determine the feasibility of the control specifications even for the most fundamental stabilization problem. We shall first define a general form of the fixed-order control problem in the next section, and then suggest an approach to tackle this problem in later sections.

11.2.1 Problem Formulation

Consider the following feasibility problem consisting of two convex sets and a nonconvex coupling constraint:

The Dual LMI Problem:

Find a matrix pair $(\mathbf{X}, \mathbf{Y})$ such that

$$\mathbf{X} \in \mathcal{X} \triangleq \{\mathbf{X} : \ \mathbf{\Phi}(\mathbf{X}) > \mathbf{0}, \ \ \mathbf{X} \in \mathcal{P}_s\} \tag{11.3}$$

$$\mathbf{Y} \in \mathcal{Y} \triangleq \{\mathbf{Y} : \ \mathbf{\Omega}(\mathbf{Y}) > \mathbf{0}, \ \ \mathbf{Y} \in \mathcal{P}_s\} \tag{11.4}$$

$$\mathbf{X} = \mathbf{Y}^{-1} > \mathbf{0}$$

where $\mathbf{\Phi}(\cdot)$ and $\mathbf{\Omega}(\cdot)$ are affine mappings on the set of real symmetric matrices, and $\mathcal{P}_s$ is the set of real symmetric matrices with a block diagonal structure.

We have shown in Chapters 6 to 9 that many fixed-order controller design problems with stability, performance and robustness specifications for both continuous-time and discrete-time systems can be reduced to the Dual LMI Problem. For example, the linear time-invariant continuous-time system

$$\dot{\mathbf{x}}_p = \mathbf{A}_p \mathbf{x}_p + \mathbf{B}_p \mathbf{u}, \quad \mathbf{z} = \mathbf{M}_p \mathbf{x}_p \tag{11.5}$$

is stabilizable via dynamic output feedback controller of order n_c if and only if there exists a matrix pair $(\mathbf{X}, \mathbf{Y})$ such that

$$\boldsymbol{\Phi}(\mathbf{X}) \triangleq -\mathbf{B}^{\perp}(\mathbf{AX} + \mathbf{XA}^T)\mathbf{B}^{\perp T} > \mathbf{0} \tag{11.6}$$

$$\boldsymbol{\Omega}(\mathbf{Y}) \triangleq -\mathbf{M}^{T\perp}(\mathbf{YA} + \mathbf{A}^T\mathbf{Y})\mathbf{M}^{T\perp T} > \mathbf{0} \tag{11.7}$$

$$\mathbf{X} = \mathbf{Y}^{-1} > \mathbf{0} \tag{11.8}$$

where

$$\mathbf{A} \triangleq \begin{bmatrix} \mathbf{A}_p & \mathbf{0} \\ \mathbf{0} & \mathbf{0} \end{bmatrix}, \quad \mathbf{B} \triangleq \begin{bmatrix} \mathbf{B}_p & \mathbf{0} \\ \mathbf{0} & \mathbf{I}_{n_c} \end{bmatrix}, \quad \mathbf{M} \triangleq \begin{bmatrix} \mathbf{M}_p & \mathbf{0} \\ \mathbf{0} & \mathbf{I}_{n_c} \end{bmatrix}. \tag{11.9}$$

The controller order is fixed by the dimension of $\mathbf{X} \in \mathcal{R}^{(n_p+n_c)\times(n_p+n_c)}$. Thus, stabilizability via a fixed-order controller can be checked by solving the Dual LMI problem with $\boldsymbol{\Phi}(\cdot)$ and $\boldsymbol{\Omega}(\cdot)$ defined as in (11.6) and (11.7), respectively, and $\mathcal{P}_s$ being the set of (unstructured) real symmetric matrices. Moreover, all stabilizing static output feedback gains can be computed using solutions $(\mathbf{X}, \mathbf{Y})$ to the Dual LMI problem by the explicit parametrization given in Theorem 2.3.12. In fact, all the control problems considered in Chapter 9 can be formulated as the Dual LMI Problems. Typically, the structure constraint $\mathcal{P}_s$ arises in robust control problems for systems with structured uncertainty. For example, the discrete-time state μ space upper bound (SSUB) μ-synthesis problem can be reduced to the Dual LMI problem with

$$\boldsymbol{\Phi}(\mathbf{X}_s) \triangleq \boldsymbol{\Gamma}^{\perp}(\mathbf{X}_s - \boldsymbol{\Theta}\mathbf{X}_s\boldsymbol{\Theta}^T)\boldsymbol{\Gamma}^{\perp T}$$

$$\boldsymbol{\Omega}(\mathbf{Y}_s) \triangleq \boldsymbol{\Lambda}^{T\perp}(\mathbf{Y}_s - \boldsymbol{\Theta}^T\mathbf{Y}_s\boldsymbol{\Theta})\boldsymbol{\Lambda}^{T\perp T}$$

$$\mathcal{P}_s \triangleq \left\{ \begin{bmatrix} \mathbf{X} & \mathbf{0} \\ \mathbf{0} & \mathbf{S} \end{bmatrix} : \mathbf{X} = \mathbf{X}^T, \quad \mathbf{S} \in \mathcal{S} \right\}$$

where $\boldsymbol{\Gamma}$, $\boldsymbol{\Lambda}$ and $\boldsymbol{\Theta}$ are the augmented matrices defined in Theorem 9.2.8, and $\mathcal{S}$ is the set of scaling matrices defined in (9.17).

11.2.2 A Minimization Approach

This section introduces an approach to address the Dual LMI problem. An algorithm based on a minimization approach is proposed and its advantages and limitations are discussed. The objective here is to provide a motivation for the XY-centering Algorithm to be presented in the next section. To this end, consider a minimization problem

$$\lambda^* \triangleq \min\{\lambda_{\max}(\mathbf{XY}) : (\mathbf{X}, \mathbf{Y}) \in \mathcal{C}\} \tag{11.10}$$

where

$$\mathcal{C} \triangleq \{(\mathbf{X}, \mathbf{Y}) : \mathbf{X} \in \mathcal{X}, \ \mathbf{Y} \in \mathcal{Y}, \ \boldsymbol{\Psi}(\mathbf{X}, \mathbf{Y}) \geq \mathbf{0}\} \tag{11.11}$$

where $\Psi(\mathbf{X}, \mathbf{Y})$ is defined in (11.1). Recall that the convex set $\mathcal{C}$ is related to the control problem with unspecified controller order (possibly equal to the plant order). For example, for the stabilization problem in section 9.1.1, $\mathcal{C} \neq \phi$ (with $\mathcal{P}_s$ being the set of unstructured real symmetric matrices) holds if and only if $(\mathbf{A}, \mathbf{B})$ is stabilizable and $(\mathbf{A}, \mathbf{C})$ is detectable. Thus, the Dual LMI problem is feasible only if the above minimization problem is feasible (i.e. $\mathcal{C} \neq \phi$). Moreover, the Dual LMI problem is feasible *if and only if* the optimal value of the above problem is $\lambda^* = 1$ and attained at an interior point of $\mathcal{C}$ (not on the boundary of $\mathcal{C}$). This can be easily seen once we notice that, if $\mathbf{X} > \mathbf{0}$, $\mathbf{Y} > \mathbf{0}$,

$$\mathbf{X} \geq \mathbf{Y}^{-1} \quad \Leftrightarrow \quad \lambda_{\min}(\mathbf{XY}) \geq 1$$

$$\lambda_{\min}(\mathbf{XY}) = \lambda_{\max}(\mathbf{XY}) = 1 \quad \Leftrightarrow \quad \mathbf{XY} = \mathbf{I}.$$

Unfortunately, the minimization problem is not convex since the objective function $\lambda_{\max}(\mathbf{XY})$ is not. Thus, it is extremely difficult to compute a globally optimal solution.

An approach to this nonconvex problem is to utilize the fact that, for each fixed $\mathbf{Y} > \mathbf{0}$, the function $\varphi(\mathbf{X}) \triangleq \lambda_{\max}(\mathbf{XY})$ is convex. Thus we can successively minimize $\lambda_{\max}(\mathbf{XY})$ over one variable while fixing the other by convex programming problems:

$$\text{Initialization: } k = 1, \quad (\mathbf{X}_0, \mathbf{Y}_1) \in \mathcal{C}$$

$$\mathbf{X}_k \triangleq \arg\min\{\lambda_{\max}(\mathbf{XY}_k) : \ (\mathbf{X}, \mathbf{Y}_k) \in \mathcal{C}\},$$

$$\mathbf{Y}_{k+1} \triangleq \arg\min\{\lambda_{\max}(\mathbf{X}_k\mathbf{Y}) : \ (\mathbf{X}_k, \mathbf{Y}) \in \mathcal{C}\}.$$

Of course, this algorithm may not give a global solution to the original problem (11.10). Note that the optimal value of each minimization problem may not be attained, i.e. $\mathbf{X}_k$ and $\mathbf{Y}_{k+1}$ lie on the boundaries of the (open) feasible domains. Specifically, it is possible that $\mathbf{X}_k \notin \mathcal{X}$ or $\mathbf{Y}_{k+1} \notin \mathcal{Y}$. Hence, the sets $\mathcal{X}$ and $\mathcal{Y}$ must be replaced by closed set inner approximations $\bar{\mathcal{X}}$ and $\bar{\mathcal{Y}}$ using a small scalar $\varepsilon > 0$ to guarantee $\mathbf{X}_k \in \mathcal{X}$ and $\mathbf{Y}_k \in \mathcal{Y}$ for all k, where

$$\bar{\mathcal{X}} \triangleq \{\mathbf{X} : \ \mathbf{\Phi}(\mathbf{X}) \geq \varepsilon\mathbf{I}, \ \mathbf{X} \in \mathcal{P}_s \}$$

$$\bar{\mathcal{Y}} \triangleq \{\mathbf{Y} : \ \Omega(\mathbf{Y}) \geq \varepsilon\mathbf{I}, \ \mathbf{Y} \in \mathcal{P}_s \}.$$

With these modifications, the feasibility of each minimization problem at any iteration is guaranteed since $\mathbf{Y}_k$ ($\mathbf{X}_{k-1}$) is a feasible point of the second (first) minimization problem. Moreover, this fact guarantees that the optimal values of the minimization problems are nonincreasing:

$$\lambda_{\max}(\mathbf{X}_k\mathbf{Y}_k) \geq \lambda_{\max}(\mathbf{X}_k\mathbf{Y}_{k+1}) \geq \lambda_{\max}(\mathbf{X}_{k+1}\mathbf{Y}_{k+1}).$$

The algorithm we shall present in the next section is obtained by modifying the above successive minimization algorithm so that an upper bound on $\lambda_{\max}(\mathbf{X}_k\mathbf{Y}_k)$ *strictly* decreases at each iteration without requiring special modifications for $\mathcal{X}$ and $\mathcal{Y}$. The idea is to replace the solution to each minimization problem in the above algorithm by analytic centers.

11.2.3 The XY-Centering Algorithm

This section presents an algorithm based on the notion of the analytic center to address the Dual LMI problem. We shall state the algorithm first, then discuss its properties and limitations.

The XY-centering Algorithm:

1 Choose a parameter θ such that $0 < \theta < 1$.

2 Find $(\hat{\mathbf{X}}, \hat{\mathbf{Y}}) \in \mathcal{C}$ and let $k = 1$ and

$$\mathbf{Y}_1 \triangleq \hat{\mathbf{Y}}, \quad \beta_1 > \lambda_{\max}(\hat{\mathbf{X}}\hat{\mathbf{Y}}).$$

3 Compute the analytic center $\mathbf{X}_k$ and update α_k:

$$\mathbf{X}_k \triangleq \mathrm{ac}\{\ \mathbf{I} < \mathbf{Y}_k^{1/2}\mathbf{X}\mathbf{Y}_k^{1/2} < \beta_k\mathbf{I}, \quad \mathbf{X} \in \mathcal{X}\ \},$$

$$\alpha_k \triangleq (1-\theta)\lambda_{\max}(\mathbf{X}_k\mathbf{Y}_k) + \theta\beta_k.$$

4 Compute the analytic center $\mathbf{Y}_{k+1}$ and update β_{k+1};

$$\mathbf{Y}_{k+1} \triangleq \mathrm{ac}\{\ \mathbf{I} < \mathbf{X}_k^{1/2}\mathbf{Y}\mathbf{X}_k^{1/2} < \alpha_k\mathbf{I}, \quad \mathbf{Y} \in \mathcal{Y}\ \},$$

$$\beta_{k+1} \triangleq (1-\theta)\lambda_{\max}(\mathbf{X}_k\mathbf{Y}_{k+1}) + \theta\alpha_k.$$

5 If $\mathbf{X}_k^{-1} \in \mathcal{Y}$ or $\mathbf{Y}_{k+1}^{-1} \in \mathcal{X}$, then stop. Otherwise let $k \leftarrow k+1$ and go to STEP 3.

In the above algorithm, an initialization parameter $(\hat{\mathbf{X}}, \hat{\mathbf{Y}}) \in \mathcal{C}$ can be found by convex programming, or it can be determined that $\mathcal{C} \neq \phi$, in which case, the Dual LMI problem is infeasible. In fact, noniterative methods to find $(\hat{\mathbf{X}}, \hat{\mathbf{Y}}) \in \mathcal{C}$ are available for many control problems. For instance, an element of $\mathcal{C}$ for the fixed-order dynamic output feedback stabilization problem can be found as follows: Find a stabilizing controller of any order (e.g. full-order LQG) and compute the closed-loop system matrix $\mathbf{A}_{c\ell}$. Choosing an arbitrary matrix $\mathbf{Q} > \mathbf{0}$, solve the Lyapunov equation for $\mathbf{X} > \mathbf{0}$

$$\mathbf{A}_{c\ell}\mathbf{X} + \mathbf{X}\mathbf{A}_{c\ell} + \mathbf{Q} = \mathbf{0}. \tag{11.12}$$

Then letting

$$\hat{\mathbf{X}} \triangleq \mathbf{X}_p, \quad \hat{\mathbf{Y}} \triangleq (\mathbf{X}_p - \mathbf{X}_{pc}\mathbf{X}_c^{-1}\mathbf{X}_{pc}^T)^{-1}, \quad \mathbf{X} \triangleq \begin{bmatrix} \mathbf{X}_p & \mathbf{X}_{pc} \\ \mathbf{X}_{pc}^T & \mathbf{X}_c \end{bmatrix}$$

yields $(\hat{\mathbf{X}}, \hat{\mathbf{Y}}) \in \mathcal{C}$. A justification of this method is left for the reader as an exercise.

The notation $\mathrm{ac}\{\cdot\}$ in STEPS 3 and 4 denotes the analytic center of the LMI $\{\cdot\}$. For example, $\mathbf{X}_k$ is the analytic center of

$$\mathbf{F}(\mathbf{X}) \triangleq \begin{bmatrix} \beta_k\mathbf{I} - \mathbf{Y}_k^{1/2}\mathbf{X}\mathbf{Y}_k^{1/2} & \mathbf{0} & \mathbf{0} \\ \mathbf{0} & \mathbf{Y}_k^{1/2}\mathbf{X}\mathbf{Y}_k^{1/2} - \mathbf{I} & \mathbf{0} \\ \mathbf{0} & \mathbf{0} & \mathbf{\Phi}(\mathbf{X}) \end{bmatrix} > \mathbf{0}$$

subject to the structure constraint $\mathbf{X} \in \mathcal{P}_s$.

The XY-centering Algorithm has the following property.

Theorem 11.2.1 *Suppose $\mathcal{C} \neq \phi$. Then the XY-centering Algorithm is well-defined, i.e. each set defined by the argument of* ac$\{\cdot\}$ *at* STEPS *3 and 4 is nonempty and bounded, and hence the analytic centers exist at any iteration $k \geq 1$. Moreover, upper bounds α_k and β_k on $\lambda_{\max}(\mathbf{X}_k\mathbf{Y}_k)$ are strictly decreasing:*

$$\beta_k > \alpha_k > \beta_{k+1} > 1, \quad \forall\, k \geq 1.$$

Proof. Since $\mathcal{C} \neq \phi$, we can find an initial pair $(\mathbf{Y}_1, \beta_1)$ as in STEP 1. Note that the set

$$\mathcal{X}_k \triangleq \{ \mathbf{X} : \mathbf{I} < \mathbf{Y}_k^{1/2}\mathbf{X}\mathbf{Y}_k^{1/2} < \beta_k \mathbf{I}, \ \ \mathbf{X} \in \mathcal{X} \}$$

is nonempty since $\hat{\mathbf{X}} \in \mathcal{X}_k$ for $k = 1$. Clearly, $\mathcal{X}_k$ is bounded, and hence the analytic center $\mathbf{X}_k$ is well-defined. Noting that

$$\lambda_{\max}(\mathbf{X}_k\mathbf{Y}_k) < \alpha_k < \beta_k,$$

the set

$$\mathcal{Y}_k \triangleq \{ \mathbf{Y} : \mathbf{I} < \mathbf{X}_k^{1/2}\mathbf{Y}\mathbf{X}_k^{1/2} < \alpha_k \mathbf{I}, \ \ \mathbf{Y} \in \mathcal{Y} \}$$

is again nonempty since $\mathbf{Y}_1 \in \mathcal{Y}_k$ for $k = 1$. β_{k+1} is determined so that

$$\lambda_{\max}(\mathbf{X}_k\mathbf{Y}_{k+1}) < \beta_{k+1} < \alpha_k.$$

Thus, in general, $\mathcal{X}_k$ and $\mathcal{Y}_k$ are nonempty for all k since $\mathbf{X}_{k-1} \in \mathcal{X}_k$, $\mathbf{Y}_{k-1} \in \mathcal{Y}_k$. □

The XY-centering Algorithm may be interpreted as a modified version of the successive minimization algorithm discussed in the previous section, except that each minimization problem is replaced by the computation of the analytic center. The strictly decreasing property of $\lambda_{\max}(\mathbf{X}_k\mathbf{Y}_k)$ may not hold for the XY-centering Algorithm, but upper bounds α_k and β_k strictly decrease. A novel aspect of the XY-centering Algorithm is that we do not require $\lambda_{\max}(\mathbf{X}_k\mathbf{Y}_k) = 1$ to have a solution to the Dual LMI problem. Intuitively, the algorithm tries to make $\mathbf{X}_k$ and $\mathbf{Y}_k^{-1}$ (or equivalently, $\mathbf{X}_k^{-1}$and $\mathbf{Y}_k$) closer to each other while maintaining $\mathbf{X}_k$ and $\mathbf{Y}_k$ to be located "deep inside" the sets $\mathcal{X}$ and $\mathcal{Y}$. Thus, $\mathbf{X}_k^{-1}$ and $\mathbf{Y}_k^{-1}$ are forced to move into $\mathcal{Y}$ and $\mathcal{X}$, respectively.

To compute the analytic center by the method described in the previous section, we need an initial feasible point. As discussed in the proof of Theorem 11.2.1, the analytic centers $\mathbf{X}_k$ and $\mathbf{Y}_{k+1}$ at STEP k can be used as such feasible starting points for the computation of the analytic centers $\mathbf{X}_{k+1}$ and $\mathbf{Y}_{k+2}$ at STEP $k + 1$. Thus, the result of the previous iteration is directly useful to the next iteration. This fact speeds up the algorithm. Moreover, in general, the analytic center can be computed much faster than the solution to the optimization problems in the successive minimization algorithm. Of course, the total number of iterations for the successive minimization algorithm may be less than that for the XY-centering Algorithm. However, numerical experiences suggests that the XY-centering Algorithm has an overall advantage.

Since the analytic center is uniquely determined, the XY-centering Algorithm can be considered as a point-to-point continuous mapping $(\mathbf{Y}_k, \beta_k) \rightarrow (\mathbf{Y}_{k+1}, \beta_{k+1})$ on $\mathcal{D}$ where

$$\mathcal{D} \triangleq \{ (\mathbf{Y}, \beta) : \ \mathbf{0} < \mathbf{Y} \in \mathcal{Y}, \ \ \exists \mathbf{X} \ \text{such that} \ \mathbf{I} < \mathbf{Y}^{1/2}\mathbf{X}\mathbf{Y}^{1/2} < \beta \mathbf{I}, \ \ \mathbf{X} \in \mathcal{X} \}.$$

Hence, in view of the global convergence theorem [83], we see that the sequence $(\mathbf{Y}_k, \beta_k)$ converges to the boundary of $\mathcal{D}$ or diverges. From Theorem 11.2.1, we know that the sequence β_k is strictly decreasing and bounded below by one, and hence converges. If the limit is $\beta_\infty > 1$, then $\mathbf{Y}_k$ will never converge to the solution of the Dual LMI problem. To see this, suppose $\mathbf{Y}_k$ converges to $\mathbf{Y}_\infty > \mathbf{0}, \mathbf{Y}_\infty \in \mathcal{Y}$. If $\mathbf{Y}_\infty$ was a solution of the Dual LMI problem, i.e. $\mathbf{Y}_\infty^{-1} \in \mathcal{X}$, then $(\mathbf{Y}_\infty, \beta_\infty) \in \mathcal{D}$ since a matrix $\mathbf{X}$ can be chosen as $\mathbf{Y}_\infty^{-1} + \varepsilon\mathbf{I}$ for sufficiently small $\varepsilon > 0$. Thus, $(\mathbf{Y}_\infty, \beta_\infty)$ is not on the boundary of $\mathcal{D}$ and, by contradiction, we conclude that $\mathbf{Y}_\infty$ is not a solution to the Dual LMI problem. If the limit is $\beta_\infty = 1$, then $\|\mathbf{X}_k\mathbf{Y}_k - \mathbf{I}\|$ can be made arbitrarily small by choosing sufficiently large k, while $\mathbf{X}_k \in \mathcal{X}$, $\mathbf{Y}_k \in \mathcal{Y}$. Note, however, that $\mathbf{X}_k$ and $\mathbf{Y}_k$ may approach the boundary of $\mathcal{X}$ and $\mathcal{Y}$, respectively, and it may still be possible that $\mathbf{X}_k^{-1} \notin \mathcal{Y}$, $\mathbf{Y}_k^{-1} \notin \mathcal{X}$ even though $\|\mathbf{X}_k\mathbf{Y}_k - \mathbf{I}\|$ is arbitrarily small. From the above discussion, the parameters α_k and β_k in the XY-centering Algorithm approach one if and only if the Dual LMI problem is feasible, provided there exist *fixed*, closed inner approximations of the sets $\mathcal{X}$ and $\mathcal{Y}$ such that $\mathbf{X}_k$ and $\mathbf{Y}_k$ belong to the closed sets for all k, and also the sequences $\mathbf{X}_k$ and $\mathbf{Y}_k$ are contained in compact subsets of the set of positive definite matrices.

Example 11.2.1 Consider the double integrator system given by (11.5) where

$$\mathbf{A}_p := \begin{bmatrix} 0 & 1 \\ 0 & 0 \end{bmatrix}, \quad \mathbf{B}_p := \begin{bmatrix} 0 \\ 1 \end{bmatrix}, \quad \mathbf{M}_p := \begin{bmatrix} 1 & 0 \end{bmatrix}.$$

This system is not stabilizable via static output feedback, but stabilizable by a dynamic controller of order $n_c \geq 1$. We apply the XY-centering Algorithm for designing stabilizing controllers of order $n_c = 0$ and $n_c = 1$. Here, we use $\mathbf{\Phi}(\cdot)$ and $\mathbf{\Omega}(\cdot)$ defined in (11.6) and (11.7), respectively.

Figure 11.1 shows the behavior of the upper bounds α_k and β_k on $\lambda_{\max}(\mathbf{X}_k\mathbf{Y}_k)$, and the minimum eigenvalues of $\Phi(\mathbf{Y}_k^{-1})$ and $\Omega(\mathbf{X}_k^{-1})$ for the case $n_c = 0$. The parameter θ is chosen to be $\theta = 0.2$ (see the discussion about the effect of the choice of θ on the speed of convergence given in [7]). The initial $\mathbf{Y}_1$ is determined by the method described in section 11.2.3 where the free matrix $\mathbf{Q} > \mathbf{0}$ in (11.12) is chosen to be identity. To visualize the convergence property, β_k^{-1} is plotted instead of β_k; in this way, the distance between the two curves corresponding to α_k and β_k^{-1} can be considered as a measure for the "distance" between $\mathbf{X}_k$ and $\mathbf{Y}_k^{-1}$. Similarly, the curves $-\lambda_{\min}(\mathbf{\Phi}(\mathbf{Y}_k^{-1}))$ and $-\lambda_{\min}(\mathbf{\Omega}(\mathbf{X}_k^{-1}))$ express the "distance" between $\mathcal{X}$ and $\mathbf{Y}_k^{-1}$, and $\mathcal{Y}$ and $\mathbf{X}_k^{-1}$, respectively. For example, $\mathbf{Y}_k^{-1} \in \mathcal{X}$ if and only if $-\lambda_{\min}(\mathbf{\Phi}(\mathbf{Y}_k^{-1})) < 0$.

Interestingly, the sequence $\mathbf{X}_k\mathbf{Y}_k$ approaches $\mathbf{I}$ (or equivalently, $\alpha_k \to 1$) even though the system is not stabilizable for $n_c = 0$. This is not a violation of the stabilizability conditions, since $\mathbf{Y}_k^{-1}$ approaches $\mathcal{X}$, but never reaches (belongs to) $\mathcal{X}$ (i.e. $\lambda_{\min}(\mathbf{\Phi}(\mathbf{Y}_k^{-1})) \leq 0$ for all k). After 33 iterations, we have

$$\mathbf{X} = \begin{bmatrix} 1.5470 & -4.4991 \times 10^{-5} \\ -4.4991 \times 10^{-5} & 4.7298 \end{bmatrix}$$

$$\mathbf{\Phi}(\mathbf{X}) = 9.00 \times 10^{-5}, \quad \mathbf{\Omega}(\mathbf{X}^{-1}) = -1.23 \times 10^{-5}$$

$$\mathbf{Y} = \begin{bmatrix} 0.64646 & -5.1534 \times 10^{-6} \\ -5.1534 \times 10^{-6} & 0.21144 \end{bmatrix}$$

$$\mathbf{\Phi}(\mathbf{Y}^{-1}) = -7.54 \times 10^{-5}, \quad \mathbf{\Omega}(\mathbf{Y}) = 1.03 \times 10^{-5}$$

$$\lambda(\mathbf{XY}) = 1.00008, \quad 1.00002.$$

Thus $\mathbf{X} \in \mathcal{X}$ and $\mathbf{Y} \in \mathcal{Y}$ but $\mathbf{X}^{-1} \notin \mathcal{Y}$ and $\mathbf{Y}^{-1} \notin \mathcal{X}$. Note that, since $\mathbf{X}_k$ and $\mathbf{Y}_k$ are approaching the boundaries of $\mathcal{X}$ and $\mathcal{Y}$, respectively, the fact that $\alpha_k \to 1$ does not imply that the Dual LMI problem is feasible (i.e. the system is not stabilizable via static output feedback).

For the fixed-order ($n_c = 1$) dynamic controller case, the parameter is chosen as before; $\theta = 0.2$. The initial matrix $\mathbf{Y}_1$ is also generated by the same procedure as before, but

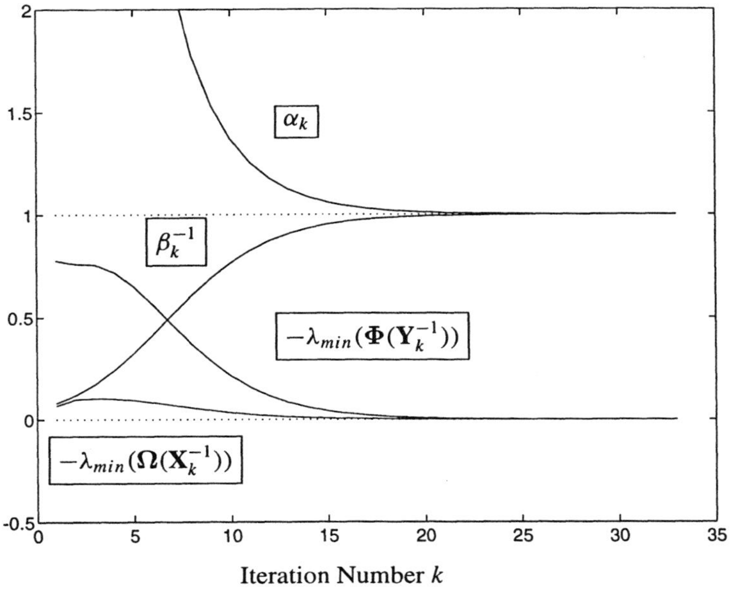

Figure 11.1 Behavior of the XY-centering Algorithm ($n_c = 0$).

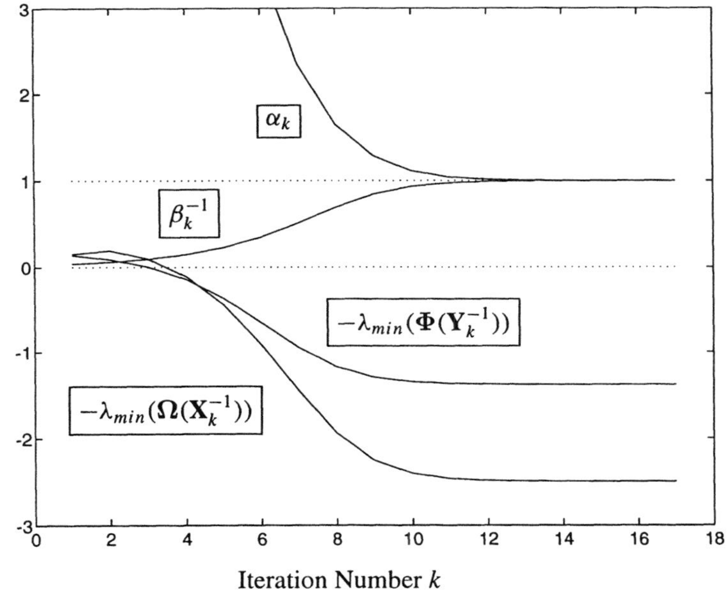

Figure 11.2 Behavior of the XY-centering Algorithm ($n_c = 1$).

$\mathbf{Q} > \mathbf{0}$ in (11.12) is chosen randomly (using the Matlab command "rand"). It can be proved that if $\mathbf{Y}_1$ is chosen to be block diagonal (two blocks with dimensions n_p and n_c where n_p is the plant order), then the XY-centering Algorithm always fails even if the Dual LMI problem is feasible, when the system matrices have the structure as in (11.9). However, for all of the (nonblock diagonal) initial conditions we have tried, the XY-centering Algorithm was successful. Figure 11.2 shows the behavior of the algorithm. To design a stabilizing controller, 4 iterations suffice since $\mathbf{Y}_k^{-1} \in \mathcal{X}$ and $\mathbf{X}_k^{-1} \in \mathcal{Y}$ at $k = 4$. If the algorithm continues to run, $\mathbf{X}_k^{-1}$ and $\mathbf{Y}_k^{-1}$ move "deep inside" the sets $\mathcal{Y}$ and $\mathcal{X}$, respectively. This is a typical behavior of the XY-centering Algorithm. The result is

$$\mathbf{Y}^{-1} \cong \mathbf{X} = \begin{bmatrix} 1.1458 & -0.6910 & -1.6325 \\ -0.6910 & 1.7966 & 2.0712 \\ -1.6325 & 2.0712 & 3.6253 \end{bmatrix}, \quad \lambda(\mathbf{XY}) = \begin{cases} 1.000048 \\ 1.000048 \\ 1.000048 \end{cases}.$$

11.2.4 Extension to Optimal Control

The Dual LMI problem defined in section 11.2.1 is a *feasibility* problem. Hence, it naturally applies to *suboptimal* control problems rather than *optimal* control problems. To make this point clear, let us take an example: the continuous-time LQR problem discussed in section 9.1.3. Using Theorem 9.1.3, one can derive the following: there exists a fixed-order dynamic output feedback controller which solves the LQR problem if and only if there exists a matrix pair $(\mathbf{X}, \mathbf{Y})$ such that $\mathbf{X} = \mathbf{Y}^{-1} > \mathbf{0}$ and

$$\boldsymbol{\Phi}(\mathbf{X}) \triangleq -\begin{bmatrix} \mathbf{B} \\ \mathbf{H} \end{bmatrix}^{\perp} \begin{bmatrix} \mathbf{AX} + \mathbf{XA}^T & \mathbf{XC}^T \\ \mathbf{CX} & -\mathbf{I} \end{bmatrix} \begin{bmatrix} \mathbf{B} \\ \mathbf{H} \end{bmatrix}^{\perp T} > \mathbf{0} \tag{11.13}$$

$$\boldsymbol{\Omega}(\mathbf{Y}, \gamma) \triangleq \begin{bmatrix} -\mathbf{M}^{T\perp}(\mathbf{YA} + \mathbf{A}^T\mathbf{Y} + \mathbf{C}^T\mathbf{C})\mathbf{M}^{T\perp T} & \mathbf{0} \\ \mathbf{0} & \gamma\mathbf{I} - \mathbf{D}^T\mathbf{YD} \end{bmatrix} > \mathbf{0}. \tag{11.14}$$

Thus, the existence of a suboptimal controller which guarantees the LQ performance bound γ can be examined by solving the Dual LMI problem with the above definitions for $\boldsymbol{\Phi}$ and $\boldsymbol{\Omega}$ *for each fixed* $\gamma > 0$.

The purpose of this section is to extend the XY-centering Algorithm which solves suboptimal control problems to handle optimal control problems: Find a controller which minimizes a specified performance measure. To this end, we shall define the following:

The Dual LMI Optimization Problem:

Solve

$$\gamma^* \triangleq \inf\{\, \gamma : \mathcal{L}(\gamma) \neq \phi \,\}$$

where

$$\mathcal{L}(\gamma) \triangleq \{\, (\mathbf{X}, \mathbf{Y}) : \mathbf{X} = \mathbf{Y}^{-1} > \mathbf{0}, \;\; \mathbf{X} \in \mathcal{X}(\gamma), \;\; \mathbf{Y} \in \mathcal{Y}(\gamma) \,\}$$

$$\mathcal{X}(\gamma) \triangleq \{\, \mathbf{X} : \boldsymbol{\Phi}(\mathbf{X}, \gamma) > \mathbf{0}, \;\; \mathbf{X} \in \mathcal{P}_s \,\}$$

$$\mathcal{Y}(\gamma) \triangleq \{\, \mathbf{Y} : \boldsymbol{\Omega}(\mathbf{Y}, \gamma) > \mathbf{0}, \;\; \mathbf{Y} \in \mathcal{P}_s \,\}$$

where $\boldsymbol{\Phi}(\cdot, \cdot)$ and $\boldsymbol{\Omega}(\cdot, \cdot)$ are affine with respect to the first arguments, and $\mathcal{P}_s$ is the set of structured real symmetric matrices.

In general, the set $\mathcal{L}(\gamma)$ corresponding to a well-posed optimal control problem has the following property:

$$\mathcal{L}(\gamma_0) \neq \phi \;\Rightarrow\; \mathcal{L}(\gamma) \neq \phi, \;\forall\, \gamma \geq \gamma_0$$

since the condition $\mathcal{L}(\gamma) \neq \phi$ is equivalent to the existence of a suboptimal controller which achieves the performance level γ. Each of the sets $\mathcal{X}(\gamma)$ and $\mathcal{Y}(\gamma)$ also has a similar property:

$$\mathcal{X}(\gamma_0) \neq \phi \;\Rightarrow\; \mathcal{X}(\gamma) \neq \phi, \;\forall\, \gamma \geq \gamma_0 \tag{11.15}$$

$$\mathcal{Y}(\gamma_0) \neq \phi \;\Rightarrow\; \mathcal{Y}(\gamma) \neq \phi, \;\forall\, \gamma \geq \gamma_0. \tag{11.16}$$

These properties are crucial to develop an algorithm to solve the Dual LMI Optimization Problem.

We shall propose the following algorithm to solve the Dual LMI Optimization Problem. To state the result, let us define the set $\mathcal{C}(\gamma)$ as in (11.11) by replacing $\mathcal{X}$ and $\mathcal{Y}$ by $\mathcal{X}(\gamma)$ and $\mathcal{Y}(\gamma)$, respectively.

The XY-centering Optimization Algorithm:

1 Choose parameters $0 < \theta_\lambda < 1$ and $0 < \theta_\gamma < 1$.

2 Let $\gamma_1 > 0$ be a sufficiently large number, and find $(\hat{\mathbf{X}}, \hat{\mathbf{Y}}) \in \mathcal{C}(\gamma_1)$. Then initialize $k = 1$ and

$$\mathbf{Y}_1 \triangleq \hat{\mathbf{Y}}, \quad \alpha_1 > \lambda_{\max}(\hat{\mathbf{X}}\hat{\mathbf{Y}}).$$

3 Compute the analytic centers;

$$\mathbf{X}_k \triangleq \text{ac}\{\, \mathbf{I} < \mathbf{Y}_k^{1/2}\mathbf{X}\mathbf{Y}_k^{1/2} < \alpha_k\mathbf{I}, \;\; \mathbf{X} \in \mathcal{X}(\gamma_k) \,\},$$

$$\mathbf{Y}_{k+1} \triangleq \text{ac}\{\, \mathbf{I} < \mathbf{X}_k^{1/2}\mathbf{Y}\mathbf{X}_k^{1/2} < \alpha_k\mathbf{I}, \;\; \mathbf{Y} \in \mathcal{Y}(\gamma_k) \,\}.$$

4 If $\mathbf{X}_k^{-1} \in \mathcal{Y}(\gamma_k)$ or $\mathbf{Y}_{k+1}^{-1} \in \mathcal{X}(\gamma_k)$, then

$$\gamma_{k+1} \triangleq (1-\theta_\gamma)\psi(\mathbf{X}_k, \mathbf{Y}_{k+1}) + \theta_\gamma\gamma_k, \quad \alpha_{k+1} \triangleq \alpha_k,$$

where

$$\psi(\mathbf{X}, \mathbf{Y}) \triangleq \min\{\gamma : \mathbf{\Phi}(\mathbf{X}, \gamma) \geq \mathbf{0}, \;\; \mathbf{\Omega}(\mathbf{Y}, \gamma) \geq \mathbf{0}\}. \tag{11.17}$$

Otherwise, let

$$\alpha_{k+1} \triangleq (1-\theta_\lambda)\lambda_{\max}(\mathbf{X}_k\mathbf{Y}_{k+1}) + \theta_\lambda\alpha_k, \quad \gamma_{k+1} \triangleq \gamma_k.$$

5 If $\gamma_k - \psi(\mathbf{X}_k, \mathbf{Y}_{k+1}) < \varepsilon$ for sufficiently small $\varepsilon > 0$, then stop. Otherwise let $k \leftarrow k+1$ and go to STEP 3.

As in the XY-centering Algorithm for the Dual LMI (feasibility) Problem, the initialization parameter $(\hat{\mathbf{X}}, \hat{\mathbf{Y}}) \in \mathcal{C}(\gamma_1)$ can be found by convex programming or some other noniterative methods. For the (optimal) LQR problem, for instance, we can use the following method similar to the one described in Section 11.1.3. First, design a stabilizing controller of any order (possibly equal to the plant order). Then compute the closed-loop system matrices $\mathbf{A}_{c\ell}$, $\mathbf{B}_{c\ell}$ and $\mathbf{C}_{c\ell}$. Choosing an arbitrary positive definite matrix $\mathbf{Q} > \mathbf{0}$, solve the Lyapunov equation for $\mathbf{Y} > \mathbf{0}$;

$$\mathbf{Y}\mathbf{A}_{c\ell} + \mathbf{A}_{c\ell}^T\mathbf{Y} + \mathbf{C}_{c\ell}^T\mathbf{C}_{c\ell} + \mathbf{Q} = \mathbf{0}. \tag{11.18}$$

Then letting

$$\hat{\mathbf{Y}} \triangleq \mathbf{Y}_p, \quad \hat{\mathbf{X}} \triangleq (\mathbf{Y}_p - \mathbf{Y}_{pc}\mathbf{Y}_c^{-1}\mathbf{Y}_{pc}^T)^{-1}, \quad \mathbf{Y} \triangleq \begin{bmatrix} \mathbf{Y}_p & \mathbf{Y}_{pc} \\ \mathbf{Y}_{pc}^T & \mathbf{Y}_c \end{bmatrix}$$

yields $(\hat{\mathbf{X}}, \hat{\mathbf{Y}}) \in \mathcal{C}(\gamma_1)$ for a given γ_1 such that

$$\|\mathbf{B}_{c\ell}^T \mathbf{Y} \mathbf{B}_{c\ell}\| < \gamma_1.$$

Note that the above condition on the initial parameter γ_1 may not be sufficient to guarantee $\mathcal{L}(\gamma_1) \neq \phi$. In this case, the only available information is that $\mathcal{L}(\gamma) \neq \phi$ for sufficiently large $\gamma > 0$ if and only if the system is stabilizable via a fixed-order dynamic output feedback controller. For the $\mathcal{H}_\infty$ control problem, a similar procedure can be applied to find $(\hat{\mathbf{X}}, \hat{\mathbf{Y}}) \in \mathcal{C}(\gamma_1)$ where we replace the above Lyapunov equation (11.18) by the $\mathcal{H}_\infty$ Riccati equation.

The idea of the XY-centering optimization algorithm is the following. If $\mathbf{X}_k^{-1} \in \mathcal{Y}(\gamma_k)$ or $\mathbf{Y}_{k+1}^{-1} \in \mathcal{X}(\gamma_k)$, then there exists a fixed-order dynamic output feedback controller which yields the cost function bounded above by γ_k. Indeed, this bound γ_k is not tight since the analytic centers $\mathbf{X}_k$ and $\mathbf{Y}_{k+1}$ are interior points, and the best (smallest) bound that can be obtained from this information is $\psi(\mathbf{X}_k, \mathbf{Y}_{k+1})$. Thus we can tighten the performance bound as

$$\psi(\mathbf{X}_k, \mathbf{Y}_{k+1}) < \gamma_{k+1} < \gamma_k.$$

Feasibility of any γ_{k+1} satisfying the above inequality is guaranteed by the properties (11.15) and (11.16). In many cases, the value of $\psi(\mathbf{X}_k, \mathbf{Y}_{k+1})$ can be evaluated without iteratively solving the minimization problem (11.17). For the LQR problem, for instance,

$$\psi(\mathbf{X}, \mathbf{Y}) = \|\mathbf{D}^T \mathbf{Y} \mathbf{D}\|.$$

Now, if $\mathbf{X}_k^{-1} \notin \mathcal{Y}(\gamma_k)$ and $\mathbf{Y}_{k+1}^{-1} \notin \mathcal{X}(\gamma_k)$, then do the feasibility iteration; tighten the bound α_k on $\lambda_{\max}(\mathbf{X}_k \mathbf{Y}_{k+1})$. Then, as in the XY-centering Algorithm, $\mathbf{X}_k$ and $\mathbf{Y}_k^{-1}$ get closer to each other by feasibility iterations and, when $\mathbf{X}_k^{-1} \in \mathcal{Y}(\gamma_k)$ or $\mathbf{Y}_{k+1}^{-1} \in \mathcal{X}(\gamma_k)$, we switch to the optimality iteration as described above.

Example 11.2.2 Consider the longitudinal motion of the VTOL helicopter [69] described by (9.10) where

$$\mathbf{A}_p := \begin{bmatrix} -0.0366 & 0.0271 & 0.0188 & -0.4555 \\ 0.0482 & -1.0100 & 0.0024 & -4.0208 \\ 0.1002 & 0.3681 & -0.7070 & 1.4200 \\ 0 & 0 & 1 & 0 \end{bmatrix}$$

$$\mathbf{B}_p := \begin{bmatrix} 0.4422 & 0.1761 \\ 3.5446 & -7.5922 \\ -5.5200 & 4.4900 \\ 0 & 0 \end{bmatrix}, \quad \mathbf{D}_p := \mathbf{I}_4,$$

$$\mathbf{C}_p := \begin{bmatrix} \mathbf{I}_2 & 0 \\ 0 & 0 \end{bmatrix}, \quad \mathbf{B}_y := \begin{bmatrix} 0 \\ \mathbf{I}_2 \end{bmatrix}, \quad \mathbf{M}_p := \begin{bmatrix} 0 & 1 & 0 & 0 \end{bmatrix}.$$

We apply the XY-centering optimization Algorithm for designing an LQ (optimal) static output feedback controller. The system is open-loop unstable, and the LQ optimal state

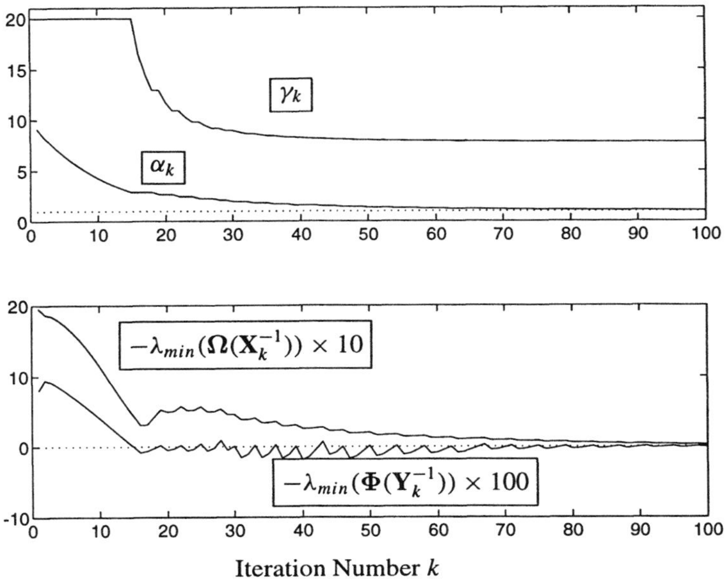

Figure 11.3 Behavior of the XY-centering optimization Algorithm.

feedback cost is $J(\mathbf{G}_{sf}) = 1.5039$ and the (full-order) LQG cost $J(\mathbf{G}_{fo}) = 7.7062$ where the LQG controller is designed with the process noise intensity $\mathbf{I}$ and the measurement noise intensity $\varepsilon\mathbf{I}$ for sufficiently small $\varepsilon > 0$. For the XY-centering optimization algorithm, the parameters θ_λ and θ_γ are chosen as $\theta_\lambda = 0.6$ and $\theta_\gamma = 0.3$. These values are chosen after a few trials, and may be refined for faster convergence. The initial matrix $\mathbf{Y}_1$ is determined by the method described above, where the LQG controller is chosen as the stabilizing controller and the arbitrary matrix $\mathbf{Q} > \mathbf{0}$ in (11.18) is set to identity. The initial cost bound γ_1 is chosen as $\gamma_1 = 1.2 \times \|\mathbf{D}_p^T\mathbf{Y}_1\mathbf{D}_p\|$. Figure 11.3 shows the behavior of the algorithm. It can be seen that the first 18 iterations are feasibility iterations where we update α_k. At $k = 18$, $-\lambda_{\min}(\boldsymbol{\Phi}(\mathbf{Y}_k^{-1}))$ becomes negative and γ_k is updated. Then, for $k > 18$, the feasibility and optimality iterations alternate. As a result, the bound α_k on $\lambda_{\max}(\mathbf{X}_k\mathbf{Y}_k)$ converged to one and the cost bound γ_k converged to 7.77561. Thus we have

$$\mathbf{X}^{-1} \cong \mathbf{Y} = \begin{bmatrix} 5.7066 & -1.3896 & -2.2371 & -1.9297 \\ -1.3896 & 1.0544 & 0.7844 & 0.5429 \\ -2.2371 & 0.7844 & 1.2420 & 0.9183 \\ -1.9297 & 0.5429 & 0.9183 & 1.0556 \end{bmatrix}, \quad \lambda(\mathbf{XY}) = \begin{cases} 1.000010 \\ 1.000037 \\ 1.000044 \\ 1.000056 \end{cases}$$

$$\lambda_{\min}(\boldsymbol{\Phi}(\mathbf{X})) = 1.3626 \times 10^{-5}, \quad \lambda_{\min}(\boldsymbol{\Omega}(\mathbf{X}^{-1})) = -4.8012 \times 10^{-5}$$

$$\lambda_{\min}(\boldsymbol{\Phi}(\mathbf{Y}^{-1})) = 2.0180 \times 10^{-6}, \quad \lambda_{\min}(\boldsymbol{\Omega}(\mathbf{Y})) = 3.5281 \times 10^{-7}.$$

The matrix $\mathbf{X}$ satisfies $\boldsymbol{\Phi}(\mathbf{X}) > \mathbf{0}$ but $\boldsymbol{\Omega}(\mathbf{X}^{-1}) \not> \mathbf{0}$. On the other hand, the matrix $\mathbf{Y}$ satisfies $\boldsymbol{\Omega}(\mathbf{Y}) > \mathbf{0}$ and $\boldsymbol{\Phi}(\mathbf{Y}^{-1}) > \mathbf{0}$, and thus we have $(\mathbf{Y}^{-1}, \mathbf{Y}) \in \mathcal{L}(7.77561)$. Using $\mathbf{Y}^{-1}$ in place of $\mathbf{X}$ in the definition for $\boldsymbol{\Gamma}$, $\boldsymbol{\Lambda}$ and $\boldsymbol{\Theta}$ given in Theorem 9.1.1, a controller can be computed from the formula in (9.7) as

$$\mathbf{G} = \begin{bmatrix} -1.4427 \\ 8.8963 \end{bmatrix}.$$

We cannot claim that this is the optimal static output feedback controller, but we may say this is a satisfactory result since the value of the cost is very close to that for the LQG controller.

11.3 CONTROL DESIGN WITH FIXED CONTROLLER STRUCTURE

In this section, we consider a problem of designing controllers with some fixed structure. Specifically, control design specifications include the structure of the controller (e.g. decentralized structure, low controller order, etc.) as well as the usual stability, performance and robustness requirements. The corresponding computational problem is more difficult than the other two considered in the previous sections. We shall first formulate the fixed structure control design problem as a computational problem in the next section, then provide an algorithm to address this problem later.

11.3.1 Problem Formulation

Recall from Chapter 9 that many control design problems are reduced to the following type of linear algebra problems:

$$\boldsymbol{\Gamma}\mathbf{G}\boldsymbol{\Lambda} + (\boldsymbol{\Gamma}\mathbf{G}\boldsymbol{\Lambda})^T + \boldsymbol{\Theta} < \mathbf{0} \quad \text{(continuous-time case)} \tag{11.19}$$

$$(\boldsymbol{\Theta} + \boldsymbol{\Gamma}\mathbf{G}\boldsymbol{\Lambda})\mathbf{R}(\boldsymbol{\Theta} + \boldsymbol{\Gamma}\mathbf{G}\boldsymbol{\Lambda})^T < \mathbf{Q} \quad \text{(discrete-time case)} \tag{11.20}$$

where $\mathbf{G}$ is the controller parameter and the other matrices are defined in terms of the plant matrices and the Lyapunov matrix $\mathbf{X}$ or $\mathbf{Y}$, and possibly the scaling matrix $\mathbf{S}$. Our approach was to solve the above algebraic problems for the controller parameter $\mathbf{G}$, while fixing the Lyapunov matrix (and the scaling matrix). In this way, the parameter search for $(\mathbf{X}, \mathbf{G})$ can be replaced by that for $\mathbf{X}$, in which case, all controllers associated with a given Lyapunov matrix $\mathbf{X}$ are parametrized explicitly. This approach was possible if the controller has no structural constraints (although we can handle the fixed controller order case, as has been done in the previous sections). If we restrict the controller to have a certain structure $\mathbf{G} \in \mathcal{G}_s$, then the above approach does not work, and in general, we need to find $\mathbf{G} \in \mathcal{G}_s$ and $\mathbf{P} \in \mathcal{P}_s$ satisfying the above matrix inequality, where $\mathcal{P}_s$ is the set of (structured) Lyapunov matrices which possibly includes the scaling matrix as well as the original Lyapunov matrix $\mathbf{X}$.

As an example, consider the discrete-time SSUB μ-synthesis problem with a decentralized controller structure. This problem can be reduced to a search for $(\mathbf{P}, \mathbf{G}) \in \mathcal{P}_s \times \mathcal{G}_s$ satisfying

$$(\boldsymbol{\Theta} + \boldsymbol{\Gamma}\mathbf{G}\boldsymbol{\Lambda})\mathbf{P}(\boldsymbol{\Theta} + \boldsymbol{\Gamma}\mathbf{G}\boldsymbol{\Lambda})^T < \mathbf{P}$$

where

$$\mathcal{P}_s \triangleq \left\{ \begin{bmatrix} \mathbf{X} & \mathbf{0} \\ \mathbf{0} & \mathbf{S} \end{bmatrix} : \mathbf{X} > \mathbf{0}, \ \mathbf{S} \in \mathcal{S} \right\}$$

$$\mathcal{G}_s \triangleq \left\{ \text{block diag}(\mathbf{G}_1 \cdots \mathbf{G}_{n_g}) : \mathbf{G}_i \in \mathcal{R}^{n_{ai} \times n_{si}} \right\}$$

$$\left[\begin{array}{ccc}\Theta & \Gamma & \Lambda^T\end{array}\right] \triangleq \left[\begin{array}{cc|c|c} \mathbf{A} & \mathbf{D} & \mathbf{B} & \mathbf{M}^T \\ \mathbf{C} & \mathbf{F} & \mathbf{H} & \mathbf{E}^T \end{array}\right].$$

Now we state a general form of the fixed structure control design problem.

The Fixed Structure Control Problem:

For a set of structured Lyapunov matrices $\mathcal{P}_s$ and a set of fixed structure controllers $\mathcal{G}_s$, solve

$$\alpha^* \triangleq \min\{\alpha : \mathcal{M}(\alpha)\}$$

where $\mathcal{M}(\alpha) \subset \mathcal{P}_s \times \mathcal{G}_s$ is such that

$$\mathcal{P}(\mathbf{G}, \alpha) \triangleq \{\mathbf{P} : (\mathbf{P}, \mathbf{G}) \in \mathcal{M}(\alpha)\}$$

$$\mathcal{G}(\mathbf{P}, \alpha) \triangleq \{\mathbf{G} : (\mathbf{P}, \mathbf{G}) \in \mathcal{M}(\alpha)\}$$

are bounded convex sets characterized by LMIs.

In the above, the parameter α has been introduced to convert a certain feasibility problem to the above optimization problem. For instance, the discrete-time SSUB μ-synthesis problem can be converted to the above optimization problem with

$$\mathcal{M}(\alpha) \triangleq \{(\mathbf{P}, \mathbf{G}) : (\Theta + \Gamma\mathbf{G}\Lambda)\mathbf{P}(\Theta + \Gamma\mathbf{G}\Lambda)^T < \alpha\mathbf{P},\ \mathbf{P} \in \mathcal{P}_s,\ \mathbf{G} \in \mathcal{G}_s\} \quad (11.21)$$

where Θ, Γ and Λ are defined above. Note that a given pair $(\mathbf{P}, \mathbf{G})$ solves the SSUB μ-synthesis problem if and only if $(\mathbf{P}, \mathbf{G}) \in \mathcal{M}(\alpha)$ for some $\alpha \leq 1$. Thus, the problem is feasible if and only if the optimal value of the above minimization problem is $\alpha^* \leq 1$. Clearly, the set $\mathcal{P}(\mathbf{G}, \alpha)$ is characterized by an LMI and thus convex. Note that the set $\mathcal{G}(\mathbf{P}, \alpha)$ is in fact convex and can be characterized by an LMI as follows:

$$\mathcal{G}(\mathbf{P}, \alpha) = \left\{\mathbf{G} : \left[\begin{array}{cc} \alpha\mathbf{P} & \Theta + \Gamma\mathbf{G}\Lambda \\ (\Theta + \Gamma\mathbf{G}\Lambda)^T & \mathbf{P}^{-1} \end{array}\right] > \mathbf{0}\right\}.$$

The set $\mathcal{G}(\mathbf{P}, \alpha)$ is bounded if $\alpha > 0$, $\mathbf{P} > \mathbf{0}$, and Γ and Λ are of full column and row rank, respectively. These conditions are usually satisfied in practice. The set $\mathcal{P}(\mathbf{G}, \alpha)$ is not bounded in general. However, adding another constraint $tr(\mathbf{P}) \leq 1$ in $\mathcal{M}(\alpha)$, the set $\mathcal{P}(\mathbf{G}, \alpha)$ can be made bounded without loss of generality. Boundedness of $\mathcal{P}(\mathbf{G}, \alpha)$ and $\mathcal{G}(\mathbf{P}, \alpha)$ is required in the algorithm given below, to guarantee the existence of the analytic centers.

11.3.2 The VK-Centering Algorithm

This section provides an algorithm to address the Fixed Structure Control Problem defined in the previous section. The idea is very simple; noting the fact that the sets $\mathcal{P}(\mathbf{G}, \alpha)$ and $\mathcal{G}(\mathbf{P}, \alpha)$ are convex, we can alternately minimize α over $\mathbf{P}$ and $\mathbf{G}$;

$$\text{Initialize } k = 0, \mathbf{G}_0$$

$$(\mathbf{P}_k, \alpha_k) \triangleq \arg\min\{\alpha : \mathbf{P} \in \mathcal{P}(\mathbf{G}_k, \alpha)\}$$

$$(\mathbf{G}_{k+1}, \beta_{k+1}) \triangleq \arg\min\{\alpha : \mathbf{G} \in \mathcal{G}(\mathbf{P}_k, \alpha)\}.$$

In this case, each minimization problem is quasi-convex and the optimal values α_k and β_{k+1} are nonincreasing. Since the value of α is bounded below, the sequences converge.

As in the XY-centering Algorithm, we shall replace each minimization problem by the computation of the analytic center. The following function will be used:

$$\psi(\mathbf{P},\mathbf{G}) \triangleq \inf\{\alpha : (\mathbf{P},\mathbf{G}) \in \mathcal{M}(\alpha)\}.$$

The VK-centering Algorithm:

1. Choose a parameter θ such that $0 < \theta < 1$.
2. Find $\hat{\alpha}$, $\hat{\mathbf{G}}$ and $\hat{\mathbf{P}}$ such that $(\hat{\mathbf{P}},\hat{\mathbf{G}}) \in \mathcal{M}(\hat{\alpha})$ and let $\mathbf{G}_1 \triangleq \hat{\mathbf{G}}$, $\beta_1 \triangleq \hat{\alpha}$, and $k = 1$.
3. Compute the analytic center $\mathbf{P}_k$ and update α_k:

$$\mathbf{P}_k \triangleq \text{ac}\,\{\mathcal{P}(\mathbf{G}_k, \beta_k)\}$$

$$\alpha_k \triangleq (1-\theta)\psi(\mathbf{P}_k,\mathbf{G}_k) + \theta\beta_k$$

4. Compute the analytic center $\mathbf{G}_{k+1}$ and update β_{k+1}:

$$\mathbf{G}_{k+1} \triangleq \text{ac}\,\{\mathcal{G}(\mathbf{P}_k, \alpha_k)\}$$

$$\beta_{k+1} \triangleq (1-\theta)\psi(\mathbf{P}_k,\mathbf{G}_{k+1}) + \theta\alpha_k.$$

5. If $\beta_{k+1} - \psi(\mathbf{P}_k,\mathbf{G}_{k+1}) < \varepsilon$ for sufficiently small $\varepsilon > 0$, then stop. Otherwise, let $k \leftarrow k+1$ and go to STEP 3.

For the special case of discrete-time unstructured (but fixed-order) control design, we do not require iterative computations for finding the analytic center $\mathbf{G}_k$; we have the following *closed form* solution to the analytic center.

Theorem 11.3.1 *Let matrices* $\mathbf{B}$, $\mathbf{C}$, $\mathbf{D}$, $\mathbf{R}$ *and* $\mathbf{Q}$ *be given. Suppose* $\mathbf{B}^T\mathbf{B} > \mathbf{0}$, $\mathbf{C}\mathbf{C}^T > \mathbf{0}$, $\mathbf{R} > \mathbf{0}$ *and* $\mathbf{Q} > \mathbf{0}$, *and consider a matrix valued function*

$$\mathbf{F}(\mathbf{G}) \triangleq \mathbf{Q} - (\mathbf{BGC}+\mathbf{D})\mathbf{R}(\mathbf{BGC}+\mathbf{D})^T.$$

Suppose further that there exists a matrix $\mathbf{G}$ *such that* $\mathbf{F}(\mathbf{G}) > \mathbf{0}$. *Then, the optimization problem*

$$\psi \triangleq \max_{\mathbf{G}}\{\det\mathbf{F}(\mathbf{G}) : \mathbf{F}(\mathbf{G}) > \mathbf{0}\}$$

is well-posed, and has the unique maximizer

$$\mathbf{G}^* = -(\mathbf{B}^T\mathbf{\Phi}\mathbf{B})^{-1}\mathbf{B}^T\mathbf{\Phi}\mathbf{D}\mathbf{R}\mathbf{C}^T(\mathbf{C}\mathbf{R}\mathbf{C}^T)^{-1} \tag{11.22}$$

where

$$\begin{aligned}\mathbf{\Phi} &\triangleq \left(\mathbf{Q}-\mathbf{D}\mathbf{R}\mathbf{D}^T+\mathbf{D}\mathbf{R}\mathbf{C}^T\mathbf{R}_c\mathbf{C}\mathbf{R}\mathbf{D}^T\right)^{-1}\\ \mathbf{R}_c &\triangleq \left(\mathbf{C}\mathbf{R}\mathbf{C}^T\right)^{-1}.\end{aligned}$$

Moreover, the maximum value ψ *is given by*

$$\psi = \frac{\det\mathbf{\Psi}}{\det\mathbf{\Phi}\ \det\mathbf{R}_c}, \tag{11.23}$$

where

$$\mathbf{\Psi} \triangleq \mathbf{R}_c - \mathbf{R}_c\mathbf{C}\mathbf{R}\mathbf{D}^T\left(\mathbf{\Phi}-\mathbf{\Phi}\mathbf{B}\left(\mathbf{B}^T\mathbf{\Phi}\mathbf{B}\right)^{-1}\mathbf{B}^T\mathbf{\Phi}\right)\mathbf{D}\mathbf{R}\mathbf{C}^T\mathbf{R}_c.$$

Proof. Following the proof of Theorem 2.3.11, we have

$$\mathbf{F}(\mathbf{G}) = \mathbf{\Phi}^{-1} - (\mathbf{BG} + \mathbf{DRC}^T\mathbf{R}_c)\mathbf{R}_c^{-1}(\mathbf{BG} + \mathbf{DRC}^T\mathbf{R}_c)^T.$$

Then, using the determinant formula,

$$\begin{aligned} \det \mathbf{F}(\mathbf{G}) &= \frac{1}{\det \mathbf{R}_c} \det \begin{bmatrix} \mathbf{\Phi}^{-1} & \mathbf{BG} + \mathbf{DRC}^T\mathbf{R}_c \\ \mathbf{G}^T\mathbf{B}^T + \mathbf{R}_c\mathbf{CRD}^T & \mathbf{R}_c \end{bmatrix} \\ &= \frac{\det \mathbf{\Phi}^{-1}}{\det \mathbf{R}_c} \det[\mathbf{R}_c - (\mathbf{BG} + \mathbf{DRC}^T\mathbf{R}_c)^T \mathbf{\Phi}(\mathbf{BG} + \mathbf{DRC}^T\mathbf{R}_c)]. \end{aligned}$$

After expanding the term $[\cdot]$ and completing the square, we have

$$\det \mathbf{F}(\mathbf{G}) = \frac{\det[\mathbf{\Psi} - (\mathbf{G} - \mathbf{G}^*)^T(\mathbf{B}^T\mathbf{\Phi}\mathbf{B})(\mathbf{G} - \mathbf{G}^*)]}{\det \mathbf{\Phi} \ \det \mathbf{R}_c}$$

where Ψ is defined in (11.1) and $\mathbf{G}^*$ is given by (11.22). Noting that $\mathbf{\Psi} > \mathbf{0}$, $\mathbf{B}^T\mathbf{\Phi}\mathbf{B} > \mathbf{0}$ and

$$\mathbf{\Psi} > \mathbf{\Psi} - (\mathbf{G} - \mathbf{G}^*)^T(\mathbf{B}^T\mathbf{\Phi}\mathbf{B})(\mathbf{G} - \mathbf{G}^*) > \mathbf{0}$$

for all $\mathbf{G}$ such that $\mathbf{F}(\mathbf{G}) > \mathbf{0}$,[4] we conclude that the determinant of $\mathbf{F}(\mathbf{G})$ is maximum when $\mathbf{G} = \mathbf{G}^*$, and the maximum value is given by (11.23). □

Example 11.3.1 We consider a mechanical system consisting of two masses connected by a spring [148]. A state space realization of the system is given by

$$\begin{bmatrix} \dot{x}_1 \\ \dot{x}_2 \\ \ddot{x}_1 \\ \ddot{x}_2 \end{bmatrix} = \begin{bmatrix} 0 & 0 & 1 & 0 \\ 0 & 0 & 0 & 1 \\ -k/m_1 & k/m_1 & 0 & 0 \\ k/m_2 & -k/m_2 & 0 & 0 \end{bmatrix} \begin{bmatrix} x_1 \\ x_2 \\ \dot{x}_1 \\ \dot{x}_2 \end{bmatrix} + \begin{bmatrix} 0 \\ 0 \\ 1/m_1 \\ 0 \end{bmatrix} u,$$

where m_1 and m_2 are masses and k is the spring constant. The position of the mass m_i is denoted by $\mathbf{x}_i$, and u is the control force input. The sensor measures the position of m_2, that is, x_2. We choose the following values for the parameters:

$$m_1 = m_2 = 1, \quad k = 1.$$

The continuous-time system is first converted to a discrete-time system assuming the zero-order hold for the control input with sampling period $T = 1$ as follows:

$$\begin{bmatrix} \mathbf{A}_p & \mathbf{B}_p \\ \mathbf{M}_p & * \end{bmatrix} = \left[\begin{array}{cccc|c} 0.5780 & 0.4220 & 0.8492 & 0.1508 & 0.4610 \\ 0.4220 & 0.5780 & 0.1508 & 0.8492 & 0.0390 \\ -0.6985 & 0.6985 & 0.5780 & 0.4220 & 0.8492 \\ 0.6985 & -0.6985 & 0.4220 & 0.5780 & 0.1508 \\ \hline 0 & 1 & 0 & 0 & 0 \end{array}\right].$$

This system is unstable since all the eigenvalues of $\mathbf{A}_p$ are on the unit circle. Our objective here is to design an n_cth-order stabilizing controller with n_c being less than the plant order ($n_p = 4$).

To address this objective, we will try to minimize the spectral radius of the closed-loop "**A**" matrix. This is a special case of the Fixed Structure Control Problem defined above,

[4] See the proof of Theorem 2.3.11.

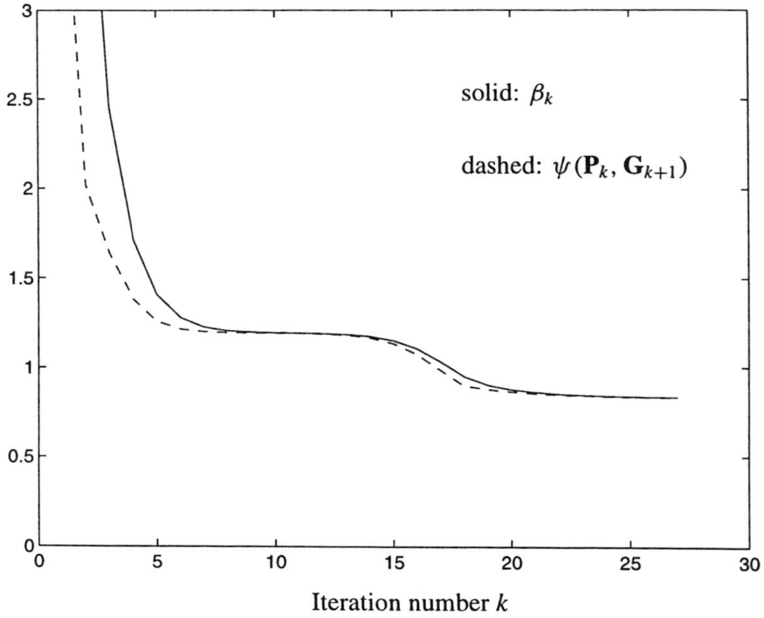

Figure 11.4 Convergence property of β_k.

with $\mathcal{M}(\alpha)$ given by (11.21) where $\mathcal{P}_s$ is the set of $(n_p + n_c) \times (n_p + n_c)$ positive definite matrices such that $\mathbf{P} \in \mathcal{P}_s$ implies $tr(\mathbf{P}) \leq 1$, $\mathcal{G}_s$ is the set of $(1+n_c)\times(1+n_c)$ unstructured matrices, and

$$[\ \mathbf{\Theta} \quad \mathbf{\Gamma} \quad \mathbf{\Lambda}^T\ ] \triangleq \left[\begin{array}{cc|cc|cc} \mathbf{A}_p & \mathbf{0} & \mathbf{B}_p & \mathbf{0} & \mathbf{M}_p^T & \mathbf{0} \\ \mathbf{0} & \mathbf{0} & \mathbf{0} & \mathbf{I}_{n_c} & \mathbf{0} & \mathbf{I}_{n_c} \end{array}\right].$$

Here, α is a (tight) upper bound on the square of the spectral radius of the closed-loop "**A**" matrix (that is, $\mathbf{A}_{c\ell} := \mathbf{\Theta}+\mathbf{\Gamma}\mathbf{G}\mathbf{\Lambda}$). Hence, the plant is stabilizable via n_cth-order controller if and only if $\alpha^* < 1$.

We choose $n_c = 2$ and apply the VK-centering Algorithm to obtain an upper bound on α^*, and corresponding controller **G** and Lyapunov matrix **P**. The algorithm generated decreasing sequences of α_k and β_k; the sequence β_k is plotted in Fig. 11.4 together with the curve for $\psi(\mathbf{P}_k, \mathbf{G}_{k+1})$.

The results are:

$$\mathbf{G} = \begin{bmatrix} \mathbf{D}_c & \mathbf{C}_c \\ \mathbf{B}_c & \mathbf{A}_c \end{bmatrix} = \left[\begin{array}{c|cc} -0.6942 & 0.3803 & -0.1565 \\ \hline 0.4832 & 0.4467 & -0.1839 \\ -0.1989 & -0.1839 & 0.0757 \end{array}\right]$$

$$\mathbf{P} = \begin{bmatrix} 9.2522 & 4.2627 & -1.6270 & -3.7541 & 8.7372 & -3.5965 \\ 4.2627 & 20.2459 & 2.2757 & -2.2217 & 6.1116 & -2.5156 \\ -1.6270 & 2.2757 & 5.7888 & -2.9487 & -2.4767 & 1.0196 \\ -3.7541 & -2.2217 & -2.9487 & 16.4480 & -11.2276 & 4.6215 \\ 8.7372 & 6.1116 & -2.4767 & -11.2276 & 18.9101 & -1.9036 \\ -3.5965 & -2.5156 & 1.0196 & 4.6215 & -1.9036 & 15.0693 \end{bmatrix} \times 10^{-2}.$$

The state space realization $\mathbf{G}$ for the controller has eigenvalues at 0.5224 and 0. The latter eigenvalue is in fact canceled by a zero at 0, and thus the controller becomes

$$\mathbf{C}(z) = -0.6942 \times \frac{z - 0.8319}{z - 0.5224}$$

which is of order one. With this controller, the eigenvalues of the resulting closed-loop system are

$$\lambda = \begin{cases} 0.7938 & (0.7938) \\ 0.7235 \pm 0.5485i & (0.9079) \\ 0.2832 \pm 0.8675i & (0.9125) \end{cases}$$

where the numbers in $(\cdot)$ indicate the magnitudes. Thus we have obtained a first-order stabilizing controller.

CHAPTER 11 SUMMARY

Control problems whose controller order is not fixed *a priori* often reduce to convex problems. This chapter gives a method for obtaining the numerical solution of each problem. The centering method goes to the center of each convex constraint set, and is quite different from the alternating projection method of Chapter 10, which goes to the *boundary* of each convex set on each iteration. As yet, there is no definitive example which can prove that one method of these two (Chapters 10 and 11) is better than the other.

Section 11.1 gives an algorithm that computes the globally optimal solution to a certain class of convex LMI problems, that includes those arising from control design with unspecified controller order. This algorithm is cited from [7]. The readers are warned that this is not necessarily the best algorithm for solving convex LMI optimization problems. Other algorithms include the potential reduction method [144], the projective method [93], positive definite programming [143], and the linear complementarity problem formulation [76].

The XY-centering Algorithm is cited from [63]. The idea for this algorithm is motivated by the min/max algorithm of [35] proposed for the fixed-order stabilization problem. The min/max algorithm has been extended for the linear quadratic suboptimal control problem in [64]. An algorithm similar to the XY-centering Algorithm (but with a different parameter space) has been proposed in [111] to deal with the constantly scaled $\mathcal{H}_\infty$ synthesis problem.

The VK-centering Algorithm is a conceptually simple way of treating control problems with design specifications related to input/output scaling and/or state similarity scaling (the Lyapunov function) "V" (while "K" is the controller). This simple idea has been used in the μ/k_m-synthesis literature (the DK iteration) [24, 116]. Other related results include [40, 57, 117].

APPENDIX ONE

Linear Algebra Basics

A.1 PARTITIONED MATRICES

Definition A.1.1 Consider the partitioned matrix

$$\mathbf{M} = \begin{bmatrix} \mathbf{A} & \mathbf{B} \\ \mathbf{C} & \mathbf{D} \end{bmatrix}.$$

Then (i) if $\mathbf{A}^{-1}$ exists, a *Schur complement* of $\mathbf{M}$ is defined as $\mathbf{D} - \mathbf{C}\mathbf{A}^{-1}\mathbf{B}$, and (ii) if $\mathbf{D}^{-1}$ exists, a *Schur complement* of $\mathbf{M}$ is defined as $\mathbf{A} - \mathbf{B}\mathbf{D}^{-1}\mathbf{C}$.

Theorem A.1.1 *Suppose* $\mathbf{A}$, $\mathbf{B}$, $\mathbf{C}$ *and* $\mathbf{D}$ *are all* $n \times n$ *matrices. Then:*

1 $\det\begin{bmatrix} \mathbf{A} & \mathbf{B} \\ \mathbf{C} & \mathbf{D} \end{bmatrix} = \det[\mathbf{A}]\det[\mathbf{D} - \mathbf{C}\mathbf{A}^{-1}\mathbf{B}]$ *provided* $\det[\mathbf{A}] \neq 0$

$\det\begin{bmatrix} \mathbf{A} & \mathbf{B} \\ \mathbf{C} & \mathbf{D} \end{bmatrix} = \det[\mathbf{D}]\det[\mathbf{A} - \mathbf{B}\mathbf{D}^{-1}\mathbf{C}]$ *provided* $\det[\mathbf{D}] \neq 0$.

2 If $\mathbf{A}$ *is an* $m \times n$ *matrix, and* $\mathbf{B}$ *is an* $n \times m$ *matrix, then*

$$\det[\mathbf{I}_m - \mathbf{A}\mathbf{B}] = \det[\mathbf{I}_n - \mathbf{B}\mathbf{A}].$$

3 If $\mathbf{A}$ *is invertible,* $\det[\mathbf{A}^{-1}] = \det[\mathbf{A}]^{-1}$.

Theorem A.1.2 *(Matrix Inversion Lemma) Suppose* $\mathbf{A}$, $\mathbf{B}$, $\mathbf{C}$ *and* $\mathbf{D}$ *are* $n \times n$, $n \times p$, $p \times p$ *and* $p \times n$ *matrices respectively. Assume* $\mathbf{A}^{-1}$ *and* $\mathbf{C}^{-1}$ *both exist. Then*

$$(\mathbf{A} + \mathbf{B}\mathbf{C}\mathbf{D})^{-1} = \mathbf{A}^{-1} - \mathbf{A}^{-1}\mathbf{B}(\mathbf{D}\mathbf{A}^{-1}\mathbf{B} + \mathbf{C}^{-1})^{-1}\mathbf{D}\mathbf{A}^{-1}. \qquad \text{(A.1.1)}$$

Theorem A.1.3 *Suppose* $\mathbf{A}$, $\mathbf{B}$, $\mathbf{C}$ *and* $\mathbf{D}$ *are* $n \times n$, $n \times p$, $p \times p$ *and* $p \times n$ *matrices respectively.*

1 Assume $\mathbf{A}^{-1}$ *exists. Then*

$$\begin{bmatrix} \mathbf{A} & \mathbf{B} \\ \mathbf{C} & \mathbf{D} \end{bmatrix}^{-1} = \begin{bmatrix} \mathbf{A}^{-1} + \mathbf{E}\mathbf{G}^{-1}\mathbf{F} & -\mathbf{E}\mathbf{G}^{-1} \\ -\mathbf{G}^{-1}\mathbf{F} & \mathbf{G}^{-1} \end{bmatrix} \qquad \text{(A.1.2)}$$

provided $\mathbf{G}^{-1}$ *exists, where*

$$\mathbf{G} = \mathbf{D} - \mathbf{CA}^{-1}\mathbf{B};\ \ \mathbf{E} = \mathbf{A}^{-1}\mathbf{B};\ \ \mathbf{F} = \mathbf{CA}^{-1}$$

2 Assume $\mathbf{D}^{-1}$ *exists. Then*

$$\begin{bmatrix} \mathbf{A} & \mathbf{B} \\ \mathbf{C} & \mathbf{D} \end{bmatrix}^{-1} = \begin{bmatrix} \mathbf{H}^{-1} & -\mathbf{H}^{-1}\mathbf{J} \\ -\mathbf{KH}^{-1} & \mathbf{D}^{-1} + \mathbf{KH}^{-1}\mathbf{J} \end{bmatrix} \tag{A.1.3}$$

provided $\mathbf{H}^{-1}$ *exists where*

$$\mathbf{H} \triangleq \mathbf{A} - \mathbf{BD}^{-1}\mathbf{C};\quad \mathbf{J} \triangleq \mathbf{BD}^{-1};\quad \mathbf{K} \triangleq \mathbf{D}^{-1}\mathbf{C}.$$

A.2 SIGN DEFINITENESS OF MATRICES

Definition A.2.1 A Hermitian matrix $\mathbf{Q} \in \mathcal{C}^{n\times n}$ is called *positive definite* if

$$\mathbf{x}^*\mathbf{Qx} > 0 \quad \textit{for all } \mathbf{x} \neq 0 \tag{A.2.1}$$

and *positive semidefinite* if

$$\mathbf{x}^*\mathbf{Qx} \geq 0 \quad \textit{for all } \mathbf{x} \tag{A.2.2}$$

and *negative definite* if

$$\mathbf{x}^*\mathbf{Qx} < 0 \quad \textit{for all } \mathbf{x} \neq 0 \tag{A.2.3}$$

and *negative semidefinite* if

$$\mathbf{x}^*\mathbf{Qx} \leq 0 \quad \textit{for all } \mathbf{x}. \tag{A.2.4}$$

Since the definiteness of the scalar $\mathbf{x}^*\mathbf{Qx}$ is a property only of the matrix $\mathbf{Q}$, we need a test for determining definiteness of a constant matrix $\mathbf{Q}$. Define a *principal submatrix* of a square matrix $\mathbf{K}$ as any square submatrix sharing some diagonal elements of $\mathbf{K}$.

Theorem A.2.1 *The constant Hermitian matrix* $\mathbf{K} \in \mathcal{C}^{n\times n}$ *is*

(a) Positive definite ($\mathbf{K} > 0$) *if either of these equivalent conditions holds:*

1 All eigenvalues of $\mathbf{K}$ *are positive*
2 All successive principal submatrices of $\mathbf{K}$ *(minors of successively increasing size) have positive determinants;*

(b) Positive semidefinite ($\mathbf{K} \geq 0$) *if either of these equivalent conditions holds:*

1 All eigenvalues of $\mathbf{K}$ *are zero or positive*
2 All principal submatrices of $\mathbf{K}$ *have zero or positive determinants;*

(c) Negative definite ($\mathbf{K} < 0$) *if either of these equivalent conditions holds:*

1 All eigenvalues of $(-\mathbf{K})$ *are positive*
2 All successive principal submatrices of $(-\mathbf{K})$ *have positive determinants;*

(d) Negative semidefinite ($\mathbf{K} \leq 0$) *if either of these equivalent conditions holds:*

1 All eigenvalues of $(-\mathbf{K})$ *are zero or positive*
2 All principal submatrices of $(-\mathbf{K})$ *have zero or positive determinants.*

Example A.2.1 Consider the matrix

$$\mathbf{A} = \begin{bmatrix} 1 & 2 & 1 \\ 2 & 4 & 2 \\ 1 & 2 & 0 \end{bmatrix}.$$

The eigenvalues of $\mathbf{A}$ are

$$\lambda_1 = -0.8541, \quad \lambda_2 = 0, \quad \lambda_3 = 5.8541.$$

Hence, $\mathbf{A}$ is an indefinite matrix (not a positive semidefinite or negative semidefinite matrix).

Note that the principal minors of $\mathbf{A}$ have the following determinants

$$1 > 0, \quad \det \begin{bmatrix} 1 & 2 \\ 2 & 4 \end{bmatrix} = 0, \quad \det(\mathbf{A}) = 0.$$

Hence, we see that non-negativeness of all successive principal submatrices is not enough to determine positive semidefiniteness of a matrix. We must check all principal submatrices. Hence

$$0 = 0, \quad \det \begin{bmatrix} 4 & 2 \\ 2 & 0 \end{bmatrix} \leq 0$$

indicates indefiniteness.

Exercise A.2.1 Show that, for any matrix $\mathbf{A}$,

(1) $\mathbf{A}^*\mathbf{A} \geq 0$
(2) $\mathbf{A}^*\mathbf{A} > 0$ if and only if $\mathbf{A}$ has linearly independent columns.

The following result provides conditions for the positive definiteness and semidefiniteness of a partitioned matrix in terms of its submatrices.

Lemma A.2.1 *The following three statements are equivalent:*

$$(i) \qquad \begin{bmatrix} \mathbf{A}_{11} & \mathbf{A}_{12} \\ \mathbf{A}_{12}^T & \mathbf{A}_{22} \end{bmatrix} > 0 \tag{A.2.5}$$

$$(ii) \quad \mathbf{A}_{22} > 0, \ \mathbf{A}_{11} - \mathbf{A}_{12}\mathbf{A}_{22}^{-1}\mathbf{A}_{12}^T > 0 \tag{A.2.6}$$

$$(iii) \quad \mathbf{A}_{11} > 0, \ \mathbf{A}_{22} - \mathbf{A}_{12}^T\mathbf{A}_{11}^{-1}\mathbf{A}_{12} > 0. \tag{A.2.7}$$

The following statements are equivalent:

$$(i) \qquad \begin{bmatrix} \mathbf{A}_{11} & \mathbf{A}_{12} \\ \mathbf{A}_{12}^T & \mathbf{A}_{22} \end{bmatrix} \geq 0$$

$$(ii) \quad \mathbf{A}_{11} \geq 0, \ \mathbf{A}_{22} - \mathbf{A}_{12}^T\mathbf{A}_{11}^{+}\mathbf{A}_{12} \geq 0, \ (\mathbf{I} - \mathbf{A}_{11}\mathbf{A}_{11}^{+})\mathbf{A}_{12} = 0$$

$$(iii) \quad \mathbf{A}_{22} \geq 0, \ \mathbf{A}_{11} - \mathbf{A}_{12}\mathbf{A}_{22}^{+}\mathbf{A}_{12}^T \geq 0, \ \mathbf{A}_{12}(\mathbf{I} - \mathbf{A}_{22}\mathbf{A}_{22}^{+}) = 0.$$

A.3 A LINEAR VECTOR SPACE

Thinking in an abstract manner, let $\mathcal{L}$ be a nonempty set of elements, and assume each pair of elements $\mathbf{x}$ and $\mathbf{y}$ in $\mathcal{L}$ can be combined by a process called *addition* to yield an element $\mathbf{z}$ in $\mathcal{L}$ denoted by

$$\mathbf{x} + \mathbf{y} = \mathbf{z}.$$

Definition A.3.1 A linear space $\mathcal{L}$ is defined by the following 8 properties:

1 $\mathbf{x} + \mathbf{y} = \mathbf{y} + \mathbf{x}$

2 $\mathbf{x} + (\mathbf{y} + \mathbf{z}) = (\mathbf{x} + \mathbf{y}) + \mathbf{z}$

3 There exists in $\mathcal{L}$ a *unique* element, denoted by $\mathbf{0}$, and called the *zero element* such that for all $\mathbf{x}$ in $\mathcal{L}$

$$\mathbf{x} + \mathbf{0} = \mathbf{x}$$

4 To each element $\mathbf{x}$ in $\mathcal{L}$ there corresponds a *unique* element in $\mathcal{L}$, denoted by $-\mathbf{x}$, and called the *negative* of $\mathbf{x}$, such that

$$\mathbf{x} + (-\mathbf{x}) = 0$$

Furthermore, each real (or complex) scalar α and each element $\mathbf{x}$ in $\mathcal{L}$ can be combined by a process called *scalar multiplication* to yield an element $\mathbf{y}$ in $\mathcal{L}$ denoted by $\mathbf{y} = \alpha\mathbf{x}$ such that:

5 $\alpha(\mathbf{x} + \mathbf{y}) = \alpha\mathbf{x} + \alpha\mathbf{y}$

Furthermore, for all scalars α and β

6 $(\alpha + \beta)\mathbf{x} = \alpha\mathbf{x} + \beta\mathbf{x}$

7 $(\alpha\beta)\mathbf{x} = \alpha(\beta\mathbf{x})$

8 $1 \cdot \mathbf{x} = \mathbf{x}$.

The algebraic system $\mathcal{L}$ which is defined by these two operations of *addition* and *scalar multiplication* which satisfy properties 1–8 is called a *linear space*. If all the scalars are *real*, then we refer to the *real* linear space $\mathcal{L}$; otherwise we refer to the *complex* linear space $\mathcal{L}$.

The set of all real n-dimensional vectors of the form

$$\mathbf{v}^T = [v_1 \ \ v_2 \ \ \cdots \ \ v_n]$$

satisfy properties 1–8 for real scalars α and so define a real linear n-dimensional space $\mathcal{L}$ which we denote by

$$\mathcal{L} = \mathcal{R}^n. \tag{A.3.1}$$

Definition A.3.2 A nonempty subset $\mathcal{S}$ of a linear space $\mathcal{L}$ is called a *linear subspace* of $\mathcal{L}$ if $\mathbf{x} + \mathbf{y}$ and $\alpha\mathbf{x}$ are in $\mathcal{S}$ whenever $\mathbf{x}$ and $\mathbf{y}$ are in $\mathcal{S}$ for any scalar α.

Since $\mathcal{S}$ is a nonempty set, $0 \cdot \mathbf{x} = \mathbf{0}$ and so the *zero element* $\mathbf{0}$ is in *any* subspace $\mathcal{S}$ of $\mathcal{L}$.

A set of elements $X = \{\mathbf{x}_1, \mathbf{x}_2, \cdots, \mathbf{x}_n\}$ is said to be a *spanning set* for a linear subspace $\mathcal{S}$ of $\mathcal{L}$ if every element $\mathbf{s}$ in $\mathcal{S}$ can be written as a linear combination of the $\mathbf{x}_k$. That is, we have

$$\mathcal{S} = \{\mathbf{s} \,\varepsilon\, \mathcal{L} \ : \ \mathbf{s} = \alpha_1\mathbf{x}_1 + \alpha_2\mathbf{x}_2 + \cdots + \alpha_n\mathbf{x}_n\} \tag{A.3.2}$$

for some scalars $\alpha_1, \alpha_2, \cdots, \alpha_n$.

A spanning set X is said to be a *basis* for $\mathcal{S}$ if no element $\mathbf{x}_k$ of the spanning set X of $\mathcal{S}$ can be written as a linear combination of the remaining elements $\mathbf{x}_1, \mathbf{x}_2, \cdots, \mathbf{x}_{k-1}, \mathbf{x}_{k+1}, \cdots, \mathbf{x}_n$, i.e. $\mathbf{x}_i$, $1 \leq i \leq n$ form a linearly independent set.

The subspace $\mathcal{S}$ in (A.3.2) is said to be of *dimension n* (or to be n-dimensional) if $\{\mathbf{x}_1, \mathbf{x}_2, \cdots, \mathbf{x}_n\}$ is a basis for $\mathcal{S}$. For any linear space $\mathcal{L}$, the *largest* subspace $\mathcal{S}$ is given by

$$\mathcal{S} = \mathcal{L}. \tag{A.3.3}$$

Example A.3.1 Suppose $\mathcal{L} = \mathcal{R}^n$. Then

$$X = \{\mathbf{e}_1, \mathbf{e}_2, \cdots, \mathbf{e}_n\}$$

where

$$\mathbf{e}_k^T = [0\,0\ \cdots\ 0\,1\,0\ \cdots\ 0]$$

has a 1 in the kth position and zeros elsewhere is a spanning set for $\mathcal{R}^n$. A three-dimensional subspace $\mathcal{S}_3$ of $\mathcal{L}$ is defined by

$$\mathcal{S} = \{\mathbf{s} \in \mathcal{R}^n \ :\ \mathbf{s} = \alpha_1\mathbf{e}_1 + \alpha_6\mathbf{e}_6 + \alpha_8\mathbf{e}_8\}$$

for any scalars α_1, α_6 and α_8.

The set of all $m \times n$ matrices of the form

$$\mathbf{A} = [a_{ij}]\ ;\quad 1 \leq i \leq m,\ 1 \leq j \leq n$$

satisfy properties 1–8 for real scalars α, and so also define a real linear space $\mathcal{L}$ which we denote by

$$\mathcal{L} = \mathcal{R}^{m\times n}. \tag{A.3.4}$$

Example A.3.2 Let $\mathcal{L}(n, n)$ denote the linear space of all real $n \times n$ matrices. Then the following sets are all linear subspaces of $\mathcal{L}(n, n)$:

1 $\quad \mathcal{S}_1 \triangleq \{\mathbf{A} \in \mathcal{L}(n, n)\ :\ \mathbf{A} = \mathbf{A}^T\}$

2 $\quad \mathcal{S}_2 \triangleq \{\mathbf{A} \in \mathcal{L}(n, n)\ :\ \mathbf{A} = -\mathbf{A}^T\}$

$\mathcal{S}_1$ is a subspace since (a) $\mathbf{0} = \mathbf{0}^T$ and so belongs to $\mathcal{S}_1$, and (b)

$$\mathbf{A} = \mathbf{A}^T,\ \ \mathbf{B} = \mathbf{B}^T \text{ implies } \mathbf{A} + \mathbf{B} = (\mathbf{A} + \mathbf{B})^T$$

For similar reasons, $\mathcal{S}_2$ is also a subspace. The following sets X are *not* subspaces of $\mathcal{L}$:

1 $\quad X = \{\mathbf{A} \in \mathcal{L}(n, n)\ :\ \det(\mathbf{A}) = 1\}$

2 $\quad X = \{\mathbf{A} \in \mathcal{L}(n, n)\ :\ \mathbf{A}^{-1} \text{ exists }\}.$

A.4 FUNDAMENTAL SUBSPACES OF MATRIX THEORY

The geometric ideas of linear vector spaces have led to the concepts of “spanning a space” and a “basis for a space”. The idea now is to define four important subspaces which are useful. The entire linear vector space of a specific problem can be decomposed into the sum of these subspaces.

A.4.1 Geometric Interpretations Definitions

Definition A.4.1 The *column space* of **A** is the space spanned by the columns of **A**, is also called the *range space* of **A**, denoted by $\mathcal{R}[\mathbf{A}]$. The *row space* of **A** is the space spanned by the rows of **A**.

Example A.4.1 Determine whether $\mathbf{y} = \begin{bmatrix} 1 \\ -1 \end{bmatrix}$ lies in the column space of

$$\mathbf{A} = \begin{bmatrix} 1 & -1 & -3 \\ 0 & 10 & 0 \end{bmatrix}.$$

Solution: The vector **y** lies in the column space of **A** if **y** can be expressed as a linear combination of the columns of **A**. In other words, we need to decide if

$$\mathbf{y} = \begin{bmatrix} 1 \\ -1 \end{bmatrix} = \begin{bmatrix} 1 \\ 0 \end{bmatrix} x_1 + \begin{bmatrix} -1 \\ 10 \end{bmatrix} x_2 + \begin{bmatrix} -3 \\ 0 \end{bmatrix} x_3 \quad \text{for some } x_1, x_2, x_3.$$

or equivalently to decide if

$$\mathbf{A}\mathbf{x} = \mathbf{y} \quad \text{for some } \mathbf{x}.$$

Since the columns of **A** span the entire two-dimensional space, the answer is *yes* for *any* y. In the present example, one can choose, for instance, $x_1 = 9/10$, $x_2 = -1/10$, $x_3 = 0$.

Example A.4.2 Does the vector $\mathbf{x} = \begin{bmatrix} 1 \\ 10 \\ 0 \end{bmatrix}$ lie within the row space of the **A** in Example A.4.1?

Solution: Note from the definition of row space that the row space of **A** is the column space of $\mathbf{A}^T$, denoted by $\mathcal{R}[\mathbf{A}^T]$. For *complex* matrices, we write $[\mathbf{A}^*]$. Thus, the question reduces to "Does **x** lie in the space spanned by the rows of **A**?" or "$\mathbf{A}^*\boldsymbol{\beta} = \mathbf{x}$ for some $\boldsymbol{\beta}$?"

$$\begin{bmatrix} 1 & 0 \\ -1 & 10 \\ -3 & 0 \end{bmatrix} \begin{bmatrix} \beta_1 \\ \beta_2 \end{bmatrix} = \begin{bmatrix} 1 \\ 10 \\ 0 \end{bmatrix}.$$

In three-dimensional space the row space of **A** is the plane spanned by the two vectors $\begin{bmatrix} 1 \\ -1 \\ -3 \end{bmatrix}$ and $\begin{bmatrix} 0 \\ 10 \\ 0 \end{bmatrix}$. The vector $\begin{bmatrix} 1 \\ 10 \\ 0 \end{bmatrix}$ is not in this plane. Hence, the answer is *no*.

Note from the scalar equations,

$$\begin{aligned} \beta_1 &= 1 \\ -\beta_1 + 10\beta_2 &= 10 \\ -3\beta_1 &= 0 \end{aligned}$$

that the first and third equations are contradictory.

Since the column rank of a matrix is the dimension of the space spanned by the columns and the row rank is the dimension of the space spanned by the rows, it is clear that the spaces $\mathcal{R}[\mathbf{A}]$ and $\mathcal{R}[\mathbf{A}^*]$ have the same dimension $r = \operatorname{rank} \mathbf{A}$.

Definition A.4.2 The right null space of $\mathbf{A}$ is the space spanned by all vectors $\mathbf{x}$ that satisfy $\mathbf{Ax} = \mathbf{0}$, and is denoted $\mathcal{N}[\mathbf{A}]$. The right null space of $\mathbf{A}$ is also called the *kernel* of $\mathbf{A}$. The left null space of $\mathbf{A}$ is the space spanned by all vectors $\mathbf{y}$ satisfying $\mathbf{y}^*\mathbf{A} = \mathbf{0}$. This space is denoted $\mathcal{N}[\mathbf{A}^*]$, since it is also characterized by all vectors $\mathbf{y}$ such that $\mathbf{A}^*\mathbf{y} = \mathbf{0}$.

Example A.4.3 Does $\mathbf{x} = \begin{bmatrix} 1 \\ 10 \\ 0 \end{bmatrix}$ lie in the right null space of the $\mathbf{A}$ in Example A.4.1?

Solution: The question reduces to deciding whether or not

$$\mathbf{Ax} = \mathbf{0}$$

Specifically,

$$\begin{bmatrix} 1 & -1 & -3 \\ 0 & 10 & 0 \end{bmatrix} \begin{bmatrix} 1 \\ 10 \\ 0 \end{bmatrix} = \begin{bmatrix} -9 \\ 100 \end{bmatrix} \neq \begin{bmatrix} 0 \\ 0 \end{bmatrix}$$

so the answer is *no*.

The dimensions of the four spaces $\mathcal{R}[\mathbf{A}]$, $\mathcal{R}[\mathbf{A}^*]$, $\mathcal{N}[\mathbf{A}]$, and $\mathcal{N}[\mathbf{A}^*]$ are to be determined now. Since $\mathbf{A}$ is $m \times n$, we have the following:

$$r \triangleq \operatorname{rank} \mathbf{A} = \text{ dimension of column space } \mathcal{R}[\mathbf{A}]$$

$$\dim \mathcal{N}[\mathbf{A}] \triangleq \text{ dimension of null space } \mathcal{N}[\mathbf{A}]$$

$$n \triangleq \text{ total number of columns of } \mathbf{A}.$$

Hence,

$$r + \dim \mathcal{N}[\mathbf{A}] = n$$

yields the dimension of the null space $\mathcal{N}[\mathbf{A}]$,

$$\boxed{\dim \mathcal{N}[\mathbf{A}] = n - r}.$$

Now, do the same for $\mathbf{A}^*$, using the fact that $\operatorname{rank} \mathbf{A} = \operatorname{rank} \mathbf{A}^*$,

$$r = \operatorname{rank}[\mathbf{A}^*] = \text{ dimension of row space } \mathcal{R}[\mathbf{A}^*]$$

$$\dim \mathcal{N}[\mathbf{A}^*] = \text{ dimension of left null space, } \mathcal{N}[\mathbf{A}^*]$$

$$m = \text{ total number of rows of } \mathbf{A}.$$

Hence,

$$r + \dim \mathcal{N}[\mathbf{A}^*] = m$$

yields

$$\boxed{\dim \mathcal{N}[\mathbf{A}^*] = m - r}.$$

These facts are summarized below:

$$\mathcal{R}[\mathbf{A}^*] = \text{ row space of } \mathbf{A}\text{: dimension } r,$$

$$\mathcal{N}[\mathbf{A}] = \text{ right null space of } \mathbf{A}\text{: dimension } n - r, \qquad \text{(A.4.1)}$$

$$\mathcal{R}[\mathbf{A}] = \text{ column space of } \mathbf{A}\text{: dimension } r,$$

$$\mathcal{N}[\mathbf{A}^*] = \text{ left null space of } \mathbf{A}\text{: dimension } m - r. \qquad \text{(A.4.2)}$$

Note from these facts that the entire n-dimensional space can be decomposed into the sum of the two subspaces $\mathcal{R}[\mathbf{A}^*]$ and $\mathcal{N}[\mathbf{A}]$. Note also that the entire m-dimensional space can be decomposed into the sum of the two subspaces $\mathcal{R}[\mathbf{A}]$ and $\mathcal{N}[\mathbf{A}^*]$.

Example A.4.4 For the $\mathbf{A}$ in Example A.4.1, express the vector $\mathbf{x} = \begin{bmatrix} 1 \\ 10 \\ 0 \end{bmatrix}$ as the sum of vectors in $\mathcal{R}[\mathbf{A}^*]$ and $\mathcal{N}[\mathbf{A}]$.

Solution:

$$\mathbf{x} = \alpha_1 \mathbf{x}^1 + \alpha_2 \mathbf{x}^2 + \alpha_3 \mathbf{x}^3$$

where $\mathbf{x}^i$, $i = 1, 2, 3$, are vectors in either $\mathcal{R}[\mathbf{A}^*]$ or $\mathcal{N}[\mathbf{A}]$. To find out how many are in each subspace, compute $\dim \mathcal{R}[\mathbf{A}^*] = \operatorname{rank} \mathbf{A} = 2$, and $\dim \mathcal{N}[\mathbf{A}] = 3 - 2 = 1$. Hence, $\mathbf{x}^1$ and $\mathbf{x}^2$ will be in $\mathcal{R}[\mathbf{A}^*]$ and $\mathbf{x}^3$ will be in $\mathcal{N}[\mathbf{A}]$. By definition, $\mathbf{x}^3$ satisfies

$$\mathbf{A}\mathbf{x}^3 = \mathbf{0} \quad \rightarrow \quad \left\{ \begin{array}{rcl} x_1^3 + x_2^3 - 3x_3^3 & = & 0 \\ 10x_2^3 & = & 0 \end{array} \right\}$$
$$\rightarrow \quad \mathbf{x}^3 = \begin{bmatrix} 3 \\ 0 \\ 1 \end{bmatrix}.$$

By definition, $\mathbf{x}^1$, $\mathbf{x}^2$ satisfy

$$\begin{aligned} \mathbf{A}^*\beta^1 &= \mathbf{x}^1 \\ \mathbf{A}^*\beta^2 &= \mathbf{x}^2 \end{aligned}$$

for some β^1, β^2. Independent $\mathbf{x}^1$ and $\mathbf{x}^2$ are obtained by the choice $\beta^1 = [1, 0]^*$, $\beta^2 = [0, 1]^*$:

$$\begin{aligned} \mathbf{A}^*\beta^1 &= \begin{bmatrix} 1 \\ -1 \\ -3 \end{bmatrix} = \mathbf{x}^1 \\ \mathbf{A}^*\beta^2 &= \begin{bmatrix} 0 \\ 10 \\ 0 \end{bmatrix} = \mathbf{x}^2. \end{aligned}$$

Now, $\mathcal{N}[\mathbf{A}]$ is one-dimensional and is spanned by $\mathbf{x}^3$, and $\mathcal{R}[\mathbf{A}^*]$ is two-dimensional and is spanned by $\mathbf{x}^1$, $\mathbf{x}^2$. Solve now for $\alpha_1, \alpha_2, \alpha_3$:

$$\mathbf{x} = \begin{bmatrix} 1 \\ 10 \\ 0 \end{bmatrix} = [\mathbf{x}^1 \mathbf{x}^2 \mathbf{x}^3] \begin{bmatrix} \alpha_1 \\ \alpha_2 \\ \alpha_3 \end{bmatrix} = \begin{bmatrix} 1 & 0 & 3 \\ 1 & 10 & 0 \\ -3 & 0 & 1 \end{bmatrix} \begin{bmatrix} \alpha_1 \\ \alpha_2 \\ \alpha_3 \end{bmatrix},$$

$$\begin{bmatrix} \alpha_1 \\ \alpha_2 \\ \alpha_3 \end{bmatrix} = \begin{bmatrix} 1 & 0 & 3 \\ 1 & 10 & 0 \\ -3 & 0 & 1 \end{bmatrix}^{-1} \begin{bmatrix} 1 \\ 10 \\ 0 \end{bmatrix} = \begin{bmatrix} \frac{1}{10} \\ \frac{99}{100} \\ \frac{3}{10} \end{bmatrix}.$$

Note from Example A.4.4 that $\mathbf{x}^3$ is orthogonal to both $\mathbf{x}^1$ and $\mathbf{x}^2$; that is

$$\mathbf{x}^{3*}\mathbf{x}^2 = 0, \quad \mathbf{x}^{3*}\mathbf{x}^1 = 0$$

The following shows that this orthogonality is no accident.

Theorem A.4.1 *$\mathcal{N}[\mathbf{A}]$ and $\mathcal{R}[\mathbf{A}^*]$ are orthogonal subspaces. This fact is denoted by $\mathcal{R}[\mathbf{A}^*]^{\perp} = \mathcal{N}[\mathbf{A}]$.*

Proof. The meaning of Theorem A.4.1 is that every vector in $\mathcal{N}[\mathbf{A}]$ is orthogonal to every vector in $\mathcal{R}[\mathbf{A}^*]$. To prove this, show that $\mathbf{x} \in \mathcal{N}[\mathbf{A}]$ satisfying

$$\mathbf{Ax} = \mathbf{0} \tag{A.4.3}$$

is orthogonal to $\mathbf{z} \in \mathcal{R}[\mathbf{A}^*]$ satisfying

$$\mathbf{A}^*\beta = \mathbf{z}. \tag{A.4.4}$$

This is accomplished as follows:

$$\mathbf{x}^*\mathbf{z} = \mathbf{x}^*\mathbf{A}^*\beta = (\mathbf{Ax})^*\beta = 0$$

where (A.4.4) is used to obtain the second equality and (A.4.3) is used to obtain the last equality. □

It should now be clear how to construct a basis for $\mathcal{R}[\mathbf{A}^*]$ and $\mathcal{N}[\mathbf{A}]$. Simply take the independent rows of $\mathbf{A}$ as a basis for $\mathcal{R}[\mathbf{A}^*]$ and take the vectors perpendicular to these independent rows of $\mathbf{A}$ as a basis for $\mathcal{N}[\mathbf{A}]$.

Example A.4.5 Graphically show the spaces $\mathcal{N}[\mathbf{A}]$ and $\mathcal{R}[\mathbf{A}^*]$ if $\mathbf{A} = \begin{bmatrix} 1 & 10 \\ 0 & 0 \end{bmatrix}$.

Solution: Since $r = \operatorname{rank} \mathbf{A} = 1$, both $\mathcal{N}[\mathbf{A}]$ and $\mathcal{R}[\mathbf{A}^*]$ are one-dimensional: $\dim \mathcal{N}[\mathbf{A}] = n - r = 2 - 1 = 1$, $\dim \mathcal{R}[\mathbf{A}^*] = r = 1$. The independent rows of $\mathbf{A}$ form a basis for $\mathcal{R}[\mathbf{A}^*]$. Any vector perpendicular to this is a basis for $\mathcal{N}[\mathbf{A}]$.

Corollary A.4.1 *$\mathcal{R}[\mathbf{A}]$ and $\mathcal{N}[\mathbf{A}^*]$ are orthogonal subspaces. This fact is denoted by $\mathcal{R}[\mathbf{A}]^{\perp} = \mathcal{N}[\mathbf{A}^*]$.*

Proof. Substitute $\mathbf{A}$ for $\mathbf{A}^*$ in Theorem A.4.1. □

Exercise A.4.1 Graphically show the spaces $\mathcal{R}[\mathbf{A}]$ and $\mathcal{N}[\mathbf{A}^*]$ if

$$\mathbf{A} = \begin{bmatrix} 1 & -10 \\ 1 & -10 \end{bmatrix}.$$

Express $\mathbf{x} = \begin{bmatrix} 1 \\ 7 \end{bmatrix}$ as the sum of vectors in $\mathcal{R}[\mathbf{A}]$ and $\mathcal{N}[\mathbf{A}^*]$. Does $\mathbf{x} = \begin{bmatrix} 1 \\ -1 \end{bmatrix}$ lie in the left null space of $\mathbf{A}$?

Exercise A.4.2 Find the set of *all* vectors $\mathbf{v}$ such that

$$\mathbf{Av} = \mathbf{0}$$

where

$$\text{(i)} \quad \mathbf{A} = \begin{bmatrix} 1 & 3 \\ 4 & 12 \end{bmatrix} \qquad \text{(ii)} \quad \mathbf{A} = \begin{bmatrix} 1 & 2 \\ 3 & 5 \\ 6 & 9 \end{bmatrix}$$

$$\text{(iii)} \quad \mathbf{A} = \begin{bmatrix} 4 & 1 & 6 \\ 0 & 1 & 3 \end{bmatrix}.$$

A.4.2 Construction of the Fundamental Subspaces by SVD

Consider an SVD of the $m \times n$ matrix $\mathbf{A}$ of rank r.

$$\mathbf{A} = [\mathbf{U}_1 \quad \mathbf{U}_2] \begin{bmatrix} \Sigma_1 & 0 \\ 0 & 0 \end{bmatrix} \begin{bmatrix} \mathbf{V}_1^* \\ \mathbf{V}_2^* \end{bmatrix} = \mathbf{U}\Sigma\mathbf{V}^* \tag{A.4.5}$$

where $\Sigma_1 > 0$ and

$$\mathbf{U}_1 \in \mathcal{C}^{m\times r}, \; \mathbf{U}_2 \in \mathcal{C}^{m\times(m-r)}, \; \Sigma_1 \in \mathcal{R}^{r\times r}$$
$$\mathbf{V}_1 \in \mathcal{C}^{n\times r}, \; \mathbf{V}_2 \in \mathcal{C}^{n\times(n-r)}.$$

Then the following interpretations of $\mathbf{U}_1$, $\mathbf{U}_2$, $\mathbf{V}_1$, $\mathbf{V}_2$ construct the four fundamental subspaces.

Theorem A.4.2 *1) $\mathbf{U}_1$ is an orthogonal basis for $\mathcal{R}[A]$, the range space of $\mathbf{A}$, and all matrices that lie in $\mathcal{R}[\mathbf{A}]$ are given by $\mathbf{U}_1\mathbf{K}_1$ for arbitrary $\mathbf{K}_1$ with r rows.*

2) $\mathbf{U}_2$ is an orthogonal basis for $\mathcal{N}[\mathbf{A}^*]$, the right null space of $\mathbf{A}^*$, and all matrices that lie in $\mathcal{N}[\mathbf{A}^*]$ are given by $\mathbf{U}_2\mathbf{K}_2$ for arbitrary $\mathbf{K}_2$ with $(m-r)$ rows.

3) $\mathbf{V}_1$ is an orthogonal basis for $\mathcal{R}[\mathbf{A}^*]$, the range space of $\mathbf{A}^*$, and all matrices that lie in $\mathcal{R}[\mathbf{A}^*]$ are given by $\mathbf{V}_1\mathbf{P}_1$ for arbitrary $\mathbf{P}_1$ with r rows.

4) $\mathbf{V}_2$ is an orthogonal basis for $\mathcal{N}[\mathbf{A}]$, the right null space of $\mathbf{A}$, and all matrices that lie in this space are given by $\mathbf{V}_2\mathbf{P}_2$ for arbitrary $\mathbf{P}_2$ with $(n-r)$ rows.

Remark A.4.1 The left null space of $\mathbf{A}$ is often denoted $\mathbf{A}^\perp$. Hence, $\mathbf{A}^\perp\mathbf{A} = \mathbf{0}$, and $\mathbf{A}^\perp = \mathbf{K}_2^*\mathbf{U}_2^*$ for any $(m-r) \times (m-r)$ matrix $\mathbf{K}_2$.

Proof.

1) The range space of $\mathbf{A}$ is spanned by the set of vectors $\mathbf{Ax}$ for arbitrary $\mathbf{x}$. Note that from the SVD $\mathbf{A} = \mathbf{U}_1\Sigma_1\mathbf{V}_1^*$, hence

$$\begin{aligned} \mathbf{Ax} &= \mathbf{U}_1\Sigma_1\mathbf{V}_1^*\mathbf{x} \\ &= \mathbf{U}_1\mathbf{z} \end{aligned}$$

for arbitrary $\mathbf{z}$ or $\mathbf{z} = \mathbf{K}_1\mathbf{y}$ for arbitrary $\mathbf{K}_1$, $\mathbf{y}$.

2) From the SVD

$$\mathbf{A}^* = [\mathbf{V}_1 \quad \mathbf{V}_2] \begin{bmatrix} \Sigma_1 & 0 \\ 0 & 0 \end{bmatrix} \begin{bmatrix} \mathbf{U}_1^* \\ \mathbf{U}_2^* \end{bmatrix}$$

and

$$\mathbf{A}^* [\mathbf{U}_1 \quad \mathbf{U}_2] = [\mathbf{V}_1 \quad \mathbf{V}_2] \begin{bmatrix} \Sigma_1 & 0 \\ 0 & 0 \end{bmatrix} = [\mathbf{V}_1\Sigma_1 \quad 0].$$

Hence

$$\mathbf{A}^*\mathbf{U}_2 = 0$$

and for arbitrary $\mathbf{K}_2$

$$\mathbf{A}^*\mathbf{U}_2\mathbf{K}_2 = 0.$$

3) The range space of $\mathbf{A}^*$ is shown to be spanned by $\mathbf{V}_1$ by applying the arguments in 1) to the SVD of $\mathbf{A}^*$.

4) From the SVD

$$\mathbf{A}\,[\mathbf{V}_1 \quad \mathbf{V}_2] = [\mathbf{U}_1 \quad \mathbf{U}_2]\begin{bmatrix} \Sigma_1 & 0 \\ 0 & 0 \end{bmatrix} = [\mathbf{U}_1\Sigma_1 \quad 0]\,.$$

Hence

$$\mathbf{A}\mathbf{V}_2\mathbf{P}_2 = 0$$

for arbitrary $\mathbf{P}_2$. □

It is clear from the unitary properties of $\mathbf{U}$ and $\mathbf{V}$

$$\mathbf{U}^*\mathbf{U} = \begin{bmatrix} \mathbf{U}_1^* \\ \mathbf{U}_2^* \end{bmatrix}[\mathbf{U}_1 \quad \mathbf{U}_2] = \begin{bmatrix} \mathbf{I} & 0 \\ 0 & \mathbf{I} \end{bmatrix}, \quad \mathbf{V}^*\mathbf{V} = \begin{bmatrix} \mathbf{V}_1^* \\ \mathbf{V}_2^* \end{bmatrix}[\mathbf{V}_1 \quad \mathbf{V}_2] = \begin{bmatrix} \mathbf{I} & 0 \\ 0 & \mathbf{I} \end{bmatrix}$$

that $\mathcal{R}[\mathbf{A}]$ and $\mathcal{N}[\mathbf{A}^*]$ are orthogonal since

$$(\mathbf{U}_1\mathbf{K}_1)^*(\mathbf{U}_2\mathbf{K}_2) = 0$$

and that $\mathcal{R}[\mathbf{A}^*]$ and $\mathcal{N}[\mathbf{A}]$ are orthogonal since

$$(\mathbf{V}_1\mathbf{P}_1)^*(\mathbf{V}_2\mathbf{P}_2) = 0.$$

Note that for any given matrix $\mathbf{B} \in \mathcal{C}^{m\times p}$ and any given unitary matrix $\mathbf{U} \in \mathcal{C}^{m\times m}$ there exists $\mathbf{K}$ such that

$$\mathbf{B} = \mathbf{U}\mathbf{K}.$$

Now suppose $\mathbf{U}$ is obtained from the SVD of any $m \times n$ matrix $\mathbf{A} \in \mathcal{C}^{m\times n}$ as in (A.4.5). Then $\mathbf{B}$ can be written, for some $\mathbf{K}_1, \mathbf{K}_2$,

$$\mathbf{B} = \mathbf{U}\mathbf{K} = [\mathbf{U}_1 \quad \mathbf{U}_2]\begin{bmatrix} \mathbf{K}_1 \\ \mathbf{K}_2 \end{bmatrix} = \mathbf{U}_1\mathbf{K}_1 + \mathbf{U}_2\mathbf{K}_2. \tag{A.4.6}$$

From Theorem 2.9.2 and (A.4.6) we can say that any $m \times p$ matrix $\mathbf{B}$ and any $m \times n$ matrix $\mathbf{A}$, (for any n) the matrix $\mathbf{B}$ can be decomposed into two orthogonal parts

$$\mathbf{B} = \mathbf{B}_1 + \mathbf{B}_2, \quad \mathbf{B}_1^*\mathbf{B}_2 = 0 \tag{A.4.7}$$

where, for some $\mathbf{K}_1, \mathbf{K}_2$,

$$\begin{aligned} \mathbf{B}_1 &= \mathbf{U}_1\mathbf{K}_1 \\ \mathbf{B}_2 &= \mathbf{U}_2\mathbf{K}_2. \end{aligned}$$

Equation (A.4.7) is often written as a direct sum of $\mathcal{R}[\mathbf{A}]$ and $\mathcal{N}[\mathbf{A}^*]$, using the notation

$$\mathbf{B} = \mathcal{R}[\mathbf{A}] \oplus \mathcal{N}[\mathbf{A}^*] \tag{A.4.8}$$

where the orthogonality of $\mathcal{R}[\mathbf{A}]$ and $\mathcal{N}[\mathbf{A}^*]$ has already been shown.

Likewise $\mathbf{B} \in \mathcal{C}^{m\times p}$ can be decomposed as the sum of the remaining two (of the four) fundamental subspaces. In the above decomposition we need not specify n, the column dimension of $\mathbf{A}$. In what follows we need not specify the row dimension. Let $\mathbf{A}$ be any $q \times m$ matrix, then any $m \times p$ matrix $\mathbf{B}$ can be written

$$\mathbf{B} = \mathbf{B}_3 + \mathbf{B}_4, \quad \mathbf{B}_3^*\mathbf{B}_4 = 0$$

where

$$\begin{aligned} \mathbf{B}_3 &= \mathbf{V}_1\mathbf{P}_1 \in \mathcal{R}[\mathbf{A}^*] \\ \mathbf{B}_4 &= \mathbf{V}_2\mathbf{P}_2 \in \mathcal{N}[\mathbf{A}] \\ \mathbf{A} &= [\mathbf{U}_1 \;\; \mathbf{U}_2] \begin{bmatrix} \Sigma_1 & 0 \\ 0 & 0 \end{bmatrix} \begin{bmatrix} \mathbf{V}_1^* \\ \mathbf{V}_2^* \end{bmatrix}. \end{aligned}$$

A.5 CONVEX SETS

In this section the basic definitions of convexity of sets and functions are provided. These concepts play a fundamental role in the computational techniques developed in the text.

Consider a finite dimensional vector space $\mathcal{L}$. Often, in this book, $\mathcal{L}$ will be a subspace of $\mathcal{C}^{n\times n}$ or $\mathcal{R}^{n\times n}$, for example the space of Hermitian or skew-Hermitian $n \times n$ matrices.

Definition A.5.1 A set $\mathcal{K}$ in $\mathcal{L}$ is said to be *convex* if for any two vectors $\mathbf{x}$ and $\mathbf{y}$ in $\mathcal{K}$ any vector of the form $(1-\lambda)\mathbf{x} + \lambda\mathbf{y}$ is also in $\mathcal{K}$ where $0 \leq \lambda \leq 1$.

This definition merely says that given two points in a convex set, the line segment between them is also in the set. Note in particular that subspaces and linear varieties (a linear variety is a translation of linear subspaces) are convex. Also the empty set is considered convex. The following facts provide important properties for convex sets. Their proofs may be found in [83].

Fact 1: Let $\mathcal{C}_i$, $i = 1, \ldots, m$, be a family of m convex sets in $\mathcal{L}$. Then the intersection $\mathcal{C}_1 \cap \mathcal{C}_2 \cap \cdots \cap \mathcal{C}_m$ is convex.

Fact 2: Let $\mathcal{C}$ be a convex set in $\mathcal{L}$ and $\mathbf{x}_0 \in \mathcal{L}$. Then the set $\{\mathbf{x}_0 + \mathbf{x} : \mathbf{x} \in \mathcal{C}\}$ is convex.

Exercise A.5.1 Is the statement provided in Fact 1 true when the intersection is replaced by union?

Exercise A.5.2 Show that sets $\mathcal{S}_1$, $\mathcal{S}_2$ defined in the Example A.3.2 are convex sets.

An important special case of a convex set is the convex cone.

Definition A.5.2 A set $\mathcal{K}$ in $\mathcal{L}$ is said to be a convex cone with vertex $\mathbf{x}_0$ if $\mathcal{K}$ is convex and $\mathbf{x} \in \mathcal{K}$ implies that $\mathbf{x}_0 + \lambda\mathbf{x} \in \mathcal{K}$ for any $\lambda \geq 0$.

An important class of convex cones for our purposes is the one defined by the positive semidefinite ordering of matrices, e.g. $\mathbf{A}_1 \geq \mathbf{A}_2 \geq \mathbf{A}_3$.

Theorem A.5.1 *Let $\mathbf{P}$ be a positive semidefinite $n \times n$ matrix. The set of matrices $\mathbf{X} \in \mathcal{C}^{n\times n}$ such that $\mathbf{X} \geq \mathbf{P}$ is a convex cone in $\mathcal{C}^{n\times n}$.*

Proof. Let $\mathbf{A}$ and $\mathbf{B}$ be $n \times n$ matrices such that $\mathbf{A} \geq \mathbf{P}$ and $\mathbf{B} \geq \mathbf{P}$. Then for any $\mathbf{x} \in \mathcal{C}^n$ we have

$$\mathbf{x}^*(\mathbf{A} - \mathbf{P})\mathbf{x} \geq 0$$

and

$$\mathbf{x}^*(\mathbf{B}-\mathbf{P})\mathbf{x} \geq 0.$$

Consider a scalar $0 \leq \lambda \leq 1$. Then

$$\begin{aligned}\mathbf{x}^*\{[(1-\lambda)\mathbf{A}+\lambda\mathbf{B}]-\mathbf{P}\}\mathbf{x} &= \mathbf{x}^*\{(1-\lambda)\mathbf{A}+\lambda\mathbf{B}-(1-\lambda)\mathbf{P}-\lambda\mathbf{P}\}\mathbf{x} \\ &= (1-\lambda)\mathbf{x}^*(\mathbf{A}-\mathbf{P})\mathbf{x}+\lambda\mathbf{x}^*(\mathbf{B}-\mathbf{P})\mathbf{x} \geq 0.\end{aligned}$$

Hence

$$(1-\lambda)\mathbf{A}+\lambda\mathbf{B} \geq \mathbf{P}$$

i.e. the set of matrices $\mathbf{X} \geq \mathbf{P}$ is convex. To show that this set is a convex cone note that for any $\lambda \geq 0$ and any $\mathbf{X} \geq \mathbf{P}$ then $\mathbf{P}+\lambda\mathbf{X} \geq \mathbf{P}$ since $\mathbf{X} \geq \mathbf{P} \geq 0$. □

Exercise A.5.3 Let $\mathbf{Y} \in \mathcal{C}^{k\times k}$ be a given *positive semidefinite* matrix and $\mathbf{C} \in \mathcal{C}^{k\times n}$, $k \leq n$, be a given matrix. Then show that the set of matrices $\mathbf{X} \in \mathcal{C}^{n\times n}$ such that $\mathbf{CXC}^* \geq \mathbf{Y}$ is a convex cone.

A.6 MATRIX INNER PRODUCTS AND THE PROJECTION THEOREM

Definition A.6.1 The *trace inner product* of two $k \times n$ matrices $\mathbf{A}$ and $\mathbf{B}$ is defined by $tr\mathbf{A}^*\mathbf{B}$ and is denoted by $\langle\mathbf{A}, \mathbf{B}\rangle$, i.e.

$$\langle\mathbf{A}, \mathbf{B}\rangle = tr\mathbf{A}^*\mathbf{B}.$$

Definition A.6.2 The *Frobenius norm* of a matrix $\mathbf{A}$ is defined by $[tr\mathbf{A}^*\mathbf{A}]^{1/2}$ and is denoted by $\|\mathbf{A}\|_F$, that is

$$\|\mathbf{A}\|_F = tr\mathbf{A}^*\mathbf{B} = [\langle\mathbf{A}, \mathbf{B}\rangle]^{1/2}.$$

Two matrices $\mathbf{A}$ and $\mathbf{B}$ will be called *orthogonal to each other* if their inner product is zero, i.e. $\langle\mathbf{A}, \mathbf{B}\rangle = tr\mathbf{A}^*\mathbf{B} = \mathbf{0}$.

Example A.6.1 Determine the conditions on α and β such that the following matrices

$$\begin{bmatrix}\alpha & 1 \\ 0 & -1 \\ \beta & \beta\end{bmatrix}, \quad \begin{bmatrix}0 & 1 \\ 5 & 1 \\ 0 & 1\end{bmatrix}$$

are orthogonal to each other with respect to the trace inner product.

The following result is the classical projection theorem onto a linear subspace.

Theorem A.6.1 *Consider a finite-dimensional vector space $\mathcal{L}$ and let $\mathcal{M}$ be a subspace of $\mathcal{L}$. For any vector $\mathbf{x}$ in $\mathcal{L}$ there exists a unique vector $\mathbf{y}_o$ in $\mathcal{M}$ such that $\|\mathbf{x}-\mathbf{y}_o\| \leq \|\mathbf{x}-\mathbf{y}\|$ for any vector $\mathbf{y}$ in $\mathcal{M}$. Furthermore, the vector $\mathbf{x}-\mathbf{y}_o$ is orthogonal to any vector in $\mathcal{M}$. The vector $\mathbf{y}_o$ is called the orthogonal projection of $\mathbf{x}$ onto the subspace $\mathcal{M}$ and is denoted by $\mathbf{y}_o = P_{\mathcal{M}}\mathbf{x}$.*

The generalized Projection Theorem onto convex sets is as follows.

Theorem A.6.2 *Consider a finite-dimensional vector space $\mathcal{L}$ and let $\mathcal{M}$ be a closed convex set in $\mathcal{L}$. For any vector $\mathbf{x}$ in $\mathcal{L}$ there exists a unique vector $\mathbf{y}_o$ in $\mathcal{M}$ such that $\|\mathbf{x} - \mathbf{y}_o\| \leq \|\mathbf{x} - \mathbf{y}\|$ for any vector $\mathbf{y}$ in $\mathcal{M}$. Furthermore, the vector $\mathbf{x} - \mathbf{y}_o$ satisfies $\langle \mathbf{x} - \mathbf{y}_o, \mathbf{y} - \mathbf{y}_o \rangle \leq 0$ for any vector in $\mathcal{M}$. The vector $\mathbf{y}_o$ is called the orthogonal projection of $\mathbf{x}$ onto the subspace $\mathcal{M}$ and is denoted by $\mathbf{y}_o = P_{\mathcal{M}}\mathbf{x}$.*

In this text we are interested in orthogonal projections of matrices onto convex matrix constraint sets. Expressions for the orthogonal projections onto some simple matrix constraint sets that are important for control design are obtained in Chapter 10.

APPENDIX TWO

Calculus of Vectors and Matrices

B.1 VECTORS

Definition B.1.1 The derivation of the real scalar valued function $f(\mathbf{v})$ of an n-dimensional real vector $\mathbf{v}$ where

$$\mathbf{v}^T = [v_1 \; v_2 \; \cdots \; v_n] \; ; \quad v_k \text{ real is defined by}$$

$$\frac{\partial f(\mathbf{v})}{\partial \mathbf{v}} \triangleq \left[\frac{\partial f(\mathbf{v})}{\partial v_1} \; \frac{\partial f(\mathbf{v})}{\partial v_2} \; \cdots \; \frac{\partial f(\mathbf{v})}{\partial v_n} \right]^T \tag{B.1.1}$$

where the partial derivations are defined by

$$\begin{aligned} \frac{\partial f(\mathbf{v})}{\partial v_k} &\triangleq \lim_{\Delta v_k \to 0} \frac{f(\mathbf{v}+\Delta\mathbf{v}) - f(\mathbf{v})}{\Delta v_k} \\ \Delta\mathbf{v}^T &\triangleq [0 \cdots \Delta v_k \cdots 0] . \end{aligned} \tag{B.1.2}$$

Exercise B.1.1 Show that

$$\frac{\partial}{\partial \mathbf{x}}(\mathbf{y}^T\mathbf{Q}\mathbf{x}) = \mathbf{Q}^T\mathbf{y} \quad \text{if } \mathbf{x}, \mathbf{y}, \mathbf{Q} \text{ are real} \tag{B.1.3}$$

$$\frac{\partial}{\partial \mathbf{x}_R}(\mathbf{y}^*\mathbf{Q}\mathbf{x}) = \mathbf{Q}^T\overline{\mathbf{y}} \quad \text{if } \mathbf{Q} \text{ is real, } \mathbf{x}, \mathbf{y} \text{ are complex (and } \mathbf{x} = \mathbf{x}_R + j\mathbf{x}_I) \tag{B.1.4}$$

$$\frac{\partial}{\partial \mathbf{x}}(\mathbf{x}^*\mathbf{Q}\mathbf{x}) = \mathbf{Q}^T\overline{\mathbf{x}} \quad \text{if } \mathbf{Q} \text{ is real, } \mathbf{x} \text{ is complex} \tag{B.1.5}$$

$$\frac{\partial}{\partial \overline{\mathbf{x}}}(\mathbf{x}^*\mathbf{Q}\mathbf{x}) = \mathbf{Q}\mathbf{x} \quad \text{if } \mathbf{Q} \text{ is real, } \mathbf{x} \text{ is complex.} \tag{B.1.6}$$

Suppose $f(\mathbf{x})$, is a real scalar function of a real vector $\mathbf{x} \in \mathcal{R}^n$. The first three terms of the Taylor's series expansion of $f(\mathbf{x})$ about $\mathbf{x}_0$ (in terms of $\delta\mathbf{x} \triangleq \mathbf{x} - \mathbf{x}_0$) are

$$\delta f(\mathbf{x}) \triangleq f(\mathbf{x}) - f(\mathbf{x}_0) = \sum_{\alpha=1}^{n} \frac{\partial f(\mathbf{x})}{\partial x_\alpha}\delta x_\alpha + \frac{1}{2}\sum_{\beta,\gamma=1}^{n} \frac{\partial^2 f(\mathbf{x})}{\partial x_\beta \partial x_\gamma}\delta x_\beta \delta x_\gamma . \tag{B.1.7}$$

By defining the gradient and variational vectors by (B.1.1),

$$\left[\frac{\partial f(\mathbf{x})}{\partial \mathbf{x}}\right]^T \triangleq \left[\frac{\partial f(\mathbf{x})}{\partial x_1} \cdots \frac{\partial f(\mathbf{x})}{\partial x_n}\right] \tag{B.1.8}$$

$$(\delta \mathbf{x})^T \triangleq [\delta x_1 \cdots \delta x_n] \tag{B.1.9}$$

and the second derivative, called the *Hessian* matrix, by

$$\frac{\partial^2 f(\mathbf{x})}{\partial \mathbf{x}^2} \triangleq \frac{\partial}{\partial \mathbf{x}}\left(\frac{\partial f}{\partial \mathbf{x}}\right)^T = \begin{bmatrix} \frac{\partial}{\partial x_1}\left(\frac{\partial f}{\partial \mathbf{x}}\right)^T \\ \vdots \\ \frac{\partial}{\partial x_n}\left(\frac{\partial f}{\partial \mathbf{x}}\right)^T \end{bmatrix} = \begin{bmatrix} \frac{\partial^2 f}{\partial x_1^2} & \cdots & \frac{\partial^2 f}{\partial x_1 \partial x_n} \\ \vdots & & \vdots \\ \frac{\partial^2 f}{\partial x_n \partial x_1} & \cdots & \frac{\partial^2 f}{\partial x_n^2} \end{bmatrix} \tag{B.1.10}$$

(B.1.7) may be written in the compact vector matrix form

$$\delta f(\mathbf{x}) = \left(\frac{\partial f}{\partial \mathbf{x}}\right)^T \delta \mathbf{x} + \frac{1}{2}\delta \mathbf{x}^T \left[\frac{\partial^2 f}{\partial \mathbf{x}^2}\right] \delta \mathbf{x}. \tag{B.1.11}$$

Suppose one wishes to choose $\mathbf{x}$ so as to minimize a scalar function $f(\mathbf{x})$. A necessary condition is that small perturbations (from the minimizing x) in $f(\mathbf{x})$ are not negative; $\delta f(\mathbf{x}) \geq 0$.

Exercise B.1.2 (i) Consider the function $f(\mathbf{x})$

$$f(\mathbf{x}) = x_1^2 + 2x_2^2 - 2x_1x_2 + x_1 + x_2 + 1$$

Construct the matrix (called the Hessian)

$$H_{ij} = \frac{\partial^2 f}{\partial x_i \partial x_j} \quad j, i = 1, 2$$

to get

$$H = \begin{bmatrix} 1 & -1 \\ -1 & 2 \end{bmatrix}.$$

Note that since

$$H_{ij} = \frac{\partial^2 f(x)}{\partial x_i \partial x_j} = \frac{\partial^2 f(x)}{\partial x_j \partial x_i} = H_{ji}$$

the Hessian is always symmetric for any twice differentiable $f(\mathbf{x})$.

(ii) Repeat this exercise (i) in vector notation by writing $f(\mathbf{x}) = \mathbf{x}^*\mathbf{Q}\mathbf{x} + \mathbf{b}^*\mathbf{x} + c$ (find $\mathbf{Q}$, $\mathbf{b}$, c first).

B.2 MATRICES

The previous section shows how to differentiate a scalar function of a vector with respect to the vector. This section shows how to differentiate a scalar function of a matrix with respect to the matrix.

The derivative of a scalar $f(\mathbf{A})$ with respect to a matrix $\mathbf{A} = [\mathbf{a}_1, \cdots, \mathbf{a}_n]$ is defined by

$$\frac{\partial f}{\partial \mathbf{A}} = \left[\frac{\partial f}{\partial \mathbf{a}_1}, \cdots, \frac{\partial f}{\partial \mathbf{a}_n}\right] = \begin{bmatrix} \frac{\partial f}{\partial A_{11}} & \cdots & \frac{\partial f}{\partial A_{1n}} \\ \vdots & \cdots & \vdots \\ \frac{\partial f}{\partial A_{n1}} & \cdots & \frac{\partial f}{\partial A_{nn}} \end{bmatrix}. \tag{B.2.1}$$

Note that

$$tr\mathbf{AB} = \sum_{\alpha=1}^{k}\sum_{\beta=1}^{n} A_{\alpha\beta}B_{\alpha\beta} = \sum_{\beta=1}^{n}\sum_{\alpha=1}^{k} B_{\beta\alpha}A_{\alpha\beta} = tr\mathbf{BA} \tag{B.2.2}$$

proves the identity

$$tr\mathbf{AB} = tr\mathbf{BA}. \tag{B.2.3}$$

Exercise B.2.1 Prove another useful identity

$$tr\mathbf{AB} = tr\mathbf{A}^T\mathbf{B}^T. \tag{B.2.4}$$

Now, the right-hand side of (B.2.3) can be expanded in terms of the columns of $\mathbf{A}$,

$$tr\mathbf{BA} = tr\begin{bmatrix} \mathbf{b}_1^* \\ \vdots \\ \mathbf{b}_n^* \end{bmatrix}[\mathbf{a}_1, \cdots, \mathbf{a}_n] = \sum_{\beta=1}^{n}\mathbf{b}_\beta^*\mathbf{a}_\beta \tag{B.2.5}$$

where $\mathbf{b}_i^*$ is defined as the ith row of $\mathbf{B}$. Equation (B.2.5) readily leads to the conclusions

$$\left[\frac{\partial(tr\mathbf{AB})}{\partial \mathbf{a}_\alpha}\right]^T = \mathbf{b}_\alpha^* \quad \text{or} \quad \frac{\partial(tr\mathbf{AB})}{\partial \mathbf{a}_\alpha} = \overline{\mathbf{b}}_\alpha. \tag{B.2.6}$$

See from (B.1.10) and (B.2.7) that the structure of $\partial(tr\mathbf{AB})/\partial\mathbf{A}$ is

$$\begin{aligned} \frac{\partial(tr\mathbf{AB})}{\partial \mathbf{A}} &= \left[\frac{\partial(tr\mathbf{AB})}{\partial \mathbf{a}_1}, \cdots, \frac{\partial(tr\mathbf{AB})}{\partial \mathbf{a}_n}\right] \\ &= [\overline{\mathbf{b}}_1, \cdots, \overline{\mathbf{b}}_n] = \mathbf{B}^T. \end{aligned}$$

This proves an identity worth remembering:

$$\frac{\partial(tr\mathbf{AB})}{\partial \mathbf{A}} = \mathbf{B}^T. \tag{B.2.7}$$

Exercise B.2.2 Prove that any $\mathbf{A} \in \mathcal{C}^{k\times n}$, $\mathbf{B} \in \mathcal{C}^{n\times k}$

$$\frac{\partial(tr\mathbf{AB})}{\partial \mathbf{B}} = \mathbf{A}^T. \tag{B.2.8}$$

Example B.2.1 Derive (B.2.5) by appealing directly to (B.2.2).
Solution

$$\begin{aligned} \left[\frac{\partial(tr\mathbf{AB})}{\partial \mathbf{A}}\right]_{\alpha\beta} &= \frac{\partial(tr\mathbf{AB})}{\partial A_{\alpha\beta}} = \frac{\partial}{\partial A_{\alpha\beta}}\left(\sum_{\omega=1}^{k}\sum_{\gamma=1}^{n} A_{\omega\gamma}B_{\gamma\omega}\right) \\ &= \sum_{\omega=1}^{k}\sum_{\gamma=1}^{n}\left(\frac{\partial A_{\omega\gamma}}{\partial A_{\alpha\beta}}\right)B_{\gamma\omega} = \sum_{\omega=1}^{k}\sum_{\gamma=1}^{n}\delta_{\alpha\omega}\delta_{\gamma\beta}B_{\gamma\omega}. \end{aligned}$$

Hence,

$$\left[\frac{\partial(tr\mathbf{AB})}{\partial\mathbf{A}}\right]_{\alpha\beta} = B_{\beta\alpha} \Rightarrow \frac{\partial(tr\mathbf{AB})}{\partial\mathbf{A}} = \mathbf{B}^T. \tag{B.2.9}$$

TRACE IDENTITIES

For convenient reference the following identities are recorded. Each identity can be derived by repeated application of (B.2.3), (B.2.4) and (B.2.7). The elements of **A** are assumed to be independent:

$$\frac{\partial(tr\mathbf{AB})}{\partial\mathbf{A}} = \frac{\partial(tr\mathbf{A}^T\mathbf{B}^T)}{\partial\mathbf{A}} = \frac{\partial(tr\mathbf{B}^T\mathbf{A}^T)}{\partial\mathbf{A}} = \frac{\partial(tr\mathbf{BA})}{\partial\mathbf{A}} = \mathbf{B}^T \tag{B.2.10}$$

$$\begin{aligned}\frac{\partial(tr\mathbf{BAC})}{\partial\mathbf{A}} &= \frac{\partial(tr\mathbf{B}^T\mathbf{C}^T\mathbf{A}^T)}{\partial\mathbf{A}} = \frac{\partial(tr\mathbf{C}^T\mathbf{A}^T\mathbf{B}^T)}{\partial\mathbf{A}} = \frac{\partial(tr\mathbf{ACB})}{\partial\mathbf{A}} \\ &= \frac{\partial(tr\mathbf{CBA})}{\partial\mathbf{A}} = \frac{\partial(tr\mathbf{A}^T\mathbf{B}^T\mathbf{C}^T)}{\partial\mathbf{A}} = \mathbf{B}^T\mathbf{C}^T\end{aligned} \tag{B.2.11}$$

$$\frac{\partial(tr\mathbf{A}^T\mathbf{BA})}{\partial\mathbf{A}} = \frac{\partial(tr\mathbf{BAA}^T)}{\partial\mathbf{A}} = \frac{\partial(tr\mathbf{AA}^T\mathbf{B})}{\partial\mathbf{A}} = (\mathbf{B}+\mathbf{B}^T)\mathbf{A}. \tag{B.2.12}$$

Using these basic ideas, a list of matrix calculus results are given below.

$$\begin{aligned}
\frac{\partial}{\partial\mathbf{X}}tr[\mathbf{AX}^T] &= \mathbf{A} \\
\frac{\partial}{\partial\mathbf{X}}tr[\mathbf{AXB}] &= \mathbf{A}^T\mathbf{B}^T \\
\frac{\partial}{\partial\mathbf{X}}tr[\mathbf{AX}^T\mathbf{B}] &= \mathbf{BA} \\
\frac{\partial}{\partial\mathbf{X}^T}tr[\mathbf{AX}] &= \mathbf{A} \\
\frac{\partial}{\partial\mathbf{X}^T}tr[\mathbf{AX}^T] &= \mathbf{A}^T \\
\frac{\partial}{\partial\mathbf{X}^T}tr[\mathbf{AXB}] &= \mathbf{BA} \\
\frac{\partial}{\partial\mathbf{X}^T}tr[\mathbf{AX}^T\mathbf{B}] &= \mathbf{A}^T\mathbf{B}^T \\
\frac{\partial}{\partial\mathbf{X}}tr[\mathbf{XX}] &= 2\mathbf{X}^T \\
\frac{\partial}{\partial\mathbf{X}}tr[\mathbf{XX}^T] &= 2\mathbf{X} \\
\frac{\partial}{\partial\mathbf{X}}tr[\mathbf{X}^n] &= n(\mathbf{X}^{n-1})^T \\
\frac{\partial}{\partial\mathbf{X}}tr[\mathbf{AX}^n] &= \left(\sum_{i=0}^{n-1}\mathbf{X}^i\mathbf{AX}^{n-1-i}\right)^T \\
\frac{\partial}{\partial\mathbf{X}}tr[\mathbf{AXBX}] &= \mathbf{A}^T\mathbf{X}^T\mathbf{B}^T + \mathbf{B}^T\mathbf{X}^T\mathbf{A}^T
\end{aligned}$$

$$\begin{aligned}
\frac{\partial}{\partial \mathbf{X}} tr[\mathbf{AXBX}^T] &= \mathbf{A}^T\mathbf{XB}^T + \mathbf{AXB} \\
\frac{\partial}{\partial \mathbf{X}} tr[\mathbf{X}^{-1}] &= -(\mathbf{X}^{-1}\mathbf{X}^{-1})^T = -(\mathbf{X}^{-2})^T \\
\frac{\partial}{\partial \mathbf{X}} tr[\mathbf{AX}^{-1}\mathbf{B}] &= -(\mathbf{X}^{-1}\mathbf{BAX}^{-1})^T \\
\frac{\partial}{\partial \mathbf{X}} \log \det[\mathbf{X}] &= (\mathbf{X}^{-1})^T \\
\frac{\partial}{\partial \mathbf{X}} \det[\mathbf{X}^T] &= \frac{\partial}{\partial \mathbf{X}} \det[\mathbf{X}] = (\det[\mathbf{X}])(\mathbf{X}^{-1})^T \\
\frac{\partial}{\partial \mathbf{X}} \det[\mathbf{X}^n] &= n(\det[\mathbf{X}]^n(\mathbf{X}^{-1})^T).
\end{aligned}$$

Exercise B.2.3 The elements of **A** are not always independent. Show for symmetric **A** that

$$\frac{\partial(tr\mathbf{AB})}{\partial \mathbf{A}} = \mathbf{B}^T + \mathbf{B} - \text{diag}[\mathbf{B}].$$

where $\text{diag}[\mathbf{B}] \triangleq \text{diag}[\cdots B_{ii} \cdots]$.

APPENDIX THREE

Balanced Model Reduction

Consider the H_∞ model reduction problem in Chapter 8. Obviously a suboptimal reduced-order model $\hat{\mathbf{G}}(s)$ exists to satisfy an H_∞ bound $\|\mathbf{G}(s) - \hat{\mathbf{G}}(s)\|_\infty < \gamma$ for some γ. We shall construct one such realization. Consider a realization with properties

$$
\begin{aligned}
\hat{\mathbf{G}}(s) = \mathbf{G}_b(s) &= \mathbf{C}_b(s\mathbf{I} - \mathbf{A}_b)^{-1}\mathbf{B}_b \\
\mathbf{0} &= \boldsymbol{\Sigma}^{1/2}\mathbf{A}_b^T + \mathbf{A}_b\boldsymbol{\Sigma}^{1/2} + \mathbf{B}_b\mathbf{B}_b^T \\
\mathbf{0} &= \mathbf{A}_b\boldsymbol{\Sigma}^{1/2} + \boldsymbol{\Sigma}^{1/2}\mathbf{A}_b^T + \mathbf{C}_b^T\mathbf{C}_b \\
\boldsymbol{\Sigma} &= \begin{bmatrix} \sigma_1 & & \\ & \ddots & \\ & & \sigma_{\hat{n}} \end{bmatrix}, \quad \gamma_B = 2(\sigma_{\hat{n}+1} + \ldots \sigma_n).
\end{aligned}
$$

Such a realization $(\mathbf{A}_b, \mathbf{B}_b, \mathbf{C}_b)$ is called a balanced realization. The following is an algorithm to compute a balanced realization of a stabilizable, detectable model $(\mathbf{A}, \mathbf{B}, \mathbf{C})$:

STEP 1: Solve for $\mathbf{X}$ from

$$\mathbf{0} = \mathbf{X}\mathbf{A}^T + \mathbf{A}\mathbf{X} + \mathbf{B}\mathbf{B}^T. \tag{C.1}$$

This is possible to do uniquely if and only if $\mathbf{A}$ has no eigenvalues that are symmetric about the j axis $(\lambda_j + \lambda_i \neq 0 \ \forall i, j)$ and if the controllable modes are stable.

STEP 2: Find the singular value decomposition of $\mathbf{X}$:

$$\mathbf{X} = \begin{bmatrix} \mathbf{U}_{11} & \mathbf{U}_{12} \end{bmatrix} \begin{bmatrix} \boldsymbol{\Sigma}_1 & \mathbf{0} \\ \mathbf{0} & \mathbf{0} \end{bmatrix} \begin{bmatrix} \mathbf{U}_{11}^T \\ \mathbf{U}_{12}^T \end{bmatrix} = \mathbf{U}_{11}\boldsymbol{\Sigma}_1\mathbf{U}_{11}^T. \tag{C.2}$$

Now, the columns of $\mathbf{U}_{11}$ span the controllable subspace and the columns of $\mathbf{U}_{12}$ span the uncontrollable subspace, and

$$\boldsymbol{\Sigma}_1 = \text{diag}\,\{\sigma_{11}, \ldots, \sigma_{1n_c}\}, \quad \mathbf{U}_{11} \in \mathcal{R}^{n_c \times n_c}. \tag{C.3}$$

STEP 3: Solve for $\mathbf{K}$:

$$\mathbf{0} = \mathbf{K}\mathbf{A} + \mathbf{A}^T\mathbf{K} + \mathbf{C}^T\mathbf{C}. \tag{C.4}$$

This is possible if $(\lambda_i + \lambda_j \neq \forall i, j)$ holds and if the observable modes are stable.

STEP 4: Find the singular value decomposition

$$\mathbf{T}_1^T\mathbf{K}\mathbf{T}_1 = \begin{bmatrix} \mathbf{U}_{21} & \mathbf{U}_{22} \end{bmatrix}\begin{bmatrix} \mathbf{\Sigma}_2 & \mathbf{0} \\ \mathbf{0} & \mathbf{0} \end{bmatrix}\begin{bmatrix} \mathbf{U}_{21}^T \\ \mathbf{U}_{22}^T \end{bmatrix} = \mathbf{U}_{21}\mathbf{\Sigma}_2\mathbf{U}_{22}^T, \quad \text{(C.5)}$$

where $\mathbf{T}_1 = \mathbf{U}_{11}\mathbf{\Sigma}_1^{1/2}$. Now, $\mathbf{U}_{21}$ columns span the subspace that is both controllable and observable; $\mathbf{U}_{22}$ columns span the controllable unobservable subspace; and

$$\mathbf{\Sigma} = \text{diag}\,\{\sigma_{21}, \ldots, \sigma_{2n_{c0}}\}, \quad \mathbf{U}_{21} \in \mathcal{R}^{n_c \times n_{c0}}. \quad \text{(C.6)}$$

A balanced realization is constructed as follows:

$$\begin{aligned} \dot{\mathbf{x}}_b &= \mathbf{A}_b\mathbf{x}_b + \mathbf{B}_b\mathbf{u}, \quad \mathbf{x}_b \in \mathcal{R}^{n_c \times n_{c0}} \\ \mathbf{y} &= \mathbf{C}_b\mathbf{x}_b \\ \mathbf{A}_b &= (\mathbf{\Sigma}_2^{1/4}\mathbf{U}_{21}^T\mathbf{\Sigma}_1^{-1/2}\mathbf{U}_{11}^T)\mathbf{A}(\mathbf{U}_{11}\mathbf{\Sigma}_1^{1/2}\mathbf{U}_{21}\mathbf{\Sigma}_2^{-1/4}) \\ \mathbf{B}_b &= (\mathbf{\Sigma}_2^{1/4}\mathbf{U}_{21}^T\mathbf{\Sigma}_1^{-1/2}\mathbf{U}_{11}^T)\mathbf{B} \\ \mathbf{C}_b &= \mathbf{C}(\mathbf{U}_{11}\mathbf{\Sigma}_1^{1/2}\mathbf{U}_{21}\mathbf{\Sigma}_2^{-1/4}). \end{aligned} \quad \text{(C.7)}$$

Note from STEP 2 that $\mathbf{\Sigma}_1$ contains all of the nonzero singular values of $\mathbf{X}$ and n_c is the number of controllable states (i.e. the controllable subspace is spanned by $\mathbf{U}_{11}$). Note from STEP 4 that $\mathbf{\Sigma}_2$ contains all of the nonzero singular values of $\mathbf{T}_1^T\mathbf{K}\mathbf{T}_1$, and hence n_{c0} is the dimension of the controllable states that are also observable (i.e. controllable observable subspace is spanned by $\mathbf{U}_{21}$).

Exercise C.1 If the original system $(\mathbf{A}, \mathbf{B}, \mathbf{C})$ is observable and controllable (i.e. if $n_{c0} = n_x$), show that $\mathbf{A}_b = \mathbf{E}_b^{-1}\mathbf{A}\mathbf{E}_b$, $\mathbf{B}_b = \mathbf{E}_b^{-1}\mathbf{B}$, $\mathbf{C}_b = \mathbf{C}\mathbf{E}_b$, where $\mathbf{E}_b = \mathbf{T}_1\mathbf{T}_2$, $\mathbf{T}_1 = \mathbf{U}_{11}\mathbf{\Sigma}_1^{1/2}$, and $\mathbf{T}_2 = \mathbf{U}_{21}\mathbf{\Sigma}_2^{-1/4}$.

Exercise C.2 Verify that (C.7) is both controllable and observable even if the triple $(\mathbf{A}, \mathbf{B}, \mathbf{C})$ is not controllable and observable.

References

[1] ANDERSON, B. D. O. and MOORE, J. B. (1990) *Optimal Control: Linear Quadratic Methods.* Prentice Hall, Englewood Cliffs, New Jersey, second edition.

[2] BARMISH, B. R. (1983) Stabilization of uncertain systems via linear control. *IEEE Trans. Automat. Control,* AC-28(8):848–850.

[3] BEN-ISRAEL, A. and GREVILLE, T. N. (1974) *Generalized Inverses: Theory and Applications.* Wiley, New York.

[4] BERNSTEIN, D. S. and HADDAD, W. M. (1990) Robust stability and performance analysis for state-space systems via quadratic Lyapunov bounds. *SIAM J. Matrix Anal. Appl.*, 11(2):239–271.

[5] BERNSTEIN, D. S. and HYLAND, D. C. (1985) Optimal projection maximum entropy stochastic modelling and reduced order design synthesis. *IFAC Workshop on Model Errors and Compensation.*

[6] BOYD, S. and BARRATT, C. H. (1991) *Linear Controller Design: Limits of Performance.* Prentice Hall, Englewood Cliffs, New Jersey.

[7] BOYD, S. and EL GHAOUI, L. (1993) Method of centers for minimizing generalized eigenvalues. *Linear Algebra Appl.*, 188,189:63–111.

[8] BOYD, S., EL GHAOUI, L., FERON, E. and BALAKRISHNAN, V. (1994) *Linear Matrix Inequalities in System and Control Theory.* SIAM Studies in Applied Mathematics.

[9] BOYLE, J. P. and DYKSTRA, R. L. (1986) A method for finding projections onto the intersection of convex sets in Hilbert space. *Lecture Notes in Statistics*, 37:28–47.

[10] BRYSON, A. E. and HO, Y.-C. (1975) *Applied Optimal Control.* Hemisphere, New York.

[11] CHEN, C.-T. (1984) *Linear System Theory and Design.* Holt, Rinehart and Winston, New York.

[12] CHENEY, W. and GOLDSTEIN, A. A. (1959) Proximity maps for convex set. *Proc. Amer. Math. Society*, 12(2):448–450.

[13] COHN, A. (1914) über die anzahl der wurzeln einer algebraischen gleichung in einem krreise. *Math. Zeit.*, 14:110–148.

[14] COLLINS, E. and SKELTON, R. (1987) A theory of state covariance assignment for discrete systems. *IEEE Trans. Automat. Contr.*, AC-32(1):35–41.

[15] CORLESS, M. (1992) Robust stability analysis and controller design with quadratic Lyapunov functions. *IEEE Workshop on Variable Structure and Lyapunov Control of Uncertain Dynamical Systems.*

[16] CORLESS, M. (1993) Control of uncertain nonlinear systems. *ASME J. Dymanic Sys. Meas. Control*, 115(2(B)):362–372.

[17] CORLESS, M. and GLIELMO, L. (1992) On the exponential stability of singularly perturbed systems. *SIAM J. Contr. Opt.*, 30(6):1338–1360.

[18] CORLESS, M. and LEITMANN, G. (1993) Bounded controllers for robust exponential convergence. *J. Optimization Theory Appl.*, 76(1):1–12.

[19] CORLESS, M., ZHU, G. and SKELTON, R. E. (1989) Robustness of covariance controllers. *Proc. IEEE Conf. Decision Contr.*, pages 2667–2672.

[20] DAVIS, C., KAHAN, W. M. and WEINBERGER, H. F. (1982) Norm-preserving dilation and their applications to optimal error bounds. *SIAM J. Numerical Analysis*, 19(3):445–469.

[21] DE VILLEMAGNE, C. and SKELTON, R. E. (1987) Model reductions using a projection formulation. *Int. J. Contr.*, 46:2141–2169.

[22] DE VILLEMAGNE, C. and SKELTON, R. E. (1988) Controller reduction using canonical interactions. *IEEE Trans. Automat. Contr.*, 33(AC-8):740–750.

[23] DOYLE, J. C. (1982) Analysis of feedback systems with structured uncertainties. *IEE Proc.*, 129, Part D(6):242–250.

[24] DOYLE, J. C. (1983) Synthesis of robust controllers and filters. *Proc. IEEE Conf. Decision Contr.*, pages 109–114.

[25] DOYLE, J. C., GLOVER, K., KHARGONEKAR, P. P. and FRANCIS, B. A. (1989) State-space solutions to standard H_2 and H_∞ control problems. *IEEE Trans. Automat. Contr.*, AC-34(8):831–847, August.

[26] DOYLE, J. C., PACKARD, A. and ZHOU, K. (1991) Review of LFTs, LMIs, and μ. *Proc. IEEE Conf. Decision Contr.*, pages 1227–1232.

[27] ENNS, D. (1984) Model reduction with balanced realizations: An error bound and a frequency weighted generalization. *Proc. IEEE Conf. Decision Contr.*, pages 127–132.

[28] FINSLER, P. (1937) Uber das vorkommen definiter und semidefiniter formen in scharen quadraticher formen. *Commentarii Mathematici Helvetici*, 1:19–28.

[29] FRANCIS, B. A. (1987) *A Course in H_∞ Control Theory*. Springer-Verlag, New York.

[30] FUJIOKA, H. and HARA, S. (1994) State covariance assignment problem with measurement noise: A unified approach based on a symmetric matrix equation. *Linear Algebra Appl.*, 203-204:579–605.

[31] FUJIWARA, M. (1926) über die algebraische gleichungen duren wurzeln in einem kreise oder in einer halbebene liegen. *Math. Zeit.*, pages 160–169.

[32] FURUTA, K. and PHOOJARUENCHANACHAI, S. (1990) An algebraic approach to discrete-time H_∞ control problems. *Proc. American Contr. Conf.*, 20:3067–3072.

[33] FURUTA, K. and WONGSAISUWAN, M. (1993) Closed-form solution to discrete-time LQ optimal control and disturbance attenuation. *Sys. Control Lett.*, 20:427–437.

[34] GAHINET, P. and APKARIAN, P. (1994) A linear matrix inequality approach to $\mathcal{H}_\infty$ control. *Int. J. Robust Nonlin. Contr.*, 4:421–448.

[35] GEROMEL, J. C., DE SOUZA, C. C. and SKELTON, R. E. (1994) LMI numerical solution for output feedback stabilization. *Proc. Amer. Contr. Conf.*, pages 40–44.

[36] GEVERS, M. and LI, G. (1993) *Parametrizations in Control, Estimation and Filtering Problems*. Springer-Verlag, New York.

[37] GLOVER, K. (1984) All optimal Hankel norm approximations of linear multivariable systems and L_∞-error bounds. *Int. J. Contr.*, 26:1115–1193.

[38] GLOVER, K. and DOYLE, J. (1988) State-space formulae for all stabilizing controllers that satisfy an H_∞ norm bound and relations to risk sensitivity. *Sys. Contr. Lett.*, 11:167–172.

[39] GLOVER, K. and LIMEBEER, D. (1983) Robust multivariable control system design using optimal reduced order plant models. *Proc. Amer. Contr. Conf.*, pages 644–648.

[40] GOH, K. C., TURAN, L., SAFONOV, M. G., PAPAVASSILOPOULOS, G. and LY, J. (1994) Biaffine matrix inequality properties and computational methods. *Proc. American Contr. Conf.*, pages 850–855.

[41] GOLUB, G. H. and VAN LOAN, C. F. (1989) *Matrix Computations*. Johns Hopkins, Baltimore.

[42] GRIGORIADIS, K. M. and SKELTON, R. E. (1993) Minimum energy covariance controllers. *Proc. IEEE Conf. Decision Contr.*, pages 823–824, December.

[43] GRIGORIADIS, K. M. and SKELTON, R. E. (1994) Alternating convex projection methods for covariance control design. *Int. J. Contr.*, 60(6):1083–1106.

[44] GRIGORIADIS, K. M. and SKELTON, R. E. (1996) Alternating convex projection methods for discrete-time covariance control design. *J. Optimization Theory Appl.*, 88(2):399–432, February.

[45] GRIGORIADIS, K. M. and SKELTON, R. E. (1996) Low-order control design for LMI problems using alternating projection methods. *Automatica*, 32(8):1117–1125.

[46] GUBIN, L. G., POLYAK, B. T. and RAIK, E. V. (1967) The method of projections for finding the common point of convex sets. *USSR Comp. Math. Phys.*, 7:1–24.

[47] HADDAD, W. M. and BERNSTEIN, D. S. (1991) Robust stabilization with positive real uncertainty: Beyond the small gain theorem. *Sys. Contr. Lett.*, 17:191–208.

[48] HAN, S. P. (1988) A successive projection method. *Math. Program.*, 40:1–14.

[49] HIGHAM, N. J. (1988) Computing the nearest symmetric positive semidefinite matrix. *Linear Algebra Appl.*, 103:103–118.

[50] HORN, R. A. and JOHNSON, C. R. (1990) *Matrix Analysis.* Cambridge University Press, Cambridge.

[51] HOROWITZ, I. (1963) *Synthesis of Feedback Systems.* Academic Press, New York.

[52] HOTZ, A. and SKELTON, R. E. (1987) Covariance control theory. *Int. J. Contr.*, 46(1):13–32.

[53] HSIEH, C. and SKELTON, R. E. (1990) All covariance controllers for linear discrete-time systems. *IEEE Trans. Automat. Contr.*, AC-35(8):908–915.

[54] HURWITZ, A. (1895) über die bedingungen unter welchen eine gleichung nur wurzeln mit negativen reellen teilen besitz. *Math. Annalen.*, 46:273–284.

[55] IWASAKI, T. (1993) A unified matrix inequality approach to linear control design. *Ph.D Dissertation, Purdue University, West Lafayette, IN 47907*, December.

[56] Iwasaki, T. (1996) Robust performance analysis for systems with norm-bounded time-varying structured uncertainty. *Int. J. Robust Nonlinear Contr.*, 6:85–99.

[57] IWASAKI, T. and ROTEA, M. A. (1995) Fixed order scaled $\mathcal{H}_\infty$ synthesis. *Optimal Control Theory and Applications*. Submitted.

[58] IWASAKI, T. and SKELTON, R. E. (1993) A complete solution to the the general H_∞ control problem: LMI existence conditions and state space formulas. *Proc. American Contr. Conf.*, pages 605–609.

[59] IWASAKI, T. and SKELTON, R. E. (1994) All controllers for the general H_∞ control problem: LMI existence conditions and state space formulas. *Automatica*, 30(8):1307–1317.

[60] IWASAKI, T. and SKELTON, R. E. (1994) On the observer-based structure of covariance controllers. *Sys. Contr. Lett.*, 22:17–25.

[61] IWASAKI, T. and SKELTON, R. E. (1994) A unified approach to fixed order controller design via linear matrix inequalities. *Proc. American Contr. Conf.*, pages 35–39.

[62] IWASAKI, T. and SKELTON, R. E. (1995) Parametrization of all stabilizing controllers via quadratic Lyapunov functions. *J. Optimiz. Theory Appl.*, 85:291–307.

[63] IWASAKI, T. and SKELTON, R. E. (1995) The XY-centering algorithm for the dual LMI problem: a new approach to fixed order control design. *Int. J. Contr.*, 62:1257–1272.

[64] IWASAKI, T., SKELTON, R. E. and GEROMEL, J. C. (1994) Linear quadratic suboptimal control with static output feedback. *Sys. Contr. Lett.*, 23:421–430.

[65] JACOBSON, D. H. (1977) *Extensions of Linear-Quadratic Control, Optimization and Matrix Theory.* Academic Press, New York.

[66] JURY, E. I. (1974) *Inners and Stability of Dynamic Systems.* Wiley-Interscience, New York.

[67] KAILATH, T. (1980) *Linear Systems.* Prentice Hall, New Jersey.

[68] KALMAN, R. E. and BERTRAM, J. E. (1960) Control system analysis and design via the second method of lyapunov i: continuous-time systems. *J. Basic Engineering*, 82:371–393, June.

[69] KEEL, L. H., BHATTACHARYYA, S. P. and Howze, J. W. (1988) Robust control with structured perturbations. *IEEE Trans. Automat. Contr.*, AC-33(1):68–78, January.

[70] KHARGONEKAR, P. P., PETERSEN, I. R. and ROTEA, M. A. (1988) H_∞ optimal control with state feedback. *IEEE Trans. Automat. Contr.*, AC-33(8):786–788.

[71] KHARGONEKAR, P. P., PETERSEN, I. R. and ZHOU, K. (1990) Robust stabilization of uncertain linear systems: Quadratic stabilizability and H_∞ control theory. *IEEE Trans. Automat. Contr.*, AC-35(3):356–361.

[72] KHATRI, C. G. and MITRA, S. K. (1976) Hermitian and nonnegative definite solutions of linear matrix equations. *SIAM J. Appl. Math.*, 14(4):579–585.

[73] KIMURA, T. (1994) A fault tolerant controller design using alternating projection techniques. *M.S. Thesis, Purdue University, West Lafayette, IN 47907*, December.

[74] KING, A. M., DESAI, U. B. and SKELTON, R. E. (1988) A generalized approach to q-markov covariance equivalent realizations of discrete systems. *Automatica*, pages 507–515.

[75] KLEMA, V. C. and LAUB, A. J. (1980) The singular value decomposition: Its computation and some applications. *IEEE Trans. Automat. Contr.*, 25(2):164–176.

[76] KOJIMA, K., SHINDOH, S. and HARA, S. (1994) Interior-point methods for the monotone linear complementarity problem in symmetric matrices. *Tech. Rep., Dept. Info. Sci., Tokyo Inst. Tech.*

[77] KWAKERNAAK, H. and SIVAN, R. (1972) *Linear Optimal Control Systems.* Wiley-Interscience, New York.

[78] LASALLE, J. P. (1986) *The Stability and Control of Discrete Processes.* Springer-Verlag, Berlin.

[79] LEITMANN, G. (1979) Guaranteed asymptotic stability for some linear systems with bounded uncertainties. *J. Dyn. Sys., Meas. Contr.*, 101:202–216.

[80] LIÈENARD, A. and CHIPART, M. H. (1914) Sur le signe de la partie rèele des racines d'une equation albebrique. *J. Math Pures et Appl.*, 10:291–346.

[81] LIU, Y. (1989) Frequency-weighted controller and model order reduction in linear systems design. *Ph.D. dissertation, Australian National University.*

[82] LU, W. M. and DOYLE, J. C. (1992) H_∞ control of LFT systems: An LMI approach. *Proc. IEEE Conf. Decision Contr.*, pages 1997–2001.

[83] LUENBERGER, D. G. (1968) *Optimization by Vector Space Methods.* John Wiley, New York.

[84] LYAPUNOV, A. M. (1907) Problème gènèral de la stabilitè du mouvement. *Ann. Frac. Sci. Toulouse*, 9:203–474.

[85] MADIWALE, A., HADDAD, W. and BERNSTEIN, D. (1989) Robust H_∞ control design for systems with structured parametric uncertainty. *Sys. Contr. Lett.*, 12:393–407.

[86] MAGNUS, J. R. (1983) L-structured matrices and linear matrix equations. *Lin. Multilin. Algebra*, 14:67–88.

[87] MANSOUR, M. (1965) Stability criteria of linear systems and the second method of Lyapunov. *Scientia Electrica*, XI(Fasc. 3):87–96.

[88] MASSERA, J. L. (1956) Contributions to stability theory. *Ann. Math.*, 64:182–206.

[89] MEGRETSKI, A. (1993) Necessary and sufficient conditions of stability: a multiloop generalization of the circle criterion. *IEEE Trans. Automat. Contr.*, AC-38(5):753–756.

[90] MOORE, B. (1981) Principal component analysis in linear systems: Controllability, observability and model reduction. *IEEE Trans. Automat. Contr.*, AC-26(1):17–31.

[91] NAEIJE, W. J. and BOSGRA, O. H. (1977) The design of dynamic compensators for linear multivariable systems. *IFAC, Fredericton, Canada*, pages 205–212.

[92] NAGAYASU, M. (1977) Realization of prescribed state covariance for linear state feedback control systems with disturbances. *Technical Report of National Aerospace Laboratory*, TR-492:1–15. In Japanese.

[93] NEMIROVSKII, A. and GAHINET, P. (1994) The projective method for solving linear matrix inequalities. *Proc. American Contr. Conf.*, pages 840–844.

[94] OHARA, A. and KITAMORI, T. (1990) Geometric structures of stable state feedback systems. *Proc. IEEE Conf. Decision Contr.*, pages 2494–2499.

[95] OHARA, A. and KITAMORI, T. (1993) Geometric structures of stable state feedback systems. *IEEE Trans. Automat. Contr.*, 38:1579–1583.

[96] PACKARD, A. and DOYLE, J. (1993) The complex structured singular value. *Automatica*, 29(1):71–109.

[97] PACKARD, A. and DOYLE, J. C. (1990) Quadratic stability with real and complex perturbations. *IEEE Trans. Automat. Contr.*, AC-35(2):198–201.

[98] PACKARD, A., ZHOU, K., PANDEY, P. and BECKER, G. (1991) A collection of robust control problems leading to LMIs. *Proc. IEEE Conf. Decision Contr.*, pages 1245–1250.

[99] PARKS, P. C. (1963) Further comment on Ralston (1962). *IEEE Trans. Automatic Control*, AC-8(3):270–271.

[100] PARKS, P. C. (1963) A new proof of the Hurwitz stability criterion by the second method of Lyapunov with applications to optimum transfer functions. *IEEE Trans. Automatic Control*, AC-9(3):319–322.

[101] PARKS, P. C. (1964) Lyapunov and the Schur-Cohn stability criterion. *IEEE Trans. Automatic Control*, AC-9(1):121.

[102] PARROTT, S. (1978) On a quotient norm and the Sz-Nagy Foias lifting theorem. *J. Functional Analysis*, 30:311–328.

[103] PENROSE, R. A. (1955) A generalized inverso for matrices. *Proc. Cambridge Phil. Soc.*, 52:17–19.

[104] PETERSEN, I. R. (1987) Disturbance attenuation and H_∞ optimization: A design method based on the algebraic Riccati equation. *IEEE Trans. Automat. Contr.*, AC-32(5):427–429, May.

[105] PETERSEN, I. R. and HOLLOT, C. V. (1986) A Riccati equation approach to the stabilization of uncertain linear systems. *Automatica*, 22(4):397–411.

[106] RALSTON, A. (1962) A symmetric matrix formulation of the Hurwitz-Routh criterion. *IEEE Trans. Automatic Control*, AC-7:50–51.

[107] RAN, A. C. M. and VREUGDENHIL, R. (1988) Existence and comparison theorems for algebraic Riccati equations for continuous and discrete time systems. *Linear Algebra Appl.*, 99:63–83.

[108] RAO, C. R. and MITRA, S. K. (1971) *Generalized inverse of matrices and its applications*. Wiley, New York.

[109] ROTEA, M. A. (1993) The generalized H_2 control problem. *Automatica*, 29(2):373–386.

[110] ROTEA, M. A., CORLESS, M., DA, D. and PETERSEN, I. R. (1993) Systems with structured uncertainty: relations between quadratic and robust stability. *IEEE Trans. Automat. Contr.*, AC-38(5):799–803.

[111] ROTEA, M. A. and IWASAKI, T. (1994) An alternative to the D-K iteration? *Proc. American Contr. Conf.*, pages 53–57.

[112] ROTEA, M. A. and KHARGONEKAR, P. P. (1991) H_2-optimal control with an H_∞-constraint: The state feedback case. *Automatica*, 27(2):307–316.

[113] ROUTH, E. J. (1905) *Dynamics of a System of Rigid Bodies*. 6th edition, Macmillan, London.

[114] RUDIN, W. (1976) *Principles of Mathematical Analysis*. McGraw-Hill, New York.

[115] SAFONOV, M. G. 1982 Stability margins of diagonally perturbed multivariable feedback systems. *IEE Proc.*, 129, Part D(6):251–256, November.

[116] SAFONOV, M. G. (1983) L_∞-optimal sensitivity vs. stability margin. *Proc. IEEE Conf. Decision Contr.*, pages 115–118.

[117] SAFONOV, M. G., GOH, K. C. and Ly, J. H. (1994) Control system synthesis via bilinear matrix inequalities. *Proc. American Contr. Conf.*, pages 45–49.

[118] SAMPEI, M., MITA, T. and NAKAMICHI, M. (1990) An algebraic approach to H_∞ output feedback control problems. *Sys. Contr. Lett.*, 14:13–24.

[119] SCHERER, C. (1992) H_∞-control by state feedback for plants with zeros on the imaginary axis. *SIAM J. Contr. Opt.*, 30(1):123–142.

[120] SCHERER, C. (1992) H_∞-optimization without assumptions on finite or infinite zeros. *SIAM J. Contr. Opt.*, 30(1):143–166.

[121] SCHONEMANN, S. (1966) A generalized solution to the orthogonal Procrustes problem. *Psychometrica*, 31:1–10.

[122] SCHUR, I. (1917/1918) über potenzreihen, die im innern des einheitskreises beschränkt sind. *Crelle's J.*, 147 and 148:205–232, 122–145.

[123] SCHWARZ, H. (1956) Ein verfahren zur stabilitätsfrage bie matrizen-eigenwertproblemen. *Zeit F. Angew Math. u. Physik*, pages 473–500.

[124] SHAMMA, J. S. (1994) Robust stability with time-varying structured uncertainty. *IEEE Trans. Automat. Contr.*, AC-39(4):714–724.

[125] SKELTON, R. E. (1988) *Dynamic Systems Control.* Wiley, New York.

[126] SKELTON, R. E. (1989) Model error concepts in control design. *Int. J. Contr.*, 49(5):1725–1753.

[127] SKELTON, R. E. (1994) Increased roles of linear algebra in control education. *Proc. American Contr. Conf.*, pages 393–397.

[128] SKELTON, R. E. (1996) Model validation for control design. *Mathematical Modelling of Systems.*

[129] SKELTON, R. E. and IKEDA, M. (1989) Covariance controllers for linear continuous-time systems. *Int. J. Contr.*, 49(5):1773–1785.

[130] SKELTON, R. E. and IWASAKI, T. (1993) Liapunov and covariance controllers. *Int. J. Contr.*, 57(3):519–536.

[131] SREERAM, V. and AGATHOKLIS, P. (1991) The generation of q-Markov covers via the inverse solution of lyapunov equation. *30th IEEE CDC*, 138(6).

[132] SREERAM, V. and AGATHOKLIS, P. (1991) On the theory of state-covariance assignment for linear SISO systems. *Proceedings of the 1991 ACC.*

[133] SREERAM, V. and AGATHOKLIS, P. (1991) Solution of Lyapunov equation with system matrix in companion form. *IEE Proceedings-D*, 138(6):529–534.

[134] SREERAM, V. and AGATHOKLIS, P. (1992) On covariance control theory for linear continuous systems. *Proc. Conf. Decision Contr.*, pages 213–217.

[135] SREERAM, V. and AGATHOKLIS, P. (1992) On the theory of state-covariance assignment for single-input linear discrete systems. *Proc. American Contr. Conf.*, pages 785–790.

[136] STENGEL, R. F. (1986) *Stochastic Optimal Control: Theory and Applications.* Wiley, New York.

[137] STOORVOGEL, A. A. (1991) The singular H_∞ control problem with dynamic measurement feedback. *SIAM J. Contr. Opt.*, 29(1):160–184.

[138] STOORVOGEL, A. A. (1992) *The $\mathcal{H}_\infty$ Control Problem: A State Space Approach.* Prentice Hall, New York.

[139] STOORVOGEL, A. A. (1993) The robust H_2 control problem: a worst case design. *IEEE Trans. Automat. Contr.*, AC-38(9):1358–1370.

[140] STOORVOGEL, A. A. and TRENTELMAN, H. L. (1990) The quadratic matrix inequality in singular H_∞ control with state feedback. *SIAM J. Contr. Opt.*, 28(5):1190–1208.

[141] SUN, W., KHARGONEKAR, P. and SHIM, D. (1994) Solution to the positive real control problem for linear time-invariant systems. *IEEE Trans. Automat. Contr.*, AC-39(10):2034–2046, October.

[142] TIKKU, A. and POOLLA, K. (1993) Robust performance against slowly-varying structured perturbations. *Proc. IEEE Conf. Decision Contr.*, pages 990–995.

[143] VANDENBERGHE, L. and BOYD, S. (1994) Semidefinite programming. *SIAM Review*, 38:48–95.

[144] VANDENBERGHE, L. and BOYD, S. (1995) Primal-dual potential reduction methods for problems involving matrix inequalities. *Math. Programm.*, 69:205–236.

[145] WAGIE, D. (1986) Model reduction and controller synthesis in the presence of parameter uncertainty. *Automatica*, 2:295–308.

[146] WALL, H. (1945) Polynomials whose zeros have negative real parts. *Amer. Math. Monthly*, 52:308–322.

[147] WICKS, M. A. and DECARLO, R. A. (1990) Gramian assignment based on the Lyapunov equation. *IEEE Trans. Automat. Contr.*, AC-35(4):465–468.

[148] WIE, B. and BERNSTEIN, D. S. (1990) A benchmark problem for robust control design. *Proc. American Contr. Conf.*, pages 961–962.

[149] WILE, H. (1959) A stability criterion for numerical integration. *J. Assoc. Comp. Mach.*, 6:363.

[150] WILLEMS, J. C. (1971) Least squares stationary optimal control and the algebraic Riccati equation. *IEEE Trans. Automat. Contr.*, AC-16:621–634.

[151] WILLEMS, J. L. (1970) *Stability Theory of Dynamical Systems*. Wiley, New York.

[152] WILLIAMSON, D. (1991) *Digital Control and Implementation: Finite Wordlength Considerations*. Prentice Hall, New Jersey.

[153] WILLIAMSON, D. and SKELTON, R. E. *Linear Algebra with Engineering Applications*. Book in preparation.

[154] WILSON, D. A. (1989) Convolution and Hankel operator norms for linear systems. *IEEE Trans. Automat. Contr.*, AC-34(1):94–98.

[155] XU, J. H. and SKELTON, R. E. (1991) Plant covariance equivalent controller reduction for discrete systems. *Proc. IEEE Conf. Decision Contr.*, pages 2668–2669.

[156] YASUDA, K. and SKELTON, R. E. (1990) Assigning controllability and observability Gramians in feedback control. *AIAA J. Guidance*, 14(5):878–885.

[157] YASUDA, K., SKELTON, R. E. and GRIGORIADIS, K. M. (1993) Covariance controllers: A new parametrization of the class of all stabilizing controllers. *Automatica*, 29(3):785–788.

[158] YOULA, D. C., JABR, H. A. and BONGIORNO, J. J. (1976) Modern Wiener-Hopf design of optimal controllers: Part 2. *IEEE Trans. Automat. Contr.*, AC-21:319–338.

[159] ZAMES, G. (1981) Feedback and optimal sensitivity: Model reference transformations, multiplicative seminorms, and approximate inverses. *IEEE Trans. Automat. Contr.*, AC-26:301–320.

[160] ZHOU, K. and KHARGONEKAR, P. P. (1988) An algebraic Riccati equation approach to H_∞ optimization. *Sys. Contr. Lett.*, 11:85–92.

[161] ZHU, G., GRIGORIADIS, K. M. and SKELTON, R. E. (1995) Covariance control design for the Hubble space telescope. *J. Guidance, Control and Dynamics*, 18(2):230–236.

[162] ZHU, G., ROTEA, M. A. and SKELTON, R. E. A convergent algorithm for the output covariance constrained control problem. Submitted to Automatica.

[163] ZHU, G. and SKELTON, R. E. (1991) Mixed L_2 and L_∞ problems by weight selection in quadratic optimal control. *Int. J. Contr.*, 53(5):1161–1176.

Index